P9-CLB-987

Installation of the BPS 2.0 for Windows 95/98/NT

The BPS SETUP program will copy onto your computer's hard disk the components necessary to run your software. Please note that the CD must be present in your CD drive after installation to allow the software to perform properly.

Step 1: Quit all open applications and save your work before beginning the installation. To begin the installation process, double-click on the SETUP icon at the root level of the CD and follow the on-screen instructions.

Step 2: After the installation is complete, quit the Installer. Restart your computer.

Step 3: Your CD requires Apple's QuickTime 3.0 [or later] system software as well as Adobe's Acrobat reader. Run the QuickTime for Windows installer application ("QuickTime30.exe") located within the \ QT302WIN directory on the CD. Run the Acrobat for Windows installer application ("AR32E301.EXE") located within the \ ACROREAD directory on the CD.

Installation of the BPS 2.0 for Mac OS

The BPS INSTALLER program will copy onto your computer's hard disk the components necessary to run your software. Please note that the CD must be present in your CD drive after installation to allow the software to perform properly.

Step 1: Quit all open applications and save your work before beginning the installation. To begin the installation process, double-click the BPS 2.0 CD-ROM Icon on your desktop, double-click on the INSTALL BPS 2.0 icon and follow the on-screen instructions.

Step 2: After the installation is complete, quit the Installer.

Step 3: Your CD requires Apple's QuickTime 3.0 system software as well as Adobe's Acrobat reader. Run the QuickTime installer application ("INSTALLER") located within the QUICKTIME 3 INSTALLER folder on the CD. Run the Acrobat installer application ("READER 3.01 INSTALLER") located within the ACROBAT READER 3.01 folder on the CD.

WITHDRAWN

Please remember that this is a library book, and that it belongs only temporarily to each person who uses it. Be considerate. Do not write in this, or any, library book.

Table entry for z is the area under the standard normal curve left of z.

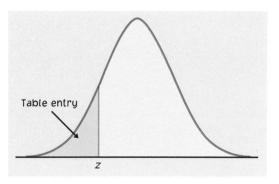

Table entry

z

TABLE A Standard normal probabilities

z	.00	.01	.02	.03	.04	.05	.06	.07	.08	.09
−3.4	.0003	.0003	.0003	.0003	.0003	.0003	.0003	.0003	.0003	.0002
−3.3	.0005	.0005	.0005	.0004	.0004	.0004	.0004	.0004	.0004	.0003
−3.2	.0007	.0007	.0006	.0006	.0006	.0006	.0006	.0005	.0005	.0005
−3.1	.0010	.0009	.0009	.0009	.0008	.0008	.0008	.0008	.0007	.0007
−3.0	.0013	.0013	.0013	.0012	.0012	.0011	.0011	.0011	.0010	.0010
−2.9	.0019	.0018	.0018	.0017	.0016	.0016	.0015	.0015	.0014	.0014
−2.8	.0026	.0025	.0024	.0023	.0023	.0022	.0021	.0021	.0020	.0019
−2.7	.0035	.0034	.0033	.0032	.0031	.0030	.0029	.0028	.0027	.0026
−2.6	.0047	.0045	.0044	.0043	.0041	.0040	.0039	.0038	.0037	.0036
−2.5	.0062	.0060	.0059	.0057	.0055	.0054	.0052	.0051	.0049	.0048
−2.4	.0082	.0080	.0078	.0075	.0073	.0071	.0069	.0068	.0066	.0064
−2.3	.0107	.0104	.0102	.0099	.0096	.0094	.0091	.0089	.0087	.0084
−2.2	.0139	.0136	.0132	.0129	.0125	.0122	.0119	.0116	.0113	.0110
−2.1	.0179	.0174	.0170	.0166	.0162	.0158	.0154	.0150	.0146	.0143
−2.0	.0228	.0222	.0217	.0212	.0207	.0202	.0197	.0192	.0188	.0183
−1.9	.0287	.0281	.0274	.0268	.0262	.0256	.0250	.0244	.0239	.0233
−1.8	.0359	.0351	.0344	.0336	.0329	.0322	.0314	.0307	.0301	.0294
−1.7	.0446	.0436	.0427	.0418	.0409	.0401	.0392	.0384	.0375	.0367
−1.6	.0548	.0537	.0526	.0516	.0505	.0495	.0485	.0475	.0465	.0455
−1.5	.0668	.0655	.0643	.0630	.0618	.0606	.0594	.0582	.0571	.0559
−1.4	.0808	.0793	.0778	.0764	.0749	.0735	.0721	.0708	.0694	.0681
−1.3	.0968	.0951	.0934	.0918	.0901	.0885	.0869	.0853	.0838	.0823
−1.2	.1151	.1131	.1112	.1093	.1075	.1056	.1038	.1020	.1003	.0985
−1.1	.1357	.1335	.1314	.1292	.1271	.1251	.1230	.1210	.1190	.1170
−1.0	.1587	.1562	.1539	.1515	.1492	.1469	.1446	.1423	.1401	.1379
−0.9	.1841	.1814	.1788	.1762	.1736	.1711	.1685	.1660	.1635	.1611
−0.8	.2119	.2090	.2061	.2033	.2005	.1977	.1949	.1922	.1894	.1867
−0.7	.2420	.2389	.2358	.2327	.2296	.2266	.2236	.2206	.2177	.2148
−0.6	.2743	.2709	.2676	.2643	.2611	.2578	.2546	.2514	.2483	.2451
−0.5	.3085	.3050	.3015	.2981	.2946	.2912	.2877	.2843	.2810	.2776
−0.4	.3446	.3409	.3372	.3336	.3300	.3264	.3228	.3192	.3156	.3121
−0.3	.3821	.3783	.3745	.3707	.3669	.3632	.3594	.3557	.3520	.3483
−0.2	.4207	.4168	.4129	.4090	.4052	.4013	.3974	.3936	.3897	.3859
−0.1	.4602	.4562	.4522	.4483	.4443	.4404	.4364	.4325	.4286	.4247
−0.0	.5000	.4960	.4920	.4880	.4840	.4801	.4761	.4721	.4681	.4641

Table entry for z is the area under the standard normal curve to the left of z.

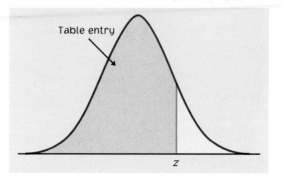

Table entry

z

TABLE A Standard normal probabilities (*continued*)

z	.00	.01	.02	.03	.04	.05	.06	.07	.08	.09
0.0	.5000	.5040	.5080	.5120	.5160	.5199	.5239	.5279	.5319	.5359
0.1	.5398	.5438	.5478	.5517	.5557	.5596	.5636	.5675	.5714	.5753
0.2	.5793	.5832	.5871	.5910	.5948	.5987	.6026	.6064	.6103	.6141
0.3	.6179	.6217	.6255	.6293	.6331	.6368	.6406	.6443	.6480	.6517
0.4	.6554	.6591	.6628	.6664	.6700	.6736	.6772	.6808	.6844	.6879
0.5	.6915	.6950	.6985	.7019	.7054	.7088	.7123	.7157	.7190	.7224
0.6	.7257	.7291	.7324	.7357	.7389	.7422	.7454	.7486	.7517	.7549
0.7	.7580	.7611	.7642	.7673	.7704	.7734	.7764	.7794	.7823	.7852
0.8	.7881	.7910	.7939	.7967	.7995	.8023	.8051	.8078	.8106	.8133
0.9	.8159	.8186	.8212	.8238	.8264	.8289	.8315	.8340	.8365	.8389
1.0	.8413	.8438	.8461	.8485	.8508	.8531	.8554	.8577	.8599	.8621
1.1	.8643	.8665	.8686	.8708	.8729	.8749	.8770	.8790	.8810	.8830
1.2	.8849	.8869	.8888	.8907	.8925	.8944	.8962	.8980	.8997	.9015
1.3	.9032	.9049	.9066	.9082	.9099	.9115	.9131	.9147	.9162	.9177
1.4	.9192	.9207	.9222	.9236	.9251	.9265	.9279	.9292	.9306	.9319
1.5	.9332	.9345	.9357	.9370	.9382	.9394	.9406	.9418	.9429	.9441
1.6	.9452	.9463	.9474	.9484	.9495	.9505	.9515	.9525	.9535	.9545
1.7	.9554	.9564	.9573	.9582	.9591	.9599	.9608	.9616	.9625	.9633
1.8	.9641	.9649	.9656	.9664	.9671	.9678	.9686	.9693	.9699	.9706
1.9	.9713	.9719	.9726	.9732	.9738	.9744	.9750	.9756	.9761	.9767
2.0	.9772	.9778	.9783	.9788	.9793	.9798	.9803	.9808	.9812	.9817
2.1	.9821	.9826	.9830	.9834	.9838	.9842	.9846	.9850	.9854	.9857
2.2	.9861	.9864	.9868	.9871	.9875	.9878	.9881	.9884	.9887	.9890
2.3	.9893	.9896	.9898	.9901	.9904	.9906	.9909	.9911	.9913	.9916
2.4	.9918	.9920	.9922	.9925	.9927	.9929	.9931	.9932	.9934	.9936
2.5	.9938	.9940	.9941	.9943	.9945	.9946	.9948	.9949	.9951	.9952
2.6	.9953	.9955	.9956	.9957	.9959	.9960	.9961	.9962	.9963	.9964
2.7	.9965	.9966	.9967	.9968	.9969	.9970	.9971	.9972	.9973	.9974
2.8	.9974	.9975	.9976	.9977	.9977	.9978	.9979	.9979	.9980	.9981
2.9	.9981	.9982	.9982	.9983	.9984	.9984	.9985	.9985	.9986	.9986
3.0	.9987	.9987	.9987	.9988	.9988	.9989	.9989	.9989	.9990	.9990
3.1	.9990	.9991	.9991	.9991	.9992	.9992	.9992	.9992	.9993	.9993
3.2	.9993	.9993	.9994	.9994	.9994	.9994	.9994	.9995	.9995	.9995
3.3	.9995	.9995	.9995	.9996	.9996	.9996	.9996	.9996	.9996	.9997
3.4	.9997	.9997	.9997	.9997	.9997	.9997	.9997	.9997	.9997	.9998

The Basic Practice of Statistics

SECOND EDITION

DAVID S. MOORE

Purdue University

W. H. Freeman and Company
New York

Acquisitions editor: Patrick Farace

Publisher: Michelle Russel Julet

Marketing manager: Kimberly Manzi

Development editor: Don Gecewicz

New media producer: Brian J. McCooey

Project editor: Diane Cimino Maass

Cover and text designer: Vicki Tomaselli

Cover and interior illustrations: Mark Chickinelli

Production coordinator: Julia DeRosa

Illustration and composition: Publication Services

Manufacturing: RR Donnelley & Sons Company

"M&M'S" CHOCOLATE CANDIES IS A REGISTERED TRADEMARK OF MARS, INCORPORATED
AND USED WITH PERMISSION OF THE OWNER. COPYRIGHT ©MARS, INC.

Library of Congress Cataloging-in-Publication Data

Moore, David S.
    The basic practice of statistics / David S. Moore—2d ed.
        p. cm.
    Includes index.
    ISBN 0-7167-3627-6
    1. Statistics.   1. Title.
QA276.12.M648   2000
519.5–dc21   99-18033

                                          CIP

©1995, 2000 by W H. Freeman and Company. All rights reserved.

No part of this book may be reproduced by any mechanical, photographic, or electronic process, or in the form of a phonographic recording, nor may it be stored in a retrieval system, transmitted, or otherwise copied for public or private use, without written permission from the publisher.

Printed in the United States of America

Third printing 2000

CONTENTS

BRIEF CONTENTS

*starred material is optional

---

*starred material is optional

---

*starred material is optional

---

*starred material is optional

*starred material is optional

---
*starred material is optional

Ties

Limitations of nonparametric tests

Section 12.1 Exercises

12.2 The Wilcoxon Signed Rank Test

The normal approximation

Ties

Section 12.2 Exercises

12.3 The Kruskal-Wallis Test

Hypotheses and assumptions

The Kruskal-Wallis test

Section 12.3 Exercises

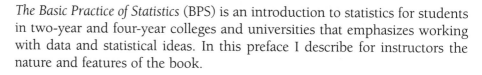

The Basic Practice of Statistics (BPS) is an introduction to statistics for students in two-year and four-year colleges and universities that emphasizes working with data and statistical ideas. In this preface I describe for instructors the nature and features of the book.

Guiding principles

Once upon a time, basic statistics courses emphasized only probability and inference. BPS reflects the current standard, in which data analysis and the design of data production join probability-based inference as major content areas. Statisticians have reached a broad consensus on the nature of first courses for general college audiences. As Richard Schaffer says in discussing a survey paper of mine, "With regard to the content of an introductory statistics course, statisticians are in closer agreement today than at any previous time in my career."[1] Figure 1 is an outline of the consensus as summarized by the joint curriculum committee of the American Statistical Association and the Mathematical Association of America.[2]

I was a member of the ASA/MAA committee, and I agree with their conclusions. Fostering active learning is the business of the teacher (though an emphasis on working with data helps). BPS is guided by the first two recommendations. Although the book is elementary in the level of mathematics

1. **Emphasize the elements of statistical thinking:**

   (a) the need for data;

   (b) the importance of data production;

   (c) the omnipresence of variability;

   (d) the measuring and modeling of variability.

2. **Incorporate more data and concepts, fewer recipes and derivations. Wherever possible, automate computations and graphics.** An introductory course should:

   (a) rely heavily on *real* (not merely realistic) data;

   (b) emphasize *statistical* concepts, e.g., causation vs. association, experimental vs. observational and longitudinal vs. cross-sectional studies;

   (c) rely on computers rather than computational recipes;

   (d) treat formal derivations as secondary in importance.

3. **Foster active learning,** through the following alternatives to lecturing:

   (a) group problem solving and discussion;

   (b) laboratory exercises;

   (c) demonstrations based on class-generated data;

   (d) written and oral presentations;

   (e) projects, either group or individual.

Figure 1 *Recommendations of the ASA/MAA Joint Curriculum Committee*

required and in the statistical procedures presented, it aims to give students both an understanding of the main ideas of statistics and useful skills for working with data. Examples and exercises, though intended for beginners, use real data and give enough background to allow students to consider the meaning of their calculations. I often ask for conclusions that are more than a number (or "reject H_0"). Some exercises require judgment in addition to right-or-wrong calculations and conclusions. I hope that teachers will encourage discussion of results in class.

Chapters 1 and 2 present the methods and unifying ideas of data analysis. Students appreciate the usefulness of data analysis; that they can actually do it relieves a bit of their anxiety about statistics. I hope that they will grow accustomed to examining data and will continue to do so even when formal inference to answer a specific question is the ultimate goal. Chapter 3 discusses random sampling and randomized comparative experiments. These are among the most important ideas in statistics and are often unjustly neglected in beginning instruction. Chapter 4 builds on the ideas of Chapter 3 and the data-analytic tools of Chapter 1 to present the central idea of a sampling distribution and (informally) the language and most important facts of probability theory. The optional Chapter 5 presents additional material for use in courses that require a greater emphasis on formal probability. Chapter 6, which describes the reasoning of statistical inference, is the cornerstone of the rest of the book. The remaining chapters present methods of inference for various settings, with a strong emphasis on practical aspects of using these methods. Chapters 7 and 8 discuss one-sample and two-sample procedures. Chapters 9, 10, and 11 (which can be read independently of each other in any order) offer a choice of somewhat more advanced topics. Working with data in real settings is a theme throughout the text.

Technology

Automating calculations increases the students' ability to complete problems, reduces their frustration, and helps them concentrate on ideas and problem recognition rather than mechanics. *All students should have at least a "two-variable statistics" calculator* with functions for correlation and the least-squares regression line as well as for the mean and standard deviation. Graphing calculators offer considerably more capability. Because students have calculators, the text doesn't discuss "computing formulas" for the sample standard deviation or the least-squares regression line. *Statistical software* has considerable advantages over calculators: easier data entry and editing and much better graphics. I encourage the use of software whenever facilities permit. BPS does not, however, assume that students will use software. Instructors will find using software easier because the CD-ROM that accompanies each copy of the book contains the data sets for exercises and tables in several common formats, along with several other features (see supplement section).

Output from several technologies, from graphing calculators through spreadsheets to statistical packages such as Minitab, appears in the text. The variety is deliberate: students see that the methods they meet in BPS are universal and that basic knowledge makes output from any technology easy to read and use.

Accessibility

The intent of BPS is to be modern *and* accessible. In comparison with the longer *Introduction to the Practice of Statistics* (IPS),[3] BPS has more concise discussions, less optional material, and more stopping places for the reader. BPS requires less advanced reading and study skills than IPS, but retains the essential content and the emphasis on working with data and thinking about statistical problems. The short Apply Your Knowledge exercise sections that follow every major new idea allow a quick check of basic mastery. Each chapter ends with Statistics in Summary, an explicit list of goals and often a pictorial review of main ideas. Please see the "Guided Book Tour" which follows the Preface.

Why did you do that?

There is no single best way to organize our presentation of statistics to beginners. That said, my choices are based on more than a whim. Here are comments on several "frequently asked questions" about the order and selection of material in BPS.

Why does the distinction between population and sample not appear in Chapters 1 and 2? This is a sign that there is more to statistics than inference. In fact, statistical inference is appropriate only in rather special circumstances. Chapters 1 and 2 present tools and tactics for describing data—any data. These tools and tactics do not depend on the idea of inference from sample to population. Many data sets in these chapters (for example, the several sets of data about the 50 states) do not lend themselves to inference because they represent an entire population. John Tukey of Bell Labs and Princeton, the philosopher of modern data analysis, insists that the population/sample distinction be avoided when it is not relevant. He uses the word "batch" for data sets in general. I see no need for a special word, but I think Tukey is right.

Why not begin with data production? It is certainly reasonable to do so—the natural flow of a planned study is from design to data analysis to inference. I choose to place the design of data production (Chapter 3) after data analysis (Chapters 1 and 2) to emphasize that data-analytic techniques apply to any data. One of the primary purposes of statistical designs for producing data is to make inference possible, so the discussion in Chapter 3 is a natural transition to inference ideas.

Why do normal distributions appear in Chapter 1? Density curves such as the normal curves are just another tool to describe the distribution of a quantitative variable, along with stemplots, histograms, and boxplots. It is becoming common for software to offer to make density curves from data just as it offers histograms. I prefer not to suggest that this material is essentially tied to probability, as the traditional order does. I also want students to think about the flow from graphs to numerical summaries to mathematical models, and density curves are the common mathematical model for the overall pattern of a distribution. Finally, I would like to break up the indigestible lump of probability that troubles students so much. Meeting normal distributions early does this and strengthens the "probability distributions are like data distributions" way of approaching probability.

Why not delay correlation and regression until late in the course, as is traditional? BPS begins by offering experience working with data and gives a conceptual structure for this non-mathematical but very important part of statistics. Students profit from more experience with data and from seeing the conceptual structure worked out in relations among variables as well as in describing single-variable data. Moreover, correlation and regression as descriptive tools (Chapter 2) have wider scope than an emphasis on inference (Chapter 11) allows. The very important discussion of lurking variables, for example, fits poorly with inference. I consider Chapter 2 essential and Chapter 11 optional.

Where did all the probability go? Much of it went to the optional Chapter 5. Students don't mind skipping a chapter and are sometimes bothered by optional material within chapters, so in this new edition all of Chapter 4 is required and all of Chapter 5 is optional.

Experienced teachers recognize that students find probability difficult. Research on learning confirms our experience. Even students who can do formally posed probability problems often have a very fragile conceptual grasp of probability ideas. Attempting to present a substantial introduction to probability in a data-oriented statistics course for students who are not mathematically trained is difficult to do. Formal probability does not help these students master the ideas of inference (at least not as much as we teachers imagine), and it depletes reserves of mental energy that might better be applied to essentially statistical ideas.

I have therefore presented little formal probability in the core Chapter 4. The chapter follows a straight line from the idea of probability as long term regularity, through concrete ways of assigning probabilities, to the central idea of the sampling distribution of a statistic. The law of large numbers and the central limit theorem appear in the context of discussing the sampling distribution of a sample mean. What is omitted here is mostly "general probability rules," including conditional probability, and combinatorics. The general rules appear in Chapter 5. Combinatorics is a different (and even harder) topic.

Why didn't you cover Topic X? Introductory texts ought not to be encyclopedic. Including each reader's favorite topic (control charts, nonparametric tests, geometric distributions, and so on) results in a text that is formidable in size and intimidating to students. I chose topics on two grounds: they are the most commonly used in practice, and they are suitable vehicles for learning broader statistical ideas. There are studies of usage in many fields of application. For example, Emerson and Colditz[4] report that just descriptive statistics, t procedures, and two-way tables would give full access to 73% of the articles in the *New England Journal of Medicine*. That suggests a reasonable semester course from BPS: Chapters 1 to 4 and 6 to 9.

Acknowledgments

I am grateful to many colleagues from two-year and four-year colleges and universities who commented on successive drafts of the manuscript:

Elizabeth Applebaum,
Park College

Edgar Avelino,
Langara College

Smiley Cheng,
University of Manitoba

James Curl,
Modesto Junior College

David Gurney,
Southeastern Louisiana University

Donald Harden,
Georgia State University

Sue Holt,
Cabrillo College

Elizabeth Houseworth,
University of Oregon

T. Henry Jablonski, Jr.,
Eastern Tennessee State University

Tom Kaupe,
Shoreline Community College

James Lang,
Valencia Community College

Donald Loftgaarden,
University of Montana

Steve Marsden,
Glendale College

Darcy Mays,
Virginia Commonwealth University

Amy Salvati,
Adirondack Community College

N. Paul Schembari,
East Stroudsburg University

W. Robert Stephenson,
Iowa State University

Martin Tanner,
Northwestern University

Bruce Torrence,
Randolph-Macon College

Mike Turegon,
Oklahoma City Community College

Jean Werner,
Mansfield University of Pennsylvania

Dex Whittinghill,
Rowan University

Rodney Wong,
University of California at Davis

I would also like to thank reviewers of the first edition:

Douglas M. Andrews
Wittenberg University

Rebecca Busam
The Ohio State University

Michael Butler
College of the Redwoods

Carolyn Pilers Dobler
Gustavus Adolphus College

Joel B. Greenhouse
Carnegie Mellon University

Larry Griffey
*Florida Community College
at Jacksonville*

Brenda Gunderson
*The University
of Michigan*

Catherine Cummins Hayes
University of Mobile

Tim Hesterberg
Franklin & Marshall College

Ronald La Porte
Macomb Community College

Ken McDonald
Northwest Missouri State University

William Notz
The Ohio State University

Mary Parker
Austin Community College

Kenneth Ross
University of Oregon

Calvin Schmall
Solano Community College

Frank Soler
De Anza College

Linda Sorenson
Algoma University

Tom Sutton
Mohawk College

Sue Holt of Cabrillo College deserves special thanks for careful comments based on long use of BPS in a two-year college. I am particularly grateful to Patrick Farace, Diane Cimino Maass and the other editorial and design professionals who have contributed greatly to the attractiveness of this book.

Finally, I am indebted to many statistics teachers with whom I have discussed the teaching of our subject over many years; to the people from diverse fields with whom I have worked to understand data; and especially to students whose compliments and complaints have changed and improved my teaching. Working with teachers, colleagues in other disciplines, and students constantly reminds me of the importance of hands-on experience with data and of statistical thinking in an era when computer routines quickly handle statistical details.

David S. Moore

A full range of supplements is available to help teachers and students use *The Basic Practice of Statistics*. **Complimentary supplements** available for the instructor include:

- **Instructor CD-ROM** contains the Instructor's version of EESEE with solutions, Presentation Manager Pro software to develop presentations from pre-loaded text figures as well as the user's own content (saved locally or found on the World Wide Web) and masters to the text figures to create presentations with the user's own presentation software. ISBN: 0-7167-3612-8

- **Instructor's Guide with Solutions** by Darryl Nester of Bluffton College and David Moore. The Instructor's Guide includes teaching suggestions, chapter comments, and worked-out solutions to all exercises. Solutions include graphics and statistical software output from various statistical packages where appropriate. ISBN: 0-7167-3616-0

- **Test Bank** by William Notz and Michael Fligner of The Ohio State University. Hundreds of page-referenced multiple-choice questions allow instructors to generate quizzes and tests. Available in a printed version as well as electronic versions for Mac and Windows platforms. The electronic version publishes texts in print, or on a LAN system, or on-line. ISBNs: Printed: 0-7167-3618-7, Windows: 0-7167-3619-5, Mac: 0-7167-3620-9

And for students. . . .

- **Student CD-ROM** developed by Sumanas Multimedia Development Services. Packaged with every copy of the text, the CD contains a number of components that illuminate and expand the themes of the text:
  - The Encyclopedia of Statistical Examples and Exercises (EESEE), developed by faculty of the Department of Statistics at Ohio State University, is a rich repository of case studies that apply the concepts of *The Basic Practice of Statistics*. Many of the diverse examples are illustrated by photos or video clips. Each case is accompanied by practice problems and a majority include data sets, which are available in Data Desk, Minitab, and JMP formats.
  - Most of the textbook's data sets are included in the various formats: ASCII, Excel, and Minitab.
  - Q&A self-quizzes for each chapter give students feedback for correct and incorrect answers. A printable report card records students' performances on the quizzes.
  - A CD/Web update section links to new and updated material as well as providing the latest statistics news via the World Wide Web.

- Additional text sections covering control charts and nonparametric tests are included on the CD.

Student Study Guide by William Notz and Michael Fligner, Ohio State University, and Rebecca Busam, University of Texas, Austin. The study guide helps students learn and review the basic concepts of the textbook in a printed format. It explains crucial concepts in each section of the text and provides solutions to key text problems and step-through models of important statistical techniques. ISBN: 0-7167-3617-9

Minitab Manual by Betsy Greenberg, University of Texas. Written specifically for students who are using Minitab with *The Basic Practice of Statistics,* this manual offers careful illustrations and exercises that allow the student to unlock the power of Minitab for statistical analysis. Appendices include lists of Minitab functions, commands, and macros for easy reference. Both windows and session commands are covered. ISBN: 0-7167-3613-X

New Excel Manual by Fred Hoppe, McMaster University. Providing exercises and applications for each chapter, this manual demonstrates how Excel's ability to organize data into spreadsheets allows easy analysis and graphic exploration. Each chapter focuses on the manner in which Excel displays and analyses data, the basic themes of *The Basic Practice of Statistics.* ISBN: 0-7167-3611-X

New TI-83 Graphing Calculator Manual by David Neal, Western Kentucky University. Offering detailed instructions on the use of the TI-83 graphing calculator with *The Basic Practice of Statistics*, this manual references text exercises to demonstrate solutions with TI-83 functions. Figures illustrate the calculator's output. A detailed list of TI-83 statistical functions is included. ISBN: 0-7167-3614-4

New SPSS Manual by Paul Stephenson, Justine Ritchie, Neal Rogness, and Patricia Stephenson, Grand Valley State University. Written specifically for students who are using SPSS with *The Basic Practice of Statistics*, this manual demonstrates the software's ability to perform a wide variety of statistical techniques ranging from descriptive statistics to complex multivariate procedures. ISBN: 0-7167-3610-1

New SAS Manual by Mike Evans, University of Toronto. This manual explores the uses of SAS analysis with *The Basic Practice of Statistics* In-depth illustrations and exercises allow the student to unlock the power of SAS for use in statistical analysis. Appendices include lists of SAS functions, and commands for easy reference. ISBN: 0-7167-3609-8

GUIDED BOOK TOUR

New Redesigned Part and Chapter Openers introduce the major concepts and themes of the three parts—analyzing and producing data in Part I, probability and inference in Part II, and more advanced inference in Part III. Chapter openers engage students by presenting capsule biographies of men and women who have been important in the development of statistics.

New Four-Color Design presents information in an appealing and stylish format making the text more accessible to students and easier to use. Statistical graphs can now be easily interpreted by using different colors, while four-color spot photographs enliven the pages with images from real applications of statistics.

"A Bit More" Think Pieces appear in the margin of each chapter and are accompanied by a light illustration. These brief notes offer interesting and often amusing anecdotes about statistics and its impact.

◀ **New and Revised Exercises and Examples** include a wide variety of real data in real settings from business and economics, demographics, public health and medicine, agriculture, biological sciences, history, psychology, and more.

◀ **Step-by-Step Examples** work through procedures and join techniques and concepts together in an easy-to-follow format.

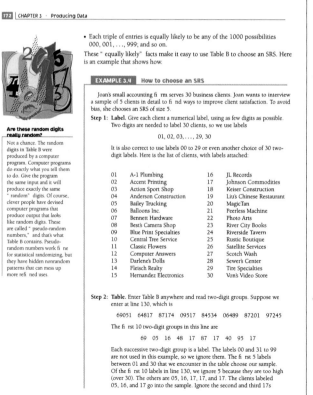

- Each triple of entries is equally likely to be any of the 1000 possibilities 000, 001, ..., 999; and so on.

These "equally likely" facts make it easy to use Table B to choose an SRS. Here is an example that shows how.

EXAMPLE 3.4 How to choose an SRS

Joan's small accounting firm serves 30 business clients. Joan wants to interview a sample of 5 clients in detail to find ways to improve client satisfaction. To avoid bias, she chooses an SRS of size 5.

Step 1: Label. Give each client a numerical label, using as few digits as possible. Two digits are needed to label 30 clients, so we use labels

01, 02, 03, ..., 29, 30

It is also correct to use labels 00 to 29 or even another choice of 30 two-digit labels. Here is the list of clients, with labels attached:

01	A-1 Plumbing	16	JL Records
02	Accent Printing	17	Johnson Commodities
03	Action Sport Shop	18	Keiser Construction
04	Anderson Construction	19	Liu's Chinese Restaurant
05	Bailey Trucking	20	MagicTan
06	Balloons Inc.	21	Peerless Machine
07	Bennett Hardware	22	Photo Arts
08	Best's Camera Shop	23	River City Books
09	Blue Print Specialties	24	Riverside Tavern
10	Central Tree Service	25	Rustic Boutique
11	Classic Flowers	26	Satellite Services
12	Computer Answers	27	Scotch Wash
13	Darlene's Dolls	28	Sewer's Center
14	Fleisch Realty	29	Tire Specialties
15	Hernandez Electronics	30	Von's Video Store

Step 2: Table. Enter Table B anywhere and read two-digit groups. Suppose we enter at line 130, which is

69051 64817 87174 09517 84534 06489 87201 97245

The first 10 two-digit groups in this line are

69 05 16 48 17 87 17 40 95 17

Each successive two-digit group is a label. The labels 00 and 31 to 99 are not used in this example, so we ignore them. The first 5 labels between 01 and 30 that we encounter in the table choose our sample. Of the first 10 labels in line 130, we ignore 5 because they are too high (over 30). The others are 05, 16, 17, 17, and 17. The clients labeled 05, 16, and 17 go into the sample. Ignore the second and third 17s

Are these random digits really random?

Not a chance. The random digits in Table B were produced by a computer program. Computer programs do exactly what you tell them to do. Give the program the same input and it will produce exactly the same "random" digits. Of course, clever people have devised computer programs that produce output that *looks* like random digits. These are called "pseudo-random numbers," and that's what Table B contains. Pseudo-random numbers work fine for statistical randomizing, but they have hidden nonrandom patterns that can mess up more refined uses.

New "Statistics in Summary" ▶
sections at the end of each chapter review key concepts, often in easy-to-remember pictorial form. A detailed outline of chapter learning objectives follows.

Review Exercises **205**

STATISTICS IN SUMMARY
Simple Random Sample

Population → All samples of size n are equally likely → Sample data $x_1, x_2, \ldots, x_n$

STATISTICS IN SUMMARY
Randomized Comparative Experiment

Random allocation → Group 1 n_1 subjects → Treatment 1 → Compare response

Random allocation → Group 2 n_2 subjects → Treatment 2 → Compare response

case should show the sizes of the groups, the specific treatments, and the response variable.

5. Use Table B of random digits to carry out the random assignment of subjects to groups in a completely randomized experiment.

6. Recognize the placebo effect. Recognize when the double-blind technique should be used.

7. Explain why a randomized comparative experiment can give good evidence for cause-and-effect relationships.

CHAPTER 3 **Review Exercises**

3.60 **Repairing knees in comfort (EESEE).** Injured knees are routinely repaired by arthroscopic surgery that does not require opening up the knee. Can we reduce patient discomfort by giving them a nonsteroidal anti-inflammatory drug (NSAID)? Eighty-three patients were placed in three groups. Group A received the NSAID both before and after surgery. Group B was given a placebo before and the NSAID after. Group C received a placebo both before and after surgery. The patients recorded a pain score by answering questions one day after the surgery.[17]

194 CHAPTER 3 · Producing Data

We hope to see a difference in the responses so large that it is unlikely to happen just because of chance variation. We can use the laws of probability, which give a mathematical description of chance behavior, to learn if the treatment effects are larger than we would expect to see if only chance were operating. If they are, we call them *statistically significant*.

STATISTICAL SIGNIFICANCE

An observed effect so large that it would rarely occur by chance is called **statistically significant**.

If we observe statistically significant differences among the groups in a comparative randomized experiment, we have good evidence that the treatments actually caused these differences. You will often see the phrase "statistically significant" in reports of investigations in many fields of study. The great advantage of randomized comparative experiments is that they can produce data that give good evidence for a cause-and-effect relationship between the explanatory and response variables. We know that in general a strong association does not imply causation. A statistically significant association in data from a well-designed experiment does imply causation.

APPLY YOUR KNOWLEDGE

3.38 **Conserving energy.** Example 3.13 describes an experiment to learn whether providing households with electronic indicators or charts will reduce their electricity consumption. An executive of the electric company objects to including a control group. He says, "It would be simpler to just compare electricity use last year (before the indicator or chart was provided) with consumption in the same period this year. If households use less electricity this year, the indicator or chart must be working." Explain clearly why this design is inferior to that in Example 3.13.

3.39 **Exercise and heart attacks.** Does regular exercise reduce the risk of a heart attack? Here are two ways to study this question. Explain clearly why the second design will produce more trustworthy data.

1. A researcher finds 2000 men over 40 who exercise regularly and have not had heart attacks. She matches each with a similar man who does not exercise regularly, and she follows both groups for 5 years.

2. Another researcher finds 4000 men over 40 who have not had heart attacks and are willing to participate in a study. She assigns 2000 of the men to a regular program of supervised exercise. The

Scratch my furry ears

Rats and rabbits, specially bred to be uniform in their inherited characteristics, are the subjects in many experiments. It turns out that animals, like people, are quite sensitive to how they are treated. This creates some amusing opportunities for hidden bias. For example, human affection can change the cholesterol level of rabbits. Choose some rabbits at random and regularly remove them from their cages to have their heads scratched by friendly people. Leave other rabbits unloved. All the rabbits eat the same diet, but the rabbits that receive affection have lower cholesterol.

▲

Titles for Examples and Exercises pique student interest by displaying the ties between statistics and a very wide variety of issues and topics.

CD-ROM Icon references exercises in BPS which are based on data from an EESEE case study on the CD-ROM.

▲

"Apply Your Knowledge" Exercises immediately follow each new idea. These straightforward exercises embedded in the text allow rapid feedback and reinforcement of basic concepts and skills.

APPLICATIONS

The Basic Practice of Statistics presents a wide variety of applications from diverse disciplines. The list below indicates the number of exercises and examples which relate to different fields.

Exercises by application:

Agriculture: 19

Biological and environmental sciences: 69

Business and economics: 145

Education: 63

History and public policy: 13

Paleontology: 2

Physical science: 31

People and places: 120

Psychology and behavioral sciences: 24

Public health and medicine: 111

Sports: 17

Examples by application:

Agriculture: 14

Biological and environmental sciences: 18

Business and economics: 32

Education: 13

History and public policy: 5

Physical science: 15

People and places: 20

Psychology and behavioral sciences: 8

Public health and medicine: 29

Sports: 8

For a complete breakdown list of examples and exercises by chapter and number, please see the *Instructor's Guide* or W.H. Freeman's Web site: www.whfreeman.com/statistics

How to Tell the Facts from the Artifacts

S tatistics is about data. Data are numbers, but they are not "just numbers." *Data are numbers with a context.* The number 10.5, for example, carries no information by itself. But if we hear that a friend's new baby weighed 10.5 pounds at birth, we congratulate her on the healthy size of the child. The context engages our background knowledge and allows us to make judgments. We know that a baby weighing 10.5 pounds is quite large, and that a human baby can't weigh 10.5 ounces or 10.5 kilograms. The context makes the number informative.

Statistics uses data to gain insight and to draw conclusions. Our tools are graphs and calculations, but the tools are guided by ways of thinking that amount to educated common sense. Let's begin our study of statistics with an informal look at some principles of statistical thinking.

Data Illuminate

What percent of the American population do you think is black? What percent do you think is white? When white Americans were asked these questions, their average answers were 23.8% black and 49.9% white. In fact, the Census Bureau tells us that 11.8% of Americans are black and 74% are white.[1]

Race remains a central social issue in the United States. It is illuminating to see that whites think (wrongly) that they are a minority. The census data—what really is true about the U.S. population—are also illuminating. We wonder if knowing the facts might help change attitudes.

Data Beat Anecdotes

An anecdote is a striking story that sticks in our minds exactly because it is striking. Anecdotes humanize an issue, but they can be misleading.

Does living near power lines cause leukemia in children? The National Cancer Institute spent 5 years and $5 million gathering data on the question. Result: no connection between leukemia and exposure to magnetic fields of the kind produced by power lines. The editorial that accompanied the study report in the *New England Journal of Medicine* thundered, "It is time to stop wasting our research resources" on the question.[2]

Now compare the effectiveness of a television news report of a 5-year, $5 million investigation against a televised interview with an articulate mother whose child has leukemia and who happens to live near a power line. In the public mind, the anecdote wins every time. A statistically literate person, however, knows that data are more reliable than anecdotes because they systematically describe an overall picture rather than focusing on a few incidents.

Beware the Lurking Variable

Air travelers would like their flights to arrive on time. Airlines collect data about on-time arrivals and report them to the Department of Transportation. Here are one month's data for flights from several western cities for two airlines:

	On time	Delayed
Alaska Airlines	3274	501
America West	6438	787

You can see that the percentages of late flights were:

$$\text{Alaska Airlines } \frac{501}{3775} = 13.3\%$$

$$\text{America West } \frac{787}{7225} = 10.9\%$$

It appears that America West does better.

This isn't the whole story, however. Almost all relationships between two variables are influenced by other variables lurking in the background. We have data on two variables, the airline and whether or not the flight was late. Let's add data on a third variable, which city the flight left from.[3]

	Alaska Airlines		America West	
	On time	Delayed	On time	Delayed
Los Angeles	497	62	694	117
Phoenix	221	12	4840	415
San Diego	212	20	383	65
San Francisco	503	102	320	129
Seattle	1841	305	201	61
Total	3274	501	6438	787

The "Total" row shows that the new table describes the same flights as the earlier table. Look again at the percentages of late flights, first for Los Angeles:

$$\text{Alaska Airlines} \quad \frac{62}{559} = 11.1\%$$

$$\text{America West} \quad \frac{117}{811} = 14.4\%$$

Alaska Airlines wins. The percentages delayed for Phoenix are:

$$\text{Alaska Airlines} \quad \frac{12}{233} = 5.2\%$$

$$\text{America West} \quad \frac{415}{5255} = 7.9\%$$

Alaska Airlines wins again. In fact, as Figure 1 shows, Alaska Airlines has a lower percentage of late flights at *every one* of these cities.

How can it happen that Alaska Airlines wins at every city but America West wins when we combine all the cities? Look at the data: America West flies most often from sunny Phoenix, where there are few delays. Alaska Airlines flies most often from Seattle, where fog and rain cause frequent delays. What city we fly from has a major influence on the chance of a delay, so including the city data reverses our conclusion. The message is worth repeating: almost all relationships between two variables are influenced by other variables lurking in the background.

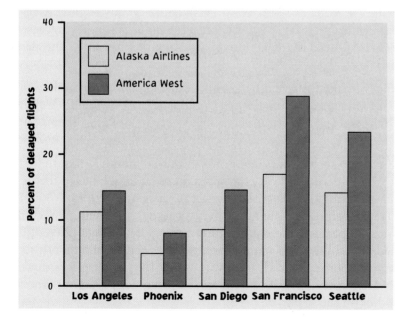

Figure 1 *Comparing the percents of delayed flights for two airlines at five airports.*

Where the Data Come from Is Important

The advice columnist Ann Landers once asked her readers, "If you had it to do over again, would you have children?" A few weeks later, her column was headlined "70% OF PARENTS SAY KIDS NOT WORTH IT." Indeed, 70% of the nearly 10,000 parents who wrote in said they would not have children if they could make the choice again. Do you believe that 70% of all parents regret having children?

You shouldn't. The people who took the trouble to write Ann Landers are not representative of all parents. Their letters showed that many of them were angry at their children. All we know from these data is that there are some unhappy parents out there. A statistically designed poll, unlike Ann Landers's appeal, targets specific people chosen in a way that gives all parents the same chance to be asked. Such a poll showed that 91% of parents *would* have children again. The lesson: if you are careless about how you get your data, you may announce 70% "No" when the truth is close to 90% "Yes."

Variation Is Everywhere

The company's sales reps file into their monthly meeting. The sales manager rises. "Congratulations! Our sales were up 2% last month, so we're all drinking champagne this morning. You remember that when sales were down 1% last month I fired half of our reps." This picture is only slightly exaggerated. Many managers overreact to small short-term variations in key figures. Here is Arthur Nielsen, head of the country's largest market research firm, describing his experience:

> *Too many business people assign equal validity to all numbers printed on paper. They accept numbers as representing Truth and find it difficult to work with the concept of probability. They do not see a number as a kind of shorthand for a range that describes our actual knowledge of the underlying condition.*[4]

Business data such as sales and prices vary from month to month for reasons ranging from the weather to a customer's financial difficulties to the inevitable errors in gathering the data. The manager's challenge is to say when there is a real pattern behind the variation. Statistical tools can help. Sometimes it is enough to simply plot the data. Figure 2 plots the average price of a gallon of unleaded gasoline each month for a decade. Behind the month-to-month variation you can see a gradual upward trend, the higher prices during each summer driving season, the upward spike during the 1990 Gulf War, and the sudden drop in the spring of 1998.[5]

Variation is everywhere. Individuals vary; repeated measurements on the same individual vary; almost everything varies over time. One reason we need to know some statistics is that statistics helps us deal with variation.

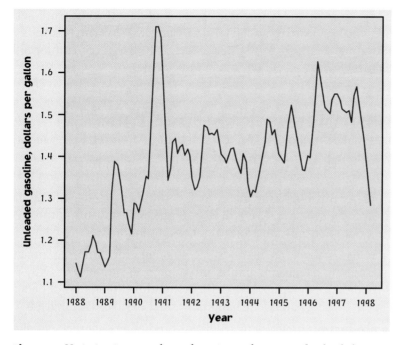

Figure 2 *Variation is everywhere: the price at the pump of unleaded gasoline, 1988 to mid-1998.*

Conclusions Are Not Certain

Most women who reach middle age have regular mammograms to detect breast cancer. Do mammograms really reduce the risk of dying of breast cancer? To seek answers, doctors rely on "randomized clinical trials" that compare different ways of screening for breast cancer. We will see later that data from randomized comparative experiments are as good as it gets. The conclusion from 13 such trials is that mammograms reduce the risk of death in women aged 50 to 64 years by 26%.[6]

On the average, then, women who have regular mammograms are less likely to die of breast cancer. But because variation is everywhere, the results are different for different women. Some women who have mammograms every year die of breast cancer, and some who never have mammograms live to 100 and die when they crash their motorcycles. Can we be sure that mammograms reduce risk on the average? No, we can't be sure. *Because variation is everywhere, conclusions are uncertain.*

Statistics gives us a language for talking about uncertainty that is used and understood by statistically literate people everywhere. In the case of mammograms, the doctors use that language to tell us that "mammography reduces the risk of dying of breast cancer by 26 percent (95 percent confidence interval, 17 to 34 percent)." That 26% is, in Arthur Nielsen's words, "a shorthand for a range that describes our actual knowledge of the underlying condition." The range is 17% to 34%, and we are 95 percent confident that the truth lies in that range. We will soon learn to understand this language. We can't escape variation and uncertainty. Learning statistics enables us to live more comfortably with these realities.

Statistical Thinking and You

What Lies Ahead in This Book

The purpose of this book is to give you a working knowledge of the ideas and tools of practical statistics. We will divide practical statistics into three parts that reflect our short introduction to statistical thinking:

1. **Data analysis** concerns methods and strategies for exploring, organizing, and describing data using graphs and numerical summaries. Only organized data can illuminate. Only thoughtful exploration of data can defeat the lurking variable. Chapters 1 and 2 discuss data analysis.

2. **Data production** provides methods for producing data that can give clear answers to specific questions. Where the data come from really is important—basic concepts about how to select samples and design experiments are the most influential ideas in statistics. These concepts are the subject of Chapter 3.

3. **Statistical inference** moves beyond the data in hand to draw conclusions about a wider universe, taking into account that variation is everywhere and that conclusions are uncertain. To describe variation and uncertainty, inference uses the language of probability, which we introduce in Chapter 4 and the optional Chapter 5. Because we are concerned with practice rather than theory, we can function with a quite limited knowledge of probability. Chapter 6 discusses the reasoning of statistical inference, and Chapters 7 and 8 present inference as used in practice in several simple settings. Chapters 9, 10, and 11 offer brief introductions to inference in some more complex settings.

Because data are numbers with a context, doing statistics means more than manipulating numbers. *The Basic Practice of Statistics* is full of data, and each set of data has some brief background to help you understand what the data say. Examples and exercises usually express briefly some understanding gained from the data. In practice, you would know much more about the background of the data you work with and about the questions you hope the data will answer. No textbook can be fully realistic. But it is important to form the habit of asking "What do the data tell me?" rather than just concentrating on making graphs and doing calculations. This book tries to encourage good habits.

Nonetheless, statistics involves lots of calculating and graphing. The text presents the techniques you need, but you should use a calculator or computer software to automate calculations and graphs as much as possible. There are many kinds of statistical software, from spreadsheets to large programs for advanced users of statistics. The kind of computing available to learners varies a great deal from place to place—but the big ideas of statistics don't depend on any particular level of access to computing. We encourage use of statistical software, but this book does not require software and is not tied to any specific software.

This book does require that you have a calculator with some built-in statistical functions. Specifically, you need a calculator that will find means and standard deviations and calculate correlations and regression lines. Look for a calculator that claims to do "two-variable statistics" or mentions "regression." Advanced calculators will do much more, including some statistical graphs, but large screens and easy data editing make computers more suitable for elaborate statistical analyses.

Calculators and computers can follow recipes for graphs and calculations both more quickly and more accurately than humans can. Because graphing and calculating are automated in statistical practice, the most important assets you can gain from the study of statistics are an understanding of the big ideas and the beginnings of good judgment in working with data. Ideas and judgment can't (at least yet) be automated. They guide you in telling the computer what to do and in interpreting its output. This book tries to explain the most important ideas of statistics, not just teach methods. Some examples of big ideas that you will meet (one from each of the three areas of statistics) are "always plot your data," "randomized comparative experiments," and "statistical significance."

You learn statistics by doing statistical problems. This book offers three levels of problems, arranged to help you learn. Short "Apply Your Knowledge" problem sets appear after each major idea. These are straightforward exercises that help you solidify the main points as you read. Pause for a few of these exercises before going on. The Section Exercises at the end of each numbered section help you combine all the ideas of the section. Finally, the Chapter Review Exercises look back over the entire chapter. At each step you are given less advance knowledge of exactly what statistical ideas and skills the problems will require, so each step requires more understanding. Each chapter ends with a Statistics in Summary section that includes a detailed list of specific things you should now be able to do. Go through that list, and be sure you can say "I can do that" to each item. Then try some chapter exercises.

The basic principle of learning is persistence. The main ideas of statistics, like the main ideas of any important subject, took a long time to discover and take some time to master. The gain will be worth the pain.

Understanding Data

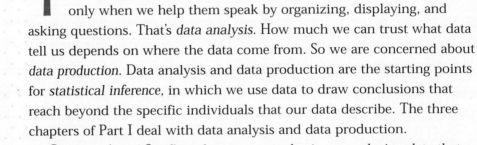

The first step in understanding data is to hear what the data say, to "let the numbers speak for themselves." Numbers speak clearly only when we help them speak by organizing, displaying, and asking questions. That's *data analysis.* How much we can trust what data tell us depends on where the data come from. So we are concerned about *data production.* Data analysis and data production are the starting points for *statistical inference,* in which we use data to draw conclusions that reach beyond the specific individuals that our data describe. The three chapters of Part I deal with data analysis and data production.

Chapters 1 and 2 reflect the strong emphasis on exploring data that characterizes modern statistics. Although careful exploration of data is essential if we are to trust the results of inference, data analysis isn't just preparation for inference. To think about inference, we carefully distinguish between the data we actually have and the larger universe we want conclusions about. The Bureau of Labor Statistics, for example, has data about employment in the 50,000 households contacted by its Current Population Survey. The bureau wants to draw conclusions about employment in all 100 million U.S. households. That's a complex problem. From the viewpoint of data analysis, things are simpler. We want simply to explore and understand the data in hand. The distinctions that inference requires don't concern us in Chapters 1 and 2. What does concern us is a systematic strategy for examining data and the tools that we use to carry out that strategy.

Of course, we often do want to use data to reach general conclusions. Whether we can do that depends most of all on how the data were produced. Good data rarely "just happen." They are products of human effort, like video games and nylon sweaters. Chapter 3 tells how to

"Tonight, we're going to let the statistics speak for themselves."

(DRAWING BY KOREN © 1974 THE NEW YORKER MAGAZINE, INC.)

produce good data yourself and how to decide whether to trust data produced by others.

The study of data analysis and data production equips you with tools and ideas that are immediately useful whenever you deal with numbers. Inference is more specialized and more subtle. Inference requires more attention in a textbook, but that doesn't mean it is more important. Statistics is the science of data, and the three chapters of Part I deal directly with data.

Florence Nightingale

Using Statistics to Save Lives

(THE GRANGER COLLECTION, NEW YORK.)

Florence Nightingale (1820–1910) won fame as a founder of the nursing profession and as a reformer of health care. As chief nurse for the British army during the Crimean War, from 1854 to 1856, she found that lack of sanitation and disease killed large numbers of soldiers hospitalized by wounds. Her reforms reduced the death rate at her military hospital from 42.7% to 2.2%, and she returned from the war famous. She at once began a fight to reform the entire military health care system, with considerable success.

One of the chief weapons Florence Nightingale used in her efforts was data.

One of the chief weapons Florence Nightingale used in her efforts was data. She had the facts, because she reformed record keeping as well as medical care. She was a pioneer in using graphs to present data in a vivid form that even generals and members of Parliament could understand. Her inventive graphs are a landmark in the growth of the new science of statistics. She considered statistics essential to understanding any social issue and tried to introduce the study of statistics into higher education.

In beginning our study of statistics, we will follow Florence Nightingale's lead. This chapter and the next will stress the analysis of data as a path to understanding. Like her, we will start with graphs to see what data can teach us. Along with the graphs we will present numerical summaries, just as Florence Nightingale calculated detailed death rates and other summaries. Data for Florence Nightingale were not dry or abstract, because they showed her, and helped her show others, how to save lives. That remains true today.

Examining Distributions

(OIL PAINTING, ABOUT 1856 BY JERRY BARRETT FROM THE GRANGER COLLECTION, NEW YORK.)

Florence Nightingale in a hospital in Turkey during the Crimean War.

Introduction

Statistics is the science of data. We therefore begin our study of statistics by mastering the art of examining data. Any set of data contains information about some group of *individuals*. The information is organized in *variables*.

> ### INDIVIDUALS AND VARIABLES
>
> **Individuals** are the objects described by a set of data. Individuals may be people, but they may also be animals or things.
>
> A **variable** is any characteristic of an individual. A variable can take different values for different individuals.

A college's student data base, for example, includes data about every currently enrolled student. The students are the individuals described by the data set. For each individual, the data contain the values of variables such as date of birth, gender (female or male), choice of major, and grade point average. In practice, any set of data is accompanied by background information that helps us understand the data. When you plan a statistical study or explore data from someone else's work, ask yourself the following questions:

1. **Who?** What **individuals** do the data describe? **How many** individuals appear in the data?
2. **What?** How many **variables** do the data contain? What are the **exact definitions** of these variables? In what **units of measurement** is each variable recorded? Weights, for example, might be recorded in pounds, in thousands of pounds, or in kilograms.
3. **Why?** What **purpose** do the data have? Do we hope to answer some specific questions? Do we want to draw conclusions about individuals other than the ones we actually have data for?

Some variables, like gender and college major, simply place individuals into categories. Others, like height and grade point average, take numerical values for which we can do arithmetic. It makes sense to give an average income for a company's employees, but it does not make sense to give an "average" gender. We can, however, count the numbers of female and male employees and do arithmetic with these counts.

> ### CATEGORICAL AND QUANTITATIVE VARIABLES
>
> A **categorical variable** places an individual into one of several groups or categories.
>
> A **quantitative variable** takes numerical values for which arithmetic operations such as adding and averaging make sense.
>
> The **distribution** of a variable tells us what values it takes and how often it takes these values.

EXAMPLE 1.1   **A corporate data set**

Here is a small part of the data set in which CyberStat Corporation records information about its employees:

	A	B	C	D	E	F
1	Name	Age	Gender	Race	Salary	Job Type
2	Fleetwood, Delores	39	Female	White	62,100	Management
3	Perez, Juan	27	Male	White	47,350	Technical
4	Wang, Lin	22	Female	Asian	18,250	Clerical
5	Johnson, LaVerne	48	Male	Black	77,600	Management

Enter   NUM

case

The *individuals* described are the employees. Each row records data on one individual. You will often see each row of data called a **case.** Each column contains the values of one *variable* for all the individuals. In addition to the person's name, there are 5 variables. Gender, race, and job type are categorical variables. Age and salary are quantitative variables. You can see that age is measured in years and salary in dollars.

spreadsheet

Most data tables follow this format—each row is an individual, and each column is a variable. This data set appears in a **spreadsheet** program that has rows and columns ready for your use. Spreadsheets are commonly used to enter and transmit data.

APPLY YOUR KNOWLEDGE ·

1.1   Here is a small part of a data set that describes the fuel economy (in miles per gallon) of 1998 model motor vehicles:

Make and model	Vehicle type	Transmission type	Number of cylinders	City MPG	Highway MPG
⋮					
BMW 318I	Subcompact	Automatic	4	22	31
BMW 318I	Subcompact	Manual	4	23	32
Buick Century	Midsize	Automatic	6	20	29
Chevrolet Blazer	Four-wheel drive	Automatic	6	16	20
⋮					

(a) What are the individuals in this data set?

(b) For each individual, what variables are given? Which of these variables are categorical and which are quantitative?

1.2 Data from a medical study contain values of many variables for each of the people who were the subjects of the study. Which of the following variables are categorical and which are quantitative?

(a) Gender (female or male)

(b) Age (years)

(c) Race (Asian, black, white, or other)

(d) Smoker (yes or no)

(e) Systolic blood pressure (millimeters of mercury)

(f) Level of calcium in the blood (micrograms per milliliter)

1.1 Displaying Distributions with Graphs

exploratory data analysis

Statistical tools and ideas help us examine data in order to describe their main features. This examination is called **exploratory data analysis.** Like an explorer crossing unknown lands, we want first to simply describe what we see. Here are two basic strategies that help us organize our exploration of a set of data:

- Begin by examining each variable by itself. Then move on to study the relationships among the variables.

- Begin with a graph or graphs. Then add numerical summaries of specific aspects of the data.

We will follow these principles in organizing our learning. This chapter presents methods for describing a single variable. We study relationships among several variables in Chapter 2. Within each chapter, we begin with graphical displays, then add numerical summaries for more complete description.

Categorical variables: bar graphs and pie charts

The values of a categorical variable are labels for the categories, such as "male" and "female." The distribution of a categorical variable lists the categories and gives either the **count** or the **percent** of individuals who fall in each category. For example, here is the distribution of marital status for all Americans aged 18 and over:[1]

Marital status	Count (millions)	Percent
Never married	43.9	22.9
Married	116.7	60.9
Widowed	13.4	7.0
Divorced	17.6	9.2

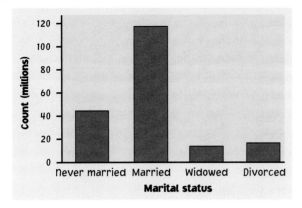

Figure 1.1(a) *Bar graph of the marital status of U.S. adults.*

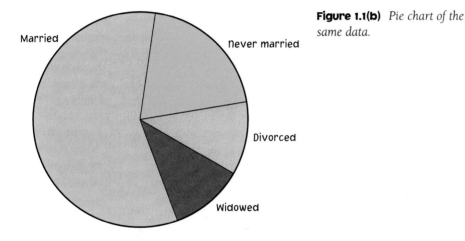

Figure 1.1(b) *Pie chart of the same data.*

The graphs in Figure 1.1 display these data. The **bar graph** in Figure 1.1(a) quickly compares the sizes of the four marital status groups. The heights of the four bars show the counts in the four categories. The **pie chart** in Figure 1.1(b) helps us see what part of the whole each group forms. For example, the "married" slice makes up 61% of the pie because 61% of adults are married. To make a pie chart, you must include all the categories that make up a whole. Bar graphs are more flexible. For example, you can use a bar graph to compare the numbers of students at your college majoring in biology, business, and political science. A pie chart cannot make this comparison because not all students fall into one of these three majors.

Bar graphs and pie charts help an audience grasp the distribution quickly. They are, however, of limited use for data analysis because it is easy to understand categorical data on a single variable such as marital status without a graph. We will move on to quantitative variables, where graphs are essential tools.

bar graph

pie chart

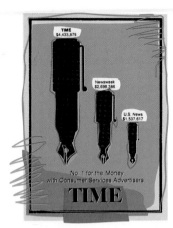

How to lie with a bar graph

This looks like a bar graph, but it isn't. *Time* used the graph to show that it leads in consumer services advertising. Instead of dull bars, the artist drew pictures of pens. *Time's* advertising take is double *Newsweek's*, but the artist made both the height and the width of *Time's* pen double those of *Newsweek's* pen. Our eyes react to the area of the pens, and *Time's* is four times as large. The pictures mislead us. Bars may be dull, but they are truthful.

histogram

APPLY YOUR KNOWLEDGE

1.3 **Female doctorates.** Here are data on the percent of females among people earning doctorates in 1994 in several fields of study (from the 1997 *Statistical Abstract of the United States*):

Computer science	15.4%	Life sciences	40.7%
Education	60.8%	Physical sciences	21.7%
Engineering	11.1%	Psychology	62.2%

(a) Present these data in a well-labeled bar graph.

(b) Would it also be correct to use a pie chart to display these data? Explain your answer.

1.4 **Accidental deaths.** In 1995 there were 90,402 deaths from accidents in the United States. Among these were 43,363 deaths from motor vehicle accidents, 10,483 from falls, 9072 from poisoning, 4350 from drowning, and 4235 from fires. (Data from the Centers for Disease Control Web site, www.cdc.gov.)

(a) Find the percent of accidental deaths from each of these causes, rounded to the nearest percent. What percent of accidental deaths were due to other causes?

(b) Make a well-labeled bar graph of the distribution of causes of accidental deaths. Be sure to include an "other causes" bar.

(c) Would it also be correct to use a pie chart to display these data? Explain your answer.

Quantitative variables: histograms

Quantitative variables often take many values. A graph of the distribution is clearer if nearby values are grouped together. The most common graph of the distribution of one quantitative variable is a **histogram.**

EXAMPLE 1.2 **How to make a histogram**

Table 1.1 presents the percent of residents aged 65 years and over in each of the 50 states. To make a histogram of this distribution, proceed as follows.

Step 1. Divide the range of the data into classes of equal width. The data in Table 1.1 range from 5.2 to 18.5, so we choose as our classes

$$5.0 < \text{percent over } 65 \le 6.0$$

$$6.0 < \text{percent over } 65 \le 7.0$$

$$\vdots$$

$$18.0 < \text{percent over } 65 \le 19.0$$

Be sure to specify the classes precisely so that each observation falls into exactly one class. A state with 6.0% of its residents aged 65 or older would fall into the first class, but 6.1% falls into the second.

Step 2. Count the number of observations in each class. Here are the counts:

Class	Count	Class	Count	Class	Count
5.1 to 6.0	1	10.1 to 11.0	4	15.1 to 16.0	4
6.1 to 7.0	0	11.1 to 12.0	8	16.1 to 17.0	0
7.1 to 8.0	0	12.1 to 13.0	13	17.1 to 18.0	0
8.1 to 9.0	1	13.1 to 14.0	12	18.1 to 19.0	1
9.1 to 10.0	1	14.1 to 15.0	5		

Step 3. Draw the histogram. First, on the horizontal axis mark the scale for the variable whose distribution you are displaying. That's "percent of state residents aged 65 and over" in this example. The scale runs from 5 to 19 because that is the span of the classes we chose. The vertical axis contains the scale of counts. Each bar represents a class. The base of the bar covers the class, and the bar height is the class count. There is no horizontal space between the bars unless a class is empty, so that its bar has height zero. Figure 1.2 is our histogram.

Table 1.1 **Percent of population 65 years old and over, by state (1996)**

State	Percent	State	Percent	State	Percent
Alabama	13.0	Louisiana	11.4	Ohio	13.4
Alaska	5.2	Maine	13.9	Oklahoma	13.5
Arizona	13.2	Maryland	11.4	Oregon	13.4
Arkansas	14.4	Massachusetts	14.1	Pennsylvania	15.9
California	10.5	Michigan	12.4	Rhode Island	15.8
Colorado	11.0	Minnesota	12.4	South Carolina	12.1
Connecticut	14.3	Mississippi	12.3	South Dakota	14.4
Delaware	12.8	Missouri	13.8	Tennessee	12.5
Florida	18.5	Montana	13.2	Texas	10.2
Georgia	9.9	Nebraska	13.8	Utah	8.8
Hawaii	12.9	Nevada	11.4	Vermont	12.1
Idaho	11.4	New Hampshire	12.0	Virginia	11.2
Illinois	12.5	New Jersey	13.8	Washington	11.6
Indiana	12.6	New Mexico	11.0	West Virginia	15.2
Iowa	15.2	New York	13.4	Wisconsin	13.3
Kansas	13.7	North Carolina	12.5	Wyoming	11.2
Kentucky	12.6	North Dakota	14.5		

Source: Statistical Abstract of the United States, 1997.

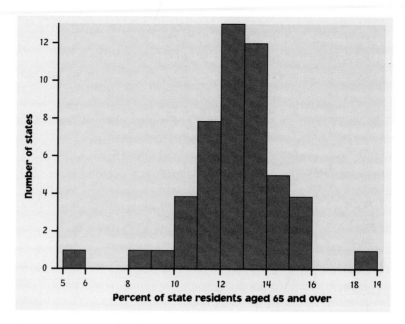

Figure 1.2 *Histogram of the percent of residents aged 65 and over in the 50 states, from Table 1.1.*

The bars of a histogram should cover the entire range of values of a variable. When the possible values of a variable have gaps between them, extend the bases of the bars to meet halfway between two adjacent possible values. For example, in a histogram of the ages in years of university faculty, the bars representing 25 to 29 years and 30 to 34 years would meet at 29.5.

Our eyes respond to the *area* of the bars in a histogram.[2] Because the classes are all the same width, area is determined by height and all classes are fairly represented. There is no one right choice of the classes in a histogram. Too few classes will give a "skyscraper" graph, with all values in a few classes with tall bars. Too many will produce a "pancake" graph, with most classes having one or no observations. Neither choice will give a good picture of the shape of the distribution. You must use your judgment in choosing classes to display the shape. Statistics software will choose the classes for you. The computer's choice is usually a good one, but you can change it if you want.

APPLY YOUR KNOWLEDGE ·

1.5 **Automobile fuel economy.** Environmental Protection Agency regulations require automakers to give the city and highway gas mileages for each model of car. Table 1.2 gives the highway

Table 1.2 Highway gas mileage for 1998 model midsize cars

Model	MPG	Model	MPG
Acura 3.5RL	25	Lexus GS300	23
Audi A6 Quattro	26	Lexus LS400	25
Buick Century	29	Lincoln Mark VIII	26
Cadillac Catera	24	Mazda 626	33
Cadillac Eldorado	26	Mercedes-Benz E320	29
Chevrolet Lumina	29	Mercedes-Benz E420	26
Chrysler Cirrus	30	Mitsubishi Diamante	24
Dodge Stratus	28	Nissan Maxima	28
Ford Taurus	28	Oldsmobile Aurora	26
Honda Accord	29	Rolls-Royce Silver Spur	16
Hyundai Sonata	27	Saab 900S	25
Infiniti I30	28	Toyota Camry	30
Infiniti Q45	23	Volvo S70	25

mileages (miles per gallon) for 26 midsize 1998 car models.[3] Make a histogram of the highway mileages of these cars.

Interpreting histograms

Making a statistical graph is not an end in itself. The purpose of the graph is to help us understand the data. After you make a graph, always ask, "What do I see?" Once you have displayed a distribution, you can see its important features as follows.

EXAMINING A DISTRIBUTION

In any graph of data, look for the **overall pattern** and for striking **deviations** from that pattern.

You can describe the overall pattern of a histogram by its **shape**, **center**, and **spread**.

An important kind of deviation is an **outlier**, an individual value that falls outside the overall pattern.

We will learn how to describe center and spread numerically in Section 1.2. For now, we can describe the center of a distribution by its *midpoint,* the value with roughly half the observations taking smaller values and half taking larger values. We can describe the spread of a distribution by giving the *smallest and largest values.*

EXAMPLE 1.3 **Describing a distribution**

Look again at the histogram in Figure 1.2. **Shape**: The distribution is roughly *symmetric* and has a *single peak*. **Center**: The midpoint of the distribution is close to the single peak, at about 13%. **Spread**: The spread is about 10% to 16% if we ignore the four most extreme observations.

Outliers: Two states stand out in the histogram of Figure 1.2. You can find them in the table once the histogram has called attention to them. Florida has 18.5% of its residents over age 65, and Alaska has only 5.2%. Once you have spotted outliers, look for an explanation. Some outliers are due to mistakes, such as typing 5.0 as 50. Other outliers point to the special nature of some observations. Florida, with its many retired people, has many residents over 65, and Alaska, the northern frontier, has few.

When you describe a distribution, concentrate on the main features. Look for major peaks, not for minor ups and downs in the bars of the histogram. Look for clear outliers, not just for the smallest and largest observations. Look for rough *symmetry* or clear *skewness*.

SYMMETRIC AND SKEWED DISTRIBUTIONS

A distribution is **symmetric** if the right and left sides of the histogram are approximately mirror images of each other.

A distribution is **skewed to the right** if the right side of the histogram (containing the half of the observations with larger values) extends much farther out than the left side. It is **skewed to the left** if the left side of the histogram extends much farther out than the right side.

In mathematics, symmetry means that the two sides of a figure like a histogram are exact mirror images of each other. Data are almost never exactly symmetric, so we are willing to call histograms like that in Figure 1.2 approximately symmetric as an overall description. Here are more examples.

EXAMPLE 1.4 **Lightning flashes and Shakespeare**

Figure 1.3 comes from a study of lightning storms in Colorado. It shows the distribution of the hour of the day during which the first lightning flash for that day occurred. The distribution has a single peak at noon and falls off on either side of this peak. The two sides of the histogram are roughly the same shape, so we call the distribution symmetric.

Figure 1.4 shows the distribution of lengths of words used in Shakespeare's plays.[4] This distribution also has a single peak but is skewed to the right. That is, there are many short words (3 and 4 letters) and few very long words (10, 11, or 12 letters), so that the right tail of the histogram extends out much farther than the left tail.

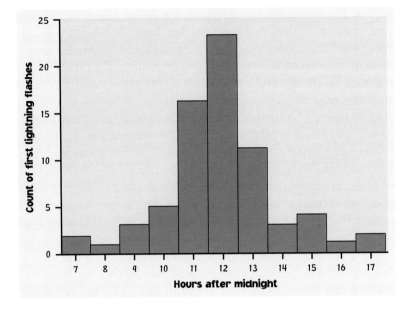

Figure 1.3 *The distribution of the time of the first lightning flash each day at a site in Colorado.*

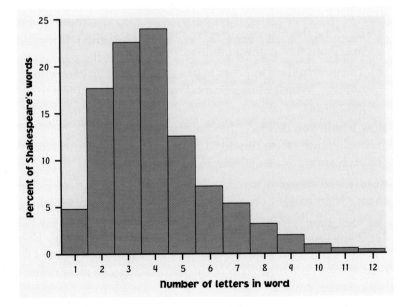

Figure 1.4 *The distribution of lengths of words used in Shakespeare's plays.*

The vital few

Skewed distributions can show us where to concentrate our efforts. Ten percent of the cars on the road account for half of all carbon dioxide emissions. A histogram of CO_2 emissions would show many cars with small or moderate values and a few with very high values. Cleaning up or replacing these cars would reduce pollution at a cost much lower than that of programs aimed at all cars. Statisticians who work at improving quality in industry make a principle of this: distinguish "the vital few" from "the trivial many."

Notice that the vertical scale in Figure 1.4 is not the *count* of words but the *percent* of all of Shakespeare's words that have each length. A histogram of percents rather than counts is convenient when the counts are very large or when we want to compare several distributions. Different kinds of writing have different distributions of word lengths, but all are right-skewed because short words are common and very long words are rare.

The overall shape of a distribution is important information about a variable. Some types of data regularly produce distributions that are symmetric or skewed. For example, the sizes of living things of the same species (like lengths of cockroaches) tend to be symmetric. Data on incomes (whether of individuals, companies, or nations) are usually strongly skewed to the right. There are many moderate incomes, some large incomes, and a few very large incomes. Do remember that many distributions have shapes that are neither symmetric nor skewed. Some data show other patterns. Scores on an exam, for example, may have a cluster near the top of the scale if many students did well. Or they may show two distinct peaks if a tough problem divided the class into those who did and didn't solve it. Use your eyes and describe what you see.

APPLY YOUR KNOWLEDGE

1.6 **Automobile fuel economy.** Table 1.2 (page 11) gives data on the fuel economy of 1998 model midsize cars. Based on a histogram of these data:

(a) Describe the main features (shape, center, spread, outliers) of the distribution of highway mileage.

(b) The government imposes a "gas guzzler" tax on cars with low gas mileage. Which of these cars do you think are subject to the gas guzzler tax?

1.7 How would you describe the center and spread of the distribution of first lightning flash times in Figure 1.3? Of the distribution of Shakespeare's word lengths in Figure 1.4?

1.8 **Returns on common stocks.** The total return on a stock is the change in its market price plus any dividend payments made. Total return is usually expressed as a percent of the beginning price. Figure 1.5 is a histogram of the distribution of total returns for all 1528 stocks listed on the New York Stock Exchange in one year.[5] Like Figure 1.4, it is a histogram of the percents in each class rather than a histogram of counts.

(a) Describe the overall shape of the distribution of total returns.

(b) What is the approximate center of this distribution? (For now, take the center to be the value with roughly half the stocks having lower returns and half having higher returns.)

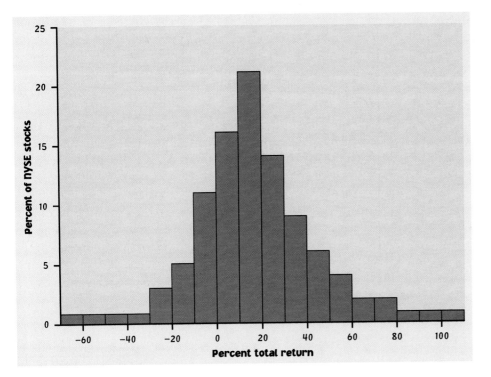

(c) Approximately what were the smallest and largest total returns? (This describes the spread of the distribution.)

(d) A return less than zero means that an owner of the stock lost money. About what percent of all stocks lost money?

Quantitative variables: stemplots

Histograms are not the only graphical display of distributions. For small data sets, a *stemplot* is quicker to make and presents more detailed information.

STEMPLOT

To make a **stemplot**:

1. Separate each observation into a **stem** consisting of all but the final (rightmost) digit and a **leaf**, the final digit. Stems may have as many digits as needed, but each leaf contains only a single digit.

2. Write the stems in a vertical column with the smallest at the top, and draw a vertical line at the right of this column.

3. Write each leaf in the row to the right of its stem, in increasing order out from the stem.

```
 5 | 2
 6 |
 7 |
 8 | 8
 9 | 9
10 | 2 5
11 | 0 0 2 2 4 4 4 4 6
12 | 0 1 1 3 4 4 5 5 5 6 6 8 9
13 | 0 2 2 3 4 4 4 5 7 8 8 8 9
14 | 1 3 4 4 5
15 | 2 2 8 9
16 |
17 |
18 | 5
```

Figure 1.6 *Stemplot of the percent of residents age 65 and over in the states. Compare the histogram of these data in Figure 1.2.*

EXAMPLE 1.5 Stemplot of the "65 and over" data

For the "65 and over" percents in Table 1.1, the whole-number part of the observation is the stem and the final digit (tenths) is the leaf. The Alabama entry, 13.0, has stem 13 and leaf 0. Stems can have as many digits as needed, but each leaf must consist of only a single digit. Figure 1.6 is the stemplot for the data in Table 1.1.

A stemplot looks like a histogram turned on end. The stemplot in Figure 1.6 resembles the histogram in Figure 1.2. The two graphs differ slightly because the classes chosen for the histogram are not the same as the stems in the stemplot. The stemplot, unlike the histogram, preserves the actual value of each observation. We interpret stemplots like histograms, looking for the overall pattern and for any outliers.

rounding

You can choose the classes in a histogram. The classes (the stems) of a stemplot are given to you. You can get more flexibility by **rounding** the data so that the final digit after rounding is suitable as a leaf. Do this when the data have too many digits. For example, data like

$$3.468 \quad 2.567 \quad 2.981 \quad 1.095 \quad \dots$$

would have too many stems if we took the first three digits as the stem and the final digit as the leaf. You should round these data to

$$3.5 \quad 2.6 \quad 3.0 \quad 1.1 \quad \dots$$

before making a stemplot.

splitting stems

You can also **split stems** to double the number of stems when all the leaves would otherwise fall on just a few stems. Each stem then appears twice. Leaves 0 to 4 go on the upper stem and leaves 5 to 9 go on the lower stem. If you split the stems in the stemplot of Figure 1.6, for example, the 12 and 13 stems become

```
12 | 011344
12 | 5556689
13 | 0223444
13 | 578889
```

Rounding and splitting stems are matters for judgment, like choosing the classes in a histogram. The stemplot in Figure 1.6 does not need either change. Stemplots work well for small sets of data. When there are more than 100 observations, a histogram is almost always a better choice.

APPLY YOUR KNOWLEDGE

1.9 **Students' attitudes.** The Survey of Study Habits and Attitudes (SSHA) is a psychological test that evaluates college students' motivation, study habits, and attitudes toward school. A private college gives the SSHA to 18 of its incoming first-year women students. Their scores are

154	109	137	115	152	140	154	178	101
103	126	126	137	165	165	129	200	148

Make a stemplot of these data. The overall shape of the distribution is irregular, as often happens when only a few observations are available. Are there any outliers? About where is the center of the distribution (the score with half the scores above it and half below)? What is the spread of the scores (ignoring any outliers)?

Time plots

Many variables are measured at intervals over time. We might, for example, measure the height of a growing child or the price of a stock at the end of each month. In these examples, our main interest is change over time. To display change over time, make a *time plot*.

TIME PLOT

A **time plot** of a variable plots each observation against the time at which it was measured. Always put time on the horizontal scale of your plot and the variable you are measuring on the vertical scale. Connecting the data points by lines helps emphasize any change over time.

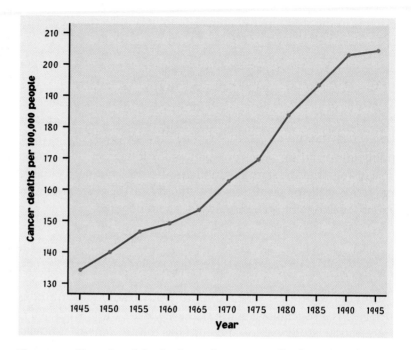

Figure 1.7 *Time plot of the death rate from cancer (deaths per 100,000 people) from 1945 to 1995.*

EXAMPLE 1.6 **Deaths from cancer**

Here are data on the rate of deaths from cancer (deaths per 100,000 people) in the United States over the 50-year period 1945 to 1995:

Year	1945	1950	1955	1960	1965	1970	1975	1980	1985	1990	1995
Deaths	134.0	139.8	146.5	149.2	153.5	162.8	169.7	183.9	193.3	203.2	204.7

Figure 1.7 is a time plot of these data. It shows the steady increase in the cancer death rate during the past half-century. This increase does not mean that we have made no progress in treating cancer. Because cancer is primarily a disease of old age, the death rate from cancer increases when people live longer even if treatment improves. In fact, if we adjust for the increasing age of the U.S. population, the rate of deaths from cancer has been dropping since 1992.

trend When you examine a time plot, look once again for an overall pattern and for strong deviations from the pattern. One common overall pattern is a **trend**, a long-term upward or downward movement over time. Figure 1.7 shows an upward trend in the cancer death rate, with no striking deviations such as short-term drops.

1.10   **Yields of money market funds.** Many people invest in "money market funds." These are mutual funds that attempt to maintain a constant price of $1 per share while paying monthly interest. Here are the average annual interest rates (in percent) paid by all taxable money market funds since 1973, the first full year in which such funds were available.[6]

Year	Rate	Year	Rate	Year	Rate	Year	Rate
1973	7.60	1979	10.92	1985	7.77	1991	5.70
1974	10.79	1980	12.88	1986	6.30	1992	3.31
1975	6.39	1981	17.16	1987	6.17	1993	2.62
1976	5.11	1982	12.55	1988	7.09	1994	3.65
1977	4.92	1983	8.69	1989	8.85	1995	5.37
1978	7.25	1984	10.21	1990	7.81	1996	4.80

(a) Make a time plot of the interest paid by money market funds for these years.

(b) Interest rates, like many economic variables, show **cycles**, clear but irregular up-and-down movements. In which years did the interest rate cycle reach temporary peaks?

cycles

(c) A time plot may show a consistent trend underneath cycles. When did interest rates reach their overall peak during these years? Has there been a general trend downward since that year?

SECTION 1.1 S u m m a r y

A data set contains information on a number of **individuals**. Individuals may be people, animals, or things. For each individual, the data give values for one or more **variables**. A variable describes some characteristic of an individual, such as a person's height, gender, or salary.

Some variables are **categorical** and others are **quantitative**. A categorical variable places each individual into a category, like male or female. A quantitative variable has numerical values that measure some characteristic of each individual, like height in centimeters or salary in dollars per year.

Exploratory data analysis uses graphs and numerical summaries to describe the variables in a data set and the relations among them.

The **distribution** of a variable describes what values the variable takes and how often it takes these values.

To describe a distribution, begin with a graph. **Bar graphs** and **pie charts** describe the distribution of a categorical variable. **Histograms** and **stemplots** graph the distributions of quantitative variables.

When examining any graph, look for an **overall pattern** and for notable **deviations** from the pattern.

Shape, center, and **spread** describe the overall pattern of a distribution. Some distributions have simple shapes, such as **symmetric** and **skewed**. Not all distributions have a simple overall shape, especially when there are few observations.

Outliers are observations that lie outside the overall pattern of a distribution. Always look for outliers and try to explain them.

When observations on a variable are taken over time, make a **time plot** that graphs time horizontally and the values of the variable vertically. A time plot can reveal **trends** or other changes over time.

SECTION 1.1 Exercises

1.11   **Athletes' salaries.** Here is a small part of a data set that describes major league baseball players as of opening day of the 1998 season:

Player	Team	Position	Age	Salary
⋮				
Perez, Eduardo	Reds	First base	28	300
Perez, Neifi	Rockies	Shortstop	23	210
Pettitte, Andy	Yankees	Pitcher	25	3750
Piazza, Mike	Dodgers	Catcher	29	8000
⋮				

(a) What individuals does this data set describe?

(b) In addition to the player's name, how many variables does the data set contain? Which of these variables are categorical and which are quantitative?

(c) Based on the data in the table, what do you think are the units of measurement for each of the quantitative variables?

1.12   **How young people die.** The number of deaths among persons aged 15 to 24 years in the United States in 1997 due to the seven leading causes of death for this age group were: accidents, 12,958; homicide, 5793; suicide, 4146; cancer, 1583; heart disease, 1013; congenital defects, 383; AIDS, 276.[7]

(a) Make a bar graph to display these data.

(b) What additional information do you need to make a pie chart?

1.13   **The statistics of writing style.** Numerical data can distinguish different types of writing, and sometimes even individual authors.

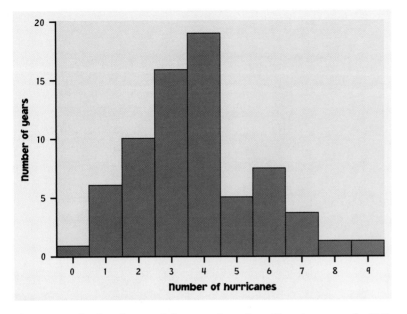

Figure 1.8 *The distribution of the annual number of hurricanes on the U.S. east coast over a 70-year period, for Exercise 1.14.*

Here are data on the percent of words of 1 to 15 letters used in articles in *Popular Science* magazine:[8]

Length	1	2	3	4	5	6	7	8	9	10	11	12	13	14	15
Percent	3.6	14.8	18.7	16.0	12.5	8.2	8.1	5.9	4.4	3.6	2.1	0.9	0.6	0.4	0.2

(a) Make a histogram of this distribution. Describe its shape, center, and spread.

(b) How does the distribution of lengths of words used in *Popular Science* compare with the similar distribution in Figure 1.4 for Shakespeare's plays? Look in particular at short words (2, 3, and 4 letters) and very long words (more than 10 letters).

1.14 **Hurricanes.** The histogram in Figure 1.8 shows the number of hurricanes reaching the east coast of the United States each year over a 70-year period.[9] Give a brief description of the overall shape of this distribution. About where does the center of the distribution lie?

1.15 **Batting averages.** Figure 1.9 displays the distribution of batting averages for all 167 American League baseball players who batted at least 200 times in the 1980 season. (The outlier is the .390 batting average of George Brett, the highest batting average in the major leagues since Ted Williams hit .406 in 1941.)

Figure 1.4 *The distribution of batting averages of American League players in 1980, for Exercise 1.15.*

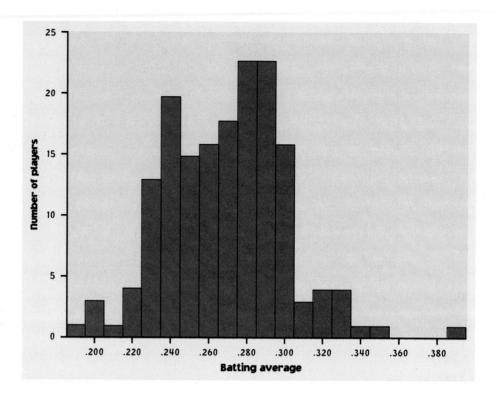

(a) Is the overall shape (ignoring the outlier) roughly symmetric or clearly skewed or neither?

(b) What was the approximate batting average of a typical American League player? About what were the highest and lowest batting averages, leaving out George Brett?

1.16 Sketch a histogram for a distribution that is skewed to the left. Suppose that you and your friends emptied your pockets of coins and recorded the year marked on each coin. The distribution of dates would be skewed to the left. Explain why.

1.17 **The changing age distribution of the United States.** The distribution of the ages of a nation's population has a strong influence on economic and social conditions. Table 1.3 shows the age distribution of U.S. residents in 1950 and 2075, in millions of persons. The 1950 data come from that year's census. The 2075 data are projections made by the Census Bureau.

(a) Because the total population in 2075 is much larger than the 1950 population, comparing percents in each age group is clearer than comparing counts. Make a table of the percent of the total population in each age group for both 1950 and 2075.

(b) Make a histogram of the 1950 age distribution (in percents). Then describe the main features of the distribution. In particular, look at the percent of children relative to the rest of the population.

Table 1.3 Age distribution in the United States, 1950 and 2075 (in millions of persons)

Age group	1950	2075
Under 10 years	29.3	34.9
10 to 19 years	21.8	35.7
20 to 29 years	24.0	36.8
30 to 39 years	22.8	38.1
40 to 49 years	19.3	37.8
50 to 59 years	15.5	37.5
60 to 69 years	11.0	34.5
70 to 79 years	5.5	27.2
80 to 89 years	1.6	18.8
90 to 99 years	0.1	7.7
100 to 109 years	—	1.7
Total	151.1	310.6

(c) Make a histogram of the projected age distribution for the year 2075. Use the same scales as in (b) for easy comparison. What are the most important changes in the U.S. age distribution projected for the 125-year period between 1950 and 2075?

1.18 **Babe Ruth's home runs.** Here are the numbers of home runs that Babe Ruth hit in his 15 years with the New York Yankees, 1920 to 1934:

54 59 35 41 46 25 47 60 54 46 49 46 41 34 22

(NATIONAL BASEBALL HALL OF FAME AND MUSEUM, INC., COOPERSTOWN, N.Y.)

Make a stemplot for these data. Is the distribution roughly symmetric, clearly skewed, or neither? About how many home runs did Ruth hit in a typical year? Is his famous 60 home runs in 1927 an outlier?

1.19 **Back-to-back stemplot.** The current major league single-season home run record is held by Mark McGwire of the St. Louis Cardinals. Here are McGwire's home run counts for 1987 to 1998:

49 32 33 39 22 42 9 9 39 52 58 70

A **back-to-back stemplot** helps us compare two distributions. Write the stems as usual, but with a vertical line both to their left and to their right. On the right, put leaves for Ruth (Exercise 1.18). On the left, put leaves for McGwire. Arrange the leaves on each stem in increasing order out from the stem. Now write a brief comparison of Ruth and McGwire as home run hitters. (McGwire was injured in 1993 and there was a baseball strike in 1994. How do these events appear in the data?)

back-to-back stemplot

Table 1.4 Size and length of bear markets

Year	Decline (percent)	Duration (months)	Year	Decline (percent)	Duration (months)
1940–1942	42	28	1966	22	8
1946	27	5	1968–1970	36	18
1950	14	1	1973–1974	48	21
1953	15	8	1981–1982	26	19
1955	10	1	1983–1984	14	10
1956–1957	22	15	1987	34	3
1959–1960	14	15	1990	20	3
1962	26	6			

1.20   **Bear markets.** Investors speak of a "bear market" when stock prices drop substantially. Table 1.4 gives data on all declines of at least 10% in the Standard & Poor's 500-stock index between 1940 and 1997. The data show how far the index fell from its peak and how long the decline in stock prices lasted.

(a) Make a stemplot of the percent declines in stock prices during these bear markets. Make a second stemplot, splitting the stems. Which graph do you prefer? Why?

(b) The shape of this distribution is irregular, but we could describe it as somewhat skewed. Is the distribution skewed to the right or to the left?

(c) Describe the center and spread of the data. What would you tell an investor about how far stocks fall in a bear market?

Table 1.5 Women's winning times in the Boston Marathon

Year	Time	Year	Time	Year	Time
1972	190	1981	147	1990	145
1973	186	1982	150	1991	144
1974	167	1983	143	1992	144
1975	162	1984	149	1993	145
1976	167	1985	154	1994	142
1977	168	1986	145	1995	145
1978	165	1987	146	1996	147
1979	155	1988	145	1997	146
1980	154	1989	144	1998	143

1.21   **The Boston Marathon.** Women were allowed to enter the Boston Marathon in 1972. The times (in minutes, rounded to the nearest minute) for the winning woman from 1972 to 1998 appear in Table 1.5.

(a) Make a time plot of the winning times.

(b) Give a brief description of the pattern of Boston Marathon winning times over these years. Have times stopped improving in recent years?

1.22   **The influenza epidemic of 1918 (EESEE).** In 1918 and 1919 a worldwide outbreak of influenza killed more than 25 million people. Here are data on the number of new influenza cases and the number of deaths from the epidemic in San Francisco week by week from October 5, 1918, to January 25, 1919. The date given is the last day of the week.[10]

Date	Oct. 5	Oct. 12	Oct. 19	Oct. 26	Nov. 2	Nov. 9
Cases	36	531	4233	8682	7164	2229
Deaths	0	0	130	552	738	414

Date	Nov. 16	Nov. 23	Nov. 30	Dec. 7	Dec. 14	Dec. 21
Cases	600	164	57	722	1517	1828
Deaths	198	90	56	50	71	137

Date	Dec. 28	Jan. 4	Jan. 11	Jan. 18	Jan. 25
Cases	1539	2416	3148	3465	1440
Deaths	178	194	290	310	149

(a) Make a time plot of weekly new cases. Based on your plot, describe the progress of the epidemic.

(b) We would like to compare the patterns over time of number of new cases and number of deaths. To make the two variables similar in size for easier comparison, plot the number of deaths against time for October 5 to January 25, then plot the number of cases divided by 10 on the same graph using a different color. What do you see? In particular, about how long is the lag between changes in the number of cases and corresponding changes in deaths?

1.23   **Watch those scales!** The impression that a time plot gives depends on the scales you use on the two axes. If you stretch the vertical axis and compress the time axis, change appears to be more rapid. Compressing the vertical axis and stretching the time axis make change appear slower. Make two more time plots of the data in Example 1.6,

Table 1.6 Education and related data for the states

State	Region*	Population (1,000)	SAT verbal	SAT math	Percent taking	Percent no HS	Teachers' pay ($1,000)
AL	ESC	4,273	565	558	8	33.1	31.3
AK	PAC	607	521	513	47	13.4	49.6
AZ	MTN	4,428	525	521	28	21.3	32.5
AR	WSC	2,510	566	550	6	33.7	29.3
CA	PAC	31,878	495	511	45	23.8	43.1
CO	MTN	3,823	536	538	30	15.6	35.4
CT	NE	3,274	507	504	79	20.8	50.3
DE	SA	725	508	495	66	22.5	40.5
DC	SA	543	489	473	50	26.9	43.7
FL	SA	14,400	498	496	48	25.6	33.3
GA	SA	7,353	484	477	63	29.1	34.1
HI	PAC	1,184	485	510	54	19.9	35.8
ID	MTN	1,189	543	536	15	20.3	30.9
IL	ENC	11,847	564	575	14	23.8	40.9
IN	ENC	5,841	494	494	57	24.4	37.7
IA	WNC	2,852	590	600	5	19.9	32.4
KS	WNC	2,572	579	571	9	18.7	35.1
KY	ESC	3,884	549	544	12	35.4	33.1
LA	WSC	4,351	559	550	9	31.7	26.8
ME	NE	1,243	504	498	68	21.2	32.9
MD	SA	5,072	507	504	64	21.6	41.2
MA	NE	6,092	507	504	80	20.0	42.9
MI	ENC	9,594	557	565	11	23.2	44.8
MN	WNC	4,658	582	593	9	17.6	36.9
MS	ESC	2,716	569	557	4	35.7	27.7

*The census regions are East North Central, East South Central, Mid-Atlantic, Mountain, New England, Pacific, South Atlantic, West North Central, and West South Central.

one that makes cancer death rates appear to increase very rapidly and one that shows only a gentle increase. The moral of this exercise is: pay close attention to the scales when you look at a time plot.

Table 1.6 presents data about the individual states that relate to education. Study of a data set with many variables begins by

Table 1.6 Education and related data for the states (*continued*)

State	Region*	Population (1,000)	SAT verbal	SAT math	Percent taking	Percent no HS	Teachers' pay ($1,000)
MO	WNC	5,359	570	569	9	26.1	33.3
MT	MTN	879	546	547	21	19.0	29.4
NE	WNC	1,652	567	568	9	18.2	31.5
NV	MTN	1,603	508	507	31	21.2	36.2
NH	NE	1,162	520	514	70	17.8	35.8
NJ	MA	7,988	498	505	69	23.3	47.9
NM	MTN	1,713	554	548	12	24.9	29.6
NY	MA	18,185	497	499	73	25.2	48.1
NC	SA	7,323	490	486	59	30.0	30.4
ND	WNC	644	596	599	5	23.3	27.0
OH	ENC	11,173	536	535	24	24.3	37.8
OK	WSC	3,301	566	557	8	25.4	28.4
OR	PAC	3,204	523	521	50	18.5	39.6
PA	MA	12,056	498	492	71	25.3	46.1
RI	NE	990	501	491	69	28.0	42.2
SC	SA	3,699	480	474	57	31.7	31.6
SD	WNC	732	574	566	5	22.9	26.3
TN	ESC	5,320	563	552	14	32.9	33.1
TX	WSC	19,128	495	500	48	27.9	32.0
UT	MTN	2,000	583	575	4	14.9	30.6
VT	NE	589	506	500	70	19.2	36.3
VA	SA	6,675	507	496	68	24.8	35.0
WA	PAC	5,533	519	519	47	16.2	38.0
WV	SA	1,826	526	506	17	34.0	32.2
WI	ENC	5,160	577	586	8	21.4	38.2
WY	MTN	481	544	544	11	17.0	31.6

examining each variable by itself. Exercises 1.24 to 1.26 concern the data in Table 1.6.

1.24 **Population of the states.** Make a stemplot of the population of the states. Briefly describe the shape, center, and spread of the distribution of population. Explain why the shape of the distribution is not surprising. Are there any states that you consider outliers?

1.25 **How many students take the SAT?** Make a stemplot of the distribution of the percent of high school seniors who take the SAT in the various states. Briefly describe the overall shape of the distribution. Find the midpoint of the data and mark this value on your stemplot. Explain why describing the center is not very useful for a distribution with this shape.

1.26 **How much are teachers paid?** Make a graph to display the distribution of average teachers' salaries for the states. Is there a clear overall pattern? Are there any outliers or other notable deviations from the pattern?

1.2 Describing Distributions with Numbers

In the summer of 1998, Mark McGwire and Sammy Sosa captured the public's imagination with their pursuit of baseball's single-season home run record. McGwire eventually set a new standard with 70 home runs. How does this accomplishment fit McGwire's career? Here are McGwire's home run counts for the years 1987 (his rookie year) to 1998:

1987	1988	1989	1990	1991	1992
49	32	33	39	22	42

1993	1994	1995	1996	1997	1998
9	9	39	52	58	70

The stemplot in Figure 1.10 shows us the *shape, center,* and *spread* of these data. The distribution is roughly symmetric with a single peak and possible outliers in both tails. The center is about 39 home runs, and the spread runs from 9 to the record 70. Shape, center, and spread provide a good description of the overall pattern of any distribution for a quantitative variable. Now we will learn specific ways to use numbers to measure the center and spread of a distribution.

```
0 | 9 9
1 |
2 | 2
3 | 2 3 9 9
4 | 2 9
5 | 2 8
6 |
7 | 0
```

Figure 1.10 *The distribution of Mark McGwire's yearly home run totals.*

Measuring center: the mean

A description of a distribution almost always includes a measure of its center or average. The most common measure of center is the ordinary arithmetic average, or *mean*.

THE MEAN $\bar{x}$

To find the **mean** of a set of observations, add their values and divide by the number of observations. If the n observations are $x_1, x_2, \ldots, x_n$, their mean is

$$\bar{x} = \frac{x_1 + x_2 + \cdots + x_n}{n}$$

or in more compact notation,

$$\bar{x} = \frac{1}{n} \sum x_i$$

The $\sum$ (capital Greek sigma) in the formula for the mean is short for "add them all up." The subscripts on the observations x_i are just a way of keeping the n observations distinct. They do not necessarily indicate order or any other special facts about the data. The bar over the x indicates the mean of all the x-values. Pronounce the mean $\bar{x}$ as "x-bar." This notation is very common. When writers who are discussing data use $\bar{x}$ or $\bar{y}$, they are talking about a mean.

EXAMPLE 1.7   **Mark McGwire's home runs**

The mean number of home runs Mark McGwire hit in his first 12 major league seasons is

$$\bar{x} = \frac{x_1 + x_2 + \cdots + x_n}{n}$$

$$= \frac{49 + 32 + \cdots + 70}{12}$$

$$= \frac{454}{12} = 37.8$$

In practice, you can key the data into your calculator and hit the mean key. You don't have to actually add and divide. But you should know that this is what the calculator is doing.

McGwire was injured in 1993 and there was a baseball strike in 1994. We might want to exclude these years as "not full seasons." Use your calculator to check that his mean home run production in his 10 full seasons is $\bar{x} = 43.6$. The two partial seasons reduced the mean home run count by almost 6 per year over the full 12 years.

resistant measure

Example 1.7 illustrates an important fact about the mean as a measure of center: it is sensitive to the influence of a few extreme observations. These may be outliers, but a skewed distribution that has no outliers will also pull the mean toward its long tail. Because the mean cannot resist the influence of extreme observations, we say that it is not a **resistant measure** of center.

· · APPLY YOUR KNOWLEDGE ·

1.27 **Students' attitudes.** Here are the scores of 18 first-year college women on the Survey of Study Habits and Attitudes (SSHA):

| 154 | 109 | 137 | 115 | 152 | 140 | 154 | 178 | 101 |
| 103 | 126 | 126 | 137 | 165 | 165 | 129 | 200 | 148 |

(a) Find the mean score from the formula for the mean. Then enter the data into your calculator and use the calculator's $\bar{x}$ button to obtain the mean. Verify that you get the same result.

(b) A stemplot (Exercise 1.9) suggests that the score 200 is an outlier. Use your calculator to find the mean for the 17 observations that remain when you drop the outlier. How does the outlier change the mean?

We don't really make that much

The American Medical Association (AMA) has long issued annual reports giving the median income of doctors in private practice. After the median income reached $177,400 in 1992, the AMA stopped releasing the data. In 1994, the AMA announced that it would again release income data but would lump doctors in private practice with doctors still in training and those who work for the government to bring the median down. "Now the physician looks less like he's gouging America," said an AMA spokesperson. There's a lesson here: it isn't enough to know the number—you must also know just what the number measures.

Measuring center: the median

In Section 1.1, we used the midpoint of a distribution as an informal measure of center. The *median* is the formal version of the midpoint, with a specific rule for calculation.

THE MEDIAN M

The **median M** is the midpoint of a distribution, the number such that half the observations are smaller and the other half are larger. To find the median of a distribution:

1. Arrange all observations in order of size, from smallest to largest.

2. If the number of observations n is odd, the median M is the center observation in the ordered list. Find the location of the median by counting $(n + 1)/2$ observations up from the bottom of the list.

3. If the number of observations n is even, the median M is the mean of the two center observations in the ordered list. The location of the median is again $(n + 1)/2$ from the bottom of the list.

Note that the formula $(n + 1)/2$ does *not* give the median, just the location of the median in the ordered list. Medians require little arithmetic, so they are easy to find by hand for small sets of data. Arranging even a moderate number of observations in order is very tedious, however, so that finding the median by hand for larger sets of data is unpleasant. Even simple calculators have an $\bar{x}$ button, but you will need computer software or a graphing calculator to automate finding the median.

EXAMPLE 1.8   **Finding the median**

To find the median number of home runs Mark McGwire hit in his first 12 seasons, first arrange the data in increasing order:

> 9   9   22   32   33   **39** / **39**   42   49   52   58   70

The count of observations $n = 12$ is even. There is no center observation, but there is a center pair. These are the two bold 39s in the list, which have 5 observations to their left in the list and 5 to their right. The median is midway between these two observations. Because both of the middle pair are 39, $M = 39$.

The rule for locating the median in the list gives

$$\text{location of } M = \frac{n + 1}{2} = \frac{13}{2} = 6.5$$

The location 6.5 means "halfway between the sixth and seventh observations in the ordered list." That agrees with what we found by eye.

How much do the two partial seasons affect the median? Drop the two 9s from the list and find the median for the remaining $n = 10$ years. It is midway between the middle pair of observations,

$$M = \frac{39 + 42}{2} = 40.5$$

We might compare McGwire with Babe Ruth, the original home run king. Here, already arranged in increasing order, are Ruth's home run counts during his 15 years with the New York Yankees, 1920 to 1934:

> 22   25   34   35   41   41   46   **46**   46   47   49   54   54   59   60

There is an odd number of observations, so there is one center observation. This is the median. It is the bold 46, which has 7 observations to its left in the list and 7 observations to its right. Although McGwire holds the single-season record, Ruth remains the king in the number of home runs hit in a typical season.

Because $n = 15$, our rule for the location of the median gives

$$\text{location of } M = \frac{n + 1}{2} = \frac{16}{2} = 8$$

That is, the median is the 8th observation in Ruth's ordered list. It is faster to use this rule than to locate the center by eye.

Comparing the mean and the median

Examples 1.7 and 1.8 illustrate an important difference between the mean and the median. The two low outliers pull McGwire's mean home run count down from 43.6 to 37.8. The median moves much less, from 40.5 to 39. The median, unlike the mean, is *resistant*. If McGwire's record 70 had been 700, his median would not change at all. The 700 just counts as one observation above the center, no matter how far above the center it lies. The mean uses the actual value of each observation and so will chase a single large observation upward.

The mean and median of a symmetric distribution are close together. If the distribution is exactly symmetric, the mean and median are exactly the same. In a skewed distribution, the mean is farther out in the long tail than is the median. For example, the distribution of house prices is strongly skewed to the right. There are many moderately priced houses and a few very expensive mansions. The few expensive houses pull the mean up but do not affect the median. The mean price of new houses sold in 1997 was $176,000, but the median price for these same houses was only $146,000. Reports about house prices, incomes, and other strongly skewed distributions usually give the median ("midpoint") rather than the mean ("arithmetic average"). However, if you are a tax assessor interested in the total value of houses in your area, use the mean. The total is the mean times the number of houses, but it has no connection with the median. The mean and median measure center in different ways, and both are useful.

Poor New York?

Is New York a rich state? New York's mean income per person ranks fourth among the states, right up there with its rich neighbors Connecticut and New Jersey, which rank first and second. But while Connecticut and New Jersey rank seventh and second in median household income, New York stands 29th, well below the national average. What's going on? Just another example of mean versus median. New York has many very highly paid people, who pull up its mean income per person. But it also has a higher proportion of poor households than do New Jersey and Connecticut, and this brings the median down. New York is not a rich state—it's a state with extremes of wealth and poverty.

APPLY YOUR KNOWLEDGE

1.28   Swiss doctors. A study in Switzerland examined the number of caesarean sections (surgical deliveries of babies) performed in a year by doctors. Here are the data for 15 male doctors:

27   50   33   25   86   25   85   31   37   44   20   36   59   34   28

(a) Make a stemplot of the data. Note the two high outliers.

(b) Find the mean and median number of operations. How do the outliers explain the difference between your two results?

(c) Find the mean and median number of operations without the two outliers. How does comparing your results in (b) and (c) illustrate the resistance of the median and the lack of resistance of the mean?

1.29   The richest 1%. The distribution of individual incomes in the United States is strongly skewed to the right. In 1997, the mean and median incomes of the top 1% of Americans were $330,000 and $675,000. Which of these numbers is the mean and which is the median? Explain your reasoning.

1.30   In Exercise 1.27 you found the mean of the SSHA scores of 18 first-year college students. Now find the median of these scores. Is the median smaller or larger than the mean? Explain why this is so.

Measuring spread: the quartiles

The mean and median provide two different measures of the center of a distribution. But a measure of center alone can be misleading. The Census Bureau reports that in 1997 the median income of American households was $37,005. Half of all households had incomes below $37,005, and half had higher incomes. But these figures do not tell the whole story. Two nations with the same median household income are very different if one has extremes of wealth and poverty and the other has little variation among households. A drug with the correct mean concentration of active ingredient is dangerous if some batches are much too high and others much too low. We are interested in the *spread* or *variability* of incomes and drug potencies as well as their centers. The simplest useful numerical description of a distribution consists of both a measure of center and a measure of spread.

One way to measure spread is to give the smallest and largest observations. For example, the number of home runs Mark McGwire has hit in a season ranges from 9 to 70. These single observations show the full spread of the data, but they may be outliers. We can improve our description of spread by also looking at the spread of the middle half of the data. The *quartiles* mark out the middle half. Count up the ordered list of observations, starting from the smallest. The *first quartile* lies one-quarter of the way up the list. The *third quartile* lies three-quarters of the way up the list. In other words, the first quartile is larger than 25% of the observations, and the third quartile is larger than 75% of the observations. The second quartile is the median, which is larger than 50% of the observations. That is the idea of quartiles. We need a rule to make the idea exact. The rule for calculating the quartiles uses the rule for the median.

THE QUARTILES Q_1 AND Q_3

To calculate the **quartiles**:

1. Arrange the observations in increasing order and locate the median M in the ordered list of observations.

2. The **first quartile** Q_1 is the median of the observations whose position in the ordered list is to the left of the location of the overall median.

3. The **third quartile** Q_3 is the median of the observations whose position in the ordered list is to the right of the location of the overall median.

Here is an example that shows how the rules for the quartiles work for both odd and even numbers of observations.

EXAMPLE 1.9   Finding the quartiles

Mark McGwire's home run counts (arranged in order) are

$$9 \quad 9 \quad \underset{\uparrow}{22} \quad \underset{Q_1}{32} \quad 33 \quad 39 \quad \underset{\uparrow}{39} \quad \underset{M}{42} \quad 49 \quad \underset{\uparrow}{52} \quad \underset{Q_3}{58} \quad 70$$

There is an even number of observations, so the median lies midway between the middle pair, the 6th and 7th in the list. The first quartile is the median of the first 6 observations, because these are the observations to the left of the location of the median. Check that $Q_1 = 27$ and $Q_3 = 50.5$. When the number of observations is even, all the observations enter into the calculation of the quartiles.

Notice that the quartiles are resistant. For example, Q_3 would have the same value if McGwire's record 70 were 700.

Babe Ruth's data, again arranged in increasing order, are

$$22 \quad 25 \quad 34 \quad \underset{\uparrow}{35} \quad \underset{Q_1}{41} \quad 41 \quad 46 \quad \underset{\uparrow}{\mathbf{46}} \quad \underset{M}{46} \quad 47 \quad 49 \quad \underset{\uparrow}{54} \quad \underset{Q_3}{54} \quad 59 \quad 60$$

There is an odd number of observations, so the median is the middle one, the bold 46 in the list. The first quartile is the median of the 7 observations to the left of the median. This is the 4th of these 7 observations, so $Q_1 = 35$. If you want, you can use the recipe for the location of the median with $n = 7$:

$$\text{location of } Q_1 = \frac{n+1}{2} = \frac{7+1}{2} = 4$$

The third quartile is the median of the 7 observations to the right of the median, $Q_3 = 54$. The overall median is left out of the calculation of the quartiles when there is an odd number of observations.

Be careful when, as in these examples, several observations take the same numerical value. Write down all of the observations and apply the rules just as if they all had distinct values. Some software packages use a slightly different rule to find the quartiles, so computer results may be a bit different from your own work. Don't worry about this. The differences will always be too small to be important.

The five-number summary and boxplots

The smallest and largest observations tell us little about the distribution as a whole, but they give information about the tails of the distribution that is missing if we know only Q_1, M, and Q_3. To get a quick summary of both center and spread, combine all five numbers.

THE FIVE-NUMBER SUMMARY

The **five-number summary** of a data set consists of the smallest observation, the first quartile, the median, the third quartile, and the largest observation, written in order from smallest to largest. In symbols, the five-number summary is

$$\text{Minimum} \quad Q_1 \quad M \quad Q_3 \quad \text{Maximum}$$

These five numbers offer a reasonably complete description of center and spread. The five-number summaries from Example 1.9 are

$$9 \quad 27 \quad 39 \quad 50.5 \quad 70$$

for McGwire and

$$22 \quad 35 \quad 46 \quad 54 \quad 60$$

for Ruth. The five-number summary of a distribution leads to a new graph, the *boxplot*. Figure 1.11 shows boxplots for the home run comparison.

BOXPLOT

A **boxplot** is a graph of the five-number summary.

- A central box spans the quartiles.
- A line in the box marks the median.
- Lines extend from the box out to the smallest and largest observations.

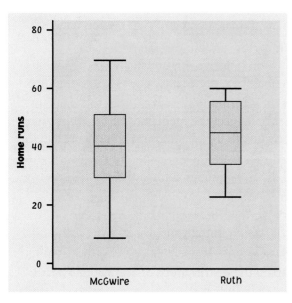

Figure 1.11 *Side-by-side boxplots comparing the number of home runs per year by Mark McGwire and Babe Ruth.*

Because boxplots show less detail than histograms or stemplots, they are best used for side-by-side comparison of more than one distribution, as in Figure 1.11. You can draw boxplots either horizontally or vertically. Be sure to include a numerical scale in the graph. When you look at a boxplot, first locate the median, which marks the center of the distribution. Then look at the spread. The quartiles show the spread of the middle half of the data, and the extremes (the smallest and largest observations) show the spread of the entire data set. We see from Figure 1.11 that Ruth is more consistent than McGwire—his home run counts are less spread out. His usual performance, as indicated by the median and spread of the middle half of the distribution, is a bit better than McGwire's.

A boxplot also gives an indication of the symmetry or skewness of a distribution. In a symmetric distribution, the first and third quartiles are equally distant from the median. In most distributions that are skewed to the right, on the other hand, the third quartile will be farther above the median than the first quartile is below it. The extremes behave the same way, but remember that they are just single observations and may say little about the distribution as a whole.

APPLY YOUR KNOWLEDGE

1.31 **Swiss doctors.** Exercise 1.28 gives the number of caesarean sections performed by 15 male doctors in Switzerland. The study also looked at 10 female doctors. The numbers of caesareans performed by these doctors (arranged in order) were

5   7   10   14   18   19   25   29   31   33

(a) Find the five-number summary for each group.

(b) Make side-by-side boxplots to compare the number of operations performed by female and male doctors. What do you conclude?

1.32 **How old are presidents?** How old are presidents at their inauguration? Was Bill Clinton, at age 46, unusually young? Table 1.7 gives the data, the ages of all U.S. presidents when they took office.

(a) Make a stemplot of the distribution of ages. From the shape of the distribution, do you expect the median to be much less than the mean, about the same as the mean, or much greater than the mean?

(b) Find the mean and the five-number summary. Verify your expectation about the median.

(c) What is the range of the middle half of the ages of new presidents? Was Bill Clinton in the youngest 25%?

1.33 **SAT scores.** Table 1.6 (page 26) contains data on education in the states. We want to compare the distributions of average SAT math

Table 1.7 Ages of the presidents at inauguration

President	Age	President	Age	President	Age
Washington	57	Buchanan	65	Harding	55
J. Adams	61	Lincoln	52	Coolidge	51
Jefferson	57	A. Johnson	56	Hoover	54
Madison	57	Grant	46	F. D. Roosevelt	51
Monroe	58	Hayes	54	Truman	60
J. Q. Adams	57	Garfield	49	Eisenhower	61
Jackson	61	Arthur	51	Kennedy	43
Van Buren	54	Cleveland	47	L. B. Johnson	55
W. H. Harrison	68	B. Harrison	55	Nixon	56
Tyler	51	Cleveland	55	Ford	61
Polk	49	McKinley	54	Carter	52
Taylor	64	T. Roosevelt	42	Reagan	69
Fillmore	50	Taft	51	Bush	64
Pierce	48	Wilson	56	Clinton	46

and verbal scores. We enter these data into a computer with the names SATM for math scores and SATV for verbal scores. Here is output from the statistical software package Minitab that gives the five-number summary along with other information. (Other software produces similar output.)

```
SATM
        N     MEAN   MEDIAN   STDEV     MIN      MAX       Q1        Q3
       51   529.27   521.00   34.83   473.00   600.00   500.00   557.00


SATV
        N     MEAN   MEDIAN   STDEV     MIN      MAX       Q1        Q3
       51   531.90   525.00   33.76   480.00   596.00   501.00   565.00
```

Use the output to make side-by-side boxplots of SAT math and verbal scores for the states. Briefly compare the two distributions in words.

Measuring spread: the standard deviation

The five-number summary is not the most common numerical description of a distribution. That distinction belongs to the combination of the mean to measure center and the *standard deviation* to measure spread. The standard deviation measures spread by looking at how far the observations are from their mean.

THE STANDARD DEVIATION s

The **variance** s^2 of a set of observations is the average of the squares of the deviations of the observations from their mean. In symbols, the variance of n observations $x_1, x_2, \ldots, x_n$ is

$$s^2 = \frac{(x_1 - \bar{x})^2 + (x_2 - \bar{x})^2 + \cdots + (x_n - \bar{x})^2}{n - 1}$$

or, more compactly,

$$s^2 = \frac{1}{n-1} \sum (x_i - \bar{x})^2$$

The **standard deviation s** is the square root of the variance s^2:

$$s = \sqrt{\frac{1}{n-1} \sum (x_i - \bar{x})^2}$$

In practice, use software or your calculator to obtain the standard deviation from keyed-in data. Doing an example step-by-step will help you understand how the variance and standard deviation work, however.

EXAMPLE 1.10   Calculating the standard deviation

A person's metabolic rate is the rate at which the body consumes energy. Metabolic rate is important in studies of weight gain, dieting, and exercise. Here are the metabolic rates of 7 men who took part in a study of dieting. (The units are calories per 24 hours. These are the same calories used to describe the energy content of foods.)

1792   1666   1362   1614   1460   1867   1439

The researchers reported $\bar{x}$ and s for these men.

First find the mean:

$$\bar{x} = \frac{1792 + 1666 + 1362 + 1614 + 1460 + 1867 + 1439}{7}$$

$$= \frac{11,200}{7} = 1600 \text{ calories}$$

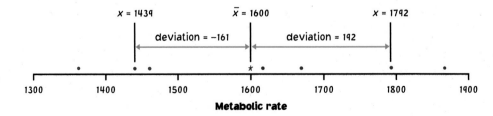

Figure 1.12 *Metabolic rates for seven men, with their mean (∗) and the deviations of two observations from the mean.*

Figure 1.12 displays the data as points above the number line, with their mean marked by an asterisk (*). The arrows mark two of the deviations from the mean. These deviations show how spread out the data are about their mean. They are the starting point for calculating the variance and the standard deviation.

Observations x_i	Deviations $x_i - \bar{x}$			Squared deviations $(x_i - \bar{x})^2$	
1792	$1792 - 1600$	=	192	$192^2 =$	36,864
1666	$1666 - 1600$	=	66	$66^2 =$	4,356
1362	$1362 - 1600$	=	-238	$(-238)^2 =$	56,644
1614	$1614 - 1600$	=	14	$14^2 =$	196
1460	$1460 - 1600$	=	-140	$(-140)^2 =$	19,600
1867	$1867 - 1600$	=	267	$267^2 =$	71,289
1439	$1439 - 1600$	=	-161	$(-161)^2 =$	25,921
	sum	=	0	sum =	214,870

The variance is the sum of the squared deviations divided by one less than the number of observations:

$$s^2 = \frac{214{,}870}{6} = 35{,}811.67$$

The standard deviation is the square root of the variance:

$$s = \sqrt{35{,}811.67} = 189.24 \text{ calories}$$

Notice that the "average" in the variance s^2 divides the sum by one fewer than the number of observations, that is, $n - 1$ rather than n. The reason is that the deviations $x_i - \bar{x}$ always sum to exactly 0, so that knowing $n - 1$ of them determines the last one. Only $n - 1$ of the squared deviations can vary freely, and we average by dividing the total by $n - 1$. The number $n - 1$ is called the **degrees of freedom** of the variance or standard deviation. Many calculators offer a choice between dividing by n and dividing by $n - 1$, so be sure to use $n - 1$.

degrees of freedom

More important than the details of hand calculation are the properties that determine the usefulness of the standard deviation:

- s measures spread about the mean and should be used only when the mean is chosen as the measure of center.

- $s = 0$ only when there is *no spread*. This happens only when all observations have the same value. Otherwise $s > 0$. As the observations become more spread out about their mean, s gets larger.

- s has the same units of measurement as the original observations. For example, if you measure metabolic rates in calories, s is also in calories.

This is one reason to prefer s to the variance s^2, which is in squared calories.

- Like the mean $\bar{x}$, s is not resistant. Strong skewness or a few outliers can greatly increase s. For example, the standard deviation of Mark McGwire's home run counts is 18.48. (Use your calculator to verify this.) If we omit the two 9s from partial seasons, the standard deviation drops to 13.99.

You may rightly feel that the importance of the standard deviation is not yet clear. We will see in the next section that the standard deviation is the natural measure of spread for an important class of symmetric distributions, the normal distributions. The usefulness of many statistical procedures is tied to distributions of particular shapes. This is certainly true of the standard deviation.

Choosing measures of center and spread

How do we choose between the five-number summary and $\bar{x}$ and s to describe the center and spread of a distribution? Because the two sides of a strongly skewed distribution have different spreads, no single number such as s describes the spread well. The five-number summary, with its two quartiles and two extremes, does a better job.

CHOOSING A SUMMARY

The five-number summary is usually better than the mean and standard deviation for describing a skewed distribution or a distribution with strong outliers. Use $\bar{x}$ and s only for reasonably symmetric distributions that are free of outliers.

Do remember that a graph gives the best overall picture of a distribution. Numerical measures of center and spread report specific facts about a distribution, but they do not describe its entire shape. Numerical summaries do not disclose the presence of multiple peaks or gaps, for example. Exercise 1.36 gives an example of a distribution for which numerical summaries alone are misleading. **Always plot your data.**

APPLY YOUR KNOWLEDGE

1.34   The level of various substances in the blood influences our health. Here are measurements of the level of phosphate in the blood of a patient, in milligrams of phosphate per deciliter of blood, made on 6 consecutive visits to a clinic:

5.6   5.2   4.6   4.9   5.7   6.4

A graph of only 6 observations gives little information, so we proceed to compute the mean and standard deviation.

(a) Find the mean from its definition. That is, find the sum of the 6 observations and divide by 6.

(b) Find the standard deviation from its definition. That is, find the deviations of each observation from the mean, square the deviations, then obtain the variance and the standard deviation. Example 1.10 shows the method.

(c) Now enter the data into your calculator and use the mean and standard deviation buttons to obtain $\bar{x}$ and s. Do the results agree with your hand calculations?

1.35 **Roger Maris.** New York Yankee Roger Maris held the single-season home run record from 1961 until 1998. Here are Maris's home run counts for his 10 years in the American League:

$$14 \quad 28 \quad 16 \quad 39 \quad 61 \quad 33 \quad 23 \quad 26 \quad 8 \quad 13$$

Maris's record 61 home runs in 1961 is an outlier in these data.

(a) Use your calculator to find the mean $\bar{x}$ and the standard deviation s.

(b) Use your calculator to find $\bar{x}$ and s for the 9 observations that remain when you leave out the outlier. How does the outlier affect the values of $\bar{x}$ and s?

1.36 **State SAT scores.** Exercise 1.33 (page 36) gives numerical summaries for the average SAT scores for the states. These numerical summaries (and the boxplots based on them) fail to show one of the most important features of the distributions. Make a stemplot of the SAT math scores from Table 1.6 (page 26). What is the overall shape of the distribution? Remember to always start with a graph of your data—numerical summaries are not a complete description.

SECTION 1.2 S u m m a r y

A numerical summary of a distribution should report its **center** and its **spread** or **variability**.

The **mean** $\bar{x}$ and the **median M** describe the center of a distribution in different ways. The mean is the arithmetic average of the observations, and the median is the midpoint of the values.

When you use the median to indicate the center of the distribution, describe its spread by giving the **quartiles**. The **first quartile** Q_1 has one-fourth of the observations below it, and the **third quartile** Q_3 has three-fourths of the observations below it.

The **five-number summary** consisting of the median, the quartiles, and the high and low extremes provides a quick overall description of a distribution. The median describes the center, and the quartiles and extremes show the spread.

Boxplots based on the five-number summary are useful for comparing several distributions. The box spans the quartiles and shows the spread of the central half of the distribution. The median is marked within the box. Lines extend from the box to the extremes and show the full spread of the data.

The **variance** s^2 and especially its square root, the **standard deviation** s, are common measures of spread about the mean as center. The standard deviation s is zero when there is no spread and gets larger as the spread increases.

A **resistant measure** of any aspect of a distribution is relatively unaffected by changes in the numerical value of a small proportion of the total number of observations, no matter how large these changes are. The median and quartiles are resistant, but the mean and the standard deviation are not.

The mean and standard deviation are good descriptions for symmetric distributions without outliers. They are most useful for the normal distributions introduced in the next section. The five-number summary is a better exploratory summary for skewed distributions.

SECTION 1.2 Exercises

1.37 Last year a small accounting firm paid each of its five clerks $22,000, two junior accountants $50,000 each, and the firm's owner $270,000. What is the mean salary paid at this firm? How many of the employees earn less than the mean? What is the median salary?

1.38 **Presidential elections.** Here are the percents of the popular vote won by the successful candidate in each of the presidential elections from 1948 to 1996:

Year	1948	1952	1956	1960	1964	1968	1972	1976	1980	1984	1988	1992	1996
Percent	49.6	55.1	57.4	49.7	61.1	43.4	60.7	50.1	50.7	58.8	53.9	43.2	49.2

(a) Make a stemplot of the winners' percents. (Round to whole numbers and use split stems.)

(b) What is the median percent of the vote won by the successful candidate in presidential elections? (Work with the unrounded data.)

(c) Call an election a landslide if the winner's percent falls at or above the third quartile. Find the third quartile. Which elections were landslides?

1.39 **How many calories in a hot dog?** Some people worry about how many calories they consume. *Consumer Reports* magazine, in a story on hot dogs, measured the calories in 20 brands of beef hot dogs, 17 brands of meat hot dogs, and 17 brands of poultry hot dogs.[11] Here is computer output describing the beef hot dogs,

```
Mean = 156.8   Standard deviation = 22.64   Min = 111   Max = 190
N = 20   Median = 152.5   Quartiles = 140, 178.5
```

the meat hot dogs,

```
Mean = 158.7   Standard deviation = 25.24   Min = 107   Max = 195
N = 17   Median = 153   Quartiles = 139, 179
```

and the poultry hot dogs,

```
Mean = 122.5   Standard deviation = 25.48   Min = 87   Max = 170
N = 17   Median = 129   Quartiles = 102, 143
```

Use this information to make side-by-side boxplots of the calorie counts for the three types of hot dogs. Write a brief comparison of the distributions. Will eating poultry hot dogs usually lower your calorie consumption compared with eating beef or meat hot dogs?

1.40 **People without high school educations.** The "Percent no HS" column in Table 1.6 (page 26) gives the percent of the adult population in each state who did not graduate from high school. We want to compare the percents of people without a high school education in the northeastern and the southern states. Take the northeastern states to be those in the MA (Mid-Atlantic) and NE (New England) regions. The southern states are those in the SA (South Atlantic) and ESC (East South Central) regions. Leave out the District of Columbia, which is a city rather than a state.

(a) List the percents without high school for the northeastern and for the southern states from Table 1.6. These are the two data sets we want to compare.

(b) Make numerical summaries and graphs to compare the two distributions. Write a brief statement of what you find.

1.41 **The density of the earth.** In 1798 the English scientist Henry Cavendish measured the density of the earth with great care. It is common practice to repeat careful measurements several times and use the mean as the final result. Cavendish repeated his work 29 times. Here are his results (the data give the density of the earth as a multiple of the density of water):[12]

5.50	5.61	4.88	5.07	5.26	5.55	5.36	5.29	5.58	5.65
5.57	5.53	5.62	5.29	5.44	5.34	5.79	5.10	5.27	5.39
5.42	5.47	5.63	5.34	5.46	5.30	5.75	5.68	5.85	

Present these measurements with a graph of your choice. Does the shape of the distribution allow the use of $\bar{x}$ and s to describe it? Find $\bar{x}$ and s. What is your estimate of the density of the earth based on these measurements?

1.42 **$\bar{x}$ and s are not enough.** The mean $\bar{x}$ and standard deviation s measure center and spread but are not a complete description of a distribution. Data sets with different shapes can have the same mean and standard deviation. To demonstrate this fact, use your calculator to find $\bar{x}$ and s for these two small data sets. Then make a stemplot of each and comment on the shape of each distribution.

Data A	9.14 8.14 8.74 8.77 9.26 8.10 6.13 3.10 9.13 7.26 4.74

Data B	6.58 5.76 7.71 8.84 8.47 7.04 5.25 5.56 7.91 6.89 12.50

1.43 Table 1.1 (page 9) records the percent of people aged 65 and over living in each of the states. Figure 1.2 (page 10) is a histogram of these data. Do you prefer the five-number summary or $\bar{x}$ and s as a brief numerical description? Why? Calculate your preferred description.

1.44 **A hot stock?** Table 1.8 gives the monthly percent returns on Philip Morris stock for the period from July 1990 to May 1997. (The return on an investment consists of the change in its price plus any cash payments made, given here as a percent of its price at the start of each month.)

(a) Make either a histogram or a stemplot of these data. How did you decide which graph to make?

(b) There is one clear outlier. What is the value of this observation? (It is explained by news of action against smoking, which depressed this tobacco company stock.) Describe the shape, center, and spread of the data after you omit the outlier.

(c) It is usual in the study of investments to use the mean and standard deviation to summarize and compare investment returns. Find the mean monthly return and the standard deviation of the returns in Table 1.8. If you invested $100 in this stock at the beginning of a month and got the mean return, how much would you have at the end of the month?

Table 1.8 Monthly percent returns on Philip Morris stock from July 1990 to May 1997

−5.7	1.2	4.1	3.2	7.3	7.5	18.6	3.7	−1.8	2.4
−6.5	6.7	9.4	−2.0	−2.8	−3.4	19.2	−4.8	0.5	−0.6
2.8	−0.5	−4.5	8.7	2.7	4.1	−10.3	4.8	−2.3	−3.1
−10.2	−3.7	−26.6	7.2	−2.9	−2.3	3.5	−4.6	17.2	4.2
0.5	8.3	−7.1	−8.4	7.7	−9.6	6.0	6.8	10.9	1.6
0.2	−2.4	−2.4	3.9	1.7	9.0	3.6	7.6	3.2	−3.7
4.2	13.2	0.9	4.2	4.0	2.8	6.7	−10.4	2.7	10.3
5.7	0.6	−14.2	1.3	2.9	11.8	10.6	5.2	13.8	−14.7
3.5	11.7	1.3							

(d) If you invested $100 in this stock at the beginning of the worst month in the data (the outlier), how much would you have at the end of the month? Find the mean and standard deviation again, this time leaving out the low outlier. How much did this one observation affect the summary measures? Would leaving out this one observation change the median? The quartiles? How do you know, without actual calculation?

1.45 **Athletes' salaries.** The Baltimore Orioles had the highest team payroll in major league baseball in 1998. Here are the salaries of the Orioles' players, in thousands of dollars. For example, 6495 stands for Mike Mussina's salary of $6,495,000.

6495	6486	6300	6269	5442	5391	3600	3600	3583
3089	2850	2500	1950	1663	1367	1333	1150	900
856	800	800	665	650	450	450	170	170

Describe this salary distribution both with a graph and with a numerical summary. Then write a brief description of the important features of the distribution.

1.46 **Household net worth.** A household's "net worth" is the total value of the household's possessions and investments less the total of its debts. In 1997, the mean and median net worth of American households were $51,000 and $212,000. Which of these numbers is the mean and which is the median? Explain your reasoning.

1.47 **Highly paid athletes.** A news article reports that of the 411 players on National Basketball Association rosters in February 1998, only 139 "made more than the league average salary" of $2.36 million. Is $2.36 million the mean or median salary for NBA players? How do you know?

1.48 **Mean or median?** Which measure of center, the mean or the median, should you use in each of the following situations?

(a) Middletown is considering imposing an income tax on citizens. The city government wants to know the average income of citizens so that it can estimate the total tax base.

(b) In a study of the standard of living of typical families in Middletown, a sociologist estimates the average family income in that city.

1.49 This is a standard deviation contest. You must choose four numbers from the whole numbers 0 to 10, with repeats allowed.

(a) Choose four numbers that have the smallest possible standard deviation.

(b) Choose four numbers that have the largest possible standard deviation.

(c) Is more than one choice possible in either (a) or (b)? Explain.

1.3 The Normal Distributions

We now have a kit of graphical and numerical tools for describing distributions. What is more, we have a clear strategy for exploring data on a single quantitative variable:

1. Always plot your data: make a graph, usually a histogram or a stemplot.
2. Look for the overall pattern (shape, center, spread) and for striking deviations such as outliers.
3. Calculate a numerical summary to briefly describe center and spread.

Here is one more step to add to this strategy:

4. Sometimes the overall pattern of a large number of observations is so regular that we can describe it by a smooth curve.

Density curves

Figure 1.13 is a histogram of the scores of all 947 seventh-grade students in Gary, Indiana, on the vocabulary part of the Iowa Test of Basic Skills.[13] Scores of many students on this national test have a quite regular distribution. The histogram is symmetric, and both tails fall off quite smoothly from a single center peak. There are no large gaps or obvious outliers, The smooth curve drawn

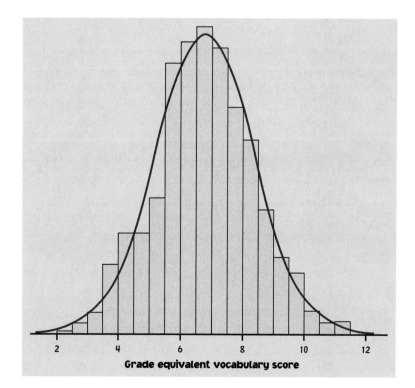

Grade equivalent vocabulary score

Figure 1.13 *Histogram of the vocabulary scores of all seventh-grade students in Gary, Indiana. The smooth curve shows the overall shape of the distribution.*

through the tops of the histogram bars in Figure 1.13 is a good description of the overall pattern of the data. The curve is a **mathematical model** for the distribution. A mathematical model is an idealized description. It gives a compact picture of the overall pattern of the data but ignores minor irregularities as well as any outliers.

mathematical model

We will see that it is easier to work with the smooth curve in Figure 1.13 than with the histogram. The reason is that the histogram depends on our choice of classes, while with a little care we can use a curve that does not depend on any choices we make. Here's how we do it.

EXAMPLE 1.11   **From histogram to density curve**

Our eyes respond to the *areas* of the bars in a histogram. The bar areas represent proportions of the observations. Figure 1.14(a) is a copy of Figure 1.13 with the leftmost bars shaded. The area of the shaded bars in Figure 1.14(a) represents the students with vocabulary scores 6.0 or lower. There are 287 such students, who make up the proportion 287/947 = 0.303 of all Gary seventh graders.

Now concentrate on the curve drawn through the bars. In Figure 1.14(b), the area under the curve to the left of 6.0 is shaded. Adjust the scale of the graph so that *the total area under the curve is exactly 1*. This area represents the proportion 1, that is, all the observations. Areas under the curve then represent proportions of the observations. The curve is now a *density curve*. The shaded area under the density curve in Figure 1.14(b) represents the proportion of students with score 6.0 or lower. This area is 0.293, only 0.010 away from the histogram result. You can see that areas under the density curve give quite good approximations of areas given by the histogram.

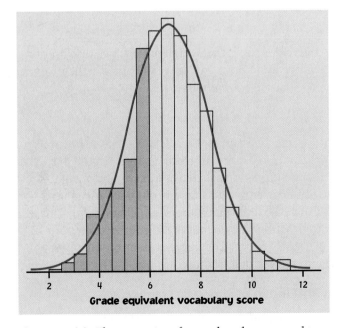

Figure 1.14(a) *The proportion of scores less than or equal to 6.0 from the histogram is 0.303.*

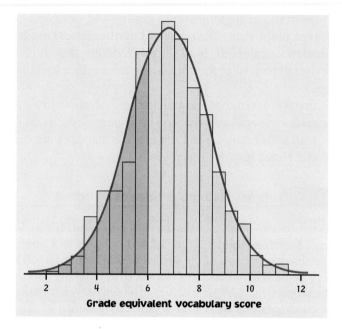

Figure 1.14(b) *The proportion of scores less than or equal to 6.0 from the density curve is 0.293.*

DENSITY CURVE

A **density curve** is a curve that

• is always on or above the horizontal axis, and

• has area exactly 1 underneath it.

A density curve describes the overall pattern of a distribution. The area under the curve and above any range of values is the proportion of all observations that fall in that range.

---

normal curve
The density curve in Figures 1.13 and 1.14 is a **normal curve.** Density curves, like distributions, come in many shapes. Figure 1.15 shows two density curves: a symmetric normal density curve and a right-skewed curve. A density curve of the appropriate shape is often an adequate description of the overall pattern of a distribution. Outliers, which are deviations from the overall pattern, are not described by the curve. Of course, no set of real data is exactly described by a density curve. The curve is an approximation that is easy to use and accurate enough for practical use.

The median and mean of a density curve

Our measures of center and spread apply to density curves as well as to actual sets of observations. The median and quartiles are easy. Areas under a density

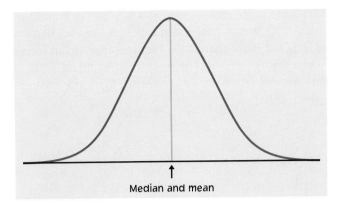

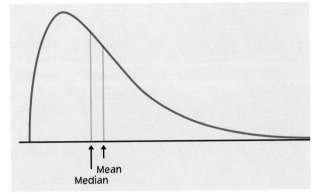

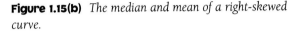

Figure 1.15(a) *The median and mean of a symmetric density curve.*

Figure 1.15(b) *The median and mean of a right-skewed curve.*

curve represent proportions of the total number of observations. The median is the point with half the observations on either side. So **the median of a density curve is the equal-areas point,** the point with half the area under the curve to its left and the remaining half of the area to its right. The quartiles divide the area under the curve into quarters. One-fourth of the area under the curve is to the left of the first quartile, and three-fourths of the area is to the left of the third quartile. You can roughly locate the median and quartiles of any density curve by eye by dividing the area under the curve into four equal parts.

Because density curves are idealized patterns, a symmetric density curve is exactly symmetric. The median of a symmetric density curve is therefore at its center. Figure 1.15(a) shows the median of a symmetric curve. It isn't so easy to spot the equal-areas point on a skewed curve. There are mathematical ways of finding the median for any density curve. We did that to mark the median on the skewed curve in Figure 1.15(b).

What about the mean? The mean of a set of observations is their arithmetic average. If we think of the observations as weights strung out along a thin rod, the mean is the point at which the rod would balance. This fact is also true of density curves. **The mean is the point at which the curve would balance if made of solid material.** Figure 1.16 illustrates this fact about the mean. A symmetric curve balances at its center because the two sides are identical. **The mean and median of a symmetric density curve are equal,** as in

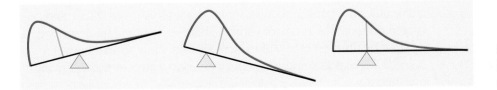

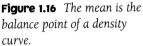

Figure 1.16 *The mean is the balance point of a density curve.*

Figure 1.15(a). We know that the mean of a skewed distribution is pulled toward the long tail. Figure 1.15(b) shows how the mean of a skewed density curve is pulled toward the long tail more than is the median. It's hard to locate the balance point by eye on a skewed curve. There are mathematical ways of calculating the mean for any density curve, so we are able to mark the mean as well as the median in Figure 1.15(b).

MEDIAN AND MEAN OF A DENSITY CURVE

The **median** of a density curve is the equal-areas point, the point that divides the area under the curve in half.

The **mean** of a density curve is the balance point, at which the curve would balance if made of solid material.

The median and mean are the same for a symmetric density curve. They both lie at the center of the curve. The mean of a skewed curve is pulled away from the median in the direction of the long tail.

We can roughly locate the mean, median, and quartiles of any density curve by eye. This is not true of the standard deviation. When necessary, we can once again call on more advanced mathematics to learn the value of the standard deviation. The study of mathematical methods for doing calculations with density curves is part of theoretical statistics. Though we are concentrating on statistical practice, we often make use of the results of mathematical study.

mean μ
standard deviation σ

Because a density curve is an idealized description of the distribution of data, we need to distinguish between the mean and standard deviation of the density curve and the mean $\bar{x}$ and standard deviation s computed from the actual observations. The usual notation for the mean of an idealized distribution is μ (the Greek letter mu). We write the standard deviation of a density curve as σ (the Greek letter sigma).

APPLY YOUR KNOWLEDGE

1.50 (a) Sketch a density curve that is symmetric but has a shape different from that of the curve in Figure 1.15(a).

(b) Sketch a density curve that is strongly skewed to the left.

1.51 Figure 1.17 displays the density curve of a *uniform distribution*. The curve takes the constant value 1 over the interval from 0 to 1 and is zero outside that range of values. This means that data described by this distribution take values that are uniformly spread between 0 and 1. Use areas under this density curve to answer the following questions.

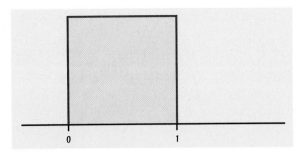

Figure 1.17 *The density curve of a uniform distribution, for Exercise 1.51.*

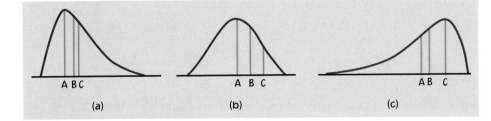

Figure 1.18 *Three density curves, for Exercise 1.52.*

(a) Why is the total area under this curve equal to 1?

(b) What percent of the observations lie above 0.8?

(c) What percent of the observations lie below 0.6?

(d) What percent of the observations lie between 0.25 and 0.75?

(e) What is the mean μ of this distribution?

1.52 Figure 1.18 displays three density curves, each with three points marked on them. At which of these points on each curve do the mean and the median fall?

Normal distributions

One particularly important class of density curves has already appeared in Figures 1.13 and 1.15(a). These density curves are symmetric, single-peaked, and bell-shaped. They are called *normal curves*, and they describe **normal distributions.** All normal distributions have the same overall shape. The exact density curve for a particular normal distribution is described by giving its mean μ and its standard deviation σ. The mean is located at the center of the symmetric curve and is the same as the median. Changing μ without changing σ moves the normal curve along the horizontal axis without changing its spread. The standard deviation σ controls the spread of a normal curve. Figure 1.19 shows two normal curves with different values of σ. The curve with the larger standard deviation is more spread out.

normal distributions

The standard deviation σ is the natural measure of spread for normal distributions. Not only do μ and σ completely determine the shape of a normal curve, but we can locate σ by eye on the curve. Here's how. Imagine that you

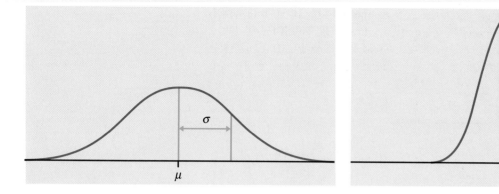

Figure 1.14 *Two normal curves, showing the mean μ and standard deviation σ.*

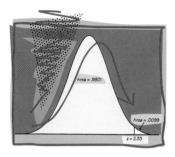

The bell curve?

Does the distribution of human intelligence follow the "bell curve" of a normal distribution? Scores on IQ tests do roughly follow a normal distribution. That is because a test score is calculated from a person's answers in a way that is designed to produce a normal distribution. To conclude that intelligence follows a bell curve, we must agree that the test scores directly measure intelligence. Many psychologists don't think there is one human characteristic that we can call "intelligence" and can measure by a single test score.

are skiing down a mountain that has the shape of a normal curve. At first, you descend at an ever-steeper angle as you go out from the peak:

Fortunately, before you find yourself going straight down, the slope begins to grow flatter rather than steeper as you go out and down:

The points at which this change of curvature takes place are located at distance σ on either side of the mean μ. You can feel the change as you run a pencil along a normal curve, and so find the standard deviation. Remember that μ and σ alone do not specify the shape of most distributions, and that the shape of density curves in general does not reveal σ. These are special properties of normal distributions.

Why are the normal distributions important in statistics? Here are three reasons. First, normal distributions are good descriptions for some distributions of *real data.* Distributions that are often close to normal include scores on tests taken by many people (such as SAT exams and many psychological tests), repeated careful measurements of the same quantity, and characteristics of biological populations (such as lengths of cockroaches and yields of corn). Second, normal distributions are good approximations to the results of many kinds of *chance outcomes,* such as tossing a coin many times. Third, and most important, we will see that many *statistical inference* procedures based on normal distributions work well for other roughly symmetric distributions. However, even though many sets of data follow a normal distribution, many do not. Most income distributions, for example, are skewed to the right and so are not normal. Nonnormal data, like nonnormal people, not only are common but are sometimes more interesting than their normal counterparts.

The 68–95–99.7 rule

Although there are many normal curves, they all have common properties. In particular, all normal distributions obey the following rule.

> **THE 68–95–99.7 RULE**
>
> In the normal distribution with mean μ and standard deviation σ:
> - **68%** of the observations fall within σ of the mean μ.
> - **95%** of the observations fall within 2σ of μ.
> - **99.7%** of the observations fall within 3σ of μ.

Figure 1.20 illustrates the 68–95–99.7 rule. By remembering these three numbers, you can think about normal distributions without constantly making detailed calculations.

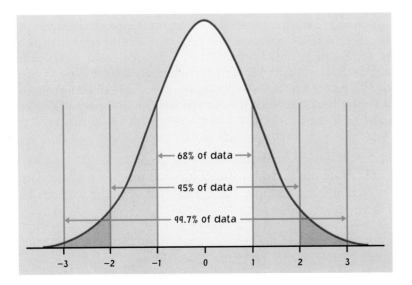

Figure 1.20 *The 68–95–99.7 rule for normal distributions.*

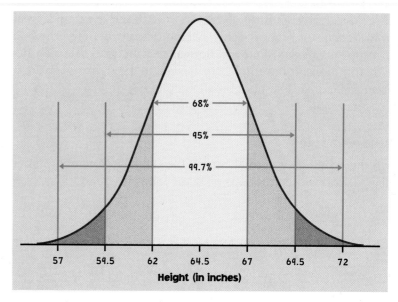

Figure 1.21 *The 68–95–99.7 rule applied to the distribution of the heights of young women. Here* $\mu = 64.5$ *and* $\sigma = 2.5$.

EXAMPLE 1.12 **Using the 68–95–99.7 rule**

The distribution of heights of young women aged 18 to 24 is approximately normal with mean $\mu = 64.5$ inches and standard deviation $\sigma = 2.5$ inches. Figure 1.21 shows what the 68–95–99.7 rule says about this distribution.

Two standard deviations is 5 inches for this distribution. The 95 part of the 68–95–99.7 rule says that the middle 95% of young women are between $64.5 - 5$ and $64.5 + 5$ inches tall, that is, between 59.5 inches and 69.5 inches. This fact is exactly true for an exactly normal distribution. It is approximately true for the heights of young women because the distribution of heights is approximately normal.

The other 5% of young women have heights outside the range from 59.5 to 69.5 inches. Because the normal distributions are symmetric, half of these women are on the tall side. So the tallest 2.5% of young women are taller than 69.5 inches.

The 99.7 part of the 68–95–99.7 rule says that almost all young women (99.7% of them) have heights between $\mu - 3\sigma$ and $\mu + 3\sigma$. This range of heights is 57 to 72 inches.

Because we will mention normal distributions often, a short notation is helpful. We abbreviate the normal distribution with mean μ and standard deviation σ as $N(\mu, \sigma)$. For example, the distribution of young women's heights is $N(64.5, 2.5)$.

APPLY YOUR KNOWLEDGE ·

1.53 **Men's heights.** The distribution of heights of adult men is approximately normal with mean 69 inches and standard deviation

2.5 inches. Draw a normal curve on which this mean and standard deviation are correctly located. (Hint: Draw the curve first, locate the points where the curvature changes, then mark the horizontal axis.)

1.54 **More on men's heights.** The distribution of heights of adult men is approximately normal with mean 69 inches and standard deviation 2.5 inches. Use the 68–95–99.7 rule to answer the following questions.

   (a) What percent of men are taller than 74 inches?

   (b) Between what heights do the middle 95% of men fall?

   (c) What percent of men are shorter than 66.5 inches?

1.55 **IQ test scores.** Scores on the Wechsler Adult Intelligence Scale (a standard "IQ test") for the 20 to 34 age group are approximately normally distributed with $\mu = 110$ and $\sigma = 25$. Use the 68–95–99.7 rule to answer these questions.

   (a) About what percent of people in this age group have scores above 110?

   (b) About what percent have scores above 160?

   (c) In what range do the middle 95% of all scores lie?

The standard normal distribution

As the 68–95–99.7 rule suggests, all normal distributions share many common properties. In fact, all normal distributions are the same if we measure in units of size σ about the mean μ as center. Changing to these units is called *standardizing*. To standardize a value, subtract the mean of the distribution and then divide by the standard deviation.

STANDARDIZING AND z-SCORES

If x is an observation from a distribution that has mean μ and standard deviation σ, the **standardized value** of x is

$$z = \frac{x - \mu}{\sigma}$$

A standardized value is often called a z-**score**.

A z-score tells us how many standard deviations the original observation falls away from the mean, and in which direction. Observations larger than the mean are positive when standardized, and observations smaller than the mean are negative.

> **EXAMPLE 1.13**   **Standardizing women's heights**
>
> The heights of young women are approximately normal with $\mu = 64.5$ inches and $\sigma = 2.5$ inches. The standardized height is
>
> $$z = \frac{\text{height} - 64.5}{2.5}$$
>
> A woman's standardized height is the number of standard deviations by which her height differs from the mean height of all young women. A woman 68 inches tall, for example, has standardized height
>
> $$z = \frac{68 - 64.5}{2.5} = 1.4$$
>
> or 1.4 standard deviations above the mean. Similarly, a woman 5 feet (60 inches) tall has standardized height
>
> $$z = \frac{60 - 64.5}{2.5} = -1.8$$
>
> or 1.8 standard deviations less than the mean height.

If the variable we standardize has a normal distribution, standardizing does more than give a common scale. It makes all normal distributions into a single distribution, and this distribution is still normal. Standardizing a variable that has any normal distribution produces a new variable that has the *standard normal distribution*.

> **STANDARD NORMAL DISTRIBUTION**
>
> The **standard normal distribution** is the normal distribution $N(0, 1)$ with mean 0 and standard deviation 1.
>
> If a variable x has any normal distribution $N(\mu, \sigma)$ with mean μ and standard deviation σ, then the standardized variable
>
> $$z = \frac{x - \mu}{\sigma}$$
>
> has the standard normal distribution.

APPLY YOUR KNOWLEDGE

1.56   **SAT versus ACT.** Eleanor scores 680 on the mathematics part of the SAT. The distribution of SAT scores in a reference population is normal, with mean 500 and standard deviation 100. Gerald takes the

American College Testing (ACT) mathematics test and scores 27. ACT scores are normally distributed with mean 18 and standard deviation 6. Find the standardized scores for both students. Assuming that both tests measure the same kind of ability, who has the higher score?

Normal distribution calculations

An area under a density curve is a proportion of the observations in a distribution. Any question about what proportion of observations lie in some range of values can be answered by finding an area under the curve. Because all normal distributions are the same when we standardize, we can find areas under any normal curve from a single table, a table that gives areas under the curve for the standard normal distribution.

EXAMPLE 1.14   **Using the standard normal distribution**

What proportion of all young women are less than 68 inches tall? This proportion is the area under the $N(64.5, 2.5)$ curve to the left of the point 68. Because the standardized height corresponding to 68 inches is

$$z = \frac{x - \mu}{\sigma} = \frac{68 - 64.5}{2.5} = 1.4$$

this area is the same as the area under the standard normal curve to the left of the point $z = 1.4$. Figure 1.22(a) shows this area.

Many calculators will give you areas under the standard normal curve. In case your calculator does not, Table A in the back of the book gives some of these areas.

THE STANDARD NORMAL TABLE

Table A is a table of areas under the standard normal curve. The table entry for each value z is the area under the curve to the left of z.

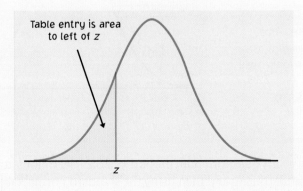

Table entry is area to left of z

z

EXAMPLE 1.15 **Using the standard normal table**

Problem: Find the proportion of observations from the standard normal distribution that are less than 1.4.

Solution: To find the area to the left of 1.40, locate 1.4 in the left-hand column of Table A, then locate the remaining digit 0 as .00 in the top row. The entry opposite 1.4 and under .00 is 0.9192. This is the area we seek. Figure 1.22(a) illustrates the relationship between the value $z = 1.40$ and the area 0.9192. Because $z = 1.40$ is the standardized value of height 68 inches, the proportion of young women who are less than 68 inches tall is 0.9192 (about 92%).

Problem: Find the proportion of observations from the standard normal distribution that are greater than -2.15.

Solution: Enter Table A under $z = -2.15$. That is, find -2.1 in the left-hand column and .05 in the top row. The table entry is 0.0158. This is the area to the *left* of -2.15. Because the total area under the curve is 1, the area lying to the *right* of -2.15 is $1 - 0.0158 = 0.9842$. Figure 1.22(b) illustrates these areas.

Figure 1.22(a) *The area under a standard normal curve to the left of the point $z = 1.4$ is 0.9192. Table A gives areas under the standard normal curve.*

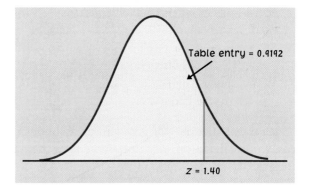

Figure 1.22(b) *Areas under the standard normal curve to the right and left of $z = -2.15$. Table A gives only areas to the left.*

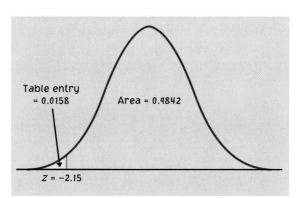

We can answer any question about proportions of observations in a normal distribution by standardizing and then using the standard normal table. Here is an outline of the method for finding the proportion of the distribution in any region.

FINDING NORMAL PROPORTIONS

1. State the problem in terms of the observed variable x.

2. Standardize x to restate the problem in terms of a standard normal variable z. Draw a picture to show the area under the standard normal curve.

3. Find the required area under the standard normal curve, using Table A and the fact that the total area under the curve is 1.

EXAMPLE 1.16   **normal distribution calculations**

The level of cholesterol in the blood is important because high cholesterol levels may increase the risk of heart disease. The distribution of blood cholesterol levels in a large population of people of the same age and sex is roughly normal. For 14-year-old boys, the mean is $\mu = 170$ milligrams of cholesterol per deciliter of blood (mg/dl) and the standard deviation is $\sigma = 30$ mg/dl.[14] Levels above 240 mg/dl may require medical attention. What percent of 14-year-old boys have more than 240 mg/dl of cholesterol?

1. *State the problem.* Call the level of cholesterol in the blood x. The variable x has the $N(170, 30)$ distribution. We want the proportion of boys with $x > 240$.

2. *Standardize.* Subtract the mean, then divide by the standard deviation, to turn x into a standard normal z:

$$x > 240$$

$$\frac{x - 170}{30} > \frac{240 - 170}{30}$$

$$z > 2.33$$

Figure 1.23 shows the standard normal curve with the area of interest shaded.

3. *Use the table.* From Table A, we see that the proportion of observations less than 2.33 is 0.9901. About 99% of boys have cholesterol levels less than 240. The area to the right of 2.33 is therefore $1 - 0.9901 = 0.0099$. This is about 0.01, or 1%. Only about 1% of boys have high cholesterol.

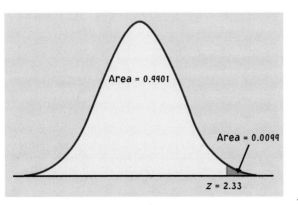

Area = 0.9901

Area = 0.0099

$z = 2.33$

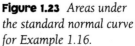

Figure 1.23 *Areas under the standard normal curve for Example 1.16.*

In a normal distribution, the proportion of observations with $x > 240$ is the same as the proportion with $x \geq 240$. There is no area under the curve and exactly over 240, so the areas under the curve with $x > 240$ and $x \geq 240$ are the same. This isn't true of the actual data. There may be a boy with exactly 240 mg/dl of blood cholesterol. The normal distribution is just an easy-to-use approximation, not a description of every detail in the actual data.

The key to using either software or Table A to do a normal calculation is to sketch the area you want, then match that area with the areas that the table or software gives you. Here is another example.

EXAMPLE 1.17 **More normal distribution calculations**

What percent of 14-year-old boys have blood cholesterol between 170 and 240 mg/dl?

1. *State the problem.* We want the proportion of boys with $170 \leq x \leq 240$.

2. *Standardize:*

$$170 \quad \leq \quad x \quad \leq \quad 240$$

$$\frac{170 - 170}{30} \leq \frac{x - 170}{30} \leq \frac{240 - 170}{30}$$

$$0 \quad \leq \quad z \quad \leq \quad 2.33$$

Figure 1.24 shows the area under the standard normal curve.

3. *Use the table.* The area between 2.33 and 0 is the area below 2.33 *minus* the area below 0. Look at Figure 1.24 to check this. From Table A,

$$\text{area between 0 and 2.33} = \text{area below 2.33} - \text{area below 0.00}$$
$$= 0.9901 - 0.5000 = 0.4901$$

About 49% of boys have cholesterol levels between 170 and 240 mg/dl.

Sometimes we encounter a value of z more extreme than those appearing in Table A. For example, the area to the left of $z = -4$ is not given directly in the

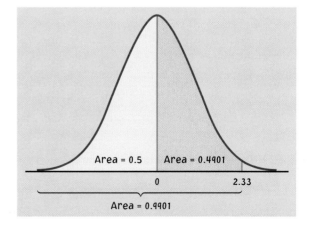

Figure 1.24 *Areas under the standard normal curve for Example 1.17.*

table. The z-values in Table A leave only area 0.0002 in each tail unaccounted for. For practical purposes, we can act as if there is zero area outside the range of Table A.

1.57 Use Table A to find the proportion of observations from a standard normal distribution that satisfies each of the following statements. In each case, sketch a standard normal curve and shade the area under the curve that is the answer to the question.

(a) $z < 2.85$

(b) $z > 2.85$

(c) $z > -1.66$

(d) $-1.66 < z < 2.85$

1.58 **How hard do locomotives pull?** An important measure of the performance of a locomotive is its "adhesion," which is the locomotive's pulling force as a multiple of its weight. The adhesion of one 4400-horsepower diesel locomotive model varies in actual use according to a normal distribution with mean $\mu = 0.37$ and standard deviation $\sigma = 0.04$.

(AL SIMPSON/VISUALS UNLIMITED.)

(a) What proportion of adhesions measured in use are higher than 0.40?

(b) What proportion of adhesions are between 0.40 and 0.50?

(c) Improvements in the locomotive's computer controls change the distribution of adhesion to a normal distribution with mean $\mu = 0.41$ and standard deviation $\sigma = 0.02$. Find the proportions in (a) and (b) after this improvement.

Finding a value given a proportion

Examples 1.16 and 1.17 illustrate the use of Table A to find what proportion of the observations satisfies some condition, such as "blood cholesterol between 170 mg/dl and 240 mg/dl." We may instead want to find the observed value with a given proportion of the observations above or below it. To do this, use Table A backward. Find the given proportion in the body of the table, read the corresponding z from the left column and top row, then "unstandardize" to get the observed value. Here is an example.

EXAMPLE 1.18 **"Backward" normal calculations**

Scores on the SAT verbal test in recent years follow approximately the $N(505, 110)$ distribution. How high must a student score in order to place in the top 10% of all students taking the SAT?

1. *State the problem.* We want to find the SAT score x with area 0.1 to its *right* under the normal curve with mean $\mu = 505$ and standard deviation $\sigma = 110$. That's

Figure 1.25 *Locating the point on a normal curve with area 0.10 to its right.*

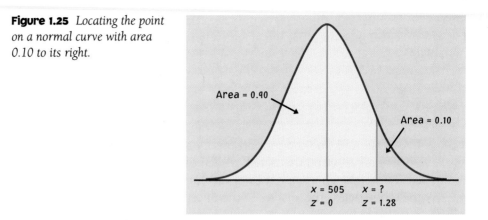

Area = 0.90

Area = 0.10

x = 505   x = ?
z = 0   z = 1.28

the same as finding the SAT score x with area 0.9 to its *left*. Figure 1.25 poses the question in graphical form. Because Table A gives the areas to the left of z-values, always state the problem in terms of the area to the left of x.

2. *Use the table.* Look in the body of Table A for the entry closest to 0.9. It is 0.8997. This is the entry corresponding to $z = 1.28$. So $z = 1.28$ is the standardized value with area 0.9 to its left.

3. *Unstandardize* to transform the solution from the z back to the original x scale. We know that the standardized value of the unknown x is $z = 1.28$. So x itself satisfies

$$\frac{x - 505}{110} = 1.28$$

Solving this equation for x gives

$$x = 505 + (1.28)(110) = 645.8$$

This equation should make sense: it says that x lies 1.28 standard deviations above the mean on this particular normal curve. That is the "unstandardized" meaning of $z = 1.28$. We see that a student must score at least 646 to place in the highest 10%.

Here is the general formula for unstandardizing a z-score. To find the value x from the normal distribution with mean μ and standard deviation σ corresponding to a given standard normal value z, use

$$x = \mu + z\sigma$$

APPLY YOUR KNOWLEDGE

1.59 Use Table A to find the value z of a standard normal variable that satisfies each of the following conditions. (Use the value of z from Table A that comes closest to satisfying the condition.) In each case, sketch a standard normal curve with your value of z marked on the axis.

(a) The point z with 25% of the observations falling below it.

(b) The point z with 40% of the observations falling above it.

1.60   **IQ test scores.** Scores on the Wechsler Adult Intelligence Scale for the 20 to 34 age group are approximately normally distributed with $\mu = 110$ and $\sigma = 25$.

(a) What percent of people aged 20 to 34 have IQ scores above 100?

(b) What IQ scores fall in the lowest 25% of the distribution?

(c) How high an IQ score is needed to be in the highest 5%?

SECTION 1.3   S u m m a r y

We can sometimes describe the overall pattern of a distribution by a **density curve.** A density curve has total area 1 underneath it. An area under a density curve gives the proportion of observations that fall in a range of values.

A density curve is an idealized description of the overall pattern of a distribution that smooths out the irregularities in the actual data. We write the mean of a density curve as μ and the standard deviation of a density curve as σ to distinguish them from the mean $\bar{x}$ and standard deviation s of the actual data.

The mean, the median, and the quartiles of a density curve can be located by eye. The **mean** μ is the balance point of the curve. The **median** divides the area under the curve in half. The **quartiles** and the median divide the area under the curve into quarters. The **standard deviation** σ cannot be located by eye on most density curves.

The mean and median are equal for symmetric density curves. The mean of a skewed curve is located farther toward the long tail than is the median.

The **normal distributions** are described by a special family of bell-shaped, symmetric density curves, called **normal curves.** The mean μ and standard deviation σ completely specify a normal distribution $N(\mu, \sigma)$. The mean is the center of the curve and σ is the distance from μ to the change-of-curvature points on either side.

To **standardize** any observation x, subtract the mean of the distribution and then divide by the standard deviation. The resulting **z-score**

$$z = \frac{x - \mu}{\sigma}$$

says how many standard deviations x lies from the distribution mean.

All normal distributions are the same when measurements are transformed to the standardized scale. In particular, all normal distributions satisfy the **68–95–99.7 rule**, which describes what percent of observations lie within one, two, and three standard deviations of the mean.

If x has the $N(\mu, \sigma)$ distribution, then the **standardized variable** $z = (x - \mu)/\sigma$ has the **standard normal distribution** $N(0, 1)$ with mean 0 and standard deviation 1. Table A gives the proportions of standard normal

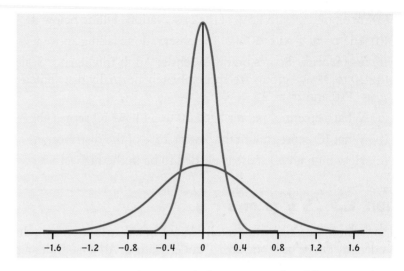

Figure 1.26 *Two normal curves with the same mean but different standard deviations, for Exercise 1.61.*

observations that are less than z for many values of z. By standardizing, we can use Table A for any normal distribution.

SECTION 1.3 Exercises

1.61   Figure 1.26 shows two normal curves, both with mean 0. Approximately what is the standard deviation of each of these curves?

1.62   **How big are soldiers' heads?** The army reports that the distribution of head circumference among male soldiers is approximately normal with mean 22.8 inches and standard deviation 1.1 inches. Use the 68–95–99.7 rule to answer the following questions.

(a) What percent of soldiers have head circumference greater than 23.9 inches?

(b) What percent of soldiers have head circumference between 21.7 inches and 23.9 inches?

1.63   **Length of pregnancies.** The length of human pregnancies from conception to birth varies according to a distribution that is approximately normal with mean 266 days and standard deviation 16 days. Use the 68–95–99.7 rule to answer the following questions.

(a) Between what values do the lengths of the middle 95% of all pregnancies fall?

(b) How short are the shortest 2.5% of all pregnancies?

1.64   **Three great hitters.** Three landmarks of baseball achievement are Ty Cobb's batting average of .420 in 1911, Ted Williams's .406 in 1941, and George Brett's .390 in 1980. These batting averages cannot be compared directly because the distribution of major league batting averages has changed over the years. The distributions are quite symmetric and (except for outliers such as Cobb, Williams, and Brett) reasonably normal. While the mean batting average has been held roughly constant by rule changes and the balance between hitting and pitching, the standard deviation has dropped over time. Here are the facts:

Decade	Mean	Std. dev.
1910s	.266	.0371
1940s	.267	.0326
1970s	.261	.0317

Compute the standardized batting averages for Cobb, Williams, and Brett to compare how far each stood above his peers.[15]

1.65 Use Table A to find the proportion of observations from a standard normal distribution that falls in each of the following regions. In each case, sketch a standard normal curve and shade the area representing the region.

(a) $z \leq -2.25$

(b) $z \geq -2.25$

(c) $z > 1.77$

(d) $-2.25 < z < 1.77$

1.66 (a) Find the number z such that the proportion of observations that are less than z in a standard normal distribution is 0.8.

(b) Find the number z such that 35% of all observations from a standard normal distribution are greater than z.

1.67 **The stock market.** The annual rate of return on stock indexes (which combine many individual stocks) is approximately normal. Since 1945, the Standard & Poor's 500 index has had a mean yearly return of 12%, with a standard deviation of 16.5%. Take this normal distribution to be the distribution of yearly returns over a long period.

(a) In what range do the middle 95% of all yearly returns lie?

(b) The market is down for the year if the return on the index is less than zero. In what proportion of years is the market down?

(c) In what proportion of years does the index gain 25% or more?

1.68 **Length of pregnancies.** The length of human pregnancies from conception to birth varies according to a distribution that is approximately normal with mean 266 days and standard deviation 16 days.

(a) What percent of pregnancies last less than 240 days (that's about 8 months)?

(b) What percent of pregnancies last between 240 and 270 days (roughly between 8 months and 9 months)?

(c) How long do the longest 20% of pregnancies last?

1.69 **Are we getting smarter?** When the Stanford-Binet "IQ test" came into use in 1932, it was adjusted so that scores for each age group of children followed roughly the normal distribution with mean $\mu = 100$ and standard deviation $\sigma = 15$. The test is readjusted from time to time to keep the mean at 100. If present-day American children took the 1932 Stanford-Binet test, their mean score would be about 120. The reasons for the increase in IQ over time are not known but probably include better childhood nutrition and more experience in taking tests.[16]

(a) IQ scores above 130 are often called "very superior." What percent of children had very superior scores in 1932?

(b) If present-day children took the 1932 test, what percent would have very superior scores? (Assume that the standard deviation $\sigma = 15$ does not change.)

1.70 The median of any normal distribution is the same as its mean. We can use normal calculations to find the quartiles for normal distributions.

(a) What is the area under the standard normal curve to the left of the first quartile? Use this to find the value of the first quartile for a standard normal distribution. Find the third quartile similarly.

(b) Your work in (a) gives the z-scores for the quartiles of any normal distribution. What are the quartiles for the lengths of human pregnancies? (Use the distribution in Exercise 1.68.)

1.71 The *deciles* of any distribution are the points that mark off the lowest 10% and the highest 10%. On a density curve, these are the points with areas 0.1 and 0.9 to their left under the curve.

(a) What are the deciles of the standard normal distribution?

(b) The heights of young women are approximately normal with mean 64.5 inches and standard deviation 2.5 inches. What are the deciles of this distribution?

STATISTICS in SUMMARY

Data analysis is the art of describing data using graphs and numerical summaries. The purpose of data analysis is to describe the most important features of a set of data. This chapter introduces data analysis by presenting statistical ideas and tools for describing the distribution of a single variable. The Statistics in Summary figure on the next page will help you organize the big ideas. The

STATISTICS In SUMMARY

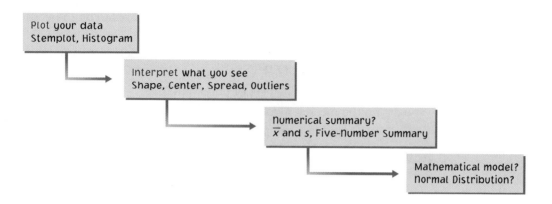

Plot your data
Stemplot, Histogram

Interpret what you see
Shape, Center, Spread, Outliers

numerical summary?
$\bar{x}$ and s, Five-number Summary

Mathematical model?
normal Distribution?

question marks at the last two stages remind us that the usefulness of numerical summaries and models such as normal distributions depends on what we find when we examine the data using graphs. Here is a review list of the most important skills you should have acquired from your study of this chapter.

A. DATA

1. Identify the individuals and variables in a set of data.
2. Identify each variable as categorical or quantitative. Identify the units in which each quantitative variable is measured.

B. DISPLAYING DISTRIBUTIONS

1. Make a bar graph of the distribution of a categorical variable. Interpret bar graphs and pie charts.
2. Make a histogram of the distribution of a quantitative variable.
3. Make a stemplot of the distribution of a small set of observations. Round leaves or split stems as needed to make an effective stemplot.

C. INSPECTING DISTRIBUTIONS (QUANTITATIVE VARIABLE)

1. Look for the overall pattern and for major deviations from the pattern.
2. Assess from a histogram or stemplot whether the shape of a distribution is roughly symmetric, distinctly skewed, or neither. Assess whether the distribution has one or more major peaks.
3. Describe the overall pattern by giving numerical measures of center and spread in addition to a verbal description of shape.
4. Decide which measures of center and spread are more appropriate: the mean and standard deviation (especially for symmetric distributions) or the five-number summary (especially for skewed distributions).
5. Recognize outliers.

D. TIME PLOTS

1. Make a time plot of data, with the time of each observation on the horizontal axis and the value of the observed variable on the vertical axis.
2. Recognize strong trends or other patterns in a time plot.

E. MEASURING CENTER

1. Find the mean $\bar{x}$ of a set of observations.
2. Find the median M of a set of observations.
3. Understand that the median is more resistant (less affected by extreme observations) than the mean. Recognize that skewness in a distribution moves the mean away from the median toward the long tail.

F. MEASURING SPREAD

1. Find the quartiles Q_1 and Q_3 for a set of observations.
2. Give the five-number summary and draw a boxplot; assess center, spread, symmetry and skewness from a boxplot.
3. Using a calculator, find the standard deviation s for a set of observations.
4. Know the basic properties of s: $s \geq 0$ always; $s = 0$ only when all observations are identical and increases as the spread increases; s has the same units as the original measurements; s is pulled strongly up by outliers or skewness.

G. DENSITY CURVES

1. Know that areas under a density curve represent proportions of all observations and that the total area under a density curve is 1.
2. Approximately locate the median (equal-areas point) and the mean (balance point) on a density curve.
3. Know that the mean and median both lie at the center of a symmetric density curve and that the mean moves farther toward the long tail of a skewed curve.

H. NORMAL DISTRIBUTIONS

1. Recognize the shape of normal curves and be able to estimate by eye both the mean and standard deviation from such a curve.
2. Use the 68–95–99.7 rule and symmetry to state what percent of the observations from a normal distribution fall between two points when both points lie at the mean or one, two, or three standard deviations on either side of the mean.
3. Find the standardized value (z-score) of an observation. Interpret z-scores and understand that any normal distribution becomes standard normal $N(0, 1)$ when standardized.
4. Given that a variable has the normal distribution with a stated mean μ and standard deviation σ, calculate the proportion of values above a

stated number, below a stated number, or between two stated numbers.

5. Given that a variable has the normal distribution with a stated mean μ and standard deviation σ, calculate the point having a stated proportion of all values above it. Also calculate the point having a stated proportion of all values below it.

CHAPTER 1 Review Exercises

1.72 Gender effects in voting. Political party preference in the United States depends in part on the age, income, and gender of the voter. A political scientist selects a large sample of registered voters. For each voter, she records gender, age, household income, and whether they voted for the Democratic or for the Republican candidate in the last congressional election. Which of these variables are categorical and which are quantitative?

1.73 Murder weapons. The 1997 *Statistical Abstract of the United States* reports FBI data on murders for 1995. In that year, 55.8% of all murders were committed with handguns, 12.4% with other firearms, 12.6% with knives, 5.9% with a part of the body (usually the hands or feet), and 4.5% with blunt objects. Make a graph to display these data. Do you need an "other methods" category?

1.74 Never on Sunday? The Canadian Province of Ontario carries out statistical studies of the working of Canada's national health care system in the province. The bar graphs in Figure 1.27 come from a study of admissions and discharges from community hospitals in Ontario.[17] They show the number of heart attack patients admitted and discharged on each day of the week during a two-year period.

(a) Explain why you expect the number of patients admitted with heart attacks to be roughly the same for all days of the week. Do the data show that this is true?

(b) Describe how the distribution of the day on which patients are discharged from the hospital differs from that of the day on which they are admitted. What do you think explains the difference?

1.75 Drive time. Professor Moore, who lives a few miles outside a college town, records the time he takes to drive to the college each morning. Here are the times (in minutes) for 42 consecutive weekdays, with the dates in order along the rows:

8.25	7.83	8.30	8.42	8.50	8.67	8.17	9.00	9.00	8.17	7.92
9.00	8.50	9.00	7.75	7.92	8.00	8.08	8.42	8.75	8.08	9.75
8.33	7.83	7.92	8.58	7.83	8.42	7.75	7.42	6.75	7.42	8.50
8.67	10.17	8.75	8.58	8.67	9.17	9.08	8.83	8.67		

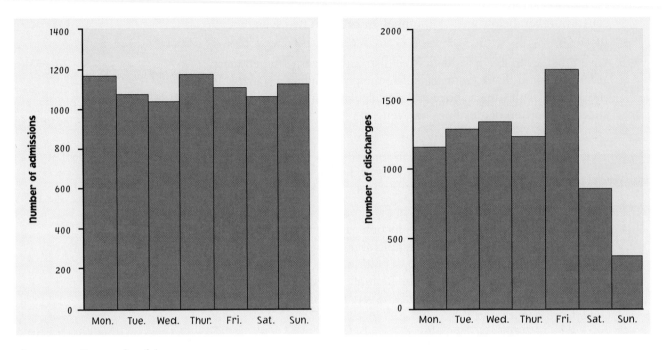

Figure 1.27 *Bar graphs of the number of heart attack victims admitted and discharged on each day of the week by hospitals in Ontario, Canada, for Exercise 1.74.*

(a) Make a stemplot of these drive times. (Round to the nearest tenth of a minute and use split stems.) Is the distribution roughly symmetric, clearly skewed, or neither? Are there any clear outliers?

(b) Make a time plot of the drive times. (Label the horizontal axis in days, 1 to 42.) The plot shows no clear trend, but it does show one unusually low drive time and two unusually high drive times. Circle these observations on your plot.

(c) All three unusual observations can be explained. The low time is the day after Thanksgiving (no traffic on campus). The two high times reflect delays due to an accident and icy roads. Remove these three observations, and find the mean $\bar{x}$ and standard deviation s of the remaining 39 drive times.

(d) Count the number of observations between $\bar{x} - s$ and $\bar{x} + s$, between $\bar{x} - 2s$ and $\bar{x} + 2s$, and between $\bar{x} - 3s$ and $\bar{x} + 3s$. Find the percent of observations in each of these intervals and compare with the 68–95–99.7 rule. As real data go, these data are reasonably close to having a normal distribution.

1.76 **Stock returns.** Table 1.8 (page 44) gives the monthly percent returns on Philip Morris stock for the period from July 1990 to May 1997. The data appear in time order reading from left to right across each row in turn, beginning with the −5.7% return in July 1990. Make a time plot of the data. This was a period of increasing action against smoking, so we might expect a trend toward lower returns. But it was also a period in which stocks in general rose sharply, which would produce an increasing trend. What does your time plot show?

1.77 **Better corn.** Corn is an important animal food. Normal corn lacks certain amino acids, which are building blocks for protein. Plant scientists have developed new corn varieties that have more of these amino acids. To test a new corn as an animal food, a group of 20 one-day-old male chicks was fed a ration containing the new corn. A control group of another 20 chicks was fed a ration that was identical except that it contained normal corn. Here are the weight gains (in grams) after 21 days:[18]

Normal corn				New corn			
380	321	366	356	361	447	401	375
283	349	402	462	434	403	393	426
356	410	329	399	406	318	467	407
350	384	316	272	427	420	477	392
345	455	360	431	430	339	410	326

(a) Compute five-number summaries for the weight gains of the two groups of chicks. Then make boxplots to compare the two distributions. What do the data show about the effect of the new corn?

(b) The researchers actually reported means and standard deviations for the two groups of chicks. What are they? How much larger is the mean weight gain of chicks fed the new corn?

1.78 **DiMaggio versus Mantle.** Joe DiMaggio played center field for the Yankees for 13 years. He was succeeded by Mickey Mantle, who played for 18 years. Here is the number of home runs hit each year by DiMaggio:

        29 46 32 30 31 30 21 25 20 39 14 32 12

Here are Mantle's home run counts:

13 23 21 27 37 52 34 42 31 40 54 30 15 35 19 23 22 18

Compute the five-number summary for each player, and make side-by-side boxplots of the home run distributions. What does your comparison show about DiMaggio and Mantle as home run hitters?

1.79 **Do SUVs waste gas?** Table 1.2 (page 11) gives the highway fuel consumption (in miles per gallon) for 26 midsize 1998 car models. Here are the highway mileages for 19 four-wheel drive 1998 sport utility vehicle models:[19]

Model	MPG	Model	MPG
Acura SLX	19	Jeep Wrangler	19
Chevrolet Blazer	20	Land Rover	16
Chevrolet Tahoe	19	Mazda MPV	19
Dodge Durango	17	Mercedes-Benz ML320	21
Ford Expedition	18	Mitsubishi Montero	20
Ford Explorer	19	Nissan Pathfinder	19
Honda Passport	20	Suzuki Sidekick	26
Infiniti QX4	19	Toyota RAV4	26
Isuzu Trooper	19	Toyota 4Runner	22
Jeep Grand Cherokee	18		

(a) Give a graphical and numerical description of highway fuel consumption for sports utility vehicles. What are the main features of the distribution?

(b) Make boxplots to compare the highway fuel consumption of midsize cars and sports utility vehicles. What are the most important differences between the two distributions?

1.80 **Guinea pig survival times.** Table 1.9 gives the survival times in days of 72 guinea pigs after they were injected with tubercle bacilli in a medical experiment.[20] Survival times, whether of machines under stress or cancer patients after treatment, usually have distributions that are skewed to the right.

(a) Graph the distribution and describe its main features. Does it show the expected right skew?

Table 1.9 Survival times (days) of guinea pigs in a medical experiment

43	45	53	56	56	57	58	66	67	73
74	79	80	80	81	81	81	82	83	83
84	88	89	91	91	92	92	97	99	99
100	100	101	102	102	102	103	104	107	108
109	113	114	118	121	123	126	128	137	138
139	144	145	147	156	162	174	178	179	184
191	198	211	214	243	249	329	380	403	511
522	598								

(b) Here is output from the statistical software Data Desk for these data:

```
Summary statistics for days

Mean 141.84722
Median 102.50000
Cases 72
StdDev 109.20863
Min 43
Max 598
25th%ile 82.250000
75th%ile 153.75000
```

(Data Desk uses "Cases" for the number of observations, and "25th%ile" for the first quartile, which is also called the 25th percentile because 25% of the data lie below it. Similarly, "75th%ile" is the third quartile.) Explain how the relationship between the mean and the median reflects the skewness of the data.

(c) Give the five-number summary and explain briefly how it reflects the skewness of the data.

1.81 **A hot stock.** The rate of return on a stock is its change in price plus any dividends paid. Rate of return is usually measured in percent of the starting value. We have data on the monthly rates of return for the stock of Wal-Mart stores for the years 1973 to 1991, the first 19 years Wal-Mart was listed on the New York Stock Exchange. There are 228 observations.

Figure 1.28 displays output from statistical software that describes the distribution of these data. The stems in the stemplot are the tens digits of the percent returns. The leaves are the ones digits. The stemplot uses split stems to give a better display. The software gives high and low outliers separately from the stemplot rather than spreading out the stemplot to include them.

(a) Give the five-number summary for monthly returns on Wal-Mart stock.

(b) Describe in words the main features of the distribution.

(c) If you had $1000 worth of Wal-Mart stock at the beginning of the best month during these 19 years, how much would your stock be worth at the end of the month? If you had $1000 worth of stock at the beginning of the worst month, how much would your stock be worth at the end of the month?

Figure 1.28 *Output from software describing the distribution of monthly returns for Wal-Mart stock, for Exercise 1.81.*

```
Mean   =  3.064
Standard deviation  =  11.49

N = 228    Median  =  3.4691
Quartiles  =  -2.950258,  8.4511

Decimal point is 1 place to the right of the colon

Low:  -34.04255  -31.25000  -27.06271  -26.61290

-1 : 985
-1 : 444443322222110000
-0 : 999988777666666665555
-0 : 4444444433333333222222222221111111100
 0 : 00000111111111111222222333333344444444
 0 : 5555555555555555555556666666666677777778888888888899999
 1 : 000000000111111111122233334444
 1 : 55566667889
 2 : 011334

High:  32.01923   41.80531   42.05607   57.89474   58.67769
```

1.82 **The 1.5 × IQR criterion.** A common criterion for detecting suspected outliers in a set of data is as follows:

interquartile range

1. Find the quartiles Q_1 and Q_3 and the **interquartile range** $IQR = Q_3 - Q_1$. The interquartile range is the spread of the central half of the data.

2. Call an observation an outlier if it falls more than $1.5 \times IQR$ above the third quartile or below the first quartile.

Find the interquartile range IQR for the Wal-Mart data in the previous exercise. Are there any outliers according to the $1.5 \times IQR$ criterion? Does it appear to you that the software uses this criterion in choosing which observations to report separately as outliers?

1.83 **Stock market returns.** Has the behavior of Wal-Mart stock changed over the 19 years 1973 to 1991? In Exercise 1.81 we saw the distribution of all 228 monthly returns. That display can't answer questions about change over time. Figure 1.29 is a variation of a time plot. Rather than plotting all 228 observations, we give side-by-side boxplots that compare the 19 years with each other. There are 12 monthly returns in each year.

(a) Is there a long term trend in the typical monthly return over time?

(b) Is there a trend in the spread of the monthly returns?

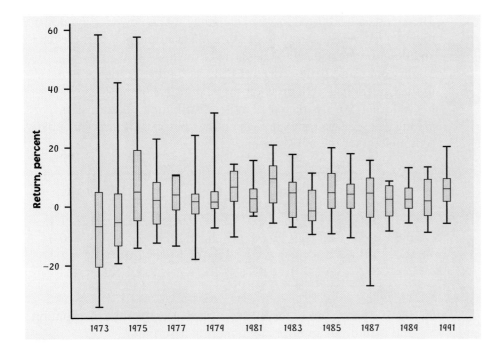

(c) The stemplot in Figure 1.28 reports several outliers. Which of these can you spot in the boxplots? In what years did they occur? Does this reinforce your conclusions from part (b)? Are there any outliers that are especially striking after taking your result from (b) into account?

1.84 Julie says, "People are living longer now, so presidents are likely to be older than in the past." John replies, "No—modern voters like youth and don't respect age, so presidents are likely to be younger now than earlier in our history."

Make a time plot of the presidential ages in Table 1.7 (page 37). Take the horizontal axis to be the order of the presidents, from 1 for Washington to 42 for Clinton. Are there any clear trends over time? Do the data support either Julie or John?

1.85 **The price of computer storage.** Computer users know that the cost of computing power falls rapidly. One measure of this is the cost per megabyte of storage on the largest hard drives available for home computers. Here are data for Macintosh hard drives:[21]

Year	1992	1993	1994	1995	1996
Cost	$5.07	$2.40	$1.14	$0.53	$0.36

These costs are adjusted for the changing buying power of the dollar to give a fair year-to-year comparison. Make a time plot of these data and describe the pattern of change over time.

1.86 **Mighty oaks and little acorns (EESEE).** Of the 50 species of oaks in the United States, 28 grow on the Atlantic coast and 11 grow in California. We are interested in the distribution of acorn sizes among oak species. Here are data on the volume of acorns (in cubic centimeters) for these 39 oak species:[22]

Atlantic							California		
1.4	3.4	9.1	1.6	10.5	2.5	0.9	4.1	5.9	17.1
6.8	1.8	0.3	0.9	0.8	2.0	1.1	1.6	2.6	0.4
0.6	1.8	4.8	1.1	3.0	1.1	1.1	2.0	6.0	7.1
3.6	8.1	3.6	1.8	0.4	1.1	1.2	5.5	1.0	

(a) Make a histogram of the 39 acorn sizes. Describe the distribution and include an appropriate numerical description.

(b) Compare the Atlantic and California region distributions with a graph and numerical summaries. What do you find?

1.87 Table 1.6 (page 26) reports data on the states. Much more information is available. Do your own exploration of differences among the states. Find in the library a recent edition of the annual *Statistical Abstract of the United States*. Look up data on either

(a) marriage rates per thousand inhabitants or

(b) death rates per thousand inhabitants

for the 50 states. Make a graph and a numerical summary to display the distribution and write a brief description of the most important characteristics. Suggest an explanation for the outliers you see.

1.88 **Mexican Americans.** The Acculturation Rating Scale for Mexican Americans (ARSMA) is a psychological test that measures the degree to which Mexican Americans are adapted to Mexican/Spanish versus Anglo/English culture. The range of possible scores is 1.0 to 5.0, with scores near 1 showing more Mexican/Spanish acculturation. The distribution of ARSMA scores in the population used to develop the test is approximately normal with mean 3.0 and standard deviation 0.8. A researcher believes that Mexicans will have an average score near 1.7 and that first-generation Mexican Americans will average about 2.1 on the ARSMA scale. What proportion of the population used to develop the test has scores below 1.7? Between 1.7 and 2.1?

1.89 **The size of soldiers' heads.** The army reports that the distribution of head circumference among soldiers is approximately normal with mean 22.8 inches and standard deviation 1.1 inches. Helmets are mass-produced for all except the smallest 5% and the largest 5% of head sizes. Soldiers in the smallest or largest 5% get custom-made helmets. What head sizes get custom-made helmets?

1.90 **Mexican Americans.** The ARSMA test is described in Exercise 1.88. How high a score on this test must a Mexican American obtain to be among the 30% of the population used to develop the test who are most Anglo/English in cultural orientation? What scores make up the 30% who are most Mexican/Spanish in their acculturation?

John W. Tukey

(AT&T ARCHIVES.)

The Philosopher of Data Analysis

He started as a chemist, became a mathematician, and was converted to statistics by what he called "the real problems experience and the real data experience" of war work during the Second World War. John W. Tukey (1915–) came to Princeton University in 1937 to study chemistry but took a doctorate in mathematics in 1939. During the war, he worked on the accuracy of range finders and of gunfire from bombers, among other problems. After the war he divided his time between Princeton and nearby Bell Labs, at that time the world's leading industrial research group.

Tukey was converted to statistics by "the real problems experience and the real data experience" during the Second World War.

Tukey devoted much of his attention to the statistical study of messy problems with lots of complex data: the safety of anesthetics used by many doctors in many hospitals on many patients, the Kinsey studies of human sexual behavior, monitoring compliance with a nuclear test ban, and air quality and environmental pollution.

From this "real problems experience and real data experience," Tukey developed exploratory data analysis. He invented some of the tools we have met, such as boxplots and stemplots. More important, he developed a philosophy for data analysis that changed the way statisticians think. In this chapter, as in Chapter 1, the approach we take in examining data follows Tukey's path.

Examining Relationships

(FRANCOIS GOHIER/PHOTO RESEARCHERS.)

Data on air quality are complex: weather, winds, pollution sources all play roles.

Introduction

A medical study finds that short women are more likely to have heart attacks than women of average height, while tall women have the fewest heart attacks. An insurance group reports that heavier cars have fewer deaths per 10,000 vehicles registered than do lighter cars. These and many other statistical studies look at the relationship between two variables. To understand such a relationship, we must often examine other variables as well. To conclude that shorter women have higher risk from heart attacks, for example, the researchers had to eliminate the effect of other variables such as weight and exercise habits. Our topic in this chapter is relationships between variables. One of our main themes is that the relationship between two variables can be strongly influenced by other variables that are lurking in the background.

Because variation is everywhere, statistical relationships are overall tendencies, not ironclad rules. They allow individual exceptions. Although smokers on the average die younger than nonsmokers, some people live to 90 while smoking three packs a day. To study a relationship between two variables, we measure both variables on the same individuals. Often, we think that one of the variables explains or influences the other.

RESPONSE VARIABLE, EXPLANATORY VARIABLE

A **response variable** measures an outcome of a study. An **explanatory variable** explains or influences changes in a response variable.

---

independent variable
dependent variable

You will often find explanatory variables called **independent variables,** and response variables called **dependent variables.** The idea behind this language is that the response variable depends on the explanatory variable. Because the words "independent" and "dependent" have other meanings in statistics that are unrelated to the explanatory-response distinction, we prefer to avoid those words.

It is easiest to identify explanatory and response variables when we actually set values of one variable in order to see how it affects another variable.

EXAMPLE 2.1 | **The effects of alcohol**

Alcohol has many effects on the body. One effect is a drop in body temperature. To study this effect, researchers give several different amounts of alcohol to mice, then measure the change in each mouse's body temperature in the 15 minutes after taking the alcohol. Amount of alcohol is the explanatory variable, and change in body temperature is the response variable.

When we don't set the values of either variable but just observe both variables, there may or may not be explanatory and response variables. Whether there are depends on how we plan to use the data.

EXAMPLE 2.2	SAT scores

Jim wants to know how the average SAT math and verbal scores in the 51 states (including the District of Columbia) are related to each other. He doesn't think that either score explains or causes the other. Jim has two related variables, and neither is an explanatory variable.

Julie looks at the same data. She asks, "Can I predict a state's SAT math score if I know its SAT verbal score?" Julie is treating the verbal score as the explanatory variable and the math score as the response variable.

In Example 2.1, alcohol actually *causes* a change in body temperature. There is no cause-and-effect relationship between SAT math and verbal scores in Example 2.2. Because the scores are closely related, we can nonetheless use a state's SAT verbal score to predict its math score. We will learn how to do the prediction in Section 2.3. Prediction requires that we identify an explanatory variable and a response variable. Some other statistical techniques ignore this distinction. Do remember that calling one variable explanatory and the other response doesn't necessarily mean that changes in one *cause* changes in the other.

Most statistical studies examine data on more than one variable. Fortunately, statistical analysis of several-variable data builds on the tools we used to examine individual variables. The principles that guide our work also remain the same:

- First plot the data, then add numerical summaries.
- Look for overall patterns and deviations from those patterns.
- When the overall pattern is quite regular, use a compact mathematical model to describe it.

APPLY YOUR KNOWLEDGE

2.1 In each of the following situations, is it more reasonable to simply explore the relationship between the two variables or to view one of the variables as an explanatory variable and the other as a response variable? In the latter case, which is the explanatory variable and which is the response variable?

(a) The amount of time a student spends studying for a statistics exam and the grade on the exam

(b) The weight and height of a person

(c) The amount of yearly rainfall and the yield of a crop

(d) A student's grades in statistics and in French

(e) The occupational class of a father and of a son

2.2 How well does a child's height at age 6 predict height at age 16? To find out, measure the heights of a large group of children at age 6, wait until they reach age 16, then measure their heights again. What are the explanatory and response variables here? Are these variables categorical or quantitative?

2.3 **Treating breast cancer.** The most common treatment for breast cancer was once removal of the breast. It is now usual to remove only the tumor and nearby lymph nodes, followed by radiation. The change in policy was due to a large medical experiment that compared the two treatments. Each treatment was given to a separate group of breast cancer patients, chosen at random. The patients were closely followed to see how long they lived following surgery. What are the explanatory and response variables? Are they categorical or quantitative variables?

2.1 Scatterplots

The most common way to display the relation between two quantitative variables is a *scatterplot.* Here is an example of a scatterplot.

EXAMPLE 2.3	State SAT scores

Some people use average SAT scores to rank state or local school systems. This is not proper, because the percent of high school students who take the SAT varies from place to place. Let us examine the relationship between the percent of a state's high school graduates who take the exam and the state average SAT mathematics score, using data from Table 1.6 on page 26.

We think that "percent taking" will help explain "average score." Therefore, "percent taking" is the explanatory variable and "average score" is the response variable. We want to see how average score changes when percent taking changes, so we put percent taking (the explanatory variable) on the horizontal axis. Figure 2.1 is the scatterplot. Each point represents a single state. In Alabama, for example, 8% take the SAT, and the average SAT math score is 558. Find 8 on the x (horizontal) axis and 558 on the y (vertical) axis. Alabama appears as the point (8, 558) above 8 and to the right of 558. Figure 2.1 shows how to locate Alabama's point on the plot.

> **SCATTERPLOT**
>
> A **scatterplot** shows the relationship between two quantitative variables measured on the same individuals. The values of one variable appear on the horizontal axis, and the values of the other variable appear on the vertical axis. Each individual in the data appears as the point in the plot fixed by the values of both variables for that individual.

Always plot the explanatory variable, if there is one, on the horizontal axis (the x axis) of a scatterplot. As a reminder, we usually call the explanatory variable x and the response variable y. If there is no explanatory-response distinction, either variable can go on the horizontal axis.

After you plot your data, think!

Abraham Wald (1902–1950), like John Tukey, worked on war problems during World War II. Wald invented some statistical methods that were military secrets until the war ended. Here is one of his simpler ideas. Asked where extra armor should be added to airplanes, Wald studied the location of enemy bullet holes in planes returning from combat. He plotted the locations on an outline of the plane. As data accumulated, most of the outline filled up. Put the armor in the few spots with no bullet holes, said Wald. That's where bullets hit the planes that didn't make it back.

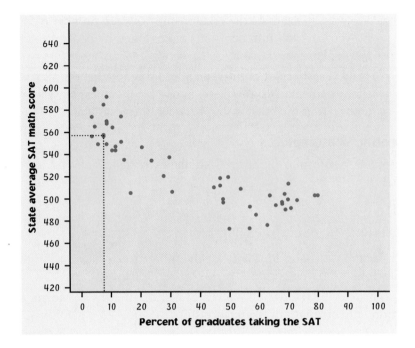

Figure 2.1 *Scatterplot of the average SAT math score in each state against the percent of that state's high school graduates who take the SAT, from Table 1.6. The dotted lines intersect at the point (8, 558), the data for Alabama.*

APPLY YOUR KNOWLEDGE

2.4 **The endangered manatee.** Manatees are large, gentle sea creatures that live along the Florida coast. Many manatees are killed or injured by powerboats. Here are data on powerboat registrations (in thousands) and the number of manatees killed by boats in Florida in the years 1977 to 1990:

(DOUGLAS FAULKNER/PHOTO RESEARCHERS)

Year	Powerboat registrations (1000)	Manatees killed	Year	Powerboat registrations (1000)	Manatees killed
1977	447	13	1984	559	34
1978	460	21	1985	585	33
1979	481	24	1986	614	33
1980	498	16	1987	645	39
1981	513	24	1988	675	43
1982	512	20	1989	711	50
1983	526	15	1990	719	47

(a) We want to examine the relationship between number of powerboats and number of manatees killed by boats. Which is the explanatory variable?

(b) Make a scatterplot of these data. (Be sure to label the axes with the variable names, not just x and y.) What does the scatterplot show about the relationship between these variables?

Interpreting scatterplots

To interpret a scatterplot, apply the strategies of data analysis learned in Chapter 1.

EXAMINING A SCATTERPLOT

In any graph of data, look for the **overall pattern** and for striking **deviations** from that pattern.

You can describe the overall pattern of a scatterplot by the **form, direction**, and **strength** of the relationship.

An important kind of deviation is an **outlier,** an individual value that falls outside the overall pattern of the relationship.

---

clusters

Figure 2.1 shows a clear *form:* there are two distinct **clusters** of states with a gap between them. In the cluster at the right of the plot, 45% or more of high school graduates take the SAT, and the average scores are low. The states in the cluster at the left have higher SAT scores and lower percents of graduates taking the test. There are no clear outliers. That is, no points fall clearly outside the clusters.

What explains the clusters? There are two widely used college entrance exams, the SAT and the American College Testing (ACT) exam. Each state favors one or the other. The left cluster in Figure 2.1 contains the ACT states, and the SAT states make up the right cluster. In ACT states, most students who take the SAT are applying to a selective college that requires SAT scores. This select group of students has a higher average score than the much larger group of students who take the SAT in SAT states.

The relationship in Figure 2.1 also has a clear *direction:* states in which a higher percent of students take the SAT tend to have lower average scores. This is a *negative association* between the two variables.

POSITIVE ASSOCIATION, NEGATIVE ASSOCIATION

Two variables are **positively associated** when above-average values of one tend to accompany above-average values of the other and below-average values also tend to occur together.

Two variables are **negatively associated** when above-average values of one tend to accompany below-average values of the other, and vice versa.

The *strength* of a relationship in a scatterplot is determined by how closely the points follow a clear form. The overall relationship in Figure 2.1 is not strong—states with similar percents taking the SAT show quite a bit of scatter in their average scores. Here is an example of a stronger relationship with a clearer form.

EXAMPLE 2.4   Heating a home

The Sanchez household is about to install solar panels to reduce the cost of heating their house. In order to know how much the solar panels help, they record their consumption of natural gas before the panels are installed. Gas consumption is higher in cold weather, so the relationship between outside temperature and gas consumption is important.

Table 2.1 gives data for 16 months.[1] The response variable y is the average amount of natural gas consumed each day during the month, in hundreds of cubic feet. The explanatory variable x is the average number of heating degree-days each day during the month. (Heating degree-days are the usual measure of demand for heating. One degree-day is accumulated for each degree a day's average temperature falls below 65°. An average temperature of 20°F, for example, corresponds to 45 degree-days.)

The scatterplot in Figure 2.2 shows a strong positive association. More degree-days means colder weather and so more gas consumed. The form of the relationship is **linear.** That is, the points lie in a straight-line pattern. It is a strong

linear relationship

Table 2.1 Average degree-days and natural gas
consumption for the Sanchez household

Month	Degree-days	Gas (100 cu. ft.)
Nov.	24	6.3
Dec.	51	10.9
Jan.	43	8.9
Feb.	33	7.5
Mar.	26	5.3
Apr.	13	4.0
May	4	1.7
June	0	1.2
July	0	1.2
Aug.	1	1.2
Sept.	6	2.1
Oct.	12	3.1
Nov.	30	6.4
Dec.	32	7.2
Jan.	52	11.0
Feb.	30	6.9

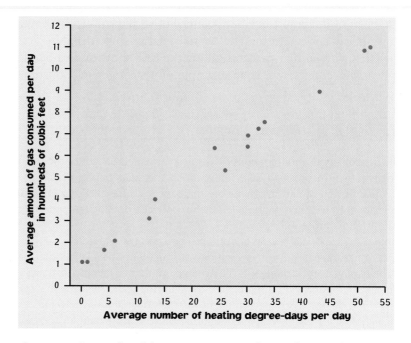

Figure 2.2 *Scatterplot of the average amount of natural gas used per day by the Sanchez household in 16 months against the average number of heating degree-days per day in those months, from Table 2.1.*

relationship because the points lie close to a line, with little scatter. If we know how cold a month is, we can predict gas consumption quite accurately from the scatterplot.

Of course, not all relationships are linear in form. What is more, not all relationships have a clear direction that we can describe as positive association or negative association. Exercise 2.6 gives an example that is not linear and has no clear direction.

APPLY YOUR KNOWLEDGE

2.5 **More on the endangered manatee.** In Exercise 2.4 you made a scatterplot of powerboats registered in Florida and manatees killed by boats.

(a) Describe the direction of the relationship. Are the variables positively or negatively associated?

(b) Describe the form of the relationship. Is it linear?

(c) Describe the strength of the relationship. Can the number of manatees killed be predicted accurately from powerboat registrations? If powerboat registrations remained constant at

716,000, about how many manatees would be killed by boats each year?

2.6 **Does fast driving waste fuel?** How does the fuel consumption of a car change as its speed increases? Here are data for a British Ford Escort. Speed is measured in kilometers per hour, and fuel consumption is measured in liters of gasoline used per 100 kilometers traveled.[2]

Speed (km/h)	Fuel used (liters/100 km)	Speed (km/h)	Fuel used (liter/100 km)
10	21.00	90	7.57
20	13.00	100	8.27
30	10.00	110	9.03
40	8.00	120	9.87
50	7.00	130	10.79
60	5.90	140	11.77
70	6.30	150	12.83
80	6.95		

(a) Make a scatterplot. (Which is the explanatory variable?)

(b) Describe the form of the relationship. Why is it not linear? Explain why the form of the relationship makes sense.

(c) It does not make sense to describe the variables as either positively associated or negatively associated. Why?

(d) Is the relationship reasonably strong or quite weak? Explain your answer.

Adding categorical variables to scatterplots

The South has long lagged behind the rest of the United States in the performance of its schools. Efforts to improve education have reduced the gap. We wonder if the South stands out in our study of state average SAT scores.

EXAMPLE 2.5 **Is the South different?**

Figure 2.3 enhances the scatterplot in Figure 2.1 by plotting the southern states in purple. (We took the South to be the states in the East South Central and South Atlantic regions.) Most of the southern states blend in with the rest of the country. Several southern states do lie at the lower edges of their clusters, along with the District of Columbia, which is a city rather than a state. Georgia, South Carolina, and West Virginia have lower SAT scores than we would expect from the percent of their high school graduates who take the examination.

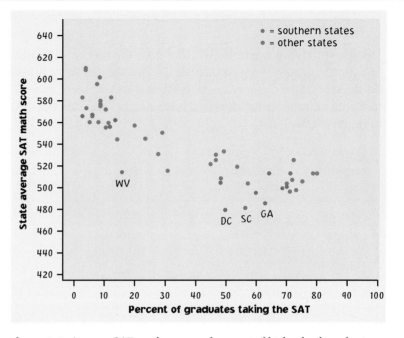

Figure 2.3 *Average SAT math score and percent of high school graduates who take the test, by state, with the southern states highlighted.*

Dividing the states into "southern" and "nonsouthern" introduces a third variable into the scatterplot. This is a categorical variable that has only two values. The two values are displayed by the two different colors. **Use different colors or symbols to plot points when you want to add a categorical variable to a scatterplot.**[3]

EXAMPLE 2.6 **Do solar panels reduce gas usage?**

After the Sanchez household gathered the information recorded in Table 2.1 and Figure 2.2, they added solar panels to their house. They then measured their natural gas consumption for 23 more months. To see how the solar panels affected gas consumption, add the degree-days and gas consumption for these months to the scatterplot. Figure 2.4 is the result. We use different colors to distinguish before from after. The "after" data form a linear pattern that is close to the "before" pattern in warm months (few degree-days). In colder months, with more degree-days, gas consumption after installing the solar panels is less than in similar months before the panels were added. The scatterplot shows the energy savings from the panels.

Our gas consumption example suffers from a common problem in drawing scatterplots that you may not notice when a computer does the work. When several individuals have exactly the same data, they occupy the same point on the scatterplot. Look at June and July in Table 2.1. Table 2.1 contains data for 16 months, but there are only 15 points in Figure 2.2. June and July both occupy the same point. You can use a different plotting symbol to call attention

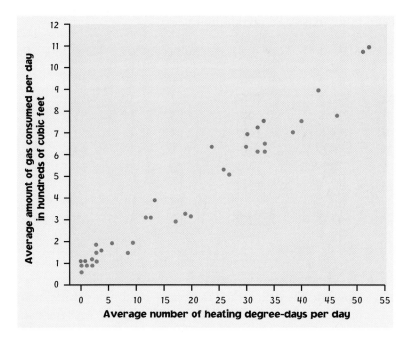

Figure 2.4 *Natural gas consumption against degree-days for the Sanchez household. The purple observations are for 16 months before installing solar panels. The blue observations are for 23 months with the panels in use.*

to points that stand for more than one individual. Some computer software does this automatically, but some does not. We recommend that you do use a different symbol for repeated observations when you plot a small number of observations by hand.

APPLY YOUR KNOWLEDGE

2.7 **Do heavier people burn more energy?** Metabolic rate, the rate at which the body consumes energy, is important in studies of weight gain, dieting, and exercise. Table 2.2 gives data on the lean body mass and resting metabolic rate for 12 women and 7 men who are subjects in a study of dieting. Lean body mass, given in kilograms, is a person's weight leaving out all fat. Metabolic rate is measured in calories burned per 24 hours, the same calories used to describe the energy content of foods. The researchers believe that lean body mass is an important influence on metabolic rate.

(a) Make a scatterplot of the data for the female subjects. Which is the explanatory variable?

(b) Is the association between these variables positive or negative? What is the form of the relationship? How strong is the relationship?

Table 2.2 Lean body mass and metabolic rate

Subject	Sex	Mass (kg)	Rate (cal)	Subject	Sex	Mass (kg)	Rate (cal)
1	M	62.0	1792	11	F	40.3	1189
2	M	62.9	1666	12	F	33.1	913
3	F	36.1	995	13	M	51.9	1460
4	F	54.6	1425	14	F	42.4	1124
5	F	48.5	1396	15	F	34.5	1052
6	F	42.0	1418	16	F	51.1	1347
7	M	47.4	1362	17	F	41.2	1204
8	F	50.6	1502	18	M	51.9	1867
9	F	42.0	1256	19	M	46.9	1439
10	M	48.7	1614				

(c) Now add the data for the male subjects to your graph, using a different color or a different plotting symbol. Does the pattern of relationship that you observed in (b) hold for men also? How do the male subjects as a group differ from the female subjects as a group?

SECTION 2.1 Summary

To study relationships between variables, we must measure the variables on the same group of individuals.

If we think that a variable x may explain or even cause changes in another variable y, we call x an **explanatory variable** and y a **response variable.**

A **scatterplot** displays the relationship between two quantitative variables measured on the same individuals. Mark values of one variable on the horizontal axis (x axis) and values of the other variable on the vertical axis (y axis). Plot each individual's data as a point on the graph.

Always plot the explanatory variable, if there is one, on the x axis of a scatterplot. Plot the response variable on the y axis.

Plot points with different colors or symbols to see the effect of a categorical variable in a scatterplot.

In examining a scatterplot, look for an overall pattern showing the **form, direction,** and **strength** of the relationship, and then for **outliers** or other deviations from this pattern.

Form: Linear relationships, where the points show a straight-line pattern, are an important form of relationship between two variables. Curved relationships and **clusters** are other forms to watch for.

Direction: If the relationship has a clear direction, we speak of either **positive association** (high values of the two variables tend to occur together) or **negative association** (high values of one variable tend to occur with low values of the other variable).

Strength: The **strength** of a relationship is determined by how close the points in the scatterplot lie to a simple form such as a line.

SECTION 2.1 Exercises

2.8 **IQ and school grades.** Do students with higher IQ test scores tend to do better in school? Figure 2.5 is a scatterplot of IQ and school grade point average (GPA) for all 78 seventh-grade students in a rural Midwest school.[4]

(a) Say in words what a positive association between IQ and GPA would mean. Does the plot show a positive association?

(b) What is the form of the relationship? Is it roughly linear? Is it very strong? Explain your answers.

(c) At the bottom of the plot are several points that we might call outliers. One student in particular has a very low GPA despite an average IQ score. What are the approximate IQ and GPA for this student?

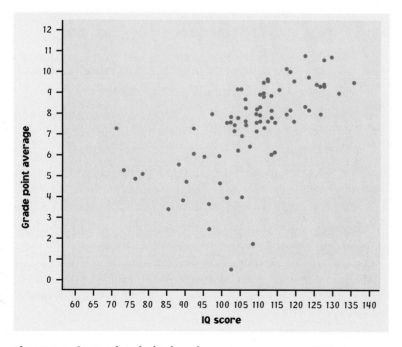

Figure 2.5 *Scatterplot of school grade point average versus IQ test score for seventh-grade students, for Exercise 2.8.*

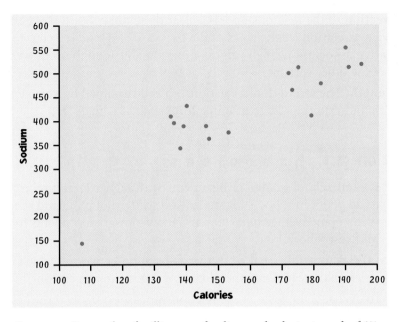

Figure 2.6 *Scatterplot of milligrams of sodium and calories in each of 17 brands of meat hot dogs, for Exercise 2.9.*

2.9   **Calories and salt in hot dogs.** Are hot dogs that are high in calories also high in salt? Figure 2.6 is a scatterplot of the calories and salt content (measured as milligrams of sodium) in 17 brands of meat hot dogs.[5]

(a) Roughly what are the lowest and highest calorie counts among these brands? Roughly what is the sodium level in the brands with the fewest and with the most calories?

(b) Does the scatterplot show a clear positive or negative association? Say in words what this association means about calories and salt in hot dogs.

(c) Are there any outliers? Is the relationship (ignoring any outliers) roughly linear in form? Still ignoring any outliers, how strong would you say the relationship between calories and sodium is?

2.10   **Rich states, poor states.** One measure of a state's prosperity is the median income of its households. Another measure is the mean personal income per person in the state. Figure 2.7 is a scatterplot of these two variables, both measured in thousands of dollars. Because both variables have the same units, the plot uses equally spaced scales on both axes.[6]

(a) We have labeled the point for New York on the scatterplot. What are the approximate values of New York's median household income and mean income per person?

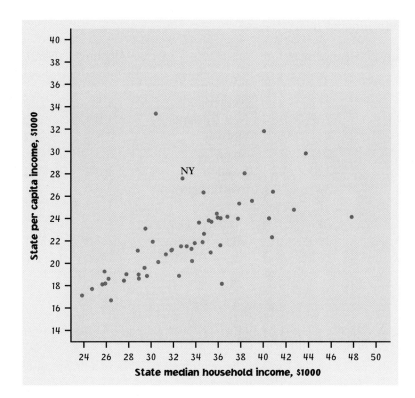

Figure 2.7 *Scatterplot of mean income per person versus median household income for the states, for Exercise 2.10.*

(b) Explain why you expect a positive association between these variables. Also explain why you expect household income to be generally higher than income per person.

(c) Nonetheless, the mean income per person in a state can be higher than the median household income. In fact, the District of Columbia has median income $30,748 per household and mean income $33,435 per person. Explain why this can happen.

(d) Alaska is the state with the highest median household income. What is the approximate median household income in Alaska? We might call Alaska and the District of Columbia outliers in the scatterplot.

(e) Describe the form, direction, and strength of the relationship, ignoring the outliers.

2.11 **Is wine good for your heart?** There is some evidence that drinking moderate amounts of wine helps prevent heart attacks. Table 2.3 gives data on yearly wine consumption (liters of alcohol from drinking wine, per person) and yearly deaths from heart disease (deaths per 100,000 people) in 19 developed nations.[7]

Table 2.3 Wine consumption and heart disease

Country	Alcohol from wine	Heart disease death rate*	Country	Alcohol from wine	Heart disease death rate*
Australia	2.5	211	Netherlands	1.8	167
Austria	3.9	167	New Zealand	1.9	266
Belgium/Lux.	2.9	131	Norway	0.8	227
Canada	2.4	191	Spain	6.5	86
Denmark	2.9	220	Sweden	1.6	207
Finland	0.8	297	Switzerland	5.8	115
France	9.1	71	United Kingdom	1.3	285
Iceland	0.8	211	United States	1.2	199
Ireland	0.7	300	West Germany	2.7	172
Italy	7.9	107			

Per 100,000 people.

(a) Make a scatterplot that shows how national wine consumption helps explain heart disease death rates.

(b) Describe the form of the relationship. Is there a linear pattern? How strong is the relationship?

(c) Is the direction of the association positive or negative? Explain in simple language what this says about wine and heart disease. Do you think these data give good evidence that drinking wine *causes* a reduction in heart disease deaths? Why?

2.12 **The professor swims.** Professor Moore swims 2000 yards regularly in a vain attempt to undo middle age. Here are his times (in minutes) and his pulse rate after swimming (in beats per minute) for 23 sessions in the pool:

Time	34.12	35.72	34.72	34.05	34.13	35.72	36.17	35.57
Pulse	152	124	140	152	146	128	136	144

Time	35.37	35.57	35.43	36.05	34.85	34.70	34.75	33.93
Pulse	148	144	136	124	148	144	140	156

Time	34.60	34.00	34.35	35.62	35.68	35.28	35.97
Pulse	136	148	148	132	124	132	139

(a) Make a scatterplot. (Which is the explanatory variable?)

(b) Is the association between these variables positive or negative? Explain why you expect the relationship to have this direction.

(c) Describe the form and strength of the relationship.

2.13 **How many corn plants are too many?** How much corn per acre should a farmer plant to obtain the highest yield? Too few plants will give a low yield. On the other hand, if there are too many plants, they will compete with each other for moisture and nutrients, and yields will fall. To find the best planting rate, plant at different rates on several plots of ground and measure the harvest. (Be sure to treat all the plots the same except for the planting rate.) Here are data from such an experiment:[8]

Plants per acre	Yield (bushels per acre)			
12,000	150.1	113.0	118.4	142.6
16,000	166.9	120.7	135.2	149.8
20,000	165.3	130.1	139.6	149.9
24,000	134.7	138.4	156.1	
28,000	119.0	150.5		

(a) Is yield or planting rate the explanatory variable?

(b) Make a scatterplot of yield and planting rate.

(c) Describe the overall pattern of the relationship. Is it linear? Is there a positive or negative association, or neither?

(d) Find the mean yield for each of the five planting rates. Plot each mean yield against its planting rate on your scatterplot and connect these five points with lines. This combination of numerical description and graphing makes the relationship clearer. What planting rate would you recommend to a farmer whose conditions were similar to those in the experiment?

2.14 **Teachers' pay.** Table 1.6 (page 26) gives data for the states. We might expect that states with less educated populations would pay their teachers less, perhaps because these states are poorer.

(a) Make a scatterplot of average teachers' pay against the percent of state residents who are not high school graduates. Take the percent with no high school degree as the explanatory variable.

(b) The plot shows a weak negative association between the two variables. Why do we say that the association is negative? Why do we say that it is weak?

(c) Circle on the plot the point for the state your school is in.

(d) There is an outlier at the upper left of the plot. Which state is this?

(e) We wonder about regional patterns. There is a relatively clear cluster of nine states at the lower right of the plot. These states have many residents who are not high school graduates and pay low salaries to teachers. Which states are these? Are they mainly from one part of the country?

transformation **2.15 Transforming data.** Data analysts often look for a **transformation** of data that simplifies the overall pattern. Here is an example of how transforming the response variable can simplify the pattern of a scatterplot. The data show the growth of Europe between 1750 and 1950.

Year	1750	1800	1850	1900	1950
Population (millions)	125	187	274	423	594

(a) Make a scatterplot of population against year. Briefly describe the pattern of Europe's growth.

(b) Now take the logarithm of the population in each year (use the `log` button on your calculator). Plot the logarithms against year. What is the overall pattern on this plot?

2.16 Categorical explanatory variable. A scatterplot shows the relationship between two quantitative variables. Here is a similar plot to study the relationship between a categorical explanatory variable and a quantitative response variable.

The presence of harmful insects in farm fields is detected by putting up boards covered with a sticky material and then examining the insects trapped on the board. Which colors attract insects best? Experimenters placed six boards of each of four colors in a field of oats and measured the number of cereal leaf beetles trapped.[9]

Board color	Insects trapped					
Lemon yellow	45	59	48	46	38	47
White	21	12	14	17	13	17
Green	37	32	15	25	39	41
Blue	16	11	20	21	14	7

(a) Make a plot of the counts of insects trapped against board color (space the four colors equally on the horizontal axis). Compute the mean count for each color, add the means to your plot, and connect the means with line segments.

(b) Based on the data, what do you conclude about the attractiveness of these colors to the beetles?

(c) Does it make sense to speak of a positive or negative association between board color and insect count?

2.2 Correlation

A scatterplot displays the form, direction, and strength of the relationship between two quantitative variables. Linear relations are particularly important
because a straight line is a simple pattern that is quite common. We say a linear relation is strong if the points lie close to a straight line, and weak if they
are widely scattered about a line. Our eyes are not good judges of how strong
a linear relationship is. The two scatterplots in Figure 2.8 depict exactly the
same data, but the lower plot is drawn smaller in a large field. The lower plot
seems to show a stronger linear relationship. Our eyes can be fooled by changing the plotting scales or the amount of white space around the cloud of points
in a scatterplot.[10] We need to follow our strategy for data analysis by using a
numerical measure to supplement the graph. *Correlation* is the measure we use.

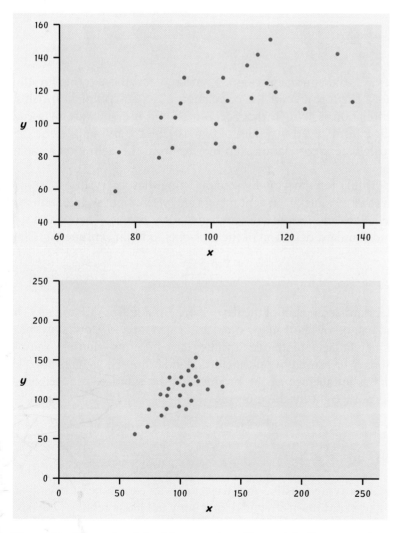

Figure 2.8 *Two scatterplots of the same data. The straight-line pattern in
the lower plot appears stronger because of the surrounding open space.*

The correlation r

CORRELATION

The **correlation** measures the direction and strength of the linear relationship between two quantitative variables. Correlation is usually written as r.

Suppose that we have data on variables x and y for n individuals. The values for the first individual are x_1 and y_1, the values for the second individual are x_2 and y_2, and so on. The means and standard deviations of the two variables are $\bar{x}$ and s_x for the x-values, and $\bar{y}$ and s_y for the y-values. The correlation r between x and y is

$$r = \frac{1}{n-1}\sum\left(\frac{x_i - \bar{x}}{s_x}\right)\left(\frac{y_i - \bar{y}}{s_y}\right)$$

As always, the summation sign $\sum$ means "add these terms for all the individuals." The formula for the correlation r is a bit complex. It helps us see what correlation is, but in practice you should use software or a calculator that finds r from keyed-in values of two variables x and y. Exercise 2.17 asks you to calculate a correlation step-by-step from the definition to solidify its meaning.

The formula for r begins by standardizing the observations. Suppose, for example, that x is height in centimeters and y is weight in kilograms and that we have height and weight measurements for n people. Then $\bar{x}$ and s_x are the mean and standard deviation of the n heights, both in centimeters. The value

$$\frac{x_i - \bar{x}}{s_x}$$

is the standardized height of the ith person, familiar from Chapter 1. The standardized height says how many standard deviations above or below the mean a person's height lies. Standardized values have no units—in this example, they are no longer measured in centimeters. Standardize the weights also. The correlation r is an average of the products of the standardized height and the standardized weight for the n people.

(JOHN D. CUNNINGHAM/VISUALS UNLIMITED.)

APPLY YOUR KNOWLEDGE

2.17 Classifying fossils. *Archaeopteryx* is an extinct beast having feathers like a bird but teeth and a long bony tail like a reptile. Only six fossil specimens are known. Because these specimens differ greatly in size, some scientists think they belong to different species. We will examine some data. If the specimens belong to the same species and differ in size because some are younger than others, there should be a positive linear relationship between the lengths of a pair of bones from all individuals. An outlier from this relationship would suggest a different species. Here are data on the lengths in centimeters of the femur (a leg bone) and the humerus (a bone in the upper arm) for the five specimens that preserve both bones:[11]

Femur	38	56	59	64	74
Humerus	41	63	70	72	84

(a) Make a scatterplot. Do you think that all five specimens come from the same species?

(b) Find the correlation r step-by-step. That is, find the mean and standard deviation of the femur lengths and of the humerus lengths. (Use your calculator for means and standard deviations.) Then find the five standardized values for each variable and use the formula for r.

(c) Now enter these data into your calculator and use the calculator's correlation function to find r. Check that you get the same result as in (b).

Facts about correlation

The formula for correlation helps us see that r is positive when there is a positive association between the variables. Height and weight, for example, have a positive association. People who are above average in height tend to also be above average in weight. Both the standardized height and the standardized weight are positive. People who are below average in height tend to also have below-average weight. Then both standardized height and standardized weight are negative. In both cases, the products in the formula for r are mostly positive and so r is positive. In the same way, we can see that r is negative when the association between x and y is negative. More detailed study of the formula gives more detailed properties of r. Here is what you need to know in order to interpret correlation.

1. Correlation makes no distinction between explanatory and response variables. It makes no difference which variable you call x and which you call y in calculating the correlation.

2. Correlation requires that both variables be quantitative, so that it makes sense to do the arithmetic indicated by the formula for r. We cannot calculate a correlation between the incomes of a group of people and what city they live in, because city is a categorical variable.

Do big skulls house smart brains?

Nineteenth-century scientists thought that the volume of a human skull might be related to the intelligence of the skull's owner. In the days before MRI (see Exercise 2.24), it was difficult to measure a skull's volume accurately, even after it was no longer attached to its owner. Paul Broca, a professor of surgery, showed that filling a skull with small lead shot, then pouring out the shot and weighing it, gave quite accurate measurements of the skull's volume.

3. Because r uses the standardized values of the observations, r does not change when we change the units of measurement of x, y, or both. Measuring height in inches rather than centimeters and weight in pounds rather than kilograms does not change the correlation between height and weight. The correlation r itself has no unit of measurement; it is just a number.

4. Positive r indicates positive association between the variables, and negative r indicates negative association.

5. The correlation r is always a number between -1 and 1. Values of r near 0 indicate a very weak linear relationship. The strength of the linear relationship increases as r moves away from 0 toward either -1 or 1. Values of r close to -1 or 1 indicate that the points in a scatterplot lie close to a straight line. The extreme values $r = -1$ and $r = 1$ occur only in the case of a perfect linear relationship, when the points lie exactly along a straight line.

6. Correlation measures the strength of only a linear relationship between two variables. Correlation does not describe curved relationships between variables, no matter how strong they are.

7. Like the mean and standard deviation, the correlation is not resistant: r is strongly affected by a few outlying observations. The correlation for Figure 2.7 (page 93) is $r = 0.634$ when all 51 observations are included, but rises to $r = 0.783$ when we omit Alaska and the District of Columbia. Use r with caution when outliers appear in the scatterplot.

The scatterplots in Figure 2.9 illustrate how values of r closer to 1 or -1 correspond to stronger linear relationships. To make the meaning of r clearer, the standard deviations of both variables in these plots are equal and the horizontal and vertical scales are the same. In general, it is not so easy to guess the value of r from the appearance of a scatterplot. Remember that changing the plotting scales in a scatterplot may mislead our eyes, but it does not change the correlation.

The real data we have examined also illustrate how correlation measures the strength and direction of linear relationships. Figure 2.2 (page 86) shows a very strong positive linear relationship between degree-days and natural gas consumption. The correlation is $r = 0.9953$. Check this on your calculator using the data in Table 2.1. Figure 2.1 (page 83) shows a clear but weaker negative association between percent of students taking the SAT and the median SAT math score in a state. The correlation is $r = -0.8581$.

Do remember that **correlation is not a complete description of two-variable data**, even when the relationship between the variables is linear. You should give the means and standard deviations of both x and y along with the correlation. (Because the formula for correlation uses the means and standard deviations, these measures are the proper choice to accompany a correlation.) Conclusions based on correlations alone may require rethinking in the light of a more complete description of the data.

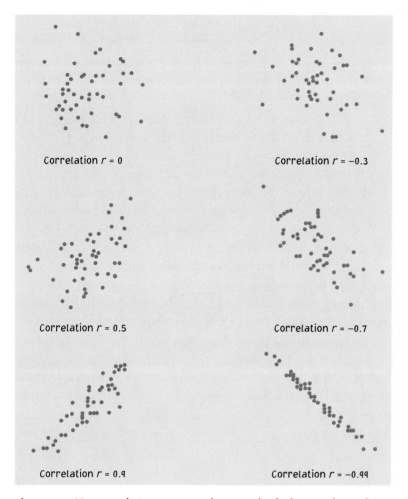

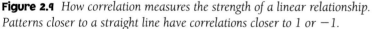

Figure 2.4 *How correlation measures the strength of a linear relationship. Patterns closer to a straight line have correlations closer to 1 or −1.*

EXAMPLE 2.7 Scoring divers

Competitive divers are scored on their form by a panel of judges who use a scale from 1 to 10. The subjective nature of the scoring often results in controversy. We have the scores awarded by two judges, Ivan and George, on a large number of dives. How well do they agree? Some calculations show that the correlation between their scores is $r = 0.9$, but the mean of Ivan's scores is 3 points lower than George's mean.

These facts do not contradict each other. They are simply different kinds of information. The mean scores show that Ivan awards much lower scores than George. But because Ivan gives *every* dive a score about 3 points lower than George, the correlation remains high. Adding or subtracting the same number to all values of either x or y does not change the correlation. If Ivan and George both rate several divers, the contest is fairly scored because Ivan and George agree on which dives are better than others. The high r shows their agreement. But if Ivan scores one diver and George another, we must add 3 points to Ivan's scores to arrive at a fair comparison.

2.18 **Thinking about correlation.** Figure 2.5 (page 91) is a scatterplot of school grade point average versus IQ score for 78 seventh-grade students.

(a) Is the correlation r for these data near -1, clearly negative but not near -1, near 0, clearly positive but not near 1, or near 1? Explain your answer.

(b) Figure 2.6 (page 92) shows the calories and sodium content in 17 brands of meat hot dogs. Is the correlation here closer to 1 than that for Figure 2.5, or closer to zero? Explain your answer.

(c) Both Figures 2.5 and 2.6 contain outliers. Removing the outliers will *increase* the correlation r in one figure and *decrease* r in the other figure. What happens in each figure, and why?

2.19 If women always married men who were 2 years older than themselves, what would be the correlation between the ages of husband and wife? (Hint: Draw a scatterplot for several ages.)

2.20 **Strong association but no correlation.** The gas mileage of an automobile first increases and then decreases as the speed increases. Suppose that this relationship is very regular, as shown by the following data on speed (miles per hour) and mileage (miles per gallon):

Speed	20	30	40	50	60
MPG	24	28	30	28	24

Make a scatterplot of mileage versus speed. Show that the correlation between speed and mileage is $r = 0$. Explain why the correlation is 0 even though there is a strong relationship between speed and mileage.

SECTION 2.2 Summary

The **correlation** r measures the strength and direction of the linear association between two quantitative variables x and y. Although you can calculate a correlation for any scatterplot, r measures only straight-line relationships.

Correlation indicates the direction of a linear relationship by its sign: $r > 0$ for a positive association and $r < 0$ for a negative association.

Correlation always satisfies $-1 \leq r \leq 1$ and indicates the strength of a relationship by how close it is to -1 or 1. Perfect correlation, $r = \pm 1$, occurs only when the points on a scatterplot lie exactly on a straight line.

Correlation ignores the distinction between explanatory and response variables. The value of r is not affected by changes in the unit of measurement of either variable. Correlation is not resistant, so outliers can greatly change the value of r.

SECTION 2.2 Exercises

2.21 **The professor swims.** Exercise 2.12 (page 94) gives data on the time to swim 2000 yards and the pulse rate after swimming for a middle-aged professor.

(a) If you did not do Exercise 2.12, do it now. Find the correlation r. Explain from looking at the scatterplot why this value of r is reasonable.

(b) Suppose that the times had been recorded in seconds. For example, the time 34.12 minutes would be 2047 seconds. How would the value of r change?

2.22 **Body mass and metabolic rate.** Table 2.2 (page 90) gives data on the lean body mass and metabolic rate for 12 women and 7 men.

(a) Make a scatterplot if you did not do so in Exercise 2.7. Use different symbols or colors for women and men. Do you think the correlation will be about the same for men and women or quite different for the two groups? Why?

(b) Calculate r for women alone and also for men alone. (Use your calculator.)

(c) Calculate the mean body mass for the women and for the men. Does the fact that the men are heavier than the women on the average influence the correlations? If so, in what way?

(d) Lean body mass was measured in kilograms. How would the correlations change if we measured body mass in pounds? (There are about 2.2 pounds in a kilogram.)

2.23 **How many calories?** A food industry group asked 3368 people to guess the number of calories in each of several common foods. Table 2.4 displays the averages of their guesses and the correct number of calories.[12]

(a) We think that how many calories a food actually has helps explain people's guesses of how many calories it has. With this in mind, make a scatterplot of these data.

(b) Find the correlation r (use your calculator). Explain why your r is reasonable based on the scatterplot.

(c) The guesses are all higher than the true calorie counts. Does this fact influence the correlation in any way? How would r change if every guess were 100 calories higher?

Table 2.4 Guessed and true calories in 10 foods

Food	Guessed calories	Correct calories
8 oz. whole milk	196	159
5 oz. spaghetti with tomato sauce	394	163
5 oz. macaroni with cheese	350	269
One slice wheat bread	117	61
One slice white bread	136	76
2-oz. candy bar	364	260
Saltine cracker	74	12
Medium-size apple	107	80
Medium-size potato	160	88
Cream-filled snack cake	419	160

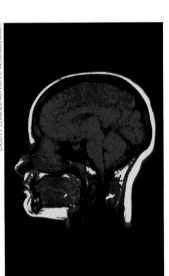

(SCOTT CAMAZINE/PHOTO RESEARCHERS)

(d) The guesses are much too high for spaghetti and snack cake. Circle these points on your scatterplot. Calculate r for the other eight foods, leaving out these two points. Explain why r changed in the direction that it did.

2.24 **Brain size and IQ score (EESEE).** Do people with larger brains have higher IQ scores? A study looked at 40 volunteer subjects, 20 men and 20 women. Brain size was measured by magnetic resonance imaging. Table 2.5 gives the data. The MRI count is the number of

Table 2.5 Brain size (MRI count) and IQ score

Men				Women			
MRI	IQ	MRI	IQ	MRI	IQ	MRI	IQ
1,001,121	140	1,038,437	139	816,932	133	951,545	137
965,353	133	904,858	89	928,799	99	991,305	138
955,466	133	1,079,549	141	854,258	92	833,868	132
924,059	135	945,088	100	856,472	140	878,897	96
889,083	80	892,420	83	865,363	83	852,244	132
905,940	97	955,003	139	808,020	101	790,619	135
935,494	141	1,062,462	103	831,772	91	798,612	85
949,589	144	997,925	103	793,549	77	866,662	130
879,987	90	949,395	140	857,782	133	834,344	83
930,016	81	935,863	89	948,066	133	893,983	88

"pixels" the brain covered in the image. IQ was measured by the Wechsler test.[13]

(a) Make a scatterplot of IQ score versus MRI count, using distinct symbols for men and women. In addition, find the correlation between IQ and MRI for all 40 subjects, for the men alone, and for the women alone.

(b) Men are larger than women on the average, so they have larger brains. How is this size effect visible in your plot? Find the mean MRI count for men and women to verify the difference.

(c) Your result in (b) suggests separating men and women in looking at the relationship between brain size and IQ. Use your work in (a) to comment on the nature and strength of this relationship for women and for men.

2.25 Changing the units of measurement can greatly alter the appearance of a scatterplot. Consider the following data:

x	-4	-4	-3	3	4	4
y	0.5	-0.6	-0.5	0.5	0.5	-0.6

(a) Draw x and y axes each extending from -6 to 6. Plot the data on these axes.

(b) Calculate the values of new variables $x^* = x/10$ and $y^* = 10y$, starting from the values of x and y. Plot y^* against x^* on the same axes using a different plotting symbol. The two plots are very different in appearance.

(c) Use your calculator to find the correlation between x and y. Then find the correlation between x^* and y^*. How are the two correlations related? Explain why this isn't surprising.

2.26 **Teaching and research.** A college newspaper interviews a psychologist about student ratings of the teaching of faculty members. The psychologist says, "The evidence indicates that the correlation between the research productivity and teaching rating of faculty members is close to zero." The paper reports this as "Professor McDaniel said that good researchers tend to be poor teachers, and vice versa." Explain why the paper's report is wrong. Write a statement in plain language (don't use the word "correlation") to explain the psychologist's meaning.

2.27 **Investment diversification.** A mutual fund company's newsletter says, "A well-diversified portfolio includes assets with low correlations." The newsletter includes a table of correlations between the returns on various classes of investments. For example, the correlation between municipal bonds and large-cap stocks is 0.50 and the correlation between municipal bonds and small-cap stocks is 0.21.[14]

(a) Rachel invests heavily in municipal bonds. She wants to diversify by adding an investment whose returns do not closely follow the returns on her bonds. Should she choose large-cap stocks or small-cap stocks for this purpose? Explain your answer.

(b) If Rachel wants an investment that tends to increase when the return on her bonds drops, what kind of correlation should she look for?

2.28 Driving speed and fuel consumption. The data in Exercise 2.20 were made up to create an example of a strong curved relationship for which, nonetheless, $r = 0$. Exercise 2.6 (page 87) gives actual data on gas used versus speed for a small car. Make a scatterplot if you did not do so in Exercise 2.6. Calculate the correlation, and explain why r is close to 0 despite a strong relationship between speed and gas used.

2.29 Sloppy writing about correlation. Each of the following statements contains a blunder. Explain in each case what is wrong.

(a) "There is a high correlation between the gender of American workers and their income."

(b) "We found a high correlation ($r = 1.09$) between students' ratings of faculty teaching and ratings made by other faculty members."

(c) "The correlation between planting rate and yield of corn was found to be $r = 0.23$ bushel."

2.3 Least-Squares Regression

Correlation measures the direction and strength of the straight-line (linear) relationship between two quantitative variables. If a scatterplot shows a linear relationship, we would like to summarize this overall pattern by drawing a line on the scatterplot. A *regression line* summarizes the relationship between two variables, but only in a specific setting: one of the variables helps explain or predict the other. That is, regression describes a relationship between an explanatory variable and a response variable.

> **REGRESSION LINE**
>
> A **regression line** is a straight line that describes how a response variable y changes as an explanatory variable x changes. We often use a regression line to predict the value of y for a given value of x.

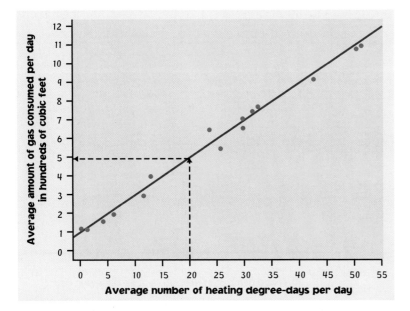

Figure 2.10 *The Sanchez household gas consumption data, with a regression line for predicting gas consumption from degree-days. The dashed lines illustrate how to use the regression line to predict gas consumption for a month averaging 20 degree-days per day.*

EXAMPLE 2.8 Predicting natural gas consumption

The scatterplot in Figure 2.10 shows that there is a strong linear relationship between the average outside temperature (measured by heating degree-days) in a month and the average amount of natural gas that the Sanchez household uses per day during the month. The correlation is $r = 0.9953$, close to the $r = 1$ of points that lie exactly on a line. The regression line drawn through the points in Figure 2.10 describes these data very well.

The Sanchez household wants to use this line to predict their natural gas consumption. "If a month averages 20 degree-days per day (that's 45° F), how much gas will we use?" To **predict** gas consumption at 20 degree-days, first locate 20 on the x axis. Then go "up and over" as in the figure to find the gas consumption y that corresponds to $x = 20$. We predict that the Sanchez household will use about 4.9 hundreds of cubic feet of gas each day in such a month.

prediction

The least-squares regression line

Different people might draw different lines by eye on a scatterplot. This is especially true when the points are more widely scattered than those in Figure 2.10. We need a way to draw a regression line that doesn't depend on our guess as to where the line should go. We will use the line to predict y from x, so the prediction errors we make are errors in y, the vertical direction in the scatterplot. If we predict 4.9 hundreds of cubic feet for a month with 20 degree-days

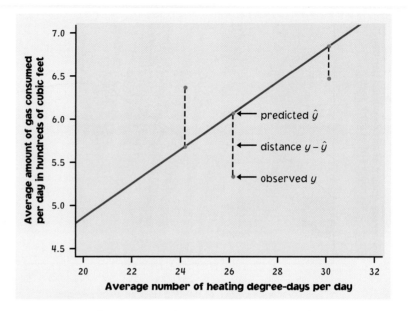

Figure 2.11 *The least-squares idea. For each observation, find the vertical distance of each point on the scatterplot from a regression line. The least-squares regression line makes the sum of the squares of these distances as small as possible.*

and the actual usage turns out to be 5.1 hundreds of cubic feet, our prediction error is

$$\text{error} = \text{observed } y - \text{predicted } y$$
$$= 5.1 - 4.9 = 0.2$$

No line will pass exactly through all the points in the scatterplot. We want the *vertical* distances of the points from the line to be as small as possible. Figure 2.11 illustrates the idea. This plot shows three of the points from Figure 2.10, along with the line, on an expanded scale. The line passes above two of the points and below one of them. The vertical distances of the data points from the line appear as vertical line segments. There are many ways to make the collection of vertical distances "as small as possible." The most common is the *least-squares* method.

> ## LEAST-SQUARES REGRESSION LINE
>
> The **least-squares regression line** of y on x is the line that makes the sum of the squares of the vertical distances of the data points from the line as small as possible.

One reason for the popularity of the least-squares regression line is that the problem of finding the line has a simple answer. We can give the recipe for the least-squares line in terms of the means and standard deviations of the two variables and their correlation.

EQUATION OF THE LEAST-SQUARES REGRESSION LINE

We have data on an explanatory variable x and a response variable y for n individuals. From the data, calculate the means $\bar{x}$ and $\bar{y}$ and the standard deviations s_x and s_y of the two variables, and their correlation r. The least-squares regression line is the line

$$\hat{y} = a + bx$$

with **slope**

$$b = r\frac{s_y}{s_x}$$

and **intercept**

$$a = \bar{y} - b\bar{x}$$

We write $\hat{y}$ (read "y hat") in the equation of the regression line to emphasize that the line gives a *predicted* response $\hat{y}$ for any x. Because of the scatter of points about the line, the predicted response will usually not be exactly the same as the actually *observed* response y. In practice, you don't need to calculate the means, standard deviations, and correlation first. Statistical software or your calculator will give the slope b and intercept a of the least-squares line from keyed-in values of the variables x and y. You can then concentrate on understanding and using the regression line.

EXAMPLE 2.9   **Using a regression line**

The line in Figure 2.10 is in fact the least-squares regression line of gas consumption on degree-days. Enter the data from Table 2.1 into your calculator and check that the equation of this line is

$$\hat{y} = 1.0892 + 0.1890x$$

The **slope** of a regression line is usually important for the interpretation of the data. The slope is the rate of change, the amount of change in $\hat{y}$ when x increases by 1. The slope $b = 0.1890$ in this example says that each additional degree-day predicts consumption of about 0.19 more hundreds of cubic feet of natural gas per day.   *slope*

The **intercept** of the regression line is the value of $\hat{y}$ when $x = 0$. Although we need the value of the intercept to draw the line, it is statistically meaningful only when x can actually take values close to zero. In our example, $x = 0$ occurs when the average outdoor temperature is at least 65° F. We predict that the Sanchez household will use an average of $a = 1.0892$ hundreds of cubic feet of gas per day when there are no degree-days. They use this gas for cooking and heating water, which continue in warm weather.   *intercept*

The equation of the regression line makes **prediction** easy. Just substitute an x-value into the equation. To predict gas consumption at 20 degree-days, substitute $x = 20$:   *prediction*

$$\hat{y} = 1.0892 + (0.1890)(20)$$
$$= 1.0892 + 3.78 = 4.869$$

plotting a line

To **plot the line** on the scatterplot, use the equation to find $\hat{y}$ for two values of x, one near each end of the range of x in the data. Plot each $\hat{y}$ above its x and draw the line through the two points.

Figure 2.12 displays the regression output for the gas consumption data from a graphing calculator and two statistical software packages. Each output

```
LinReg
  y=ax+b
  a=.1889989538
  b=1.089210843
  r²=.9905504416
  r =.995264006
```

(a)

The regression equation is
Gas Used = 1.09 + 0.189 D-days

Predictor	Coef	Stdev	t-ratio	p
Constant	1.0892	0.1389	7.84	0.000
D-days	0.188999	0.004934	38.31	0.000

s = 0.3389   R-sq = 99.1%   R-sq(adj) = 99.0%

Analysis of Variance

SOURCE	DF	SS	MS	F	p
Regression	1	168.58	168.58	1467.55	0.000
Error	14	1.61	0.11		
Total	15	170.19			

(b)

Dependent variable is: **Gas used**
No Selector
R squared = 99.1% R squared (adjusted) = 99.0%
s = 0.3389   with 16 − 2 = 14 degrees of freedom

Source	Sum of Squares	df	Mean Square	F-ratio
Regression	168.581	1	168.581	1468
Residual	1.60821	14	0.114872	

Variable	Coefficient	s.e. of Coeff	t-ratio	prob
Constant	1.08921	0.1389	7.84	≤0.0001
Degree-days	0.188999	0.0049	38.3	≤0.0001

(c)

Figure 2.12 *Least-squares regression output for the gas consumption data from a graphing calculator and two statistical software packages.* **(a)** *The TI-83 calculator.* **(b)** *Minitab.* **(c)** *Data Desk.*

records the slope and intercept of the least-squares line, calculated to more decimal places than we need. The software also provides information that we do not yet need—part of the art of using software is to ignore the extra information that is almost always present. We will make use of other parts of the output in Chapter 10.

APPLY YOUR KNOWLEDGE

2.30 Example 2.9 gives the equation of the regression line of gas consumption y on degree-days x for the data in Table 2.1 as

$$\hat{y} = 1.0892 + 0.1890x$$

Enter the data from Table 2.1 into your calculator.

(a) Use your calculator's regression function to find the equation of the least-squares regression line.

(b) Use your calculator to find the mean and standard deviation of both x and y and their correlation r. Find the slope b and intercept a of the regression line from these, using the facts in the box Equation of the Least-Squares Regression Line. Verify that in both part (a) and part (b) you get the equation in Example 2.9. (Results may differ slightly because of rounding off.)

2.31 **Acid rain.** Researchers studying acid rain measured the acidity of precipitation in a Colorado wilderness area for 150 consecutive weeks. Acidity is measured by pH. Lower pH values show higher acidity. The acid rain researchers observed a linear pattern over time. They reported that the least-squares regression line

$$pH = 5.43 - (0.0053 \times weeks)$$

fit the data well.[15]

(a) Draw a graph of this line. Is the association positive or negative? Explain in plain language what this association means.

(b) According to the regression line, what was the pH at the beginning of the study (weeks = 1)? At the end (weeks = 150)?

(c) What is the slope of the regression line? Explain clearly what this slope says about the change in the pH of the precipitation in this wilderness area.

2.32 **The endangered manatee.** Exercise 2.4 (page 83) gives data on the number of powerboats registered in Florida and the number of manatees killed by boats in the years from 1977 to 1990. The regression line for predicting manatees killed from powerboats registered is

$$killed = -41.4 + (0.125 \times boats)$$

Regression toward the mean

To "regress" means to go backward. Why are statistical methods for predicting a response from an explanatory variable called "regression"? Sir Francis Galton (1822–1911), who was the first to apply regression to biological and psychological data, looked at examples such as the heights of children versus the heights of their parents. He found that the taller-than-average parents tended to have children who were also taller than average, but not as tall as their parents. Galton called this fact "regression toward the mean" and the name came to be applied to the statistical method.

(a) Make a scatterplot and draw this regression line on the plot. Predict the number of manatees that will be killed by boats in a year when 716,000 powerboats are registered.

(b) Here are four more years of manatee data, in the same form as in Exercise 2.4:

| 1991 | 716 | 53 | 1993 | 716 | 35 |
| 1992 | 716 | 38 | 1994 | 735 | 49 |

Add these points to your scatterplot. Florida took stronger measures to protect manatees during these years. Do you see any evidence that these measures succeeded? (c) In part (a) you predicted manatee deaths in a year with 716,000 powerboat registrations. In fact, powerboat registrations remained at 716,000 for the next three years. Compare the mean manatee deaths in these years with your prediction from part (a). How accurate was your prediction?

Facts about least-squares regression

Regression is one of the most common statistical settings, and least-squares is the most common method for fitting a regression line to data. Here are some facts about least-squares regression lines.

Fact 1. The distinction between explanatory and response variables is essential in regression. Least-squares regression looks at the distances of the data points from the line only in the y direction. If we reverse the roles of the two variables, we get a different least-squares regression line.

| **EXAMPLE 2.10** | **The expanding universe** |

Figure 2.13 is a scatterplot of data that played a central role in the discovery that the universe is expanding. They are the distances from earth of 24 spiral galaxies and the speed at which these galaxies are moving away from us, reported by the astronomer Edwin Hubble in 1929.[16] There is a positive linear relationship, $r = 0.7842$, so that more distant galaxies are moving away more rapidly. Astronomers believe that there is in fact a perfect linear relationship, and that the scatter is caused by imperfect measurements.

The two lines on the plot are the two least-squares regression lines. The regression line of velocity on distance is solid. The regression line of distance on velocity is dashed. *Regression of velocity on distance and regression of distance on velocity give different lines.* In the regression setting you must know clearly which variable is explanatory.

Fact 2. There is a close connection between correlation and the slope of the least-squares line. The slope is

$$b = r\frac{s_y}{s_x}$$

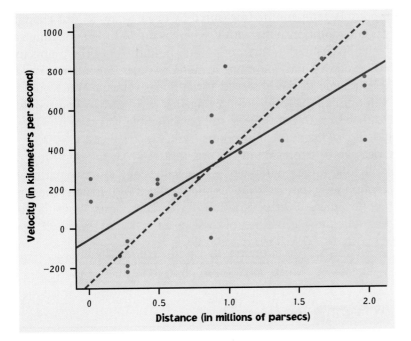

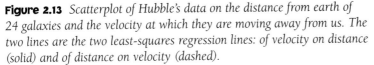

Figure 2.13 *Scatterplot of Hubble's data on the distance from earth of 24 galaxies and the velocity at which they are moving away from us. The two lines are the two least-squares regression lines: of velocity on distance (solid) and of distance on velocity (dashed).*

This equation says that along the regression line, **a change of one standard deviation in x corresponds to a change of r standard deviations in y.** When the variables are perfectly correlated ($r = 1$ or $r = -1$), the change in the predicted response $\hat{y}$ is the same (in standard deviation units) as the change in x. Otherwise, because $-1 \le r \le 1$, the change in $\hat{y}$ is less than the change in x. As the correlation grows less strong, the prediction $\hat{y}$ moves less in response to changes in x.

Fact 3. The least-squares regression line always passes through the point $(\overline{x}, \overline{y})$ on the graph of y against x. So the least-squares regression line of y on x is the line with slope rs_y/s_x that passes through the point $(\overline{x}, \overline{y})$. We can describe regression entirely in terms of the basic descriptive measures $\overline{x}$, s_x, $\overline{y}$, s_y, and r.

Fact 4. The correlation r describes the strength of a straight-line relationship. In the regression setting, this description takes a specific form: **the square of the correlation, r^2, is the fraction of the variation in the values of y that is explained by the least-squares regression of y on x.**

The idea is that when there is a linear relationship, some of the variation in y is accounted for by the fact that as x changes it pulls y along with it. Look again at Figure 2.10 on page 107. There is a lot of variation in the observed y's, the gas consumption data. They range from a low of about 1 to a high of 11. The scatterplot shows that most of this variation in y is accounted for by the fact that outdoor temperature (measured by degree-days x) was changing and pulled gas consumption along with it. There is only a little remaining variation in y, which appears in the scatter of points about the line. The points in Figure 2.13, on the other hand, are more scattered. Linear dependence on distance does explain some of the observed variation in velocity. You would guess a higher value for the velocity y knowing that $x = 2$ than you would if you were told that $x = 0$. But there is still considerable variation in y even when x is held fixed—look at the four points in Figure 2.13 with $x = 2$.

This idea can be expressed in algebra, though we won't do it. It is possible to break the total variation in the observed values of y into two parts. One part is the variation we expect as x moves and $\hat{y}$ moves with it along the regression line. The other measures the variation of the data points about the line. The squared correlation r^2 is the first of these as a fraction of the whole:

$$r^2 = \frac{\text{variation in } \hat{y} \text{ as } x \text{ pulls it along the line}}{\text{total variation in observed values of } y}$$

EXAMPLE 2.11   **Using r^2**

In Figure 2.10, $r = 0.9953$ and $r^2 = 0.9906$. Over 99% of the variation in gas consumption is accounted for by the linear relationship with degree-days. In Figure 2.13, $r = 0.7842$ and $r^2 = 0.6150$. The linear relationship between distance and velocity explains 61.5% of the variation in either variable. There are two regression lines, but just one correlation, and r^2 helps interpret both regressions.

When you report a regression, give r^2 as a measure of how successful the regression was in explaining the response. All the software outputs in Figure 2.12 include r^2, either in decimal form or as a percent. When you see a correlation, square it to get a better feel for the strength of the association. Perfect correlation ($r = -1$ or $r = 1$) means the points lie exactly on a line. Then $r^2 = 1$ and all of the variation in one variable is accounted for by the linear relationship with the other variable. If $r = -0.7$ or $r = 0.7$, $r^2 = 0.49$ and about half the variation is accounted for by the linear relationship. In the r^2 scale, correlation ± 0.7 is about halfway between 0 and ± 1.

These facts are special properties of least-squares regression. They are not true for other methods of fitting a line to data. Another reason that least squares is the most common method for fitting a regression line to data is that it has many convenient special properties.

APPLY YOUR KNOWLEDGE

2.33   **The professor swims.** Here are Professor Moore's times (in minutes) to swim 2000 yards and his pulse rate after swimming (in beats per minute) for 23 sessions in the pool:

Time	34.12	35.72	34.72	34.05	34.13	35.72	36.17	35.57
Pulse	152	124	140	152	146	128	136	144

Time	35.37	35.57	35.43	36.05	34.85	34.70	34.75	33.93
Pulse	148	144	136	124	148	144	140	156

Time	34.60	34.00	34.35	35.62	35.68	35.28	35.97
Pulse	136	148	148	132	124	132	139

(a) A scatterplot shows a moderately strong negative linear relationship. Use your calculator or software to verify that the least-squares regression line is

$$\text{pulse} = 479.9 - (9.695 \times \text{time})$$

(b) The next day's time is 34.30 minutes. Predict the professor's pulse rate. In fact, his pulse rate was 152. How accurate is your prediction?

(c) Suppose you were told only that the pulse rate was 152. You now want to predict swimming time. Find the equation of the least-squares regression line that is appropriate for this purpose. What is your prediction, and how accurate is it?

(d) Explain clearly, to someone who knows no statistics, why there are two different regression lines.

2.34 **Predicting the stock market.** Some people think that the behavior of the stock market in January predicts its behavior for the rest of the year. Take the explanatory variable x to be the percent change in a stock market index in January and the response variable y to be the change in the index for the entire year. We expect a positive correlation between x and y because the change during January contributes to the full year's change. Calculation from data for the years 1960 to 1997 gives

$$\bar{x} = 1.75\% \qquad s_x = 5.36\% \qquad r = 0.596$$
$$\bar{y} = 9.07\% \qquad s_y = 15.35\%$$

(a) What percent of the observed variation in yearly changes in the index is explained by a straight-line relationship with the change during January?

(b) What is the equation of the least-squares line for predicting full-year change from January change?

(c) The mean change in January is $\bar{x} = 1.75\%$. Use your regression line to predict the change in the index in a year in which the index rises 1.75% in January. Why could you have given this result (up to roundoff error) without doing the calculation?

(CARLYN GALATI/VISUALS UNLIMITED.)

2.35 **Beavers and beetles.** Ecologists sometimes find rather strange relationships in our environment. One study seems to show that beavers benefit beetles. The researchers laid out 23 circular plots, each four meters in diameter, in an area where beavers were cutting down cottonwood trees. In each plot, they counted the number of stumps from trees cut by beavers and the number of clusters of beetle larvae. Here are the data:[17]

Stumps	2	2	1	3	3	4	3	1	2	5	1	3
Beetle larvae	10	30	12	24	36	40	43	11	27	56	18	40

Stumps	2	1	2	2	1	1	4	1	2	1	4
Beetle larvae	25	8	21	14	16	6	54	9	13	14	50

(a) Make a scatterplot that shows how the number of beaver-caused stumps influences the number of beetle larvae clusters. What does your plot show? (Ecologists think that the new sprouts from stumps are more tender than other cottonwood growth, so that beetles prefer them.)

(b) Find the least-squares regression line and draw it on your plot.

(c) What percent of the observed variation in beetle larvae counts can be explained by straight-line dependence on stump counts?

Residuals

A regression line is a mathematical model for the overall pattern of a linear relationship between an explanatory variable and a response variable. Deviations from the overall pattern are also important. In the regression setting, we see deviations by looking at the scatter of the data points about the regression line. The vertical distances from the points to the least-squares regression line are as small as possible, in the sense that they have the smallest possible sum of squares. Because they represent "left-over" variation in the response after fitting the regression line, these distances are called *residuals*.

> **RESIDUALS**
>
> A **residual** is the difference between an observed value of the response variable and the value predicted by the regression line. That is,
>
> $$\text{residual} = \text{observed } y - \text{predicted } y$$
> $$= y - \hat{y}$$

EXAMPLE 2.12 Predicting mental ability

Does the age at which a child begins to talk predict later score on a test of mental ability? A study of the development of young children recorded the age in months at which each of 21 children spoke their first word and their Gesell Adaptive Score, the result of an aptitude test taken much later. The data appear in Table 2.6.[18]

Figure 2.14 is a scatterplot, with age at first word as the explanatory variable x and Gesell score as the response variable y. Children 3 and 13, and also Children 16 and 21, have identical values of both variables. We use a different plotting symbol to show that one point stands for two individuals. The plot shows a negative association. That is, children who begin to speak later tend to have lower test scores than early talkers. The overall pattern is moderately linear. The correlation describes both the direction and strength of the linear relationship. It is $r = -0.640$.

The line on the plot is the least-squares regression line of Gesell score on age at first word. Its equation is

$$\hat{y} = 109.8738 - 1.1270x$$

For Child 1, who first spoke at 15 months, we predict the score

$$\hat{y} = 109.8738 - (1.1270)(15) = 92.97$$

This child's actual score was 95. The residual is

$$\text{residual} = \text{observed } y - \text{predicted } y$$
$$= 95 - 92.97 = 2.03$$

The residual is positive because the data point lies above the line.

Table 2.6 Age at first word and Gesell score

Child	Age	Score	Child	Age	Score
1	15	95	11	7	113
2	26	71	12	9	96
3	10	83	13	10	83
4	9	91	14	11	84
5	15	102	15	11	102
6	20	87	16	10	100
7	18	93	17	12	105
8	11	100	18	42	57
9	8	104	19	17	121
10	20	94	20	11	86
			21	10	100

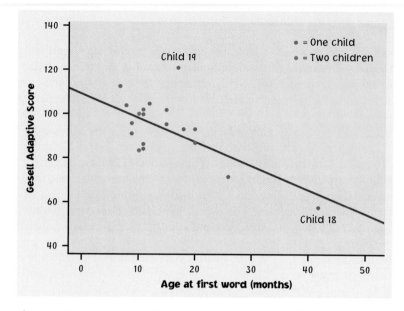

Figure 2.14 *Scatterplot of Gesell Adaptive Score versus the age at first word for 21 children, from Table 2.6. The line is the least-squares regression line for predicting Gesell score from age at first word.*

There is a residual for each data point. Finding the residuals with a calculator is a bit unpleasant, because you must first find the predicted response for every x. Statistical software gives you the residuals all at once. Here are the 21 residuals for the Gesell data, from regression software:

```
residuals:
    2.0310  -9.5721 -15.6040  -8.7309   9.0310   -0.3341    3.4120
    2.5230   3.1421   6.6659  11.0151  -3.7309  -15.6040  -13.4770
    4.5230   1.3960   8.6500  -5.5403  30.2850  -11.4770    1.3960
```

Because the residuals show how far the data fall from our regression line, examining the residuals helps assess how well the line describes the data. Although residuals can be calculated from any model fitted to the data, the residuals from the least-squares line have a special property: **the mean of the least-squares residuals is always zero.**

Compare the scatterplot in Figure 2.14 with the *residual plot* for the same data in Figure 2.15. The horizontal line at zero in Figure 2.15 helps orient us. It corresponds to the regression line in Figure 2.14.

RESIDUAL PLOTS

A **residual plot** is a scatterplot of the regression residuals against the explanatory variable. Residual plots help us assess the fit of a regression line.

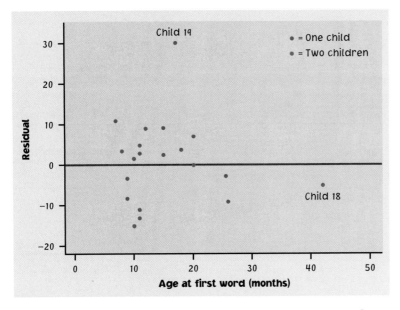

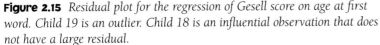

Figure 2.15 *Residual plot for the regression of Gesell score on age at first word. Child 19 is an outlier. Child 18 is an influential observation that does not have a large residual.*

If the regression line captures the overall relationship between x and y, the residuals should have no systematic pattern. The residual plot will look something like the simplified pattern in Figure 2.16(a). That plot shows a uniform scatter of the points about the fitted line, with no unusual individual observations. Here are some things to look for when you examine the residuals, using either a scatterplot of the data or a residual plot.

- **A curved pattern** shows that the relationship is not linear. Figure 2.16(b) is a simplified example. A straight line is not a good summary for such data.

- **Increasing or decreasing spread about the line** as x increases. Figure 2.16(c) is a simplified example. Prediction of y will be less accurate for larger x in that example.

- **Individual points with large residuals,** like Child 19 in Figures 2.14 and 2.15. These points are outliers in the vertical (y) direction because they lie far from the line that describes the overall pattern.

- **Individual points that are extreme in the x direction,** like Child 18 in Figures 2.14 and 2.15. Such points may not have large residuals, but they can be very important. We address such points next.

Influential observations

Children 18 and 19 are both unusual in the Gesell example. They are unusual in different ways. Child 19 lies far from the regression line. This child's Gesell

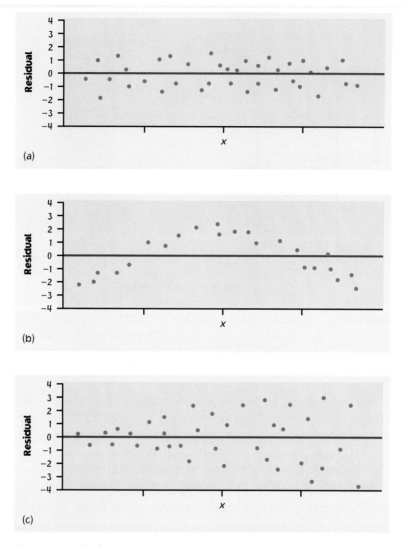

Figure 2.16 *Idealized patterns in plots of least-squares residuals. Plot (a) indicates that the regression line fits the data well. The data in plot (b) have a curved pattern, so a straight line fits poorly. The response variable y in plot (c) has more spread for larger values of the explanatory variable x, so prediction will be less accurate when x is large.*

score is so high that we should check for a mistake in recording it. In fact, the score is correct. Child 18 is close to the line but far out in the x direction. He or she began to speak much later than any of the other children. *Because of its extreme position on the age scale, this point has a strong influence on the position of the regression line.* Figure 2.17 adds a second regression line, calculated after leaving out Child 18. You can see that this one point moves the line quite a bit. We call such points *influential*.

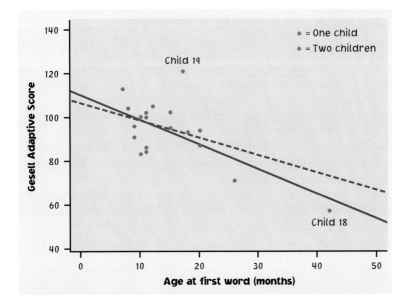

Figure 2.17 *Two least-squares regression lines of Gesell score on age at first word. The solid line is calculated from all the data. The dashed line was calculated leaving out Child 18. Child 18 is an influential observation because leaving out this point moves the regression line quite a bit.*

OUTLIERS AND INFLUENTIAL OBSERVATIONS IN REGRESSION

An **outlier** is an observation that lies outside the overall pattern of the other observations.

An observation is **influential** for a statistical calculation if removing it would markedly change the result of the calculation. Points that are outliers in the *x* direction of a scatterplot are often influential for the least-squares regression line.

Children 18 and 19 are both outliers in Figure 2.17. Child 18 is an outlier in the *x* direction and influences the least-squares line. Child 19 is an outlier in the *y* direction. It has less influence on the regression line because the many other points with similar values of *x* anchor the line well below the outlying point. Influential points often have small residuals, because they pull the regression line toward themselves. If you just look at residuals, you will miss influential points. Influential observations can greatly change the interpretation of data.

EXAMPLE 2.13 An influential observation

The strong influence of Child 18 makes the original regression of Gesell score on age at first word misleading. The original data have $r^2 = 0.41$. That is, the age at which a child begins to talk explains 41% of the variation on a later test of mental ability. This relationship is strong enough to be interesting to parents. If we leave out Child 18, r^2 drops to only 11%. The apparent strength of the association was largely due to a single influential observation.

What should the child development researcher do? She must decide whether Child 18 is so slow to speak that this individual should not be allowed to influence the analysis. If she excludes Child 18, much of the evidence for a connection between the age at which a child begins to talk and later ability score vanishes. If she keeps Child 18, she needs data on other children who were also slow to begin talking, so that the analysis no longer depends so heavily on just one child.

APPLY YOUR KNOWLEDGE

2.36 **Driving speed and fuel consumption.** Exercise 2.6 (page 87) gives data on the fuel consumption y of a car at various speeds x. Fuel consumption is measured in liters of gasoline per 100 kilometers driven and speed is measured in kilometers per hour. Software tells us that the equation of the least-squares regression line is

$$\hat{y} = 11.058 - 0.01466x$$

The residuals, in the same order as the observations, are

| 10.09 | 2.24 | −0.62 | −2.47 | −3.33 | −4.28 | −3.73 | −2.94 |
| −2.17 | −1.32 | −0.42 | 0.57 | 1.64 | 2.76 | 3.97 | |

(a) Make a scatterplot of the observations and draw the regression line on your plot.

(b) Would you use the regression line to predict y from x? Explain your answer.

(c) Check that the residuals have sum zero (up to roundoff error).

(d) Make a plot of the residuals against the values of x. Draw a horizontal line at height zero on your plot. Notice that the residuals show the same pattern about this line as the data points show about the regression line in the scatterplot in (a).

2.37 **How many calories?** Table 2.4 (page 104) gives data on the true calories in ten foods and the average guesses made by a large group of people. Exercise 2.23 explored the influence of two outlying observations on the correlation.

(a) Make a scatterplot suitable for predicting guessed calories from true calories. Circle the points for spaghetti and snack cake on your plot. These points lie outside the linear pattern of the other eight points.

(b) Use your calculator to find the least-squares regression line of guessed calories on true calories. Do this twice, first for all ten data points and then leaving out spaghetti and snack cake.

(c) Plot both lines on your graph. (Make one dashed so you can tell them apart.) Are spaghetti and snack cake, taken together, influential observations? Explain your answer.

2.38 **Influential or not?** We have seen that Child 18 in the Gesell data in Table 2.6 is an influential observation. Now we will examine the effect of Child 19, who is also an outlier in Figure 2.14.

(a) Find the least-squares regression line of Gesell score on age at first word, leaving out Child 19. Example 2.12 gives the regression line from all the children. Plot both lines on the same graph. (You do not have to make a scatterplot of all the points; just plot the two lines.) Would you call Child 19 very influential? Why?

(b) How does removing Child 19 change the r^2 for this regression? Explain why r^2 changes in this direction when you drop Child 19.

SECTION 2.3 Summary

A **regression line** is a straight line that describes how a response variable y changes as an explanatory variable x changes.

The most common method of fitting a line to a scatterplot is least squares. The **least-squares regression line** is the straight line $\hat{y} = a + bx$ that minimizes the sum of the squares of the vertical distances of the observed points from the line.

You can use a regression line to **predict** the value of y for any value of x by substituting this x into the equation of the line.

The **slope** b of a regression line $\hat{y} = a + bx$ is the rate at which the predicted response $\hat{y}$ changes along the line as the explanatory variable x changes. Specifically, b is the change in $\hat{y}$ when x increases by 1.

The **intercept** a of a regression line $\hat{y} = a + bx$ is the predicted response $\hat{y}$ when the explanatory variable $x = 0$. This prediction is of no statistical use unless x can actually take values near 0.

The least-squares regression line of y on x is the line with slope $r s_y/s_x$ and intercept $a = \bar{y} - b\bar{x}$. This line always passes through the point $(\bar{x}, \bar{y})$.

Correlation and regression are closely connected. The correlation r is the slope of the least-squares regression line when we measure both x and y in standardized units. The square of the correlation r^2 is the fraction of the variance of one variable that is explained by least-squares regression on the other variable.

You can examine the fit of a regression line by studying the **residuals**, which are the differences between the observed and predicted values of y. Be on the lookout for outlying points with unusually large residuals and also for nonlinear patterns and uneven variation about the line.

Also look for **influential observations**, individual points that substantially change the regression line. Influential observations are often outliers in the x direction, but they need not have large residuals.

SECTION 2.3 Exercises

2.39 **Review of straight lines.** Fred keeps his savings in his mattress. He began with $500 from his mother and adds $100 each year. His total savings y after x years are given by the equation

$$y = 500 + 100x$$

(a) Draw a graph of this equation. (Choose two values of x, such as 0 and 10. Compute the corresponding values of y from the equation. Plot these two points on graph paper and draw the straight line joining them.)

(b) After 20 years, how much will Fred have in his mattress?

(c) If Fred had added $200 instead of $100 each year to his initial $500, what is the equation that describes his savings after x years?

2.40 **Review of straight lines.** During the period after birth, a male white rat gains exactly 40 grams (g) per week. (This rat is unusually regular in his growth, but 40 g per week is a realistic rate.)

(a) If the rat weighed 100 g at birth, give an equation for his weight after x weeks. What is the slope of this line?

(b) Draw a graph of this line between birth and 10 weeks of age.

(c) Would you be willing to use this line to predict the rat's weight at age 2 years? Do the prediction and think about the reasonableness of the result. (There are 454 grams in a pound. To help you assess the result, note that a large cat weighs about 10 pounds.)

2.41 **IQ and school GPA.** Figure 2.5 (page 91) plots school grade point average (GPA) against IQ test score for 78 seventh-grade students. Calculation shows that the mean and standard deviation of the IQ scores are

$$\bar{x} = 108.9 \qquad s_x = 13.17$$

For the grade point averages,

$$\bar{y} = 7.447 \qquad s_y = 2.10$$

The correlation between IQ and GPA is $r = 0.6337$.

(a) Find the equation of the least-squares line for predicting GPA from IQ.

(b) What percent of the observed variation in these students' GPAs can be explained by the linear relationship between GPA and IQ?

(c) One student has an IQ of 103 but a very low GPA of 0.53. What is the predicted GPA for a student with IQ = 103? What is the residual for this particular student?

2.42 **Take me out to the ball game.** What is the relationship between the price charged for a hot dog and the price charged for a 16-ounce soda in major league baseball stadiums? Here are some data:[19]

Team	Hot dog	Soda	Team	Hot dog	Soda	Team	Hot dog	Soda
Angels	2.50	1.75	Giants	2.75	2.17	Rangers	2.00	2.00
Astros	2.00	2.00	Indians	2.00	2.00	Red Sox	2.25	2.29
Braves	2.50	1.79	Marlins	2.25	1.80	Rockies	2.25	2.25
Brewers	2.00	2.00	Mets	2.50	2.50	Royals	1.75	1.99
Cardinals	3.50	2.00	Padres	1.75	2.25	Tigers	2.00	2.00
Dodgers	2.75	2.00	Phillies	2.75	2.20	Twins	2.50	2.22
Expos	1.75	2.00	Pirates	1.75	1.75	White Sox	2.00	2.00

(a) Make a scatterplot appropriate for predicting soda price from hot dog price. Describe the relationship that you see. Are there any outliers?

(b) Find the correlation between hot dog price and soda price. What percent of the variation in soda price does a linear relationship account for?

(c) Find the equation of the least-squares line for predicting soda price from hot dog price. Draw the line on your scatterplot. Based on your findings in (b), explain why it is not surprising that the line is nearly horizontal (slope near zero).

(d) Circle the observation that is potentially the most influential. What team is this? Find the least-squares line without this one observation and draw it on your scatterplot. Was the observation in fact influential?

2.43 **Keeping water clean.** Keeping water supplies clean requires regular measurement of levels of pollutants. The measurements are indirect—a typical analysis involves forming a dye by a chemical reaction with the dissolved pollutant, then passing light through the solution and measuring its "absorbence." To calibrate such measurements, the laboratory measures known standard solutions and uses regression to relate absorbence to pollutant concentration. This is usually done every day. Here is one series of data on the absorbence for different levels of nitrates. Nitrates are measured in milligrams per liter of water.[20]

Nitrates	50	50	100	200	400	800	1200	1600	2000	2000
Absorbence	7.0	7.5	12.8	24.0	47.0	93.0	138.0	183.0	230.0	226.0

(a) Chemical theory says that these data should lie on a straight line. If the correlation is not at least 0.997, something went wrong and the calibration procedure is repeated. Plot the data and find the correlation. Must the calibration be done again?

(b) What is the equation of the least-squares line for predicting absorbence from concentration? If the lab analyzed a specimen with 500 milligrams of nitrates per liter, what do you expect the absorbence to be? Based on your plot and the correlation, do you expect your predicted absorbence to be very accurate?

2.44 **A growing child.** Sarah's parents are concerned that she seems short for her age. Their doctor has the following record of Sarah's height:

Age (months)	36	48	51	54	57	60
Height (cm)	86	90	91	93	94	95

(a) Make a scatterplot of these data. Note the strong linear pattern.

(b) Using your calculator, find the equation of the least-squares regression line of height on age.

(c) Predict Sarah's height at 40 months and at 60 months. Use your results to draw the regression line on your scatterplot.

(d) What is Sarah's rate of growth, in centimeters per month? Normally growing girls gain about 6 cm in height between ages 4 (48 months) and 5 (60 months). What rate of growth is this in centimeters per month? Is Sarah growing more slowly than normal?

2.45 **Investing at home and overseas.** Investors ask about the relationship between returns on investments in the United States and on investments overseas. Table 2.7 gives the total returns on U.S. and overseas common stocks over a 26-year period. (The total return is change in price plus any dividends paid, converted into U.S. dollars. Both returns are averages over many individual stocks.)[21]

Table 2.7 Annual total return on overseas and U.S. stocks

Year	Overseas % return	U.S. % return	Year	Overseas % return	U.S. % return
1971	29.6	14.6	1984	7.4	6.1
1972	36.3	18.9	1985	56.2	31.6
1973	−14.9	−14.8	1986	69.4	18.6
1974	−23.2	−26.4	1987	24.6	5.1
1975	35.4	37.2	1988	28.5	16.8
1976	2.5	23.6	1989	10.6	31.5
1977	18.1	−7.4	1990	−23.0	−3.1
1978	32.6	6.4	1991	12.8	30.4
1979	4.8	18.2	1992	−12.1	7.6
1980	22.6	32.3	1993	32.9	10.1
1981	−2.3	−5.0	1994	6.2	1.3
1982	−1.9	21.5	1995	11.2	37.6
1983	23.7	22.4	1996	6.4	23.0
			1997	2.1	33.4

(a) Make a scatterplot suitable for predicting overseas returns from U.S. returns.

(b) Find the correlation and r^2. Describe the relationship between U.S. and overseas returns in words, using r and r^2 to make your description more precise.

(c) Find the least-squares regression line of overseas returns on U.S. returns. Draw the line on the scatterplot.

(d) In 1997, the return on U.S. stocks was 33.4%. Use the regression line to predict the return on overseas stocks. The actual overseas return was 2.1%. Are you confident that predictions using the regression line will be quite accurate? Why?

(e) Circle the point that has the largest residual (either positive or negative). What year is this? Are there any points that seem likely to be very influential?

2.46 **Always plot your data!** Table 2.8 presents four sets of data prepared by the statistician Frank Anscombe to illustrate the dangers of calculating without first plotting the data.[22]

(a) Without making scatterplots, find the correlation and the least-squares regression line for all four data sets. What do you notice? Use the regression line to predict y for $x = 10$.

(b) Make a scatterplot for each of the data sets and add the regression line to each plot.

Table 2.8 Four data sets for exploring correlation and regression

Data Set A

x	10	8	13	9	11	14	6	4	12	7	5
y	8.04	6.95	7.58	8.81	8.33	9.96	7.24	4.26	10.84	4.82	5.68

Data Set B

x	10	8	13	9	11	14	6	4	12	7	5
y	9.14	8.14	8.74	8.77	9.26	8.10	6.13	3.10	9.13	7.26	4.74

Data Set C

x	10	8	13	9	11	14	6	4	12	7	5
y	7.46	6.77	12.74	7.11	7.81	8.84	6.08	5.39	8.15	6.42	5.73

Data Set D

x	8	8	8	8	8	8	8	8	8	8	19
y	6.58	5.76	7.71	8.84	8.47	7.04	5.25	5.56	7.91	6.89	12.50

Source: Frank J. Anscombe, "Graphs in statistical analysis," The American Statistician, 27 (1973), pp. 17–21.

(c) In which of the four cases would you be willing to use the regression line to describe the dependence of y on x? Explain your answer in each case.

2.47 **What's my grade?** In Professor Friedman's economics course the correlation between the students' total scores prior to the final examination and their final examination scores is $r = 0.6$. The pre-exam totals for all students in the course have mean 280 and standard deviation 30. The final exam scores have mean 75 and standard deviation 8. Professor Friedman has lost Julie's final exam but knows that her total before the exam was 300. He decides to predict her final exam score from her pre-exam total.

(a) What is the slope of the least-squares regression line of final exam scores on pre-exam total scores in this course? What is the intercept?

(b) Use the regression line to predict Julie's final exam score.

(c) Julie doesn't think this method accurately predicts how well she did on the final exam. Calculate r^2 and use the value you get to argue that her actual score could have been much higher (or much lower) than the predicted value.

2.48 **A nonsense prediction.** Use the least-squares regression line for the data in Exercise 2.44 to predict Sarah's height at age 40 years (480 months). Your prediction is in centimeters. Convert it to inches using the fact that a centimeter is 0.3937 inch.

The prediction is impossibly large. It is not reasonable to use data for 36 to 60 months to predict height at 480 months.

2.49 **Investing at home and overseas.** Exercise 2.45 examined the relationship between returns on U.S. and overseas stocks. Investors also want to know what typical returns are and how much year-to-year variability (called *volatility* in finance) there is. Regression and correlation do not answer these questions.

(a) Find the five-number summaries for both U.S. and overseas returns, and make side-by-side boxplots to compare the two distributions.

(b) Were returns generally higher in the United States or overseas during this period? Explain your answer.

(c) Were returns more volatile (more variable) in the United States or overseas during this period? Explain your answer.

2.50 A study of class attendance and grades among first-year students at a state university showed that in general students who attended a higher percent of their classes earned higher grades. Class attendance explained 16% of the variation in grade index among the students. What is the numerical value of the correlation between percent of classes attended and grade index?

2.51 **Will I bomb the final?** We expect that students who do well on the midterm exam in a course will usually also do well on the final exam. Gary Smith of Pomona College looked at the exam scores of all 346 students who took his statistics class over a 10-year period.[23] The least-squares line for predicting final exam score from midterm exam score was $\hat{y} = 46.6 + 0.41x$.

Octavio scores 10 points above the class mean on the midterm. How many points above the class mean do you predict that he will score on the final? (Hint: Use the fact that the least-squares line passes through the point $(\bar{x}, \bar{y})$ and the fact that Octavio's midterm score is $\bar{x} + 10$. This is an example of the phenomenon that gave "regression" its name: students who do well on the midterm will on the average do less well, but still above average, on the final.)

2.52 **Predicting enrollments.** The mathematics department of a large state university would like to use the number of freshmen entering the university x to predict the number of students y who will sign up for freshman-level math courses in the fall semester. Here are data for recent years:[24]

Year	1991	1992	1993	1994	1995	1996	1997	1998
x	4595	4827	4427	4258	3995	4330	4265	4351
y	7364	7547	7099	6894	6572	7156	7232	7450

Software gives the correlation $r = 0.8333$ and the least-squares regression line

$$\hat{y} = 2492.69 + 1.0663x$$

The software also gives a table of the residuals:

Year	1991	1992	1993	1994	1995	1996	1997	1998
Residual	−28.44	−92.83	−114.30	−139.09	−180.65	46.13	191.44	317.74

(a) Make a scatterplot and draw the regression line on it. The regression line does not predict very accurately. What percent of variation in class enrollment is explained by the linear relationship with the count of freshmen?

(b) Check that the residuals have sum zero (at least up to roundoff error).

(c) Plots of the residuals against other variables are often revealing. Plot the residuals against year. One of the schools in the university recently changed its program to require that entering students take another mathematics course. How does the residual plot show the change? In what year was the change effective?

2.4 Cautions about Correlation and Regression

Correlation and regression are powerful tools for describing the relationship between two variables. When you use these tools, you must be aware of their limitations, beginning with the fact that **correlation and regression describe only linear relationships**. Also remember that **the correlation r and the least-squares regression line are not resistant**. One influential observation or incorrectly entered data point can greatly change these measures. Always plot your data before interpreting regression or correlation. Here are some other cautions to keep in mind when you apply correlation and regression or read accounts of their use.

Extrapolation

Suppose that you have data on a child's growth between 3 and 8 years of age. You find a strong linear relationship between age x and height y. If you fit a regression line to these data and use it to predict height at age 25 years, you will predict that the child will be 8 feet tall. Growth slows down and stops at maturity, so extending the straight line to adult ages is foolish. Few relationships are linear for all values of x. So don't stray far from the range of x that actually appears in your data.

> **EXTRAPOLATION**
>
> **Extrapolation** is the use of a regression line for prediction far outside the range of values of the explanatory variable x that you used to obtain the line. Such predictions are often not accurate.

APPLY YOUR KNOWLEDGE

2.53 **The declining farm population.** The number of people living on American farms has declined steadily during this century. Here are data on the farm population (millions of persons) from 1935 to 1980:

Year	1935	1940	1945	1950	1955	1960	1965	1970	1975	1980
Population	32.1	30.5	24.4	23.0	19.1	15.6	12.4	9.7	8.9	7.2

(a) Make a scatterplot of these data and find the least-squares regression line of farm population on year.

(b) According to the regression line, how much did the farm population decline each year on the average during this period? What percent of the observed variation in farm population is accounted for by linear change over time?

(c) Use the regression equation to predict the number of people living on farms in 1990. Is this result reasonable? Why?

Using averaged data

Many regression or correlation studies work with averages or other measures that combine information from many individuals. You should note this carefully and resist the temptation to apply the results of such studies to individuals. Figure 2.2 (page 86) shows a strong relationship between outside temperature and the Sanchez household's natural gas consumption. Each point on the scatterplot represents a month. Both degree-days and gas consumed are averages over all the days in the month. The data for individual days would form a cloud around each month's averages—they would have more scatter and lower correlation. Averaging over an entire month smooths out the day-to-day variation due to doors left open, house guests using more gas to heat water, and so on. **Correlations based on averages are usually too high when applied to** individuals. This is another reminder that it is important to note exactly what variables were measured in a statistical study.

Did the vote counters cheat?

Republican Bruce Marks was ahead of Democrat William Stinson when the voting machines were tallied in their Pennsylvania election. But Stinson was ahead after absentee ballots were counted by the Democrats, who controlled the electoral board. A court fight followed. The court called in a statistician, who used regression with data from past elections to predict the counts of absentee ballots from the voting-machine results. Marks's lead of 564 votes from the machines predicted that he would get 133 more absentee votes than Stinson. In fact, Stinson got 1023 more absentee votes than Marks. Did the vote counters cheat?

2.54 **Stock market indexes.** The Standard & Poor's 500-stock index is an average of the price of 500 stocks. There is a moderately strong correlation (roughly $r = 0.6$) between how much this index changes in January and how much it changes during the entire year. If we looked instead at data on all 500 individual stocks, we would find a quite different correlation. Would the correlation be higher or lower? Why?

Lurking variables

Correlation and regression describe the relationship between two variables. Often the relationship between two variables is strongly influenced by other variables. More advanced statistical methods allow the study of many variables together, so that we can take other variables into account. Sometimes, however, the relationship between two variables is influenced by other variables that we did not measure or even think about. Because these variables are lurking in the background, we call them *lurking variables*.

> **LURKING VARIABLE**
>
> A **lurking variable** is a variable that has an important effect on the relationship among the variables in a study but is not included among the variables studied.

A lurking variable can falsely suggest a strong relationship between x and y, or it can hide a relationship that is really there. Here are examples of each of these effects.

EXAMPLE 2.14 **Discrimination in medical treatment?**

Studies show that men who complain of chest pain are more likely to get detailed tests and aggressive treatment such as bypass surgery than are women with similar complaints. Is this association between gender and treatment due to discrimination?

Perhaps not. Men and women develop heart problems at different ages—women are on the average between 10 and 15 years older than men. Aggressive treatments are more risky for older patients, so doctors may hesitate to advise them. Lurking variables—the patient's age and condition—may explain the relationship between gender and doctors' decisions. As the author of one study of the issue said, "When men and women are otherwise the same and the only difference is gender, you find that treatments are very similar."[25]

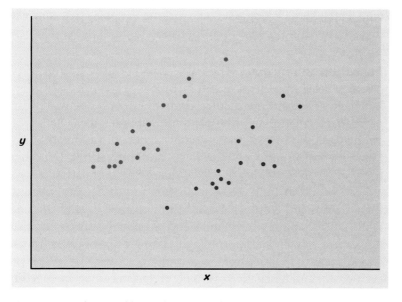

Figure 2.18 *The variables in this scatterplot have a small correlation even though there is a strong correlation within each of the two clusters.*

EXAMPLE 2.15 **Measuring inadequate housing**

A study of housing conditions in the city of Hull, England, measured several variables for each of the city's wards. Figure 2.18 is a simplified version of one of the study's findings. The variable x measures how crowded the ward is. The variable y measures the fraction of housing units that have no indoor toilet. We expect inadequate housing to be crowded (high x) and lack toilets (high y). That is, we expect a strong correlation between x and y. In fact the correlation was only $r = 0.08$. How can this be?

Figure 2.18 shows that there are two clusters of wards. The wards in the lower cluster have a lot of public housing. These wards are crowded (high values of x), but they have low values of y because public housing always includes indoor toilets. The wards in the upper cluster lack public housing and have high values of both x and y. Within wards of each type, there is a strong positive association between x and y. In Figure 2.18, $r = 0.85$ and $r = 0.91$ in the two clusters. However, because similar values of x correspond to quite different values of y in the two clusters, x alone is of little value for predicting y. Analyzing all wards together ignores the lurking variable—amount of public housing—and hides the nature of the relationship between x and y.[26]

APPLY YOUR KNOWLEDGE

2.55 **TV watching and school grades.** Children who watch many hours of television get lower grades in school on the average than those

Do left handers die early?

Yes, said a study of 1000 deaths in California. Left-handed people died at an average age of 66 years; right-handers, at 75 years of age. Should left-handed people fear an early death? No—the lurking variable has struck again. Older people grew up in an era when many natural left-handers were forced to use their right hands. So right-handers are more common among older people, and left-handers are more common among the young. When we look at deaths, the left-handers who die are younger on the average because left-handers in general are younger. Mystery solved.

who watch less TV. Suggest some lurking variables that may explain this relationship because they contribute to both heavy TV viewing and poor grades.

2.56 **Education and income.** There is a strong positive correlation between years of education and income for economists employed by business firms. In particular, economists with doctorates earn more than economists with only a bachelor's degree. There is also a strong positive correlation between years of education and income for economists employed by colleges and universities. But when all economists are considered, there is a *negative* correlation between education and income. The explanation for this is that business pays high salaries and employs mostly economists with bachelor's degrees, while colleges pay lower salaries and employ mostly economists with doctorates. Sketch a scatterplot with two groups of cases (business and academic) illustrating how a strong positive correlation within each group and a negative overall correlation can occur together. (Hint: Begin by studying Figure 2.18.)

Association is not causation

When we study the relationship between two variables, we often hope to show that changes in the explanatory variable *cause* changes in the response variable. But a strong association between two variables is not enough to draw conclusions about cause and effect. Sometimes an observed association really does reflect cause and effect. The Sanchez household uses more natural gas in colder months because cold weather requires burning more gas to stay warm. In other cases, an association is explained by lurking variables, and the conclusion that x causes y is either wrong or not proved. Here are several examples.

EXAMPLE 2.16 **Does television extend life?**

Measure the number of television sets per person x and the average life expectancy y for the world's nations. There is a high positive correlation: nations with many TV sets have higher life expectancies.

The basic meaning of causation is that by changing x we can bring about a change in y. Could we lengthen the lives of people in Rwanda by shipping them TV sets? No. Rich nations have more TV sets than poor nations. Rich nations also have longer life expectancies because they offer better nutrition, clean water, and better health care. There is no cause-and-effect tie between TV sets and length of life.

Correlations such as that in Example 2.16 are sometimes called "nonsense correlations." The correlation is real. What is nonsense is the conclusion that changing one of the variables causes changes in the other. A lurking variable—such as national wealth in Example 2.16—that influences both x and y can create a high correlation even though there is no direct connection between x and y.

EXAMPLE 2.17   Safety of anesthetics

The National Halothane Study was a major investigation of the safety of the anesthetics used in surgery. Records of over 850,000 operations performed in 34 major hospitals showed the following death rates for four common anesthetics:[27]

Anesthetic	A	B	C	D
Death rate	1.7%	1.7%	3.4%	1.9%

There is a clear association between the anesthetic used and the death rate of patients. Anesthetic C appears dangerous. But there are obvious lurking variables: the age and condition of the patient and the seriousness of the surgery. In fact, anesthetic C was more often used in serious operations on older patients in poor condition. The death rate would be higher among these patients no matter what anesthetic they received. After measuring the lurking variables and adjusting for their effect, the apparent relationship between anesthetic and death rate is very much weaker.

These and other examples lead us to the most important caution about correlation, regression, and statistical association between variables in general.

> **ASSOCIATION DOES NOT IMPLY CAUSATION**
>
> An association between an explanatory variable x and a response variable y, even if it is very strong, is not by itself good evidence that changes in x actually cause changes in y.

experiment

The best way to get good evidence that x causes y is to do an **experiment** in which we change x and keep lurking variables under control. We will discuss experiments in Chapter 3. When experiments cannot be done, finding the explanation for an observed association is often difficult and controversial. Many of the sharpest disputes in which statistics plays a role involve questions of causation that cannot be settled by experiment. Does smoking cause lung cancer? What about secondhand smoke? Does living near power lines cause leukemia? Has increased free trade widened the gap between the incomes of more educated and less educated American workers? All of these questions have become public issues. All concern associations among variables. And all have this in common: they try to pinpoint cause and effect in a setting involving complex relations among many interacting variables.

EXAMPLE 2.18 **Does smoking cause lung cancer?**

Despite the difficulties, it is sometimes possible to build a strong case for causation in the absence of experiments. The evidence that smoking causes lung cancer is about as strong as nonexperimental evidence can be.

Doctors had long observed that most lung cancer patients were smokers. Comparison of smokers and "similar" nonsmokers showed a very strong association between smoking and death from lung cancer. Could the association be explained by lurking variables? Might there be, for example, a genetic factor that predisposes people both to nicotine addiction and to lung cancer? Smoking and lung cancer would then be positively associated even if smoking had no direct effect on the lungs. How were these objections overcome?

Let's answer this question in general terms: What are the criteria for establishing causation when we cannot do an experiment?

- *The association is strong.* The association between smoking and lung cancer is very strong.

- *The association is consistent.* Many studies of different kinds of people in many countries link smoking to lung cancer. That reduces the chance that a lurking variable specific to one group or one study explains the association.

- *Higher doses are associated with stronger responses.* People who smoke more cigarettes per day or who smoke over a longer period get lung cancer more often. People who stop smoking reduce their risk.

- *The alleged cause precedes the effect in time.* Lung cancer develops after years of smoking. The number of men dying of lung cancer rose as smoking became more common, with a lag of about 30 years. Lung cancer kills more men than any other form of cancer. Lung cancer was rare among women until women began to smoke. Lung cancer in women rose along with smoking, again with a lag of about 30 years, and has now passed breast cancer as the leading cause of cancer death among women.

- *The alleged cause is plausible.* Experiments with animals show that tars from cigarette smoke do cause cancer.

Medical authorities do not hesitate to say that smoking causes lung cancer. The U.S. Surgeon General has long stated that cigarette smoking is "the largest avoidable cause of death and disability in the United States."[28] The evidence for causation is overwhelming—but it is not as strong as the evidence provided by well-designed experiments.

APPLY YOUR KNOWLEDGE

2.57 **Do firefighters make fires worse?** Someone says, "There is a strong positive correlation between the number of firefighters at a fire and the amount of damage the fire does. So sending lots of firefighters just causes more damage." Explain why this reasoning is wrong.

2.58 **How's your self-esteem?** People who do well tend to feel good about themselves. Perhaps helping people feel good about themselves will help them do better in school and life. Raising self-esteem became for a time a goal in many schools. California even created a state commission to advance the cause. Can you think of explanations for the association between high self-esteem and good school performance other than "Self-esteem causes better work in school"?

2.59 **Are big hospitals bad for you?** A study shows that there is a positive correlation between the size of a hospital (measured by its number of beds x) and the median number of days y that patients remain in the hospital. Does this mean that you can shorten a hospital stay by choosing a small hospital? Why?

SECTION 2.4 Summary

Correlation and regression must be **interpreted with caution. Plot the data** to be sure the relationship is roughly linear and to detect outliers and influential observations.

Avoid **extrapolation**, the use of a regression line for prediction for values of the explanatory variable far outside the range of the data from which the line was calculated.

Remember that **correlations based on averages** are usually too high when applied to individuals.

Lurking variables that you did not measure may explain the relations between the variables you did measure. Correlation and regression can be misleading if you ignore important lurking variables.

Most of all, be careful not to conclude that there is a cause-and-effect relationship between two variables just because they are strongly associated. **High correlation does not imply causation.** The best evidence that an association is due to causation comes from an **experiment** in which the explanatory variable is directly changed and other influences on the response are controlled.

SECTION 2.4 Exercises

2.60 **Is math the key to success in college?** Here is the opening of a newspaper account of a College Board study of 15,941 high school graduates:

> Minority students who take high school algebra and geometry succeed in college at almost the same rate as whites, a new study says.

The link between high school math and college graduation is "almost magical," says College Board President Donald Stewart, suggesting "math is the gatekeeper for success in college."

"These findings," he says, "justify serious consideration of a national policy to ensure that all students take algebra and geometry."[29]

What lurking variables might explain the association between taking several math courses in high school and success in college? Explain why requiring algebra and geometry may have little effect on who succeeds in college.

2.61   **Shoe size and reading score.** A study of elementary school children, ages 6 to 11, finds a high positive correlation between shoe size x and score y on a test of reading comprehension. What explains this correlation?

2.62   **Do artificial sweeteners cause weight gain?** People who use artificial sweeteners in place of sugar tend to be heavier than people who use sugar. Does this mean that artificial sweeteners cause weight gain? Give a more plausible explanation for this association.

2.63   **SAT math and verbal scores.** Table 1.6 (page 26) gives education data for the states. The correlation between the average SAT math scores and the average SAT verbal scores for the states is $r = 0.970$.

(a) Find r^2 and explain in simple language what this number tells us.

(b) If you calculated the correlation between the SAT math and verbal scores of a large number of individual students, would you expect the correlation to be about 0.97 or quite different? Explain your answer.

2.64   **Does herbal tea help nursing home residents?** A group of college students believes that herbal tea has remarkable powers. To test this belief, they make weekly visits to a local nursing home, where they visit with the residents and serve them herbal tea. The nursing home staff reports that after several months many of the residents are healthier and more cheerful. We should commend the students for their good deeds but doubt that herbal tea helped the residents. Identify the explanatory and response variables in this informal study. Then explain what lurking variables account for the observed association.

2.65   **The benefits of foreign language study.** Members of a high school language club believe that study of a foreign language improves a student's command of English. From school records, they obtain the scores on an English achievement test given to all seniors. The mean score of seniors who studied a foreign language for at least two years is much higher than the mean score of seniors who studied no foreign language. These data are not good evidence that language study strengthens English skills. Identify the explanatory and response variables in this study. Then explain what lurking variable prevents the conclusion that language study improves students' English scores.

2.66 **Education and income.** There is a strong positive correlation between years of schooling completed x and lifetime earnings y for American men. One possible reason for this association is causation: more education leads to higher-paying jobs. But lurking variables may explain some of the correlation. Suggest some lurking variables that would explain why men with more education earn more.

2.67 **Do power lines cause cancer?** It has been suggested that electromagnetic fields of the kind present near power lines can cause leukemia in children. Experiments with children and power lines are not ethical. Careful studies have found no association between exposure to electromagnetic fields and childhood leukemia.[30] Suggest several lurking variables that you would want information about in order to investigate the claim that living near power lines is associated with cancer.

(PHIL MOUGHMER/INDEX STOCK IMAGERY.)

2.5 Relations in Categorical Data*

We have concentrated on relationships in which at least the response variable is quantitative. Now we will shift to describing relationships between two or more categorical variables. Some variables—such as gender, race, and occupation— are categorical by nature. Other categorical variables are created by grouping values of a quantitative variable into classes. Published data often appear in grouped form to save space. To analyze categorical data, we use the *counts* or *percents* of individuals that fall into various categories.

EXAMPLE 2.19 **Age and education**

Table 2.9 presents Census Bureau data on the years of school completed by Americans of different ages. Many people under 25 years of age have not completed their education, so they are left out of the table. Both variables, age and education, are grouped into categories. This is a **two-way table** because it describes two categorical variables. Education is the **row variable** because each row in the table describes people with one level of education. Age is the **column variable** because each column describes one age group. The entries in the table are the counts of persons in each age-by-education class. Although both age and education in this table are categorical variables, both have a natural order from least to most. The order of the rows and the columns in Table 2.9 reflects the order of the categories.

two-way table

row and column variables

Marginal distributions

How can we best grasp the information contained in Table 2.9? First, *look at the distribution of each variable separately*. The distribution of a categorical variable

---

*This material is important in statistics, but it is needed later in this book only for Chapter 9. You may omit it if you do not plan to read Chapter 9 or delay reading it until you reach Chapter 9.

Table 2.9 Years of school completed, by age, 1995 (thousands of persons)

Education	Age Group			Total
	25 to 34	35 to 54	55 and over	
Did not complete high school	5,325	9,152	16,035	30,512
Completed high school	14,061	24,070	18,320	56,451
College, 1 to 3 years	11,659	19,926	9,662	41,247
College, 4 or more years	10,342	19,878	8,005	38,225
Total	41,388	73,028	52,022	166,438

says how often each outcome occurred. The "Total" column at the right of the table contains the totals for each of the rows. These row totals give the distribution of education level (the row variable) among all people over 25 years of age: 30,512,000 did not complete high school, 56,451,000 finished high school but did not attend college, and so on. In the same way, the "Total" row at the bottom of the table gives the age distribution. If the row and column totals are missing, the first thing to do in studying a two-way table is to calculate them. The distributions of education alone and age alone are called **marginal distributions** because they appear at the right and bottom margins of the two-way table.

marginal distribution

If you check the row and column totals in Table 2.9, you will notice a few discrepancies. For example, the sum of the entries in the "25 to 34" column is 41,387. The entry in the "Total" row for that column is 41,388. The explanation is **roundoff error**. The table entries are in thousands of persons, and each is rounded to the nearest thousand. The Census Bureau obtained the "Total" entry by rounding the exact number of people aged 25 to 34 to the nearest thousand. The result was 41,388,000. Adding the row entries, each of which is already rounded, gives a slightly different result.

roundoff error

Percents are often more informative than counts. We can display the marginal distribution of education level in terms of percents by dividing each row total by the table total and converting to a percent.

EXAMPLE 2.20 **Calculating a marginal distribution**

The percent of people over 25 years of age who have at least 4 years of college is

$$\frac{\text{total with 4 years of college}}{\text{table total}} = \frac{38,225}{166,438} = 0.230 = 23.0\%$$

Do three more such calculations to obtain the marginal distribution of education level in percents. Here it is:

	Did not complete high school	Completed high school	1 to 3 years of college	≥ 4 years of college
Percent	18.3	33.9	24.8	23.0

The total is 100% because everyone is in one of the four education categories.

Each marginal distribution from a two-way table is a distribution for a single categorical variable. As we saw in Chapter 1 (page 6), we can use a bar graph or a pie chart to display such a distribution. Figure 2.19 is a bar graph of the distribution of years of schooling. We see that people with at least some college education make up about half of the over-25 population.

In working with two-way tables, you must calculate lots of percents. Here's a tip to help decide what fraction gives the percent you want. Ask, "What group represents the total that I want a percent of?" The count for that group is the denominator of the fraction that leads to the percent. In Example 2.20, we wanted a percent "of people over 25 years of age," so the count of people over 25 (the table total) is the denominator.

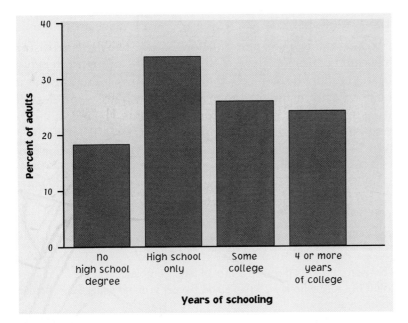

Figure 2.19 *A bar graph of the distribution of years of schooling completed among people aged 25 years and over. This is one of the marginal distributions for Table 2.9.*

2.68 The counts in the "Total" column at the right of Table 2.9 are the counts of people in each education group. Explain why the sum of these counts is not equal to 166,438, the table total that appears at the lower right corner of the table.

2.69 Give the marginal distribution of age (in percents) among people over 25 years of age, starting from the counts in Table 2.9.

2.70 **Smoking by students and their parents.** Here are data from eight high schools on smoking among students and among their parents:[31]

	Neither parent smokes	One parent smokes	Both parents smoke
Student does not smoke	1168	1823	1380
Student smokes	188	416	400

(a) How many students do these data describe?

(b) What percent of these students smoke?

(c) Give the marginal distribution of parents' smoking behavior, both in counts and in percents.

Describing relationships

Table 2.9 contains much more information than the two marginal distributions of age alone and education alone. The nature of the relationship between age and education cannot be deduced from the separate distributions but requires the full table. **Relationships among categorical variables are described by calculating appropriate percents from the counts given.** We use percents because counts are often hard to compare. For example, 19,878,000 persons aged 35 to 54 have completed college, and only 8,005,000 persons in the 55 and up age group have done so. These counts do not accurately describe the association, however, because there are many more people in the younger age group.

EXAMPLE 2.21 **How common is college education?**

What percent of people aged 25 to 34 have completed 4 years of college? This is the count of those who are 25 to 34 and have 4 years of college as a percent of the age group total:

$$\frac{10{,}342}{41{,}388} = 0.250 = 25.0\%$$

"People aged 25 to 34" is the group we want a percent of, so the count for that group is the denominator. In the same way, find the percent of people in each age group who have completed college. The comparison of all three groups is

	25 to 34	35 to 54	55 and over
Percent with 4 years of college	25.0	27.2	15.4

These percentages make it clear that a college education is less common among Americans over age 55 than among younger adults. This is an important aspect of the association between age and education.

APPLY YOUR KNOWLEDGE

2.71 Using the counts in Table 2.9, find the percent of people in each age group who did not complete high school. Draw a bar graph that compares these percents. State briefly what the data show.

2.72 **Python eggs.** How is the hatching of water python eggs influenced by the temperature of the snake's nest? Researchers assigned newly laid eggs to one of three temperatures: hot, neutral, or cold. Hot duplicates the extra warmth provided by the mother python, and cold duplicates the absence of the mother. Here are the data on the number of eggs and the number that hatched:[32]

	Cold	Neutral	Hot
Number of eggs	27	56	104
Number hatched	16	38	75

(a) Make a two-way table of temperature by outcome (hatched or not).

(b) Calculate the percent of eggs in each group that hatched. The researchers anticipated that eggs would not hatch in cold water. Do the data support that anticipation?

Conditional distributions

Example 2.21 does not compare the complete distributions of years of schooling in the three age groups. It compares only the percent who finished college. Let's look at the complete picture.

EXAMPLE 2.22 **Calculating a conditional distribution**

Information about the 25 to 34 age group occupies the first column in Table 2.9. To find the complete distribution of education in this age group, look only at that column. Compute each count as a percent of the column total, which is 41,388. Here is the distribution:

	Did not complete high school	Completed high school	1 to 3 years of college	$\geq$ 4 years of college
Percent	12.9	34.0	28.2	25.0

conditional distribution

These percents should add to 100% because all 25- to 34-year-olds fall in one of the educational categories. (In fact, they add to 100.1% because of roundoff error.) The four percents together are the **conditional distribution** of education, given that a person is 25 to 34 years of age. We use the term "conditional" because the distribution refers only to people who satisfy the condition that they are 25 to 34 years old.

Now focus in turn on the second column (people aged 35 to 54) and then the third column (people 55 and over) of Table 2.9 in order to find two more conditional distributions. Statistical software can speed the task of finding each entry in a two-way table as a percent of its column total. Figure 2.20 displays the result. The software found the row and column totals from the table entries, so they may differ slightly from those in Table 2.9.

Each cell in this table contains a count from Table 2.9 along with that count as a percent of the column total. The percents in each column form the conditional distribution of years of schooling for one age group. The percents in

```
                      TABLE OF EDU BY AGE
         EDU              AGE

         Frequency |          |          |          |
         Col Pct   |  25-34   |  35-54   | 55 over  | Total
         ----------+----------+----------+----------+
         NoHS      |   5325   |   9152   |  16035   | 30512
                   |  12.87   |  12.53   |  30.82   |
         ----------+----------+----------+----------+
         HSonly    |  14061   |  24070   |  18320   | 56451
                   |  33.97   |  32.96   |  35.22   |
         ----------+----------+----------+----------+
         SomeColl  |  11659   |  19926   |   9662   | 41247
                   |  28.17   |  27.29   |  18.57   |
         ----------+----------+----------+----------+
         Coll4yrs  |  10342   |  19878   |   8005   | 38225
                   |  24.99   |  27.22   |  15.39   |
         ----------+----------+----------+----------+
         Total        41387      73026      52022     166435
```

Figure 2.20 *SAS output of the two-way table of education by age with the three conditional distributions of education, one for each age group. The percents in each column add to 100%.*

each column add to 100% because everyone in the age group is accounted for. Comparing the conditional distributions reveals the nature of the association between age and education. The distributions of education in the two younger groups are quite similar, but higher education is less common in the 55 and over group.

Bar graphs can help make the association visible. We could make three side-by-side bar graphs, each resembling Figure 2.19, to present the three conditional distributions. Figure 2.21 shows an alternative form of bar graph. Each set of three bars compares the percents in the three age groups who have reached a specific educational level. We see at once that the "25 to 34" and "35 to 54" bars are similar for all four levels of education, and that the "55 and

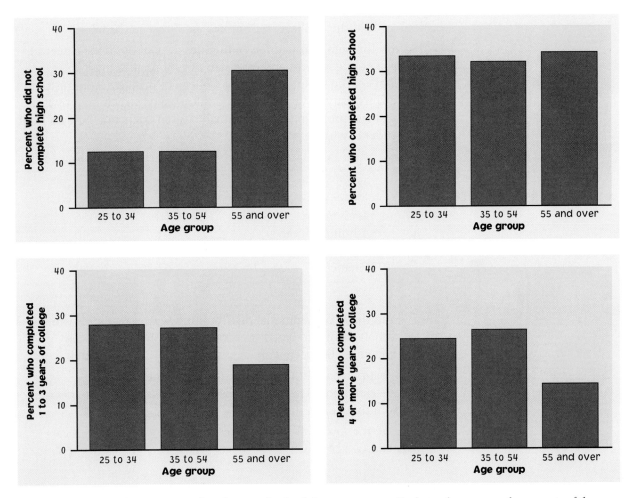

Figure 2.21 *Bar graphs to compare the education levels of three age groups. Each graph compares the percents of three groups who fall in one of the four education levels.*

over" bars show that many more people in this group did not finish high school and that many fewer have any college.

No single graph (such as a scatterplot) portrays the form of the relationship between categorical variables. No single numerical measure (such as the correlation) summarizes the strength of the association. Bar graphs are flexible enough to be helpful, but you must think about what comparisons you want to display. For numerical measures, we rely on well-chosen percents. You must decide which percents you need. Here is a hint: compare the conditional distributions of the response variable (education) for the separate values of the explanatory variable (age). That's what we did in Figure 2.20.

APPLY YOUR KNOWLEDGE

2.73 Find the conditional distribution of age among people with at least 4 years of college, starting from the counts in Table 2.9. (To do this, look only at the "College, 4 or more years" row in the table.)

2.74 **Majors for men and women in business.** A study of the career plans of young women and men sent questionnaires to all 722 members of the senior class in the College of Business Administration at the University of Illinois. One question asked which major within the business program the student had chosen. Here are the data from the students who responded:[33]

	Female	Male
Accounting	68	56
Administration	91	40
Economics	5	6
Finance	61	59

(a) Find the two conditional distributions of major, one for women and one for men. Based on your calculations, describe the differences between women and men with a graph and in words.

(b) What percent of the students did not respond to the questionnaire? The nonresponse weakens conclusions drawn from these data.

2.75 Here are the row and column totals for a two-way table with two rows and two columns:

$$
\begin{array}{cc|c}
a & b & 50 \\
c & d & 50 \\
\hline
60 & 40 & 100
\end{array}
$$

Find *two different* sets of counts a, b, c, and d for the body of the table that give these same totals. This shows that the relationship between two variables cannot be obtained from the two individual distributions of the variables.

Simpson's paradox

As is the case with quantitative variables, the effects of lurking variables can change or even reverse relationships between two categorical variables. Here is a hypothetical example that demonstrates the surprises that can await the unsuspecting user of data.

EXAMPLE 2.23 **Which hospital is safer?**

To help consumers make informed decisions about health care, the government releases data about patient outcomes in hospitals. You want to compare Hospital A and Hospital B, which serve your community. Here is a two-way table of data on the survival of patients after surgery in these two hospitals. All patients undergoing surgery in a recent time period are included. "Survived" means that the patient lived at least 6 weeks following surgery.

	Hospital A	Hospital B
Died	63	16
Survived	2037	784
Total	2100	800

Hospital A loses 3% (63/2100) of its surgery patients, and Hospital B loses only 2% (16/800). It seems that you should choose Hospital B if you need surgery.

Not all surgery cases are equally serious, however. Patients are classified as being in either "poor" or "good" condition before surgery. Here are the data broken down by patient condition. Check that the entries in the original two-way table are just the sums of the "poor" and "good" entries in this pair of tables.

Good Condition	Hospital A	Hospital B
Died	6	8
Survived	594	592
Total	600	600

Poor Condition	Hospital A	Hospital B
Died	57	8
Survived	1443	192
Total	1500	200

Hospital A beats Hospital B for patients in good condition: only 1% (6/600) died in Hospital A, compared with 1.3% (8/600) in Hospital B. Hospital A wins again for patients in poor condition, losing 3.8% (57/1500) to Hospital B's 4% (8/200). So Hospital A is safer for both patients in good condition and patients in poor condition. You should choose Hospital A if you are facing surgery.

The patient's condition is a lurking variable when we compare the death rates at the two hospitals. When we ignore the lurking variable, Hospital B seems safer, even though Hospital A does better for both classes of patients. How can A do better in both groups, yet do worse overall? Look at the data. Hospital A is a medical center that attracts seriously ill patients from a wide region. It had 1500 patients in poor condition. Hospital B had only 200 such cases. Because patients in poor condition are more likely to die, Hospital A has a higher death rate despite its superior performance for each class of patients. The original two-way table, which did not take account of the condition of the patients, was misleading. Example 2.23 illustrates *Simpson's paradox*.

SIMPSON'S PARADOX

An association or comparison that holds for all of several groups can reverse direction when the data are combined to form a single group. This reversal is called **Simpson's paradox**.

The lurking variables in Simpson's paradox are categorical. That is, they break the individuals into groups, as when surgery patients are classified as "good condition" or "poor condition." Simpson's paradox is just an extreme form of the fact that observed associations can be misleading when there are lurking variables.

APPLY YOUR KNOWLEDGE

2.76 Airline flight delays. Here are the numbers of flights on time and delayed for two airlines at five airports in one month. Overall on-time percentages for each airline are often reported in the news. The airport that flights serve is a lurking variable that can make such reports misleading.[34]

	Alaska Airlines		America West	
	On time	Delayed	On time	Delayed
Los Angeles	497	62	694	117
Phoenix	221	12	4840	415
San Diego	212	20	383	65
San Francisco	503	102	320	129
Seattle	1841	305	201	61

(a) What percent of all Alaska Airlines flights were delayed? What percent of all America West flights were delayed? These are the numbers usually reported.

(b) Now find the percent of delayed flights for Alaska Airlines at each of the five airports. Do the same for America West.

(c) America West does worse at *every one* of the five airports, yet does better overall. That sounds impossible. Explain carefully, referring to the data, how this can happen. (The weather in Phoenix and Seattle lies behind this example of Simpson's paradox.)

2.77 **Race and the death penalty.** Whether a convicted murderer gets the death penalty seems to be influenced by the race of the victim. Here are data on 326 cases in which the defendant was convicted of murder:[35]

White Defendant	White victim	Black victim
Death	19	0
Not	132	9

Black Defendant	White victim	Black victim
Death	11	6
Not	52	97

(a) Use these data to make a two-way table of defendant's race (white or black) versus death penalty (yes or no).

(b) Show that Simpson's paradox holds: a higher percent of white defendants are sentenced to death overall, but for both black and white victims a higher percent of black defendants are sentenced to death.

(c) Use the data to explain why the paradox holds in language that a judge could understand.

SECTION 2.5   Summary

A **two-way table** of counts organizes data about two categorical variables. Values of the **row variable** label the rows that run across the table, and values of the **column variable** label the columns that run down the table. Two-way tables are often used to summarize large amounts of data by grouping outcomes into categories.

The **row totals** and **column totals** in a two-way table give the **marginal distributions** of the two individual variables. It is clearer to present these distributions as percents of the table total. Marginal distributions tell us nothing about the relationship between the variables.

To find the **conditional distribution** of the row variable for one specific value of the column variable, look only at that one column in the table. Find each entry in the column as a percent of the column total.

There is a conditional distribution of the row variable for each column in the table. Comparing these conditional distributions is one way to describe the association between the row and the column variables. It is particularly useful when the column variable is the explanatory variable.

Bar graphs are a flexible means of presenting categorical data. There is no single best way to describe an association between two categorical variables.

A comparison between two variables that holds for each individual value of a third variable can be changed or even reversed when the data for all values of the third variable are combined. This is **Simpson's paradox**. Simpson's paradox is an example of the effect of lurking variables on an observed association.

SECTION 2.5 Exercises

College undergraduates. Exercises 2.78 to 2.82 are based on Table 2.10. This two-way table reports data on all undergraduate students enrolled in U.S. colleges and universities in the fall of 1995 whose age was known.[36]

2.78 (a) How many undergraduate students were enrolled in colleges and universities?

(b) What percent of all undergraduate students were 18 to 24 years old in the fall of the academic year?

(c) Find the percent of the undergraduates enrolled in each of the four types of program who were 18 to 24 years old. Make a bar graph to compare these percents.

(d) The 18 to 24 group is the traditional age group for college students. Briefly summarize what you have learned from the data about the extent to which this group predominates in different kinds of college programs.

2.79 (a) An association of two-year colleges asks: "What percent of students enrolled part-time at 2-year colleges are 25 to 39 years old?"

Table 2.10 Undergraduate college enrollment, fall 1995 (thousands of students)

Age	2-year full-time	2-year part-time	4-year full-time	4-year part-time
Under 18	41	125	75	45
18 to 24	1378	1198	4607	588
25 to 39	428	1427	1212	1321
40 and up	119	723	225	605
Total	1966	3472	6119	2559

(b) A bank that makes education loans to adults asks: "What percent of all 25- to 39-year-old students are enrolled part-time at 2-year colleges?"

2.80 (a) Find the marginal distribution of age among all undergraduate students, first in counts and then in percents. Make a bar graph of the distribution in percents.

(b) Find the conditional distribution of age (in percents) among students enrolled part-time in 2-year colleges and make a bar graph of this distribution.

(c) Briefly describe the most important differences between the two age distributions.

(d) The sum of the entries in the "2-year part-time" column is not the same as the total given for that column. Why is this?

2.81 Call students aged 40 and up "older students." Compare the presence of older students in the four types of program with numbers, a graph, and a brief summary of your findings.

2.82 With a little thought, you can extract from Table 2.10 information other than marginal and conditional distributions. The traditional college age group is ages 18 to 24 years.

(a) What percent of all undergraduates fall in this age group?

(b) What percent of students at 2-year colleges fall in this age group?

(c) What percent of part-time students fall in this group?

2.83 **Firearm deaths.** Firearms are second to motor vehicles as a cause of nondisease deaths in the United States. Here are counts from a study of all firearm-related deaths in Milwaukee, Wisconsin, between 1990 and 1994.[37] We want to compare the types of firearms used in homicides and in suicides. We suspect that long guns (shotguns and rifles) will more often be used in suicides because many people keep them at home for hunting. Make a careful comparison of homicides and suicides, with a bar graph. What do you find about long guns versus handguns?

	Handgun	Shotgun	Rifle	Unknown	Total
Homicides	468	28	15	13	524
Suicides	124	22	24	5	175

2.84 **Nonresponse in a survey.** A business school conducted a survey of companies in its state. They mailed a questionnaire to 200 small companies, 200 medium-sized companies, and 200 large companies.

The rate of nonresponse is important in deciding how reliable survey results are. Here are the data on response to this survey:

	Small	Medium	Large
Response	125	81	40
No response	75	119	160
Total	200	200	200

(a) What was the overall percent of nonresponse?

(b) Describe how nonresponse is related to the size of the business. (Use percents to make your statements precise.)

(c) Draw a bar graph to compare the nonresponse percents for the three size categories.

2.85 **Helping cocaine addicts.** Cocaine addiction is hard to break. Addicts need cocaine to feel any pleasure, so perhaps giving them an antidepressant drug will help. A 3-year study with 72 chronic cocaine users compared an antidepressant drug called desipramine with lithium and a placebo. (Lithium is a standard drug to treat cocaine addiction. A placebo is a dummy drug, used so that the effect of being in the study but not taking any drug can be seen.) One-third of the subjects, chosen at random, received each drug. Here are the results:[38]

	Desipramine	Lithium	Placebo
Relapse	10	18	20
No relapse	14	6	4
Total	24	24	24

(a) Compare the effectiveness of the three treatments in preventing relapse. Use percents and draw a bar graph.

(b) Do you think that this study gives good evidence that desipramine actually *causes* a reduction in relapses?

2.86 **Age and marital status of women.** The following two-way table describes the age and marital status of American women in 1995. The table entries are in thousands of women.

Age (years)	Marital Status				Total
	Never married	Married	Widowed	Divorced	
18 to 24	9,289	3,046	19	260	12,613
25 to 39	6,948	21,437	206	3,408	32,000
40 to 64	2,307	26,679	2,219	5,508	36,713
≥ 65	768	7,767	8,636	1,091	18,264
Total	19,312	58,931	11,080	10,266	99,588

(a) Find the sum of the entries in the "Married" column. Why does this sum differ from the "Total" entry for that column?

(b) Give the marginal distribution of marital status for all adult women (use percents). Draw a bar graph to display this distribution.

(c) Compare the conditional distributions of marital status for women aged 18 to 24 and women aged 40 to 64. Briefly describe the most important differences between the two groups of women, and back up your description with percents.

(d) You are planning a magazine aimed at women who have never been married. Find the conditional distribution of age among single women and display it in a bar graph. What age group or groups should your magazine aim to attract?

2.87 **Discrimination?** Wabash Tech has two professional schools, business and law. Here are two-way tables of applicants to both schools, categorized by gender and admission decision. (Although these data are made up, similar situations occur in reality.[39])

Business

	Admit	Deny
Male	480	120
Female	180	20

Law

	Admit	Deny
Male	10	90
Female	100	200

(a) Make a two-way table of gender by admission decision for the two professional schools together by summing entries in these tables.

(b) From the two-way table, calculate the percent of male applicants who are admitted and the percent of female applicants who are admitted. Wabash admits a higher percent of male applicants.

(c) Now compute separately the percents of male and female applicants admitted by the business school and by the law school. Each school admits a higher percent of female applicants.

(d) This is Simpson's paradox: both schools admit a higher percent of the women who apply, but overall Wabash admits a lower percent of female applicants than of male applicants. Explain carefully, as if speaking to a skeptical reporter, how it can happen that Wabash appears to favor males when each school individually favors females.

2.88 Obesity and health. Recent studies have shown that earlier reports underestimated the health risks associated with being overweight. The error was due to overlooking lurking variables. In particular, smoking tends both to reduce weight and to lead to earlier death. Illustrate Simpson's paradox by a simplified version of this situation. That is, make up tables of overweight (yes or no) by early death (yes or no) by smoker (yes or no) such that

- Overweight smokers and overweight nonsmokers both tend to die earlier than those not overweight.

- But when smokers and nonsmokers are combined into a two-way table of overweight by early death, persons who are not overweight tend to die earlier.

STATISTICS IN SUMMARY

Chapter 1 dealt with data analysis for a single variable. In this chapter, we have studied analysis of data for two or more variables. The proper analysis depends on whether the variables are categorical or quantitative and on whether one is an explanatory variable and the other a response variable.

When you have a categorical explanatory variable and a quantitative response variable, use the tools of Chapter 1 to compare the distributions of the response variable for the different categories of the explanatory variable. Make side-by-side boxplots, stemplots, or histograms and compare medians or means. If both variables are categorical, there is no satisfactory graph (though bar graphs can help). We describe the relationship numerically by comparing percents. The optional Section 2.5 explains how to do this.

Most of this chapter concentrates on relations between two quantitative variables. The Statistics in Summary figure on the next page organizes the main ideas in a way that stresses that our tactics are the same as when we faced single-variable data in Chapter 1. Here is a review list of the most important skills you should have gained from studying this chapter.

STATISTICS IN SUMMARY
Analyzing Data for Two Variables

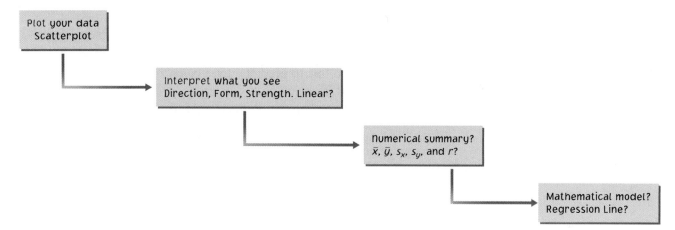

A. DATA

1. Recognize whether each variable is quantitative or categorical.
2. Identify the explanatory and response variables in situations where one variable explains or influences another.

B. SCATTERPLOTS

1. Make a scatterplot to display the relationship between two quantitative variables. Place the explanatory variable (if any) on the horizontal scale of the plot.
2. Add a categorical variable to a scatterplot by using a different plotting symbol or color.
3. Describe the form, direction, and strength of the overall pattern of a scatterplot. In particular, recognize positive or negative association and linear (straight-line) patterns. Recognize outliers in a scatterplot.

C. CORRELATION

1. Using a calculator, find the correlation r between two quantitative variables.
2. Know the basic properties of correlation: r measures the strength and direction of only linear relationships; $-1 \leq r \leq 1$ always; $r = \pm 1$ only for perfect straight-line relations; r moves away from 0 toward ± 1 as the linear relation gets stronger.

D. STRAIGHT LINES

1. Explain what the slope b and the intercept a mean in the equation $y = a + bx$ of a straight line.
2. Draw a graph of the straight line when you are given its equation.

E. REGRESSION

1. Using a calculator, find the least-squares regression line of a response variable y on an explanatory variable x from data.

2. Find the slope and intercept of the least-squares regression line from the means and standard deviations of x and y and their correlation.

3. Use the regression line to predict y for a given x. Recognize extrapolation and be aware of its dangers.

4. Use r^2 to describe how much of the variation in one variable can be accounted for by a straight-line relationship with another variable.

5. Recognize outliers and potentially influential observations from a scatterplot with the regression line drawn on it.

6. Calculate the residuals and plot them against the explanatory variable x or against other variables. Recognize unusual patterns.

F. LIMITATIONS OF CORRELATION AND REGRESSION

1. Understand that both r and the least-squares regression line can be strongly influenced by a few extreme observations.

2. Recognize that correlations based on averages of several observations are usually stronger than the correlation for individual observations.

3. Recognize possible lurking variables that may explain the observed association between two variables x and y.

4. Understand that even a strong correlation does not mean that there is a cause-and-effect relationship between x and y.

G. CATEGORICAL DATA (OPTIONAL)

1. From a two-way table of counts, find the marginal distributions of both variables by obtaining the row sums and column sums.

2. Express any distribution in percents by dividing the category counts by their total.

3. Describe the relationship between two categorical variables by computing and comparing percents. Often this involves comparing the conditional distributions of one variable for the different categories of the other variable.

4. Recognize Simpson's paradox and be able to explain it.

CHAPTER 2 Review Exercises

2.89   **Is wine good for your heart?** Table 2.3 (page 94) gives data on wine consumption and heart disease death rates in 19 countries. A scatterplot (Exercise 2.11) shows a moderately strong relationship.

(a) The correlation for these variables is $r = -0.843$. Why is the correlation negative? About what percent of the variation among countries in heart disease death rates is explained by the straight-line relationship with wine consumption?

(b) The least-squares regression line for predicting heart disease death rate from wine consumption, calculated from the data in Table 2.3, is

$$y = 260.56 - 22.969x$$

Use this equation to predict the heart disease death rate in another country where adults average 4 liters of alcohol from wine each year.

(c) The correlation in (a) and the slope of the least-squares line in (b) are both negative. Is it possible for these two quantities to have opposite signs? Explain your answer.

2.90   **Age and education in the states.**   Because older people as a group have less education than younger people, we might suspect a relationship between the percent of state residents aged 65 and over and the percent who are not high school graduates. Figure 2.22 is a scatterplot of these variables. The data appear in Tables 1.1 (page 9) and 1.6 (page 26).

(a) Say in words what a positive association between these variables would mean.

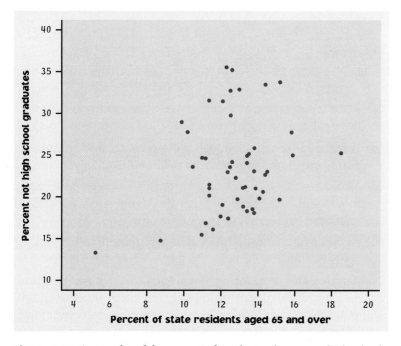

Figure 2.22 *Scatterplot of the percent of residents who are not high school graduates against the percent of residents aged 65 and over in the 50 states, for Exercise 2.90.*

(b) There are three outliers in the plot. Two of these are Alaska and Florida, which we identified as outliers in the histogram of Figure 1.2 (page 10). What is the third outlying state?

(c) If we ignore the outliers, does the relationship have a clear form and direction? Explain your answer.

(d) If we calculate the correlation with and without the three outliers, we get $r = 0.054$ and $r = 0.259$. Which of these is the correlation without the outliers? Explain your answer.

2.91 **Food poisoning.** Here are data on 18 people who fell ill from an incident of food poisoning.[40] The data give each person's age in years, the incubation period (the time in hours between eating the infected food and the first signs of illness), and whether the victim survived (S) or died (D).

Person	1	2	3	4	5	6	7	8	9
Age	29	39	44	37	42	17	38	43	51
Incubation	13	46	43	34	20	20	18	72	19
Outcome	D	S	S	D	D	S	D	S	D

Person	10	11	12	13	14	15	16	17	18
Age	30	32	59	33	31	32	32	36	50
Incubation	36	48	44	21	32	86	48	28	16
Outcome	D	D	S	D	D	S	D	S	D

(a) Make a scatterplot of incubation period against age, using different symbols for people who died and those who survived.

(b) Is there an overall relationship between age and incubation period? If so, describe it.

(c) More important, is there a relationship between either age or incubation period and whether the victim survived? Describe any relations that seem important here.

(d) Are there any unusual cases that may require individual investigation?

2.92 **Nematodes and tomatoes.** Nematodes are microscopic worms. Here are data from an experiment to study the effect of nematodes in the soil on plant growth. The experimenter prepared 16 planting pots and introduced different numbers of nematodes. Then he placed a tomato seedling in each pot and measured its growth (in centimeters) after 16 days.[41]

Nematodes	Seedling growth (cm)			
0	10.8	9.1	13.5	9.2
1,000	11.1	11.1	8.2	11.3
5,000	5.4	4.6	7.4	5.0
10,000	5.8	5.3	3.2	7.5

Analyze these data and give your conclusions about the effects of nematodes on plant growth.

2.93 **A hot stock?** It is usual in finance to describe the returns from investing in a single stock by regressing the stock's returns on the returns from the stock market as a whole. This helps us see how closely the stock follows the market. We analyzed the monthly percent total return y on Philip Morris common stock and the monthly return x on the Standard & Poor's 500-stock index, which represents the market, for the period between July 1990 and May 1997. Here are the results:

$$\bar{x} = 1.304 \qquad s_x = 3.392 \qquad r = 0.5251$$
$$\bar{y} = 1.878 \qquad s_y = 7.554$$

A scatterplot shows no very influential observations.

(a) Find the equation of the least-squares line from this information. What percent of the variation in Philip Morris stock is explained by the linear relationship with the market as a whole?

(b) Explain carefully what the slope of the line tells us about how Philip Morris stock responds to changes in the market. This slope is called "beta" in investment theory.

(c) Returns on most individual stocks have a positive correlation with returns on the entire market. That is, when the market goes up, an individual stock tends to also go up. Explain why an investor should prefer stocks with beta > 1 when the market is rising and stocks with beta < 1 when the market is falling.

2.94 **The influenza epidemic of 1918 (EESEE).** Exercise 1.22 (page 25) reported data on the great influenza outbreak of 1918. Time plots seem to show that deaths per week follow cases per week with about a one-week lag. We will examine this finding in more detail.[42]

(a) Make three scatterplots of deaths (the response variable) against each of new cases the same week, new cases one week earlier, and new cases two weeks earlier. Describe and compare the patterns you see.

(b) Find the correlations that go with your three plots.

(c) What do you conclude? Do the cases data predict deaths best with no lag, a one-week lag, or a two-week lag?

(CORBIS-BETTMANN.)

(d) Find the least-squares line for predicting weekly deaths for the choice of explanatory variable that gives the best predictions.

2.95 Women scientists. A study by the National Science Foundation[43] found that the median salary of newly graduated female engineers and scientists was only 73% of the median salary for males. When the new graduates were broken down by field, however, the picture changed. Women's median salaries as a percent of the male median in the 16 fields studied were

94%	96%	98%	95%	85%	85%	84%	100%
103%	100%	107%	93%	104%	93%	106%	100%

How can women do nearly as well as men in every field yet fall far behind men when we look at all young engineers and scientists?

2.96 Transforming data. Ecologists collect data to study nature's patterns. Table 2.11 gives data on the mean number of seeds produced in a year by several common tree species and the mean weight (in milligrams) of the seeds produced. (Some species appear twice because their seeds were counted in two locations.) We might expect that trees with heavy seeds produce fewer of them, but what is the form of the relationship?[44]

(a) Make a scatterplot showing how the weight of tree seeds helps explain how many seeds the tree produces. Describe the form, direction, and strength of the relationship.

(b) When dealing with sizes and counts, the logarithms of the original data are often the "natural" variables. Use your calculator or software to obtain the logarithms of both the seed weights and the seed counts in Table 2.11. Make a new scatterplot using these

Table 2.11 Count and weight of seeds produced by common tree species

Tree species	Seed count	Seed weight (mg)	Tree species	Seed count	Seed weight (mg)
Paper birch	27,239	0.6	American beech	463	247
Yellow birch	12,158	1.6	American beech	1,892	247
White spruce	7,202	2.0	Black oak	93	1,851
Engelmann spruce	3,671	3.3	Scarlet oak	525	1,930
Red spruce	5,051	3.4	Red oak	411	2,475
Tulip tree	13,509	9.1	Red oak	253	2,475
Ponderosa pine	2,667	37.7	Pignut hickory	40	3,423
White fir	5,196	40.0	White oak	184	3,669
Sugar maple	1,751	48.0	Chestnut oak	107	4,535
Sugar pine	1,159	216.0			

new variables. Now what are the form, direction, and strength of the relationship?

2.97 **Husbands and wives.** The mean height of American women in their early twenties is about 64.5 inches and the standard deviation is about 2.5 inches. The mean height of men the same age is about 68.5 inches, with standard deviation about 2.7 inches. If the correlation between the heights of husbands and wives is about $r = 0.5$, what is the slope of the regression line of the husband's height on the wife's height in young couples? Draw a graph of this regression line. Predict the height of the husband of a woman who is 67 inches tall.

2.98 **A computer game.** A multimedia statistics learning system includes a test of skill in using the computer's mouse. The software displays a circle at a random location on the computer screen. The subject tries to click in the circle with the mouse as quickly as possible. A new circle appears as soon as the subject clicks the old one. Table 2.12 gives data for one subject's trials, 20 with each hand. Distance is the distance from the cursor location to the center of the new circle, in units whose actual size depends on the size of the screen. Time is the time required to click in the new circle, in milliseconds.[45]

(a) We suspect that time depends on distance. Make a scatterplot of time against distance, using separate symbols for each hand.

Table 2.12 Reaction times in a computer game

Time	Distance	Hand	Time	Distance	Hand
115	190.70	right	240	190.70	left
96	138.52	right	190	138.52	left
110	165.08	right	170	165.08	left
100	126.19	right	125	126.19	left
111	163.19	right	315	163.19	left
101	305.66	right	240	305.66	left
111	176.15	right	141	176.15	left
106	162.78	right	210	162.78	left
96	147.87	right	200	147.87	left
96	271.46	right	401	271.46	left
95	40.25	right	320	40.25	left
96	24.76	right	113	24.76	left
96	104.80	right	176	104.80	left
106	136.80	right	211	136.80	left
100	308.60	right	238	308.60	left
113	279.80	right	316	279.80	left
123	125.51	right	176	125.51	left
111	329.80	right	173	329.80	left
95	51.66	right	210	51.66	left
108	201.95	right	170	201.95	left

(b) Describe the pattern. How can you tell that the subject is right-handed?

(c) Find the regression line of time on distance separately for each hand. Draw these lines on your plot. Which regression does a better job of predicting time from distance? Give numerical measures that describe the success of the two regressions.

(d) It is possible that the subject got better in later trials due to learning. It is also possible that he got worse due to fatigue. Plot the residuals from each regression against the time order of the trials (down the columns in Table 2.12). Is either of these systematic effects of time visible in the data?

2.99 Table 1.6 gives data about education in the states. Use computer software to examine the relationship between the median SAT verbal and mathematics scores as follows.

(a) You want to predict a state's SAT math score from its verbal score. Find the least-squares regression line for this purpose. You learn that a state's median verbal score the following year was 455. Use your regression line to predict its median math score.

(b) Plot the residuals from your regression against the SAT verbal score. (The software should do this for you.) One state is an outlier. What state is this? Does this state have a median math score higher or lower than would be predicted on the basis of its median verbal score?

The following exercises concern material in the optional Section 2.5.

2.100 Aspirin and heart attacks. Does taking aspirin regularly help prevent heart attacks? The Physicians' Health Study tried to find out. The subjects were 22,071 healthy male doctors at least 40 years old. Half the subjects, chosen at random, took aspirin every other day. The other half took a placebo, a dummy pill that looked and tasted like aspirin. Here are the results.[46] (The row for "None of these" is left out of the two-way table.)

	Aspirin group	Placebo group
Fatal heart attacks	10	26
Other heart attacks	129	213
Strokes	119	98
Total	11,037	11,034

What do the data show about the association between taking aspirin and heart attacks and stroke? Use percents to make your statements precise. Do you think the study provides evidence that aspirin actually reduces heart attacks (cause and effect)?

2.101 Suicides. Here is a two-way table of suicides committed in 1993, categorized by the gender of the victim and the method used. ("Hanging" also includes suffocation.) Write a brief account of differences in suicide between men and women. Use calculations and a graph to justify your statements.

	Male	Female
Firearms	16,381	2,559
Poison	3,569	2,110
Hanging	3,824	803
Other	1,641	623

2.102 Smoking and staying alive. In the mid-1970s, a medical study contacted randomly chosen people in a district in England. Here are data on the 1314 women contacted who were either current smokers or who had never smoked. The table classifies these women by their smoking status and age at the time of the survey and whether they were still alive 20 years later.[47]

Age 18 to 44	Smoker	Not
Dead	19	13
Alive	269	327

Age 45 to 64	Smoker	Not
Dead	78	52
Alive	167	147

Age 65+	Smoker	Not
Dead	42	165
Alive	7	28

(a) Make from these data a two-way table of smoking (yes or no) by dead or alive. What percent of the smokers stayed alive for 20 years? What percent of the nonsmokers survived? It seems surprising that a higher percent of smokers stayed alive.

(b) The age of the women at the time of the study is a lurking variable. Show that within each of the three age groups in the data, a higher percent of nonsmokers remained alive 20 years later. This is another example of Simpson's paradox.

(c) The study authors give this explanation: "Few of the older women (over 65 at the original survey) were smokers, but many of them had died by the time of follow-up." Compare the percent of smokers in the three age groups to verify the explanation.

Ronald A. Fisher

The Father of Statistics

The ideas and methods that we study as "statistics" were invented in the nineteenth and twentieth centuries by people working on problems that required analysis of data. Astronomy, biology, social science, and even surveying can claim a role in the birth of statistics. But if anyone can claim to be "the father of statistics," that honor belongs to Sir Ronald A. Fisher (1890–1962).

Like other statistical pioneers, Fisher was driven by the demands of practical problems.

Fisher's writings helped organize statistics as a distinct field of study whose methods apply to practical problems across many disciplines. He systematized the mathematical theory of statistics and invented many new techniques. The randomized comparative experiment is perhaps Fisher's greatest contribution.

Like other statistical pioneers, Fisher was driven by the demands of practical problems. Beginning in 1919, he worked on agricultural field experiments at Rothamsted in England. How should we arrange the planting of different crop varieties or the application of different fertilizers to get a fair comparison among them? Because fertility and other variables change as we move across a field, experimenters used elaborate checkerboard planting arrangements to obtain fair comparisons. Fisher had a better idea: "arrange the plots deliberately at random."

This chapter explores statistical designs for producing data to answer specific questions like "Which crop variety has the highest mean yield?" Fisher's innovation, the deliberate use of chance in producing data, is the central theme of the chapter and one of the most important ideas in statistics.

Producing Data

(LARRY LEFEVER/GRANT HEILMAN.)

Comparing crop varieties is a job for statistically-designed experiments.

Introduction

Exploratory data analysis seeks to discover and describe what data say by using graphs and numerical summaries. The conclusions we draw from data analysis apply to the specific data that we examine. Often, however, we want to answer questions about some large group of individuals. To get sound answers, we must produce data in a way that is designed to answer our questions.

sample

Suppose our question is "What percent of American adults agree that the United Nations should continue to have its headquarters in the United States?" To answer the question, we interview American adults. We can't afford to ask all adults, so we put the question to a **sample** chosen to represent the entire adult population. How shall we choose a sample that truly represents the opinions of the entire population? Statistical designs for choosing samples are the topic of Section 3.1.

Our goal in choosing a sample is a picture of the population, disturbed as little as possible by the act of gathering information. Sample surveys are one kind of *observational study*. In other settings, we gather data from an *experiment*. In doing an experiment, we don't just observe individuals or ask them questions. We actively impose some treatment in order to observe the response. Experiments can answer questions such as "Does aspirin reduce the chance of a heart attack?" and "Does a majority of college students prefer Pepsi to Coke when they taste both without knowing which they are drinking?" Experiments, like samples, provide useful data only when properly designed. We will discuss statistical design of experiments in Section 3.2. The distinction between experiments and observational studies is one of the most important ideas in statistics.

OBSERVATION VERSUS EXPERIMENT

An **observational study** observes individuals and measures variables of interest but does not attempt to influence the responses.

An **experiment**, on the other hand, deliberately imposes some treatment on individuals in order to observe their responses.

Observational studies are essential sources of data about topics from the opinions of voters to the behavior of animals in the wild. But an observational study, even one based on a statistical sample, is a poor way to gauge the effect of an intervention. To see the response to a change, we must actually impose the change. When our goal is to understand cause and effect, experiments are the only source of fully convincing data.

EXAMPLE 3.1 | **Helping welfare mothers find jobs**

Most adult recipients of welfare are mothers of young children. Observational studies of welfare mothers show that many are able to increase their earnings and leave the welfare system. Some take advantage of voluntary job-training programs to improve their skills. Should participation in job-training and job-search programs be required of all able-bodied welfare mothers? Observational studies cannot tell us what the effects of such a policy would be. Even if the mothers studied are a properly chosen sample of all welfare recipients, those who seek out training and find jobs may differ in many ways from those who do not. They are observed to have more education, for example, but they may also differ in values and motivation, things that cannot be observed.

To see if a required jobs program will help mothers escape welfare, such a program must actually be tried. Choose two similar groups of mothers when they apply for welfare. Require one group to participate in a job-training program, but do not offer the program to the other group. This is an experiment. Comparing the income and work record of the two groups after several years will show whether requiring training has the desired effect.

When we simply observe welfare mothers, the effect of job-training programs on success in finding work is *confounded* with (mixed up with) the characteristics of mothers who seek out training on their own.

CONFOUNDING

Two variables (explanatory variables or lurking variables) are **confounded** when their effects on a response variable cannot be distinguished from each other.

Observational studies of the effect of one variable on another often fail because the explanatory variable is confounded with lurking variables. We will see that well-designed experiments take steps to defeat confounding.

APPLY YOUR KNOWLEDGE

3.1 **The political gender gap.** There may be a "gender gap" in political party preference in the United States, with women more likely than men to prefer Democratic candidates. A political scientist interviews a large sample of registered voters, both men and women. She asks each voter whether they voted for the Democratic or the Republican candidate in the last congressional election. Is this study an experiment? Why or why not? What are the explanatory and response variables?

3.2 **Teaching reading.** An educator wants to compare the effectiveness of computer software that teaches reading with that of a standard reading curriculum. He tests the reading ability of each student in a class of fourth graders, then divides them into two groups. One group uses the computer regularly, while the other studies a standard curriculum. At the end of the year, he retests all the students and

Is it the government's business?

Statistical sampling by governments became common in the 1930s, in response to the Great Depression. Data on unemployment, consumption, and incomes did help manage the economy. But government data collection also raised concerns that are still with us. The wife of a cabinet member complained that the Consumer Purchases Study was asking about delicate products such as girdles and corsets. The U.S. Chamber of Commerce said a study of incomes that showed a highly skewed distribution was socialist propaganda.

compares the increase in reading ability in the two groups. Is this an experiment? Why or why not? What are the explanatory and response variables?

3.3 **The effects of propaganda.** In 1940, a psychologist conducted an experiment to study the effect of propaganda on attitude toward a foreign government. He administered a test of attitude toward the German government to a group of American students. After the students read German propaganda for several months, he tested them again to see if their attitudes had changed.

Unfortunately, Germany attacked and conquered France while the experiment was in progress. Explain clearly why confounding makes it impossible to determine the effect of reading the propaganda.

3.1 Designing Samples

An opinion poll wants to know what fraction of the public approves of the president's performance in office. A quality engineer must estimate what fraction of the bearings rolling off an assembly line are defective. Government economists inquire about household income. In all these situations, we want to gather information about a large group of people or things. Time, cost, and inconvenience usually forbid inspecting every bearing or contacting every household. In such cases, we gather information about only part of the group in order to draw conclusions about the whole.

> **POPULATION, SAMPLE**
>
> The entire group of individuals that we want information about is called the **population**.
>
> A **sample** is a part of the population that we actually examine in order to gather information.

Notice that the population is defined in terms of our desire for knowledge. If we wish to draw conclusions about all U.S. college students, that group is our population even if only local students are available for questioning. The sample is the part from which we draw conclusions about the whole. The **design** of a sample refers to the method used to choose the sample from the population. Poor sample designs can produce misleading conclusions.

sample design

EXAMPLE 3.2   Call-in opinion polls

Television news programs like to conduct call-in polls of public opinion. The program announces a question and asks viewers to call one telephone number to respond "Yes" and another for "No." Telephone companies charge for these calls. The ABC network program *Nightline* once asked whether the United Nations should continue to have its headquarters in the United States. More than 186,000 callers responded, and 67% said "No."

People who spend the time and money to respond to call-in polls are not representative of the entire adult population. In fact, they tend to be the same people who call radio talk shows. People who feel strongly, especially those with strong negative opinions, are more likely to call. It is not surprising that a properly designed sample showed that 72% of adults want the UN to stay.[1]

Call-in opinion polls are an example of *voluntary response sampling*. A voluntary response sample can easily produce 67% "No" when the truth about the population is close to 72% "Yes."

VOLUNTARY RESPONSE SAMPLE

A **voluntary response sample** consists of people who choose themselves by responding to a general appeal. Voluntary response samples are biased because people with strong opinions, especially negative opinions, are most likely to respond.

Voluntary response is one common type of bad sample design. Another is **convenience sampling**, which chooses the individuals easiest to reach. Here is an example of convenience sampling.

convenience sampling

EXAMPLE 3.3 **Interviewing at the mall**

Manufacturers and advertising agencies often use interviews at shopping malls to gather information about the habits of consumers and the effectiveness of ads. A sample of mall shoppers is fast and cheap. "Mall interviewing is being propelled primarily as a budget issue," one expert told the *New York Times*. But people contacted at shopping malls are not representative of the entire U.S. population. They are richer, for example, and more likely to be teenagers or retired. Moreover, mall interviewers tend to select neat, safe-looking individuals from the stream of customers. Decisions based on mall interviews may not reflect the preferences of all consumers.[2]

Both voluntary response samples and convenience samples choose a sample that is almost guaranteed not to represent the entire population. These sampling methods display *bias*, or systematic error, in favoring some parts of the population over others.

BIAS

The design of a study is **biased** if it systematically favors certain outcomes.

3.4 Sampling employed women. A sociologist wants to know the opinions of employed adult women about government funding for day care. She obtains a list of the 520 members of a local business and professional women's club and mails a questionnaire to 100 of these women selected at random. Only 48 questionnaires are returned. What is the population in this study? What is the sample from whom information is actually obtained? What is the rate (percent) of nonresponse?

3.5 What is the population? For each of the following sampling situations, identify the population as exactly as possible. That is, say what kind of individuals the population consists of and say exactly which individuals fall in the population. If the information given is not sufficient, complete the description of the population in a reasonable way.

(a) Each week, the Gallup Poll questions a sample of about 1500 adult U.S. residents to determine national opinion on a wide variety of issues.

(b) Every 10 years, the census tries to gather basic information from every household in the United States. In addition, a "long form" requesting much additional information is sent to a sample of about 17% of households.

(c) A machinery manufacturer purchases voltage regulators from a supplier. There are reports that variation in the output voltage of the regulators is affecting the performance of the finished products. To assess the quality of the supplier's production, the manufacturer sends a sample of 5 regulators from the last shipment to a laboratory for study.

3.6 Letters to Congress. You are on the staff of a member of Congress who is considering a bill that would provide government-sponsored insurance for nursing home care. You report that 1128 letters have been received on the issue, of which 871 oppose the legislation. "I'm surprised that most of my constituents oppose the bill. I thought it would be quite popular," says the congresswoman. Are you convinced that a majority of the voters oppose the bill? How would you explain the statistical issue to the congresswoman?

(AP/WIDE WORLD PHOTOS.)

Simple random samples

In a voluntary response sample, people choose whether to respond. In a convenience sample, the interviewer makes the choice. In both cases, personal choice produces bias. The statistician's remedy is to allow impersonal chance to choose the sample. A sample chosen by chance allows neither favoritism by the sampler nor self-selection by respondents. Choosing a sample by chance attacks bias by giving all individuals an equal chance to be chosen. Rich and poor, young and old, black and white, all have the same chance to be in the sample.

The simplest way to use chance to select a sample is to place names in a hat (the population) and draw out a handful (the sample). This is the idea of *simple random sampling*.

> ### SIMPLE RANDOM SAMPLE
>
> A **simple random sample (SRS)** of size *n* consists of *n* individuals from the population chosen in such a way that every set of *n* individuals has an equal chance to be the sample actually selected.

An SRS not only gives each individual an equal chance to be chosen (thus avoiding bias in the choice) but also gives every possible sample an equal chance to be chosen. There are other random sampling designs that give each individual, but not each sample, an equal chance. Exercise 3.27 describes one such design, called systematic random sampling.

The idea of an SRS is to choose our sample by drawing names from a hat. In practice, computer software can choose an SRS almost instantly from a list of the individuals in the population. If you don't use software, you can randomize by using a *table of random digits*.

> ### RANDOM DIGITS
>
> A **table of random digits** is a long string of the digits 0, 1, 2, 3, 4, 5, 6, 7, 8, 9 with these two properties:
>
> 1. Each entry in the table is equally likely to be any of the 10 digits 0 through 9.
> 2. The entries are independent of each other. That is, knowledge of one part of the table gives no information about any other part.

Table B at the back of the book is a table of random digits. You can think of Table B as the result of asking an assistant (or a computer) to mix the digits 0 to 9 in a hat, draw one, then replace the digit drawn, mix again, draw a second digit, and so on. The assistant's mixing and drawing save us the work of mixing and drawing when we need to randomize. Table B begins with the digits 19223950340575628713. To make the table easier to read, the digits appear in groups of five and in numbered rows. The groups and rows have no meaning—the table is just a long list of randomly chosen digits. Because the digits in Table B are random:

- Each entry is equally likely to be any of the 10 possibilities 0, 1, ..., 9.
- Each pair of entries is equally likely to be any of the 100 possible pairs 00, 01, ..., 99.

- Each triple of entries is equally likely to be any of the 1000 possibilities 000, 001, …, 999; and so on.

These "equally likely" facts make it easy to use Table B to choose an SRS. Here is an example that shows how.

Are these random digits really random?

Not a chance. The random digits in Table B were produced by a computer program. Computer programs do exactly what you tell them to do. Give the program the same input and it will produce exactly the same "random" digits. Of course, clever people have devised computer programs that produce output that *looks* like random digits. These are called "pseudo-random numbers," and that's what Table B contains. Pseudo-random numbers work fine for statistical randomizing, but they have hidden nonrandom patterns that can mess up more refined uses.

EXAMPLE 3.4   How to choose an SRS

Joan's small accounting firm serves 30 business clients. Joan wants to interview a sample of 5 clients in detail to find ways to improve client satisfaction. To avoid bias, she chooses an SRS of size 5.

Step 1: Label. Give each client a numerical label, using as few digits as possible. Two digits are needed to label 30 clients, so we use labels

$$01, 02, 03, \ldots, 29, 30$$

It is also correct to use labels 00 to 29 or even another choice of 30 two-digit labels. Here is the list of clients, with labels attached:

01	A-1 Plumbing	16	JL Records
02	Accent Printing	17	Johnson Commodities
03	Action Sport Shop	18	Keiser Construction
04	Anderson Construction	19	Liu's Chinese Restaurant
05	Bailey Trucking	20	MagicTan
06	Balloons Inc.	21	Peerless Machine
07	Bennett Hardware	22	Photo Arts
08	Best's Camera Shop	23	River City Books
09	Blue Print Specialties	24	Riverside Tavern
10	Central Tree Service	25	Rustic Boutique
11	Classic Flowers	26	Satellite Services
12	Computer Answers	27	Scotch Wash
13	Darlene's Dolls	28	Sewer's Center
14	Fleisch Realty	29	Tire Specialties
15	Hernandez Electronics	30	Von's Video Store

Step 2: Table. Enter Table B anywhere and read two-digit groups. Suppose we enter at line 130, which is

69051   64817   87174   09517   84534   06489   87201   97245

The first 10 two-digit groups in this line are

69   05   16   48   17   87   17   40   95   17

Each successive two-digit group is a label. The labels 00 and 31 to 99 are not used in this example, so we ignore them. The first 5 labels between 01 and 30 that we encounter in the table choose our sample. Of the first 10 labels in line 130, we ignore 5 because they are too high (over 30). The others are 05, 16, 17, 17, and 17. The clients labeled 05, 16, and 17 go into the sample. Ignore the second and third 17s because that client is already in the sample. Now run your finger across line 130 (and continue to line 131 if needed) until 5 clients are chosen.

The sample is the clients labeled 05, 16, 17, 20, 19. These are Bailey Trucking, JL Records, Johnson Commodities, MagicTan, and Liu's Chinese Restaurant.

CHOOSING AN SRS

Choose an SRS in two steps:

Step 1: Label. Assign a numerical label to every individual in the population.

Step 2: Table. Use Table B to select labels at random.

You can assign labels in any convenient manner, such as alphabetical order for names of people. Be certain that all labels have the same number of digits. Only then will all individuals have the same chance to be chosen. Use the shortest possible labels: one digit for a population of up to 10 members, two digits for 11 to 100 members, three digits for 101 to 1000 members, and so on. As standard practice, we recommend that you begin with label 1 (or 01 or 001, as needed). You can read digits from Table B in any order—across a row, down a column, and so on—because the table has no order. As standard practice, we recommend reading across rows.

APPLY YOUR KNOWLEDGE

3.7 A firm wants to understand the attitudes of its minority managers toward its system for assessing management performance. Below is a list of all the firm's managers who are members of minority groups. Use Table B at line 139 to choose 6 to be interviewed in detail about the performance appraisal system.

Agarwal	Dewald	Huang	Puri
Anderson	Fernandez	Kim	Richards
Baxter	Fleming	Liao	Rodriguez
Bowman	Gates	Mourning	Santiago
Brown	Goel	Naber	Shen
Castillo	Gomez	Peters	Vega
Cross	Hernandez	Pliego	Wang

3.8 Your class in ancient Ugaritic religion is poorly taught and wants to complain to the dean. The class decides to choose 4 of its members at random to carry the complaint. The class list appears below. Choose an SRS of 4 using the table of random digits, beginning at line 145.

Anderson	Gupta	Patnaik
Aspin	Gutierrez	Pirelli
Bennett	Harter	Rao
Bock	Henderson	Rider
Breiman	Hughes	Robertson
Castillo	Johnson	Rodriguez
Dixon	Kempthorne	Sosa
Edwards	Laskowsky	Tran
Gonzalez	Liang	Trevino
Green	Olds	Wang

3.9 You must choose an SRS of 10 of the 440 retail outlets in New York that sell your company's products. How would you label this population? Use Table B, starting at line 105, to choose your sample.

Other sampling designs

The general framework for statistical sampling is a *probability sample*.

> **PROBABILITY SAMPLE**
>
> A **probability sample** gives each member of the population a known chance (greater than zero) to be selected.

Some probability sampling designs (such as an SRS) give each member of the population an equal chance to be selected. This may not be true in more elaborate sampling designs. In every case, however, the use of chance to select the sample is the essential principle of statistical sampling.

Designs for sampling from large populations spread out over a wide area are usually more complex than an SRS. For example, it is common to sample important groups within the population separately, then combine these samples. This is the idea of a *stratified random sample*.

> **STRATIFIED RANDOM SAMPLE**
>
> To select a **stratified random sample**, first divide the population into groups of similar individuals, called **strata**. Then choose a separate SRS in each stratum and combine these SRSs to form the full sample.

Choose the strata based on facts known before the sample is taken. For example, a population of election districts might be divided into urban, suburban, and rural strata. A stratified design can produce more exact information than an SRS of the same size by taking advantage of the fact that individuals in the same stratum are similar to one another. If all individuals in each stratum are identical, for example, just one individual from each stratum is enough to completely describe the population.

EXAMPLE 3.5   **Who wrote that song?**

A radio station that broadcasts a piece of music owes a royalty to the composer. The organization of composers (called ASCAP) collects these royalties for all its members by charging stations a license fee for the right to play members' songs. ASCAP has four million songs in its catalog and collects $435 million in fees each year. How should ASCAP distribute this income among its members? By sampling: ASCAP tapes about 60,000 hours from the 53 million hours of local radio programs across the country each year.

Radio stations are stratified by type of community (metropolitan, rural), geographic location (New England, Pacific, etc.), and the size of the license fee paid to ASCAP, which reflects the size of the audience. In all, there are 432 strata. Tapes are made at random hours for randomly selected members of each stratum. The tapes are reviewed by experts who can recognize almost every piece of music ever written, and the composers are then paid according to their popularity.[3]

Multistage samples

Another common means of restricting random selection is to choose the sample in stages. This is usual practice for national samples of households or people. For example, government data on employment and unemployment are gathered by the **Current Population Survey**, which conducts interviews in about 50,000 households each month. It is not practical to maintain a list of all U.S. households from which to select an SRS. Moreover, the cost of sending interviewers to the widely scattered households in an SRS would be too high. The Current Population Survey therefore uses a **multistage sample design.** The final sample consists of clusters of nearby households. Most opinion polls and

Current Population Survey

multistage sample

Figure 3.1 *The monthly Current Population Survey is one of the most important sample surveys in the United States. You can find information and data from the survey on its Web page, http://www.bls.census.gov/cps.*

other national samples are also multistage, though interviewing in most national samples today is done by telephone rather than in person, eliminating the economic need for clustering.

The Current Population Survey sampling design is roughly as follows:

Stage 1. Divide the United States into 2007 geographical areas called Primary Sampling Units, or PSUs. Select a sample of 754 PSUs. This sample includes the 428 PSUs with the largest population and a random sample of the others.

Stage 2. Divide each PSU selected into smaller areas called "blocks." Stratify the blocks using ethnic and other information and take a stratified sample of the blocks in each PSU.

Stage 3. Sort the housing units in each block into clusters of four nearby units. Interview the households in a random sample of these clusters.

Analysis of data from sampling designs more complex than an SRS takes us beyond basic statistics. But the SRS is the building block of more elaborate designs, and analysis of other designs differs more in complexity of detail than in fundamental concepts.

APPLY YOUR KNOWLEDGE

3.10 A club has 30 student members and 10 faculty members. The students are

Abel	Fisher	Huber	Miranda	Reinmann
Carson	Ghosh	Jimenez	Moskowitz	Santos
Chen	Griswold	Jones	Neyman	Shaw
David	Hein	Kim	O'Brien	Thompson
Deming	Hernandez	Klotz	Pearl	Utts
Elashoff	Holland	Liu	Potter	Varga

The faculty members are

Andrews	Fernandez	Kim	Moore	West
Besicovitch	Gupta	Lightman	Vicario	Yang

The club can send 4 students and 2 faculty members to a convention. It decides to choose those who will go by random selection. Use Table B to choose a stratified random sample of 4 students and 2 faculty members.

3.11 Sampling by accountants. Accountants use stratified samples during audits to verify a company's records of such things as accounts receivable. The stratification is based on the dollar amount of the item and often includes 100% sampling of the largest items. One company reports 5000 accounts receivable. Of these, 100 are in amounts over $50,000; 500 are in amounts between $1000 and $50,000; and the remaining 4400 are in amounts under $1000. Using these groups as strata, you decide to verify all of the largest accounts and to sample 5% of the midsize accounts and 1% of the small accounts. How would you label the two strata from which you will sample? Use Table B, starting at line 115, to select *only the first 5* accounts from each of these strata.

3.12 What do schoolkids want (EESEE)? What are the most important goals of schoolchildren? Do girls and boys have different goals? Are goals different in urban, suburban, and rural areas? To find out, researchers wanted to ask children in the fourth, fifth, and sixth grades this question:

> *What would you most like to do at school?*
> *A. Make good grades.*
> *B. Be good at sports.*
> *C. Be popular.*

Because most children live in heavily populated urban and suburban areas, an SRS might contain few rural children. Moreover, it is too expensive to choose children at random from a large region—we must start by choosing schools rather than children. Describe a suitable sample design for this study and explain the reasoning behind your choice of design.[4]

Cautions about sample surveys

Random selection eliminates bias in the choice of a sample from a list of the population. When the population consists of human beings, however, accurate information from a sample requires much more than a good sampling design.[5] To begin, we need an accurate and complete list of the population. Because such a list is rarely available, most samples suffer from some degree of *undercoverage*. A sample survey of households, for example, will miss not only homeless people but prison inmates and students in dormitories. An opinion poll conducted by telephone will miss the 6% of American households without residential phones. The results of national sample surveys therefore have some bias if the people not covered—who most often are poor people—differ from the rest of the population.

A more serious source of bias in most sample surveys is *nonresponse*, which occurs when a selected individual cannot be contacted or refuses to cooperate. Nonresponse to sample surveys often reaches 30% or more, even with careful planning and several callbacks. Because nonresponse is higher in urban areas, most sample surveys substitute other people in the same area to avoid favoring rural areas in the final sample. If the people contacted differ from those who are rarely at home or who refuse to answer questions, some bias remains.

(RUBBER BALL PRODUCTIONS/PNI.)

Why the census matters

A *census* is an attempt to gather data from every member of a population. The Constitution of the United States requires a census of the American population every ten years. The first census was held in 1790. Although the census asks questions about many subjects, the constitutional purpose of counting the people is to divide the 435 seats in Congress among the states and to draw up congressional voting districts that are equal in population. Census counts also determine how some federal funds are divided among cities and states. Because political power and money depend on the census, its statistical details become political issues.

UNDERCOVERAGE AND NONRESPONSE

Undercoverage occurs when some groups in the population are left out of the process of choosing the sample.

Nonresponse occurs when an individual chosen for the sample can't be contacted or refuses to cooperate.

EXAMPLE 3.6   The census undercount

Even the U.S. census, backed by the resources of the federal government, suffers from undercoverage and nonresponse. The census begins by mailing forms to every household in the country. The Census Bureau's list of addresses is incomplete, resulting in undercoverage. Despite special efforts to count homeless people (who can't be reached at any address), homelessness causes more undercoverage.

In 1990, about 35% of the households that were mailed census forms did not mail them back. In New York City, 47% did not return the form. That's nonresponse. The Census Bureau sent interviewers to these households. In inner-city areas, the interviewers could not contact about one in five of the nonresponders, even after six tries.

The Census Bureau estimates that the 1990 census missed about 1.8% of the total population due to undercoverage and nonresponse. Because the undercount was greater in the poorer sections of large cities, the Census Bureau estimates that it failed to count 4.4% of blacks and 5.0% of Hispanics.[6]

For the 2000 census, the Bureau planned to replace follow-up of all nonresponders with more intense pursuit of a probability sample of nonresponding households plus a national sample of 750,000 households. The final counts would be based on comparing the national sample with the original responses. This idea was politically controversial. The Supreme Court ruled that sampling could be used for most purposes, but not for dividing seats in Congress among the states.

---

response bias   In addition, the behavior of the respondent or of the interviewer can cause **response bias** in sample results. Respondents may lie, especially if asked about illegal or unpopular behavior. The sample then underestimates the presence of such behavior in the population. An interviewer whose attitude suggests that some answers are more desirable than others will get these answers more often. The race or sex of the interviewer can influence responses to questions about race relations or attitudes toward feminism. Answers to questions that ask respondents to recall past events are often inaccurate because of faulty memory. For example, many people "telescope" events in the past, bringing them forward in memory to more recent time periods. "Have you visited a dentist in the last 6 months?" will often draw a "Yes" from someone who last visited a dentist 8 months ago.[7] Careful training of interviewers and careful supervision to avoid variation among the interviewers can greatly reduce response bias. Good interviewing technique is another aspect of a well-done sample survey.

wording effects   The **wording of questions** is the most important influence on the answers given to a sample survey. Confusing or leading questions can introduce strong bias, and even minor changes in wording can change a survey's outcome. Here are two examples.

EXAMPLE 3.7 **Should we ban disposable diapers?**

A survey paid for by makers of disposable diapers found that 84% of the sample opposed banning disposable diapers. Here is the actual question:

It is estimated that disposable diapers account for less than 2% of the trash in today's landfills. In contrast, beverage containers, third-class mail and yard wastes are esti-mated to account for about 21% of the trash in landfills. Given this, in your opinion, would it be fair to ban disposable diapers?[8]

This question gives information on only one side of an issue, then asks an opinion. That's a sure way to bias the responses. A different question that described how long disposable diapers take to decay and how many tons they contribute to landfills each year would draw a quite different response.

EXAMPLE 3.8 **Doubting the Holocaust**

An opinion poll conducted in 1992 for the American Jewish Committee asked: "Does it seem possible or does it seem impossible to you that the Nazi extermination of the Jews never happened?" When 22% of the sample said "possible," the news media wondered how so many Americans could be uncertain that the Holocaust happened. Then a second poll asked the question in different words: "Does it seem possible to you that the Nazi extermination of the Jews never happened, or do you feel certain that it happened?" Now only 1% of the sample said "possible." The complicated wording of the first question confused many respondents.[9]

Never trust the results of a sample survey until you have read the exact questions posed. The amount of nonresponse and the date of the survey are also important. Good statistical design is a part, but only a part, of a trustworthy survey.

APPLY YOUR KNOWLEDGE

3.13 **Random digit dialing.** The list of individuals from which a sample is actually selected is called the *sampling frame.* Ideally, the frame should list every individual in the population, but in practice this is often difficult. A frame that leaves out part of the population is a common source of undercoverage.

(a) Suppose that a sample of households in a community is selected at random from the telephone directory. What households are omitted from this frame? What types of people do you think are likely to live in these households? These people will probably be underrepresented in the sample.

(b) It is usual in telephone surveys to use random digit dialing equipment that selects the last four digits of a telephone number at random after being given the exchange (the first three digits). Which of the households you mentioned in your answer to (a) will be included in the sampling frame by random digit dialing?

3.14 **Ring-no-answer.** A common form of nonresponse in telephone surveys is "ring-no-answer." That is, a call is made to an active number but no one answers. The Italian National Statistical Institute looked at nonresponse to a government survey of households in Italy during the periods January 1 to Easter and July 1 to August 31. All calls were made between 7 and 10 p.m., but 21.4% gave "ring-no-answer" in one period versus 41.5% "ring-no-answer" in the other period.[10] Which period do you think had the higher rate of no answers? Why? Explain why a high rate of nonresponse makes sample results less reliable.

3.15 **Campaign contributions.** Here are two wordings for the same question. The first question was asked by presidential candidate Ross Perot, and the second by a Time/CNN poll, both in March 1993.[11]

A. Should laws be passed to eliminate all possibilities of special interests giving huge sums of money to candidates?

B. Should laws be passed to prohibit interest groups from contributing to campaigns, or do groups have a right to contribute to the candidates they support?

One of these questions drew 40% favoring banning contributions; the other drew 80% with this opinion. Which question produced the 40% and which got 80%? Explain why the results were so different.

Inference about the population

Despite the many practical difficulties in carrying out a sample survey, using chance to choose a sample does eliminate bias in the actual selection of the sample from the list of available individuals. But it is unlikely that results from a sample are exactly the same as for the entire population. Sample results, like the official unemployment rate obtained from the monthly Current Population Survey, are only estimates of the truth about the population. If we select two samples at random from the same population, we will draw different individuals. So the sample results will almost certainly differ somewhat. Two runs of the Current Population Survey would produce somewhat different unemployment rates. Properly designed samples avoid systematic bias, but their results are rarely exactly correct and they vary from sample to sample.

How accurate is a sample result like the monthly unemployment rate? We can't say for sure, because the result would be different if we took another sample. But the results of random sampling don't change haphazardly from sample to sample. Because we deliberately use chance, the results obey the laws of probability that govern chance behavior. We can say how large an error we are likely to make in drawing conclusions about the population from a sample. Results from a sample survey usually come with a margin of error that sets bounds on the size of the likely error. How to do this is part of the business of statistical inference. We will describe the reasoning in Chapter 6.

One point is worth making now: **larger random samples give more accurate results than smaller samples**. By taking a very large sample, you can be confident that the sample result is very close to the truth about the population. The Current Population Survey's sample of 50,000 households estimates the national unemployment rate very accurately. Of course, only probability samples carry this guarantee. *Nightline's* voluntary response sample is worthless even though 186,000 people called in. Using a probability sampling design and taking care to deal with practical difficulties reduce bias in a sample. The size of the sample then determines how close to the population truth the sample result is likely to fall.

APPLY YOUR KNOWLEDGE

3.16 **Ask more people.** Just before a presidential election, a national opinion polling firm increases the size of its weekly sample from the usual 1500 people to 4000 people. Why do you think the firm does this?

SECTION 3.1 Summary

We can produce data intended to answer specific questions by **observational studies** or **experiments**. **Sample surveys** that select a part of a population of interest to represent the whole are one type of observational study. **Experiments**, unlike observational studies, actively impose some treatment on the subjects of the experiment.

A sample survey selects a **sample** from the **population** of all individuals about which we desire information. We base conclusions about the population on data about the sample.

The **design** of a sample refers to the method used to select the sample from the population. **Probability sampling** designs use random selection to give each member of the population a known chance (greater than zero) to be selected for the sample.

The basic probability sample is a **simple random sample (SRS)**. An SRS gives every possible sample of a given size the same chance to be chosen.

Choose an SRS by labeling the members of the population and using a **table of random digits** to select the sample. Software can automate this process.

To choose a **stratified random sample**, divide the population into **strata**, groups of individuals that are similar in some way that is important to the response. Then choose a separate SRS from each stratum.

Failure to use probability sampling often results in **bias**, or systematic errors in the way the sample represents the population. **Voluntary response samples**, in which the respondents choose themselves, are particularly prone to large bias.

In human populations, even probability samples can suffer from bias due to **undercoverage** or **nonresponse**, from **response bias**, or from misleading results due to **poorly worded questions**. Sample surveys must deal expertly with these potential problems in addition to using a probability sampling design.

SECTION 3.1 Exercises

3.17   To study the effect of living in public housing on family stability in poverty-level households, researchers obtained a list of all applicants for public housing during the previous year. Some applicants had been accepted, while others had been turned down by the housing authority. Both groups were interviewed and compared. Was this an observational study or an experiment? Explain your answer. What are the explanatory and response variables?

3.18   Different types of writing can sometimes be distinguished by the lengths of the words used. A student interested in this fact wants to study the lengths of words used by Tom Clancy in his novels. She opens a Clancy novel at random and records the lengths of each of the first 250 words on the page. What is the population in this study? What is the sample? What is the variable measured?

3.19   **What is the population?** For each of the following sampling situations, identify the population as exactly as possible. That is, say what kind of individuals the population consists of and say exactly which individuals fall in the population. If the information given is not sufficient, complete the description of the population in a reasonable way.

(a) A business school researcher wants to know what factors affect the survival and success of small businesses. She selects a sample of 150 eating-and-drinking establishments from those listed in the telephone directory Yellow Pages for a large city.

(b) A member of Congress wants to know whether his constituents support proposed legislation on health care. His staff reports that 228 letters have been received on the subject, of which 193 oppose the legislation.

(c) An insurance company wants to monitor the quality of its procedures for handling loss claims from its auto insurance policyholders. Each month the company selects an SRS from all auto insurance claims filed that month to examine them for accuracy and promptness.

3.20   **Ann Landers takes a sample.** Advice columnist Ann Landers once asked her female readers whether they would be content with affectionate treatment by men with no sex ever. Over 90,000 women wrote in, with 72% answering "Yes." Many of the letters described unfeeling treatment at the hands of men. Explain why this sample is certainly biased. What is the likely direction of the bias? That is, is that 72% probably higher or lower than the truth about the population of all adult women?

3.21 **You call the shots.** A newspaper advertisement for *USA Today: The Television Show* once said:

> *Should handgun control be tougher? You call the shots in a special call-in poll tonight. If yes, call 1-900-720-6181. If no, call 1-900-720-6182. Charge is 50 cents for the first minute.*

Explain why this opinion poll is almost certainly biased.

3.22 **Rating the president.** A newspaper article about an opinion poll says that "43% of Americans approve of the president's overall job performance." Toward the end of the article, you read: "The poll is based on telephone interviews with 1210 adults from around the United States, excluding Alaska and Hawaii." What variable did this poll measure? What population do you think the newspaper wants information about? What was the sample? Are there any sources of bias in the sampling method used?

3.23 **Rating the police.** The Miami Police Department wants to know how black residents of Miami feel about police service. A sociologist prepares several questions about the police. A sample of 300 mailing addresses in predominantly black neighborhoods is chosen, and a uniformed black police officer goes to each address to ask the questions of an adult living there. What are the population and the sample? Why are the results likely to be biased?

3.24 A manufacturer of chemicals chooses 3 containers from each lot of 25 containers of a reagent to test for purity and potency. Below are the control numbers stamped on the bottles in the current lot. Use Table B at line 111 to choose an SRS of 3 of these bottles.

A1096	A1097	A1098	A1101	A1108
A1112	A1113	A1117	A2109	A2211
A2220	B0986	B1011	B1096	B1101
B1102	B1103	B1110	B1119	B1137
B1189	B1223	B1277	B1286	B1299

3.25 **Sampling from a census tract.** Figure 3.2 is a map of a census tract. Census tracts are small, homogeneous areas averaging 4000 in population. On the map, each block is marked with a Census Bureau identification number. An SRS of blocks from a census tract is often the next-to-last stage in a multistage sample. Use Table B beginning at line 125 to choose an SRS of 5 blocks from this census tract.

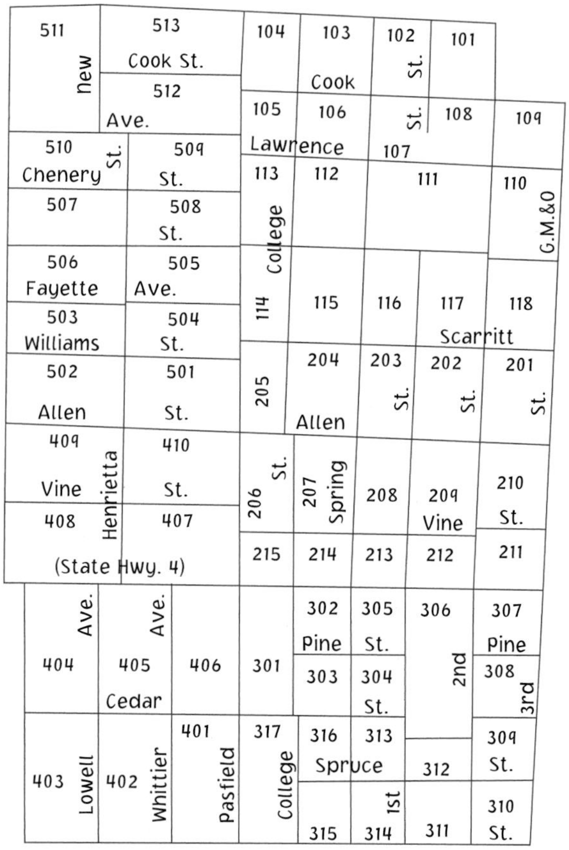

Figure 3.2 *Map of a census tract, for Exercise 3.25.*

3.26 **Random digits.** Which of the following statements are true of a table of random digits, and which are false? Briefly explain your answers.

(a) There are exactly four 0s in each row of 40 digits.

(b) Each pair of digits has chance 1/100 of being 00.

(c) The digits 0000 can never appear as a group, because this pattern is not random.

systematic random sample **3.27** **Systematic random samples.** The last stage of the Current Population Survey chooses addresses within small areas called blocks. The method used is **systematic random sampling**. An example will illustrate the idea of a systematic sample. Suppose that we must choose 4 addresses out of 100. Because 100/4 = 25, we can think of the list as four lists of 25 addresses. Choose 1 of the first 25 at random, using Table B. The sample will contain this address and the addresses 25, 50, and 75 places down the list from it. If 13 is chosen, for example, then the systematic random sample consists of the addresses numbered 13, 38, 63, and 88.

(a) Use Table B to choose a systematic random sample of 5 addresses from a list of 200. Enter the table at line 120.

(b) Like an SRS, a systematic sample gives all individuals the same chance to be chosen. Explain why this is true. Then explain carefully why a systematic sample is nonetheless *not* an SRS.

3.28 A university has 2000 male and 500 female faculty members. A stratified random sample of 50 female and 200 male faculty members gives each member of the faculty 1 chance in 10 to be chosen. This sample design gives every individual in the population the same chance to be chosen for the sample. Is it an SRS? Explain your answer.

(WILL & DENI MCINTYRE/PHOTO RESEARCHERS.)

3.29 A university has 2000 male and 500 female faculty members. The equal opportunity employment officer wants to poll the opinions of a random sample of faculty members. In order to give adequate attention to female faculty opinion, he decides to choose a stratified random sample of 200 males and 200 females. He has alphabetized lists of female and male faculty members. Explain how you would assign labels and use random digits to choose the desired sample. Enter Table B at line 122 and give the labels of the first 5 females and the first 5 males in the sample.

3.30 **Wording survey questions.** Comment on each of the following as a potential sample survey question. Is the question clear? Is it slanted toward a desired response?

(a) Which of the following best represents your opinion on gun control?

1. The government should confiscate our guns.

2. We have the right to keep and bear arms.

(b) A freeze in nuclear weapons should be favored because it would begin a much-needed process to stop everyone in the world from building nuclear weapons now and reduce the possibility of nuclear war in the future. Do you agree or disagree?

(c) In view of escalating environmental degradation and incipient resource depletion, would you favor economic incentives for recycling of resource-intensive consumer goods?

3.31 **Errors in a poll.** A *New York Times* opinion poll on women's issues contacted a sample of 1025 women and 472 men by randomly selecting telephone numbers. The *Times* publishes descriptions of its polling methods. Here is part of the description for this poll:

> In theory, in 19 cases out of 20 the results based on the entire sample will differ by no more than three percentage points in either direction from what would have been obtained by seeking out all adult Americans.
>
> The potential sampling error for smaller subgroups is larger. For example, for men it is plus or minus five percentage points.[12]

Explain why the margin of error is larger for conclusions about men alone than for conclusions about all adults.

3.2 Designing Experiments

A study is an experiment when we actually do something to people, animals, or objects in order to observe the response. Here is the basic vocabulary of experiments.

> ### EXPERIMENTAL UNITS, SUBJECTS, TREATMENT
>
> The individuals on which the experiment is done are the **experimental units**. When the units are human beings, they are called **subjects**. A specific experimental condition applied to the units is called a **treatment**.

factor

level

Because the purpose of an experiment is to reveal the response of one variable to changes in other variables, the distinction between explanatory and response variables is essential. The explanatory variables in an experiment are often called **factors**. Many experiments study the joint effects of several factors. In such an experiment, each treatment is formed by combining a specific value (often called a **level**) of each of the factors.

EXAMPLE 3.9   Absorption of a drug

Researchers studying the absorption of a drug into the bloodstream inject the drug (the treatment) into 25 people (the subjects). The response variable is the concentration of the drug in a subject's blood, measured 30 minutes after the injection. This experiment has a single factor with only one level. If three different doses of the drug are injected, there is still a single factor (the dosage of the drug), now with three levels. The three levels of the single factor are the treatments that the experiment compares.

EXAMPLE 3.10   Effects of TV advertising

What are the effects of repeated exposure to an advertising message? The answer may depend both on the length of the ad and on how often it is repeated. An experiment investigated this question using undergraduate students as subjects. All subjects viewed a 40-minute television program that included ads for a 35 mm camera. Some subjects saw a 30-second commercial; others, a 90-second version. The same commercial was repeated either 1, 3, or 5 times during the program. After viewing, all of the subjects answered questions about their recall of the ad, their attitude toward the camera, and their intention to purchase it. These are the response variables.[13]

This experiment has two factors: length of the commercial, with 2 levels, and repetitions, with 3 levels. The 6 combinations of one level of each factor form 6 treatments. Figure 3.3 shows the layout of the treatments.

		Factor B Repetitions	
Factor A Length	1 time	3 times	5 times
30 seconds	1	2	3
90 seconds	4	5	6

Figure 3.3 *The treatments in the experimental design of Example 3.10. Combinations of levels of the two factors form six treatments.*

Examples 3.9 and 3.10 illustrate the advantages of experiments over observational studies. Experimentation allows us to study the effects of the specific treatments we are interested in. Moreover, we can control the environment of the experimental units to hold constant factors that are of no interest to us, such as the specific product advertised in Example 3.10. The ideal case is a laboratory experiment in which we control all lurking variables and so see only the effect of the treatments on the response. Like most ideals, such control is not always realized in practice. Nonetheless, in principle, experiments can give good evidence for causation.

Another advantage of experiments is that we can study the combined effects of several factors simultaneously. The interaction of several factors can produce effects that could not be predicted from looking at the effect of each factor alone. Perhaps longer commercials increase interest in a product, and more commercials also increase interest, but if we both make a commercial longer and show it more often, viewers get annoyed and their interest in the product drops. The two-factor experiment in Example 3.10 will help us find out.

APPLY YOUR KNOWLEDGE

3.32 **Sickle cell disease.** Sickle cell disease is an inherited disorder of the red blood cells that in the United States affects mostly blacks. It can cause severe pain and many complications. Can the drug hydroxyurea reduce the severe pain caused by sickle cell disease? A study by the National Institutes of Health gave the drug to 150 sickle cell sufferers and a placebo (a dummy medication) to another 150. The researchers then counted the episodes of pain reported by each subject. What are the experimental units or subjects, the factors, the treatments, and the response variables?

3.33 **Sealing food packages.** A manufacturer of food products uses package liners that are sealed at the top by applying heated jaws after the package is filled. The customer peels the sealed pieces apart to open the package. What effect does the temperature of the jaws have on the force needed to peel the liner? To answer this question, engineers obtain 20 pairs of pieces of package liner. They seal five pairs at each of 250°F, 275°F, 300°F, and 325°F. Then they measure the force needed to peel each seal.

(a) What are the experimental units?

(b) There is one factor (explanatory variable). What is it, and what are its levels?

(c) What is the response variable?

3.34 **An industrial experiment.** A chemical engineer is designing the production process for a new product. The chemical reaction that produces the product may have higher or lower yield, depending on the temperature and the stirring rate in the vessel in which the reaction takes place. The engineer decides to investigate the effects of combinations of two temperatures (50°C and 60°C) and three stirring rates (60 rpm, 90 rpm, and 120 rpm) on the yield of the process. She will process two batches of the product at each combination of temperature and stirring rate.

(a) What are the experimental units and the response variable in this experiment?

(b) How many factors are there? How many treatments? Use a diagram like that in Figure 3.3 to lay out the treatments.

(c) How many experimental units are required for the experiment?

Comparative experiments

Experiments in the science laboratory often have a simple design: impose the treatment and see what happens. We can outline that design like this:

$$\text{Units} \longrightarrow \text{Treatment} \longrightarrow \text{Response}$$

In the laboratory, we try to avoid confounding by rigorously controlling the environment of the experiment so that nothing except the experimental treatment influences the response. Once we get out of the laboratory, however, there are almost always lurking variables waiting to confound us. When our experimental units are people or animals rather than electrons or chemical compounds, confounding can happen even in the controlled environment of a laboratory or medical clinic. Here is an example.

EXAMPLE 3.11 Gastric freezing to treat ulcers

"Gastric freezing" is a clever treatment for ulcers. The patient swallows a deflated balloon with tubes attached, then a refrigerated liquid is pumped through the balloon for an hour. The idea is that cooling the stomach will reduce its production of acid and so relieve ulcers. An experiment reported in the *Journal of the American Medical Association* showed that gastric freezing did reduce acid production and relieve ulcer pain. The treatment was safe and easy and was widely used for several years. The design of the experiment was

Subjects ⟶ Gastric freezing ⟶ Observe pain relief

The gastric freezing experiment was poorly designed. The patients' response may have been due to the **placebo effect**. A placebo is a dummy treatment. Many patients respond favorably to *any* treatment, even a placebo, presumably because of trust in the doctor and expectations of a cure. This response to a dummy treatment is the placebo effect.

placebo effect

A later experiment divided ulcer patients into two groups. One group was treated by gastric freezing as before. The other group received a placebo treatment in which the liquid in the balloon was at body temperature rather than freezing. The results: 34% of the 82 patients in the treatment group improved, but so did 38% of the 78 patients in the placebo group. This and other properly designed experiments showed that gastric freezing was no better than a placebo, and its use was abandoned.[14]

The first gastric freezing experiment was *biased*. It systematically favored gastric freezing because the placebo effect was confounded with the effect of the treatment. Fortunately, the remedy is simple. Experiments should *compare* treatments rather than attempt to assess a single treatment in isolation. When we compare the two groups of patients in the second gastric freezing experiment, the placebo effect and other lurking variables operate on both groups. The only difference between the groups is the actual effect of gastric freezing. The group of patients who received a sham treatment is called a **control group**, because it enables us to control the effects of lurking variables on the outcome.

control group

Randomized comparative experiments

The design of an experiment first describes the response variables, the factors (explanatory variables), and the layout of the treatments, with *comparison* as the leading principle. The second aspect of design is the rule used to assign the experimental units to the treatments. Comparison of the effects of several treatments is valid only when all treatments are applied to similar groups of experimental units. If one corn variety is planted on more fertile ground, or if one cancer drug is given to less seriously ill patients, comparisons among treatments are biased. How can we assign experimental units to treatments in a way that is fair to all the treatments?

Our answer is the same as in sampling: let impersonal chance make the assignment. The use of chance to divide experimental units into groups is called **randomization**. Groups formed by randomization don't depend on any characteristic of the experimental units or on the judgment of the experimenter. An experiment that uses both comparison and randomization is a **randomized comparative experiment**. Here is an example.

randomization

randomized comparative experiment

Figure 3.4 *Outline of a randomized comparative experiment, for Example 3.12.*

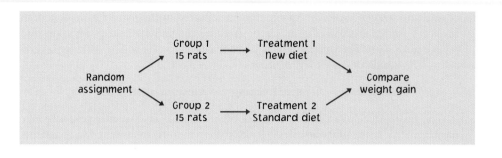

EXAMPLE 3.12 Testing a breakfast food

A food company assesses the nutritional quality of a new "instant breakfast" product by feeding it to newly weaned male white rats. The response variable is a rat's weight gain over a 28-day period. A control group of rats eats a standard diet but otherwise receives exactly the same treatment as the experimental group.

This experiment has one factor (the diet) with two levels. The researchers use 30 rats for the experiment and so must divide them into two groups of 15. To do this in an unbiased fashion, put the cage numbers of the 30 rats in a hat, mix them up, and draw 15. These rats form the experimental group and the remaining 15 make up the control group. That is, *each group is an SRS of the available rats*. Figure 3.4 outlines the design of this experiment.

We can use software or the table of random digits to randomize. Label the rats 01 to 30. Enter Table B at (say) line 130. Run your finger along this line (and continue to lines 131 and 132 as needed) until 15 rats are chosen. They are the rats labeled

05, 16, 17, 20, 19, 04, 25, 29, 18, 07, 13, 02, 23, 27, 21

These rats form the experimental group; the remaining 15 are the control group.

Completely randomized designs

The design in Figure 3.4 combines comparison and randomization to arrive at the simplest statistical design for an experiment. This "flowchart" outline presents all the essentials: randomization, the sizes of the groups and which treatment they receive, and the response variable. There are, as we will see later, statistical reasons for generally using treatment groups about equal in size. We call designs like that in Figure 3.4 *completely randomized*.

COMPLETELY RANDOMIZED DESIGN

In a **completely randomized** experimental design, all the experimental units are allocated at random among all the treatments.

Completely randomized designs can compare any number of treatments. Here is an example that compares three treatments.

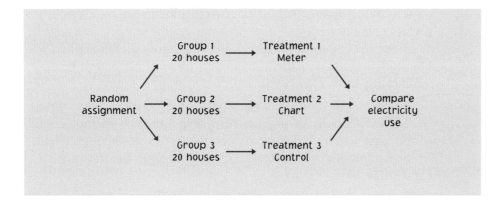

Figure 3.5 *Outline of a completely randomized design comparing three treatments, for Example 3.13.*

EXAMPLE 3.13 **Conserving energy**

Many utility companies have introduced programs to encourage energy conservation among their customers. An electric company considers placing electronic indicators in households to show what the cost would be if the electricity use at that moment continued for a month. Will indicators reduce electricity use? Would cheaper methods work almost as well? The company decides to design an experiment.

One cheaper approach is to give customers a chart and information about monitoring their electricity use. The experiment compares these two approaches (indicator, chart) and also a control. The control group of customers receives information about energy conservation but no help in monitoring electricity use. The response variable is total electricity used in a year. The company finds 60 single-family residences in the same city willing to participate, so it assigns 20 residences at random to each of the 3 treatments. Figure 3.5 outlines the design.

To carry out the random assignment, label the 60 households 01 to 60. Enter Table B (or use software) to select an SRS of 20 to receive the indicators. Continue in Table B, selecting 20 more to receive charts. The remaining 20 form the control group.

Examples 3.12 and 3.13 describe completely randomized designs that compare levels of a single factor. In Example 3.12, the factor is the diet fed to the rats. In Example 3.13, it is the method used to encourage energy conservation. Completely randomized designs can have more than one factor. The advertising experiment of Example 3.10 has two factors: the length and the number of repetitions of a television commercial. Their combinations form the six treatments outlined in Figure 3.3 (page 187). A completely randomized design assigns subjects at random to these six treatments. Once the layout of treatments is set, the randomization needed for a completely randomized design is tedious but straightforward.

APPLY YOUR KNOWLEDGE

3.35 **Treating prostate disease.** A large study used records from Canada's national health care system to compare the effectiveness of two ways to treat prostate disease. The two treatments are traditional surgery and a new method that does not require surgery. The records described many patients whose doctors had chosen each method. The study found that patients treated by the new method were significantly more likely to die within 8 years.[15]

(a) Further study of the data showed that this conclusion was wrong. The extra deaths among patients who got the new method could be explained by lurking variables. What lurking variables might be confounded with a doctor's choice of surgical or nonsurgical treatment?

(b) You have 300 prostate patients who are willing to serve as subjects in an experiment to compare the two methods. Use a diagram to outline the design of a randomized comparative experiment. (When using a diagram to outline the design of an experiment, be sure to indicate the size of the treatment groups and the response variable. The diagrams in Figures 3.4 and 3.5 are models.)

3.36 **Sealing food packages.** Use a diagram to describe a completely randomized experimental design for the package liner experiment of Exercise 3.33. (When you outline the design of an experiment, be sure to indicate the size of the treatment groups and the response variable. The diagrams in Figures 3.4 and 3.5 are models.) Use software or Table B (starting at line 120) to do the randomization required by your design.

3.37 **Does child care help recruit employees?** Will providing child care for employees make a company more attractive to women, even those who are unmarried? You are designing an experiment to answer this question. You prepare recruiting material for two fictitious companies, both in similar businesses in the same location. Company A's brochure does not mention child care. There are two versions of Company B's material, identical except that one describes the company's on-site child-care facility. Your subjects are 40 unmarried women who are college seniors seeking employment. Each subject will read recruiting material for both companies and choose the one she would prefer to work for. You will give each version of Company B's brochure to half the women. You expect that a higher percentage of those who read the description that includes child care will choose Company B.

(a) Outline an appropriate design for the experiment.

(b) The names of the subjects appear below. Use Table B, beginning at line 131, to do the randomization required by your design. List the subjects who will read the version that mentions child care.

CHAPTER 9   Inference for Two-way Tables

- Expected count for a cell in a two-way table:

$$\text{expected count} = \frac{\text{row total} \times \text{column total}}{\text{table total}}$$

- Chi-square test statistic for testing whether the row and column variables in an $r \times c$ table are unrelated (expected cell counts not too small):

$$X^2 = \sum \frac{(\text{observed count} - \text{expected count})^2}{\text{expected count}}$$

  with P-values from the chi-square distribution with $\text{df} = (r - 1) \times (c - 1)$.

- Describe the relationship using percents, comparison of observed with expected counts, and components of X^2.

CHAPTER 10   One-way Analysis of Variance: Comparing Several Means

- ANOVA F tests whether all of I populations have the same mean, based on independent SRSs from I normal populations with the same σ. P-values come from the F distribution with $I - 1$ and $N - I$ degrees of freedom, where N is the total observations in all samples (independent SRSs from I normal populations with the same σ).

- Describe the data using the I sample means and standard deviations and side-by-side graphs of the samples.

- The ANOVA test statistic (use software) is $F = \text{MSG/MSE}$, where

$$\text{MSG} = \frac{n_1(\bar{x}_1 - \bar{x})^2 + \cdots + n_I(\bar{x}_I - \bar{x})^2}{I - 1}$$

$$\text{MSE} = \frac{(n_1 - 1)s_1^2 + \cdots + (n_I - 1)s_I^2}{N - I}$$

CHAPTER 11   Inference for Regression

- The regression model: We have n observations on x and y. The response y for any fixed x has a normal distribution with mean given by the true regression line $\mu_y = \alpha + \beta x$ and standard deviation σ.

- Estimate α by the intercept a and β by the slope b of the least-squares line. Estimate σ by the standard error about the line (use software):

$$s = \sqrt{\frac{1}{n - 2} \sum \text{residual}^2}$$

- t confidence interval for regression slope β:

$$b \pm t^* \text{SE}_b \qquad t^* \text{ from } t(n - 2)$$

  where the standard error of b is

$$\text{SE}_b = \frac{s}{\sqrt{\sum (x - \bar{x})^2}}$$

- t test statistic for no linear relationship, H_0: $\beta = 0$:

$$t = \frac{b}{\text{SE}_b} \qquad P\text{-values from } t(n - 2)$$

- t confidence interval for mean response μ_y when $x = x^*$:

$$\hat{y} \pm t^* \text{SE}_{\hat{\mu}} \qquad t^* \text{ from } t(n - 2)$$

  where the standard error is

$$\text{SE}_{\hat{\mu}} = s \sqrt{\frac{1}{n} + \frac{(x^* - \bar{x})^2}{\sum (x - \bar{x})^2}}$$

- t prediction interval for an individual observation y when $x = x^*$:

$$\hat{y} \pm t^* \text{SE}_{\hat{y}} \qquad t^* \text{ from } t(n - 2)$$

  where the standard error is

$$\text{SE}_{\hat{y}} = s \sqrt{1 + \frac{1}{n} + \frac{(x^* - \bar{x})^2}{\sum (x - \bar{x})^2}}$$

TABLE A Standard normal probabilities

z	.00	.01	.02	.03	.04	.05	.06	.07	.08	.09
−3.4	.0003	.0003	.0003	.0003	.0003	.0003	.0003	.0003	.0003	.0002
−3.3	.0005	.0005	.0005	.0004	.0004	.0004	.0004	.0004	.0004	.0003
−3.2	.0007	.0007	.0006	.0006	.0006	.0006	.0006	.0005	.0005	.0005
−3.1	.0010	.0009	.0009	.0009	.0008	.0008	.0008	.0008	.0007	.0007
−3.0	.0013	.0013	.0013	.0012	.0012	.0011	.0011	.0011	.0010	.0010
−2.9	.0019	.0018	.0018	.0017	.0016	.0016	.0015	.0015	.0014	.0014
−2.8	.0026	.0025	.0024	.0023	.0023	.0022	.0021	.0021	.0020	.0019
−2.7	.0035	.0034	.0033	.0032	.0031	.0030	.0029	.0028	.0027	.0026
−2.6	.0047	.0045	.0044	.0043	.0041	.0040	.0039	.0038	.0037	.0036
−2.5	.0062	.0060	.0059	.0057	.0055	.0054	.0052	.0051	.0049	.0048
−2.4	.0082	.0080	.0078	.0075	.0073	.0071	.0069	.0068	.0066	.0064
−2.3	.0107	.0104	.0102	.0099	.0096	.0094	.0091	.0089	.0087	.0084
−2.2	.0139	.0136	.0132	.0129	.0125	.0122	.0119	.0116	.0113	.0110
−2.1	.0179	.0174	.0170	.0166	.0162	.0158	.0154	.0150	.0146	.0143
−2.0	.0228	.0222	.0217	.0212	.0207	.0202	.0197	.0192	.0188	.0183
−1.9	.0287	.0281	.0274	.0268	.0262	.0256	.0250	.0244	.0239	.0233
−1.8	.0359	.0351	.0344	.0336	.0329	.0322	.0314	.0307	.0301	.0294
−1.7	.0446	.0436	.0427	.0418	.0409	.0401	.0392	.0384	.0375	.0367
−1.6	.0548	.0537	.0526	.0516	.0505	.0495	.0485	.0475	.0465	.0455
−1.5	.0668	.0655	.0643	.0630	.0618	.0606	.0594	.0582	.0571	.0559
−1.4	.0808	.0793	.0778	.0764	.0749	.0735	.0721	.0708	.0694	.0681
−1.3	.0968	.0951	.0934	.0918	.0901	.0885	.0869	.0853	.0838	.0823
−1.2	.1151	.1131	.1112	.1093	.1075	.1056	.1038	.1020	.1003	.0985
−1.1	.1357	.1335	.1314	.1292	.1271	.1251	.1230	.1210	.1190	.1170
−1.0	.1587	.1562	.1539	.1515	.1492	.1469	.1446	.1423	.1401	.1379
−0.9	.1841	.1814	.1788	.1762	.1736	.1711	.1685	.1660	.1635	.1611
−0.8	.2119	.2090	.2061	.2033	.2005	.1977	.1949	.1922	.1894	.1867
−0.7	.2420	.2389	.2358	.2327	.2296	.2266	.2236	.2206	.2177	.2148
−0.6	.2743	.2709	.2676	.2643	.2611	.2578	.2546	.2514	.2483	.2451
−0.5	.3085	.3050	.3015	.2981	.2946	.2912	.2877	.2843	.2810	.2776
−0.4	.3446	.3409	.3372	.3336	.3300	.3264	.3228	.3192	.3156	.3121
−0.3	.3821	.3783	.3745	.3707	.3669	.3632	.3594	.3557	.3520	.3483
−0.2	.4207	.4168	.4129	.4090	.4052	.4013	.3974	.3936	.3897	.3859
−0.1	.4602	.4562	.4522	.4483	.4443	.4404	.4364	.4325	.4286	.4247
−0.0	.5000	.4960	.4920	.4880	.4840	.4801	.4761	.4721	.4681	.4641
0.0	.5000	.5040	.5080	.5120	.5160	.5199	.5239	.5279	.5319	.5359
0.1	.5398	.5438	.5478	.5517	.5557	.5596	.5636	.5675	.5714	.5753
0.2	.5793	.5832	.5871	.5910	.5948	.5987	.6026	.6064	.6103	.6141
0.3	.6179	.6217	.6255	.6293	.6331	.6368	.6406	.6443	.6480	.6517
0.4	.6554	.6591	.6628	.6664	.6700	.6736	.6772	.6808	.6844	.6879
0.5	.6915	.6950	.6985	.7019	.7054	.7088	.7123	.7157	.7190	.7224
0.6	.7257	.7291	.7324	.7357	.7389	.7422	.7454	.7486	.7517	.7549
0.7	.7580	.7611	.7642	.7673	.7704	.7734	.7764	.7794	.7823	.7852
0.8	.7881	.7910	.7939	.7967	.7995	.8023	.8051	.8078	.8106	.8133
0.9	.8159	.8186	.8212	.8238	.8264	.8289	.8315	.8340	.8365	.8389
1.0	.8413	.8438	.8461	.8485	.8508	.8531	.8554	.8577	.8599	.8621
1.1	.8643	.8665	.8686	.8708	.8729	.8749	.8770	.8790	.8810	.8830
1.2	.8849	.8869	.8888	.8907	.8925	.8944	.8962	.8980	.8997	.9015
1.3	.9032	.9049	.9066	.9082	.9099	.9115	.9131	.9147	.9162	.9177
1.4	.9192	.9207	.9222	.9236	.9251	.9265	.9279	.9292	.9306	.9319
1.5	.9332	.9345	.9357	.9370	.9382	.9394	.9406	.9418	.9429	.9441
1.6	.9452	.9463	.9474	.9484	.9495	.9505	.9515	.9525	.9535	.9545
1.7	.9554	.9564	.9573	.9582	.9591	.9599	.9608	.9616	.9625	.9633
1.8	.9641	.9649	.9656	.9664	.9671	.9678	.9686	.9693	.9699	.9706
1.9	.9713	.9719	.9726	.9732	.9738	.9744	.9750	.9756	.9761	.9767
2.0	.9772	.9778	.9783	.9788	.9793	.9798	.9803	.9808	.9812	.9817
2.1	.9821	.9826	.9830	.9834	.9838	.9842	.9846	.9850	.9854	.9857
2.2	.9861	.9864	.9868	.9871	.9875	.9878	.9881	.9884	.9887	.9890
2.3	.9893	.9896	.9898	.9901	.9904	.9906	.9909	.9911	.9913	.9916
2.4	.9918	.9920	.9922	.9925	.9927	.9929	.9931	.9932	.9934	.9936
2.5	.9938	.9940	.9941	.9943	.9945	.9946	.9948	.9949	.9951	.9952
2.6	.9953	.9955	.9956	.9957	.9959	.9960	.9961	.9962	.9963	.9964
2.7	.9965	.9966	.9967	.9968	.9969	.9970	.9971	.9972	.9973	.9974
2.8	.9974	.9975	.9976	.9977	.9977	.9978	.9979	.9979	.9980	.9981
2.9	.9981	.9982	.9982	.9983	.9984	.9984	.9985	.9985	.9986	.9986
3.0	.9987	.9987	.9987	.9988	.9988	.9989	.9989	.9989	.9990	.9990
3.1	.9990	.9991	.9991	.9991	.9992	.9992	.9992	.9992	.9993	.9993
3.2	.9993	.9993	.9994	.9994	.9994	.9994	.9994	.9995	.9995	.9995
3.3	.9995	.9995	.9995	.9996	.9996	.9996	.9996	.9996	.9996	.9997
3.4	.9997	.9997	.9997	.9997	.9997	.9997	.9997	.9997	.9997	.9998

TABLE E Chi-square distribution critical values

df	.25	.20	.15	.10	.05	.025	.02	.01	.005	.0025	.001	.0005
					Upper tail probability p							
1	1.32	1.64	2.07	2.71	3.84	5.02	5.41	6.63	7.88	9.14	10.83	12.12
2	2.77	3.22	3.79	4.61	5.99	7.38	7.82	9.21	10.60	11.98	13.82	15.20
3	4.11	4.64	5.32	6.25	7.81	9.35	9.84	11.34	12.84	14.32	16.27	17.73
4	5.39	5.99	6.74	7.78	9.49	11.14	11.67	13.28	14.86	16.42	18.47	20.00
5	6.63	7.29	8.12	9.24	11.07	12.83	13.39	15.09	16.75	18.39	20.51	22.11
6	7.84	8.56	9.45	10.64	12.59	14.45	15.03	16.81	18.55	20.25	22.46	24.10
7	9.04	9.80	10.75	12.02	14.07	16.01	16.62	18.48	20.28	22.04	24.32	26.02
8	10.22	11.03	12.03	13.36	15.51	17.53	18.17	20.09	21.95	23.77	26.12	27.87
9	11.39	12.24	13.29	14.68	16.92	19.02	19.68	21.67	23.59	25.46	27.88	29.67
10	12.55	13.44	14.53	15.99	18.31	20.48	21.16	23.21	25.19	27.11	29.59	31.42
11	13.70	14.63	15.77	17.28	19.68	21.92	22.62	24.72	26.76	28.73	31.26	33.14
12	14.85	15.81	16.99	18.55	21.03	23.34	24.05	26.22	28.30	30.32	32.91	34.82
13	15.98	16.98	18.20	19.81	22.36	24.74	25.47	27.69	29.82	31.88	34.53	36.48
14	17.12	18.15	19.41	21.06	23.68	26.12	26.87	29.14	31.32	33.43	36.12	38.11
15	18.25	19.31	20.60	22.31	25.00	27.49	28.26	30.58	32.80	34.95	37.70	39.72
16	19.37	20.47	21.79	23.54	26.30	28.85	29.63	32.00	34.27	36.46	39.25	41.31
17	20.49	21.61	22.98	24.77	27.59	30.19	31.00	33.41	35.72	37.95	40.79	42.88
18	21.60	22.76	24.16	25.99	28.87	31.53	32.35	34.81	37.16	39.42	42.31	44.43
19	22.72	23.90	25.33	27.20	30.14	32.85	33.69	36.19	38.58	40.88	43.82	45.97
20	23.83	25.04	26.50	28.41	31.41	34.17	35.02	37.57	40.00	42.34	45.31	47.50
21	24.93	26.17	27.66	29.62	32.67	35.48	36.34	38.93	41.40	43.78	46.80	49.01
22	26.04	27.30	28.82	30.81	33.92	36.78	37.66	40.29	42.80	45.20	48.27	50.51
23	27.14	28.43	29.98	32.01	35.17	38.08	38.97	41.64	44.18	46.62	49.73	52.00
24	28.24	29.55	31.13	33.20	36.42	39.36	40.27	42.98	45.56	48.03	51.18	53.48
25	29.34	30.68	32.28	34.38	37.65	40.65	41.57	44.31	46.93	49.44	52.62	54.95
26	30.43	31.79	33.43	35.56	38.89	41.92	42.86	45.64	48.29	50.83	54.05	56.41
27	31.53	32.91	34.57	36.74	40.11	43.19	44.14	46.96	49.64	52.22	55.48	57.86
28	32.62	34.03	35.71	37.92	41.34	44.46	45.42	48.28	50.99	53.59	56.89	59.30
29	33.71	35.14	36.85	39.09	42.56	45.72	46.69	49.59	52.34	54.97	58.30	60.73
30	34.80	36.25	37.99	40.26	43.77	46.98	47.96	50.89	53.67	56.33	59.70	62.16
40	45.62	47.27	49.24	51.81	55.76	59.34	60.44	63.69	66.77	69.70	73.40	76.09
50	56.33	58.16	60.35	63.17	67.50	71.42	72.61	76.15	79.49	82.66	86.66	89.56
60	66.98	68.97	71.34	74.40	79.08	83.30	84.58	88.38	91.95	95.34	99.61	102.7
80	88.13	90.41	93.11	96.58	101.9	106.6	108.1	112.3	116.3	120.1	124.8	128.3
100	109.1	111.7	114.7	118.5	124.3	129.6	131.1	135.8	140.2	144.3	149.4	153.2

© 2000 W. H. Freeman and Company

TABLES AND FORMULAS FOR MOORE,
The Basic Practice of Statistics

CHAPTER 1   Examining Distributions

- Remember overall pattern (shape, center, spread) and deviations.
- Mean (use a calculator):

$$\bar{x} = \frac{x_1 + x_2 + \cdots + x_n}{n} = \frac{1}{n}\sum x_i$$

- Standard deviation (use a calculator):

$$s = \sqrt{\frac{1}{n-1}\sum(x_i - \bar{x})^2}$$

- Median: Arrange all observations from smallest to largest. The median M is located $(n + 1)/2$ observations from the beginning of this list.
- Quartiles: The first quartile Q_1 is the median of the observations whose position in the ordered list is to the left of the location of the overall median. The third quartile Q_3 is the median of the observations to the right of the location of the overall median.
- Five-number summary:

  Minimum, Q_1, M, Q_3, Maximum

- Standardized observation from x:

$$z = \frac{x - \mu}{\sigma}$$

CHAPTER 2   Examining Relationships

- Remember overall pattern (form, direction, strength) and deviations.
- Correlation (use a calculator):

$$r = \frac{1}{n-1}\sum\left(\frac{x_i - \bar{x}}{s_x}\right)\left(\frac{y_i - \bar{y}}{s_y}\right)$$

- Least-squares regression line (use a calculator): $\hat{y} = a + bx$ with slope $b = rs_y/s_x$ and intercept $a = \bar{y} - b\bar{x}$
- Residuals:

$$\text{residual} = \text{observed } y - \text{predicted } y$$
$$= y - \hat{y}$$

CHAPTER 3   Producing Data

- Simple random sample: Choose an SRS by giving every individual in the population a numerical label and using Table B of random digits to choose the sample.
- Randomized comparative experiments:

$$\text{Random} \nearrow \quad \text{Group 1} \rightarrow \text{Treatment 1} \quad \searrow \text{Observe}$$
$$\text{Allocation} \searrow \quad \text{Group 2} \rightarrow \text{Treatment 2} \quad \nearrow \text{Response}$$

CHAPTER 4   Probability and Sampling Distributions

- Probability rules:
  - Any probability satisfies $0 \leq P(A) \leq 1$.
  - The sample space S has probability $P(S) = 1$.
  - For any event A, $P(A \text{ does not occur}) = 1 - P(A)$.
  - If events A and B are disjoint, $P(A \text{ or } B) = P(A) + P(B)$.
- Sampling distribution of a sample mean:
  - $\bar{x}$ has mean μ and standard deviation $\sigma/\sqrt{n}$.
  - $\bar{x}$ has a normal distribution if the population distribution is normal.
  - Central limit theorem: $\bar{x}$ is approximately normal when n is large.

CHAPTER 6   Introduction to Inference

- z confidence interval for a population mean (σ known, SRS from normal population):

$$\bar{x} \pm z^* \frac{\sigma}{\sqrt{n}} \qquad z^* \text{ from } N(0, 1)$$

- Sample size for desired margin of error m:

$$n = \left(\frac{z^* \sigma}{m}\right)^2$$

- z test statistic for H_0: $\mu = \mu_0$ (σ known, SRS from normal population):

$$z = \frac{\bar{x} - \mu_0}{\sigma/\sqrt{n}} \qquad P\text{-values from } N(0, 1)$$

CHAPTER 7   Inference for Distributions

- t confidence interval for a population mean (SRS from normal population):

$$\bar{x} \pm t^* \frac{s}{\sqrt{n}} \qquad t^* \text{ from } t(n - 1)$$

- t test statistic for H_0: $\mu = \mu_0$ (SRS from normal population):

$$t = \frac{\bar{x} - \mu_0}{s/\sqrt{n}} \qquad P\text{-values from } t(n - 1)$$

- Matched pairs: To compare the responses to the two treatments, apply the one-sample t procedures to the observed differences.

- Two-sample t confidence interval for $\mu_1 - \mu_2$ (independent SRSs from normal populations):

$$(\bar{x}_1 - \bar{x}_2) \pm t^* \sqrt{\frac{s_1^2}{n_1} + \frac{s_2^2}{n_2}}$$

with conservative t^* from t with df the smaller of $n_1 - 1$ and $n_2 - 1$.

- Two-sample t test statistic for H_0: $\mu_1 = \mu_2$ (independent SRSs from normal populations):

$$t = \frac{\bar{x}_1 - \bar{x}_2}{\sqrt{\frac{s_1^2}{n_1} + \frac{s_2^2}{n_2}}}$$

with conservative P-values from t with df the smaller of $n_1 - 1$ and $n_2 - 1$.

CHAPTER 8   Inference for Proportions

- Sampling distribution of a sample proportion: when the population is large and both $np \geq 10$ and $n(1 - p) \geq 10$, $\hat{p}$ is approximately normal with mean p and standard deviation $\sqrt{p(1 - p)/n}$.

- z confidence interval for p (large SRS):

$$\hat{p} \pm z^* \sqrt{\frac{\hat{p}(1 - \hat{p})}{n}} \qquad z^* \text{ from } N(0, 1)$$

- z test statistic for H_0: $p = p_0$ (large SRS):

$$z = \frac{\hat{p} - p_0}{\sqrt{\frac{p_0(1 - p_0)}{n}}} \qquad P\text{-values from } N(0, 1)$$

- Sample size for desired margin of error m:

$$n = \left(\frac{z^*}{m}\right)^2 p^*(1 - p^*)$$

where p^* is a guessed value for p or $p^* = 0.5$.

- z confidence interval for $p_1 - p_2$ (large independent SRSs):

$$(\hat{p}_1 - \hat{p}_2) \pm z^* \text{SE} \qquad z^* \text{ from } N(0, 1)$$

where the standard error of $\hat{p}_1 - \hat{p}_2$ is

$$\text{SE} = \sqrt{\frac{\hat{p}_1(1 - \hat{p}_1)}{n_1} + \frac{\hat{p}_2(1 - \hat{p}_2)}{n_2}}$$

- Two-sample z test statistic for H_0: $p_1 = p_2$ (large independent SRSs):

$$z = \frac{\hat{p}_1 - \hat{p}_2}{\sqrt{\hat{p}(1 - \hat{p})\left(\frac{1}{n_1} + \frac{1}{n_2}\right)}}$$

where $\hat{p}$ is the pooled proportion of successes.

TABLE B Random digits

Line								
101	19223	95034	05756	28713	96409	12531	42544	82853
102	73676	47150	99400	01927	27754	42648	82425	36290
103	45467	71709	77558	00095	32863	29485	82226	90056
104	52711	38889	93074	60227	40011	85848	48767	52573
105	95592	94007	69971	91481	60779	53791	17297	59335
106	68417	35013	15529	72765	85089	57067	50211	47487
107	82739	57890	20807	47511	81676	55300	94383	14893
108	60940	72024	17868	24943	61790	90656	87964	18883
109	36009	19365	15412	39638	85453	46816	83485	41979
110	38448	48789	18338	24697	39364	42006	76688	08708
111	81486	69487	60513	09297	00412	71238	27649	39950
112	59636	88804	04634	71197	19352	73089	84898	45785
113	62568	70206	40325	03699	71080	22553	11486	11776
114	45149	32992	75730	66280	03819	56202	02938	70915
115	61041	77684	94322	24709	73698	14526	31893	32592
116	14459	26056	31424	80371	65103	62253	50490	61181
117	38167	98532	62183	70632	23417	26185	41448	75532
118	73190	32533	04470	29669	84407	90785	65956	86382
119	95857	07118	87664	92099	58806	66979	98624	84826
120	35476	55972	39421	65850	04266	35435	43742	11937
121	71487	09984	29077	14863	61683	47052	62224	51025
122	13873	81598	95052	90908	73592	75186	87136	95761
123	54580	81507	27102	56027	55892	33063	41842	81868
124	71035	09001	43367	49497	72719	96758	27611	91596
125	96746	12149	37823	71868	18442	35119	62103	39244

TABLE C t distribution critical values

df	.25	.20	.15	.10	.05	.025	.02	.01	.005	.0025	.001	.0005
						Upper tail probability p						
1	1.000	1.376	1.963	3.078	6.314	12.71	15.89	31.82	63.66	127.3	318.3	636.6
2	0.816	1.061	1.386	1.886	2.920	4.303	4.849	6.965	9.925	14.09	22.33	31.60
3	0.765	0.978	1.250	1.638	2.353	3.182	3.482	4.541	5.841	7.453	10.21	12.92
4	0.741	0.941	1.190	1.533	2.132	2.776	2.999	3.747	4.604	5.598	7.173	8.610
5	0.727	0.920	1.156	1.476	2.015	2.571	2.757	3.365	4.032	4.773	5.893	6.869
6	0.718	0.906	1.134	1.440	1.943	2.447	2.612	3.143	3.707	4.317	5.208	5.959
7	0.711	0.896	1.119	1.415	1.895	2.365	2.517	2.998	3.499	4.029	4.785	5.408
8	0.706	0.889	1.108	1.397	1.860	2.306	2.449	2.896	3.355	3.833	4.501	5.041
9	0.703	0.883	1.100	1.383	1.833	2.262	2.398	2.821	3.250	3.690	4.297	4.781
10	0.700	0.879	1.093	1.372	1.812	2.228	2.359	2.764	3.169	3.581	4.144	4.587
11	0.697	0.876	1.088	1.363	1.796	2.201	2.328	2.718	3.106	3.497	4.025	4.437
12	0.695	0.873	1.083	1.356	1.782	2.179	2.303	2.681	3.055	3.428	3.930	4.318
13	0.694	0.870	1.079	1.350	1.771	2.160	2.282	2.650	3.012	3.372	3.852	4.221
14	0.692	0.868	1.076	1.345	1.761	2.145	2.264	2.624	2.977	3.326	3.787	4.140
15	0.691	0.866	1.074	1.341	1.753	2.131	2.249	2.602	2.947	3.286	3.733	4.073
16	0.690	0.865	1.071	1.337	1.746	2.120	2.235	2.583	2.921	3.252	3.686	4.015
17	0.689	0.863	1.069	1.333	1.740	2.110	2.224	2.567	2.898	3.222	3.646	3.965
18	0.688	0.862	1.067	1.330	1.734	2.101	2.214	2.552	2.878	3.197	3.611	3.922
19	0.688	0.861	1.066	1.328	1.729	2.093	2.205	2.539	2.861	3.174	3.579	3.883
20	0.687	0.860	1.064	1.325	1.725	2.086	2.197	2.528	2.845	3.153	3.552	3.850
21	0.686	0.859	1.063	1.323	1.721	2.080	2.189	2.518	2.831	3.135	3.527	3.819
22	0.686	0.858	1.061	1.321	1.717	2.074	2.183	2.508	2.819	3.119	3.505	3.792
23	0.685	0.858	1.060	1.319	1.714	2.069	2.177	2.500	2.807	3.104	3.485	3.768
24	0.685	0.857	1.059	1.318	1.711	2.064	2.172	2.492	2.797	3.091	3.467	3.745
25	0.684	0.856	1.058	1.316	1.708	2.060	2.167	2.485	2.787	3.078	3.450	3.725
26	0.684	0.856	1.058	1.315	1.706	2.056	2.162	2.479	2.779	3.067	3.435	3.707
27	0.684	0.855	1.057	1.314	1.703	2.052	2.158	2.473	2.771	3.057	3.421	3.690
28	0.683	0.855	1.056	1.313	1.701	2.048	2.154	2.467	2.763	3.047	3.408	3.674
29	0.683	0.854	1.055	1.311	1.699	2.045	2.150	2.462	2.756	3.038	3.396	3.659
30	0.683	0.854	1.055	1.310	1.697	2.042	2.147	2.457	2.750	3.030	3.385	3.646
40	0.681	0.851	1.050	1.303	1.684	2.021	2.123	2.423	2.704	2.971	3.307	3.551
50	0.679	0.849	1.047	1.299	1.676	2.009	2.109	2.403	2.678	2.937	3.261	3.496
60	0.679	0.848	1.045	1.296	1.671	2.000	2.099	2.390	2.660	2.915	3.232	3.460
80	0.678	0.846	1.043	1.292	1.664	1.990	2.088	2.374	2.639	2.887	3.195	3.416
100	0.677	0.845	1.042	1.290	1.660	1.984	2.081	2.364	2.626	2.871	3.174	3.390
1000	0.675	0.842	1.037	1.282	1.646	1.962	2.056	2.330	2.581	2.813	3.098	3.300
z^*	0.674	0.841	1.036	1.282	1.645	1.960	2.054	2.326	2.576	2.807	3.091	3.291
	50%	60%	70%	80%	90%	95%	96%	98%	99%	99.5%	99.8%	99.9%
						Confidence level C						

Abrams	Danielson	Gutierrez	Lippman	Rosen
Adamson	Durr	Howard	Martinez	Sugiwara
Afifi	Edwards	Hwang	McNeill	Thompson
Brown	Fluharty	Iselin	Morse	Travers
Cansico	Garcia	Janle	Ng	Turing
Chen	Gerson	Kaplan	Quinones	Ullmann
Cortez	Green	Kim	Rivera	Williams
Curzakis	Gupta	Lattimore	Roberts	Wong

The logic of randomized comparative experiments

Randomized comparative experiments are designed to give good evidence that differences in the treatments actually *cause* the differences we see in the response. The logic is as follows:

- Random assignment of subjects forms groups that should be similar in all respects before the treatments are applied.

- Comparative design ensures that influences other than the experimental treatments operate equally on all groups.

- Therefore, differences in average response must be due either to the treatments or to the play of chance in the random assignment of subjects to the treatments.

That "either-or" deserves more thought. In Example 3.12, we cannot say that *any* difference in the average weight gains of rats fed the two diets must be caused by a difference between the diets. There would be some difference even if both groups received the same diet, because the natural variability among rats means that some grow faster than others. Chance assigns the faster-growing rats to one group or the other, and this creates a chance difference between the groups. We would not trust an experiment with just one rat in each group, for example. The results would depend too much on which group got lucky and received the faster-growing rat. If we assign many rats to each diet, however, the effects of chance will average out and there will be little difference in the average weight gains in the two groups unless the diets themselves cause a difference. "Use enough experimental units to reduce chance variation" is the third big idea of statistical design of experiments.

PRINCIPLES OF EXPERIMENTAL DESIGN

The basic principles of statistical design of experiments are

1. **Control** of the effects of lurking variables on the response, most simply by comparing several treatments.

2. **Randomization,** the use of impersonal chance to assign experimental units to treatments.

3. **Replication** of the experiment on many units to reduce chance variation in the results.

We hope to see a difference in the responses so large that it is unlikely to happen just because of chance variation. We can use the laws of probability, which give a mathematical description of chance behavior, to learn if the treatment effects are larger than we would expect to see if only chance were operating. If they are, we call them *statistically significant*.

STATISTICAL SIGNIFICANCE

An observed effect so large that it would rarely occur by chance is called **statistically significant**.

If we observe statistically significant differences among the groups in a comparative randomized experiment, we have good evidence that the treatments actually caused these differences. You will often see the phrase "statistically significant" in reports of investigations in many fields of study. The great advantage of randomized comparative experiments is that they can produce data that give good evidence for a cause-and-effect relationship between the explanatory and response variables. We know that in general a strong association does not imply causation. A statistically significant association in data from a well-designed experiment does imply causation.

Scratch my furry ears

Rats and rabbits, specially bred to be uniform in their inherited characteristics, are the subjects in many experiments. It turns out that animals, like people, are quite sensitive to how they are treated. This creates some amusing opportunities for hidden bias. For example, human affection can change the cholesterol level of rabbits. Choose some rabbits at random and regularly remove them from their cages to have their heads scratched by friendly people. Leave other rabbits unloved. All the rabbits eat the same diet, but the rabbits that receive affection have lower cholesterol.

APPLY YOUR KNOWLEDGE

3.38 Conserving energy. Example 3.13 describes an experiment to learn whether providing households with electronic indicators or charts will reduce their electricity consumption. An executive of the electric company objects to including a control group. He says, "It would be simpler to just compare electricity use last year (before the indicator or chart was provided) with consumption in the same period this year. If households use less electricity this year, the indicator or chart must be working." Explain clearly why this design is inferior to that in Example 3.13.

3.39 Exercise and heart attacks. Does regular exercise reduce the risk of a heart attack? Here are two ways to study this question. Explain clearly why the second design will produce more trustworthy data.

1. A researcher finds 2000 men over 40 who exercise regularly and have not had heart attacks. She matches each with a similar man who does not exercise regularly, and she follows both groups for 5 years.

2. Another researcher finds 4000 men over 40 who have not had heart attacks and are willing to participate in a study. She assigns 2000 of the men to a regular program of supervised exercise. The

other 2000 continue their usual habits. The researcher follows both groups for 5 years.

3.40   **Statistical significance.** The financial aid office of a university asks a sample of students about their employment and earnings. The report says that "for academic year earnings, a significant difference was found between the sexes, with men earning more on the average. No significant difference was found between the earnings of black and white students." Explain the meaning of "a significant difference" and "no significant difference" in plain language.

Cautions about experimentation

The logic of a randomized comparative experiment depends on our ability to treat all the experimental units identically in every way except for the actual treatments being compared. Good experiments therefore require careful attention to details. For example, the subjects in both groups of the second gastric freezing experiment (Example 3.11, page 188) all got the same medical attention over the several years the studies continued. The researchers paid attention to such details as ensuring that the tube in the mouth of each subject was cold, whether or not the fluid in the balloon was refrigerated. Moreover, the study was **double-blind**—neither the subjects themselves nor the medical personnel who worked with them knew which treatment any subject had received. The double-blind method avoids unconscious bias by, for example, a doctor who doesn't think that "just a placebo" can benefit a patient.

double-blind

The most serious potential weakness of experiments is **lack of realism**. The subjects or treatments or setting of an experiment may not realistically duplicate the conditions we really want to study. Here are two examples.

lack of realism

EXAMPLE 3.14   Response to advertising

The study of television advertising in Example 3.10 showed a 40-minute video-tape to students who knew an experiment was going on. We can't be sure that the results apply to everyday television viewers. Many behavioral science experiments use as subjects students who know they are subjects in an experiment. That's not a realistic setting.

EXAMPLE 3.15   Center brake lights

Do those high center brake lights, required on all cars sold in the United States since 1986, really reduce rear-end collisions? Randomized comparative experiments with fleets of rental and business cars, done before the lights were required, showed that the third brake light reduced rear-end collisions by as much as 50%. Alas, requiring the third light in all cars led to only a 5% drop.

What happened? Most cars did not have the extra brake light when the experiments were carried out, so it caught the eye of following drivers. Now that almost all cars have the third light, they no longer capture attention.

Lack of realism can limit our ability to apply the conclusions of an experiment to the settings of greatest interest. Most experimenters want to generalize their conclusions to some setting wider than that of the actual experiment. Statistical analysis of the original experiment cannot tell us how far the results will generalize. Nonetheless, the randomized comparative experiment, because of its ability to give convincing evidence for causation, is one of the most important ideas in statistics.

APPLY YOUR KNOWLEDGE

3.41 **Does meditation reduce anxiety?** An experiment that claimed to show that meditation reduces anxiety proceeded as follows. The experimenter interviewed the subjects and rated their level of anxiety. Then the subjects were randomly assigned to two groups. The experimenter taught one group how to meditate and they meditated daily for a month. The other group was simply told to relax more. At the end of the month, the experimenter interviewed all the subjects again and rated their anxiety level. The meditation group now had less anxiety. Psychologists said that the results were suspect because the ratings were not blind. Explain what this means and how lack of blindness could bias the reported results.

3.42 Fizz Laboratories, a pharmaceutical company, has developed a new pain-relief medication. Sixty patients suffering from arthritis and needing pain relief are available. Each patient will be treated and asked an hour later, "About what percentage of pain relief did you experience?"

 (a) Why should Fizz not simply administer the new drug and record the patients' responses?

 (b) Outline the design of an experiment to compare the drug's effectiveness with that of aspirin and of a placebo.

 (c) Should patients be told which drug they are receiving? How would this knowledge probably affect their reactions?

 (d) If patients are not told which treatment they are receiving, the experiment is single-blind. Should this experiment be double-blind also? Explain.

Matched pairs designs

Completely randomized designs are the simplest statistical designs for experiments. They illustrate clearly the principles of control, randomization, and replication. However, completely randomized designs are often inferior to more elaborate statistical designs. In particular, matching the subjects in various ways can produce more precise results than simple randomization.

One common design that combines matching with randomization is the **matched pairs design**. A matched pairs design compares just two treatments. Choose pairs of subjects that are as closely matched as possible. Assign one of the treatments to each subject in a pair by tossing a coin or reading odd and even digits from Table B. Sometimes each "pair" in a matched pairs design consists of just one subject, who gets both treatments one after the other. Each subject serves as his or her own control. The *order* of the treatments can influence the subject's response, so we randomize the order for each subject, again by a coin toss.

matched pairs design

EXAMPLE 3.16   Coke versus Pepsi

Pepsi wanted to demonstrate that Coke drinkers prefer Pepsi when they taste both colas blind. The subjects, all people who said they were Coke drinkers, tasted both colas from glasses without brand markings and said which they liked better. This is a matched pairs design in which each subject compares the two colas. Because responses may depend on which cola is tasted first, the order of tasting should be chosen at random for each subject.

When more than half the Coke drinkers chose Pepsi, Coke claimed that the experiment was biased. The Pepsi glasses were marked M and Coke glasses were marked Q. Aha, said Coke, this just shows that people like the letter M better than the letter Q. A careful experiment would in fact take care to avoid any distinction other than the actual treatments.[16]

Block designs

Matched pairs designs use the principles of comparison of treatments, randomization, and replication on several experimental units or subjects. However, the randomization is not complete—we do not randomly assign all the subjects at once to the two treatments. Instead, we only randomize within each matched pair. This allows matching to reduce the effect of variation among the subjects. Matched pairs are an example of *block designs*.

> ### BLOCK DESIGN
>
> A **block** is a group of experimental units or subjects that are known before the experiment to be similar in some way that is expected to affect the response to the treatments. In a **block design**, the random assignment of units to treatments is carried out separately within each block.

A block design combines the idea of creating equivalent treatment groups by matching with the principle of forming treatment groups at random. Blocks are another form of *control*. They control the effects of some outside variables by bringing those variables into the experiment to form the blocks. Here are some typical examples of block designs.

Figure 3.6 *Outline of a block design for Example 3.17. The blocks consist of male and female subjects. The treatments are three therapies for cancer.*

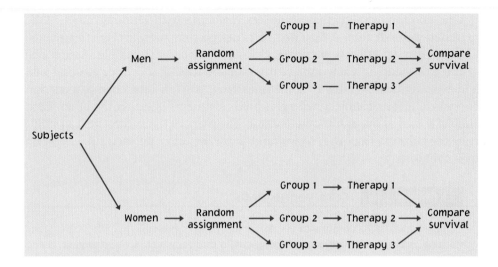

EXAMPLE 3.17   Comparing cancer treatments

The progress of a type of cancer differs in women and men. A clinical experiment to compare three therapies for this cancer therefore treats sex as a blocking variable. Two separate randomizations are done, one assigning the female subjects to the treatments and the other assigning the male subjects. Figure 3.6 outlines the design of this experiment. Note that there is no randomization involved in making up the blocks. They are groups of subjects who differ in some way (sex in this case) that is apparent before the experiment begins.

EXAMPLE 3.18   Comparing welfare policies

A social policy experiment will assess the effect on family income of several proposed new welfare systems and compare them with the present welfare system. Because the future income of a family is strongly related to its present income, the families who agree to participate are divided into blocks of similar income levels. The families in each block are then allocated at random among the welfare systems.

Blocks allow us to draw separate conclusions about each block, for example, about men and women in the cancer study in Example 3.17. Blocking also allows more precise overall conclusions, because the systematic differences between men and women can be removed when we study the overall effects of the three therapies. The idea of blocking is an important additional principle of statistical design of experiments. A wise experimenter will form blocks based on the most important unavoidable sources of variability among the experimental units. Randomization will then average out the effects of the remaining variation and allow an unbiased comparison of the treatments.

APPLY YOUR KNOWLEDGE

3.43   Comparing hand strength. Is the right hand generally stronger than the left in right-handed people? You can crudely measure hand strength by placing a bathroom scale on a shelf with the end protruding, then squeezing the scale between the thumb below and the four fingers above it. The reading of the scale shows the force exerted. Describe the design of a matched pairs experiment to compare the strength of the right and left hands, using 10 right-handed people as subjects. (You need not actually do the randomization.)

3.44 **Does charting help investors?** Some investment advisors believe that charts of past trends in the prices of securities can help predict future prices. Most economists disagree. In an experiment to examine the effects of using charts, business students trade (hypothetically) a foreign currency at computer screens. There are 20 student subjects available, named for convenience A, B, C, . . ., T. Their goal is to make as much money as possible, and the best performances are rewarded with small prizes. The student traders have the price history of the foreign currency in dollars in their computers. They may or may not also have software that highlights trends. Describe two designs for this experiment, a completely randomized design and a matched pairs design in which each student serves as his or her own control. In both cases, carry out the randomization required by the design.

3.45 **Comparing weight-loss treatments.** Twenty overweight females have agreed to participate in a study of the effectiveness of 4 weight-loss treatments: A, B, C, and D. The researcher first calculates how overweight each subject is by comparing the subject's actual weight with her "ideal" weight. The subjects and their excess weights in pounds are

Birnbaum	35	Hernandez	25	Moses	25	Smith	29
Brown	34	Jackson	33	Nevesky	39	Stall	33
Brunk	30	Kendall	28	Obrach	30	Tran	35
Cruz	34	Loren	32	Rodriguez	30	Wilansky	42
Deng	24	Mann	28	Santiago	27	Williams	22

The response variable is the weight lost after 8 weeks of treatment. Because a subject's excess weight will influence the response, a block design is appropriate.

(a) Arrange the subjects in order of increasing excess weight. Form 5 blocks of 4 subjects each by grouping the 4 least overweight, then the next 4, and so on.

(b) Use Table B to randomly assign the 4 subjects in each block to the 4 weight-loss treatments. Be sure to explain exactly how you used the table.

SECTION 3.2 Summary

In an experiment, we impose one or more **treatments** on the **experimental units** or **subjects**. Each treatment is a combination of levels of the explanatory variables, which we call **factors**.

The **design** of an experiment describes the choice of treatments and the manner in which the experimental units or subjects are assigned to the treatments.

The basic principles of statistical design of experiments are **control**, **randomization**, and **replication**.

The simplest form of control is **comparison**. Experiments should compare two or more treatments in order to avoid **confounding** of the effect of a treatment with other influences, such as lurking variables.

Randomization uses chance to assign subjects to the treatments. Randomization creates treatment groups that are similar (except for chance variation) before the treatments are applied. Randomization and comparison together prevent **bias**, or systematic favoritism, in experiments.

You can carry out randomization by giving numerical labels to the experimental units and using a **table of random digits** to choose treatment groups.

Replication of the treatments on many units reduces the role of chance variation and makes the experiment more sensitive to differences among the treatments.

Good experiments require attention to detail as well as good statistical design. Many behavioral and medical experiments are **double-blind**. **Lack of realism** in an experiment can prevent us from generalizing its results.

In addition to comparison, a second form of control is to restrict randomization by forming **blocks** of experimental units that are similar in some way that is important to the response. Randomization is then carried out separately within each block.

Matched pairs are a common form of blocking for comparing just two treatments. In some matched pairs designs, each subject receives both treatments in a random order. In others, the subjects are matched in pairs as closely as possible, and one subject in each pair receives each treatment.

SECTION 3.2 Exercises

3.46 (a) Exercise 2.64 (page 138) describes a study of the effects of herbal tea on the attitude of nursing home residents. Is this study an experiment? Why?

(b) Exercise 2.65 (page 138) describes a study of the effect of learning a foreign language on scores on an English test. Is this study an experiment? Why?

3.47 **Marketing to children.** If children are given more choices within a class of products, will they tend to prefer that product to a competing product that offers fewer choices? Marketers want to know. An experiment prepared three sets of beverages. Set 1 contained two milk drinks and two fruit drinks. Set 2 had two fruit drinks and four milk drinks. Set 3 contained four fruit drinks but only two milk drinks.

The researchers divided 210 children aged 4 to 12 years into 3 groups at random. They offered each group one of the sets. As each child chose a beverage to drink from the set presented, the researchers noted whether the choice was a milk drink or a fruit drink.

(a) What are the experimental subjects?

(b) What is the factor and what are its levels? What is the response variable?

(c) Use a diagram to outline a completely randomized design for the study.

(d) Explain how you would assign labels to the subjects. Use Table B at line 125 to choose *only the first 5* subjects assigned to the first treatment.

3.48 **Aspirin and heart attacks.** Can aspirin help prevent heart attacks? The Physicians' Health Study, a large medical experiment involving 22,000 male physicians, attempted to answer this question. One group of about 11,000 physicians took an aspirin every second day, while the rest took a placebo. After several years the study found that subjects in the aspirin group had significantly fewer heart attacks than subjects in the placebo group.

(a) Identify the experimental subjects, the factor and its levels, and the response variable in the Physicians' Health Study.

(b) Use a diagram to outline a completely randomized design for the Physicians' Health Study.

3.49 **Prayer and meditation.** You read in a magazine that "nonphysical treatments such as meditation and prayer have been shown to be effective in controlled scientific studies for such ailments as high blood pressure, insomnia, ulcers, and asthma." Explain in simple language what the article means by "controlled scientific studies" and why such studies can show that meditation and prayer are effective treatments for some medical problems.

3.50 **Reducing health care spending.** Will people spend less on health care if their health insurance requires them to pay some part of the cost themselves? An experiment on this issue asked if the percent of medical costs that are paid by health insurance has an effect either on the amount of medical care that people use or on their health. The treatments were four insurance plans. Each plan paid all medical costs above a ceiling. Below the ceiling, the plans paid 100%, 75%, 50%, or 0% of costs incurred.

(a) Outline the design of a randomized comparative experiment suitable for this study.

(b) Describe briefly the practical and ethical difficulties that might arise in such an experiment.

3.51 **Treating drunk drivers.** Once a person has been convicted of drunk driving, one purpose of court-mandated treatment or punishment is to prevent future offenses of the same kind. Suggest three different treatments that a court might require. Then outline the design of an experiment to compare their effectiveness. Be sure to specify the response variables you will measure.

3.52 **Sickle cell disease.** Exercise 3.32 (page 187) describes a medical study of a new treatment for sickle cell disease.

(a) Outline the design of this experiment.

(b) Use of a placebo is considered ethical if there is no effective standard treatment to give the control group. It might seem humane to give all the subjects hydroxyurea in the hope that it will help them. Explain clearly why this would not provide information about the effectiveness of the drug. (In fact, the experiment was stopped ahead of schedule because the hydroxyurea group had only half as many pain episodes as the control group. Ethical standards required stopping the experiment as soon as significant evidence became available.)

3.53 **Calcium and blood pressure.** Some medical researchers suspect that added calcium in the diet reduces blood pressure. You have available 40 men with high blood pressure who are willing to serve as subjects.

(a) Outline an appropriate design for the experiment, taking the placebo effect into account.

(b) The names of the subjects appear below. Use Table B, beginning at line 119, to do the randomization required by your design, and list the subjects to whom you will give the drug.

Alomar	Denman	Han	Liang	Rosen
Asihiro	Durr	Howard	Maldonado	Solomon
Bennett	Edwards	Hruska	Marsden	Tompkins
Bikalis	Farouk	Imrani	Moore	Townsend
Chen	Fratianna	James	O'Brian	Tullock
Clemente	George	Kaplan	Ogle	Underwood
Cranston	Green	Krushchev	Plochman	Willis
Curtis	Guillen	Lawless	Rodriguez	Zhang

3.54 You decide to use a completely randomized design in the two-factor experiment on response to advertising described in Example 3.10 (page 186). You have 36 students who will serve as subjects. Outline the design. Then use Table B at line 130 to randomly assign the subjects to the 6 treatments.

3.55 **The placebo effect.** A survey of physicians found that some doctors give a placebo to a patient who complains of pain for which the physician can find no cause. If the patient's pain improves, these doctors conclude that it had no physical basis. The medical school researchers who conducted the survey claimed that these doctors do not understand the placebo effect. Why?

3.56 **Advertising to men and women.** Return to the advertising experiment of Example 3.10 (page 186). You have 36 subjects: 24 women and 12 men. Men and women often react differently to advertising. You therefore decide to use a block design with the two genders as blocks. You must assign the 6 treatments at random within each block separately.

(a) Outline the design with a diagram.

(b) Use Table B, beginning at line 140, to do the randomization. Report your result in a table that lists the 24 women and 12 men and the treatment you assigned to each.

3.57 **Temperature and work performance.** An expert on worker performance is interested in the effect of room temperature on the performance of tasks requiring manual dexterity. She chooses temperatures of 70°F and 90°F as treatments. The response variable is the number of correct insertions, during a 30-minute period, in a peg-and-hole apparatus that requires the use of both hands simultaneously. Each subject is trained on the apparatus and then asked to make as many insertions as possible in 30 minutes of continuous effort.

(a) Outline a completely randomized design to compare dexterity at 70° and 90°. Twenty subjects are available.

(b) Because individuals differ greatly in dexterity, the wide variation in individual scores may hide the systematic effect of temperature unless there are many subjects in each group. Describe in detail the design of a matched pairs experiment in which each subject serves as his or her own control.

3.58 **The culture of Mexican Americans.** There are several psychological tests that measure the extent to which Mexican Americans are oriented toward Mexican/Spanish or Anglo/English culture. Two such tests are the Bicultural Inventory (BI) and the Acculturation Rating Scale for Mexican Americans (ARSMA). To study the correlation between the scores on these two tests, researchers will give both tests to a group of 22 Mexican Americans.

(a) Briefly describe a matched pairs design for this study. In particular, how will you use randomization in your design?

(b) You have an alphabetized list of the subjects (numbered 1 to 22). Carry out the randomization required by your design and report the result.

3.59 **More on calcium and blood pressure.** You are participating in the design of a medical experiment to investigate whether a calcium supplement in the diet will reduce the blood pressure of middle-aged men. Preliminary work suggests that calcium may be effective and that the effect may be greater for black men than for white men.

    (a) Outline in graphical form the design of an appropriate experiment.

    (b) Choosing the sizes of the treatment groups requires more statistical expertise. We will learn more about this aspect of design in later chapters. Explain in plain language the advantage of using larger groups of subjects.

STATISTICS in SUMMARY

Designs for producing data are essential parts of statistics in practice. The Statistics in Summary figure on the next page displays the big ideas visually. Random sampling and randomized comparative experiments are perhaps the most important statistical inventions in this century. Both were slow to gain acceptance, and you will still see many voluntary response samples and uncontrolled experiments. This chapter has explained good techniques for producing data and has also explained why bad techniques often produce worthless data.

The deliberate use of chance in producing data is a central idea in statistics. It allows use of the laws of probability to analyze data, as we will see in the following chapters. Here are the major skills you should have now that you have studied this chapter.

A. SAMPLING

1. Identify the population in a sampling situation.
2. Recognize bias due to voluntary response samples and other inferior sampling methods.
3. Use software or Table B of random digits to select a simple random sample (SRS) from a population.
4. Recognize the presence of undercoverage and nonresponse as sources of error in a sample survey. Recognize the effect of the wording of questions on the responses.
5. Use random digits to select a stratified random sample from a population when the strata are identified.

B. EXPERIMENTS

1. Recognize whether a study is an observational study or an experiment.
2. Recognize bias due to confounding of explanatory variables with lurking variables in either an observational study or an experiment.
3. Identify the factors (explanatory variables), treatments, response variables, and experimental units or subjects in an experiment.
4. Outline the design of a completely randomized experiment using a diagram like those in Figures 3.4 and 3.5. The diagram in a specific

STATISTICS IN SUMMARY
Simple Random Sample

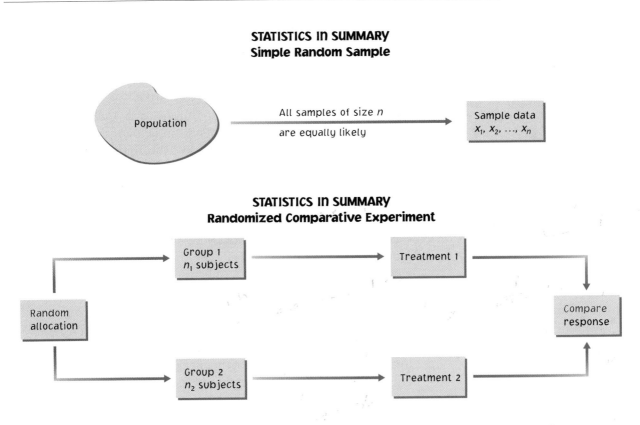

STATISTICS IN SUMMARY
Randomized Comparative Experiment

case should show the sizes of the groups, the specific treatments, and the response variable.

5. Use Table B of random digits to carry out the random assignment of subjects to groups in a completely randomized experiment.

6. Recognize the placebo effect. Recognize when the double-blind technique should be used.

7. Explain why a randomized comparative experiment can give good evidence for cause-and-effect relationships.

CHAPTER 3 Review Exercises

3.60 **Repairing knees in comfort (EESEE).** Injured knees are routinely repaired by arthroscopic surgery that does not require opening up the knee. Can we reduce patient discomfort by giving them a nonsteroidal anti-inflammatory drug (NSAID)? Eighty-three patients were placed in three groups. Group A received the NSAID both before and after surgery. Group B was given a placebo before and the NSAID after. Group C received a placebo both before and after surgery. The patients recorded a pain score by answering questions one day after the surgery.[17]

(BOB DAEMMRICH STOCK BOSTON.)

(a) Outline the design of this experiment. You do not need to do the randomization that your design requires.

(b) You read that "the patients, physicians, and physical therapists were blinded" during the study. What does this mean?

(c) You also read that "the pain scores for Group A were significant lower than Group C but not significantly lower than Group B." What does this mean? What does this finding lead you to conclude about the use of NSAIDs?

3.61 **Physical fitness and leadership.** A study of the relationship between physical fitness and leadership uses as subjects middle-aged executives who have volunteered for an exercise program. The executives are divided into a low-fitness group and a high-fitness group on the basis of a physical examination. All subjects then take a psychological test designed to measure leadership, and the results for the two groups are compared. Is this an observational study or an experiment? Explain your answer.

3.62 **Treating breast cancer.** What is the preferred treatment for breast cancer that is detected in its early stages? The most common treatment was once removal of the breast. It is now usual to remove only the tumor and nearby lymph nodes, followed by radiation. To study whether these treatments differ in their effectiveness, a medical team examines the records of 25 large hospitals and compares the survival times after surgery of all women who have had either treatment.

(a) What are the explanatory and response variables?

(b) Explain carefully why this study is not an experiment.

(c) Explain why confounding will prevent this study from discovering which treatment is more effective. (The current treatment was in fact recommended after a large randomized comparative experiment.)

3.63 **Canada's national health care.** The Ministry of Health in the Canadian Province of Ontario wants to know whether the national health care system is achieving its goals in the province. Much information about health care comes from patient records, but that source doesn't allow us to compare people who use health services with those who don't. So the Ministry of Health conducted the Ontario Health Survey, which interviewed a random sample of 61,239 people who live in the Province of Ontario.[18]

(a) What is the population for this sample survey? What is the sample?

(b) The survey found that 76% of males and 86% of females in the sample had visited a general practitioner at least once in the past year. Do you think these estimates are close to the truth about the entire population? Why?

3.64 **Daytime running lights.** Canada requires that cars be equipped with "daytime running lights," headlights that automatically come on at a low level when the car is started. Some manufacturers are now equipping cars sold in the United States with running lights. Will running lights reduce accidents by making cars more visible?

(a) Briefly discuss the design of an experiment to help answer this question. In particular, what response variables will you examine?

(b) Example 3.15 (page 195) discusses center brake lights. What cautions do you draw from that example that apply to an experiment on the effects of running lights?

3.65 **How much do students earn?** A university's financial aid office wants to know how much it can expect students to earn from summer employment. This information will be used to set the level of financial aid. The population contains 3478 students who have completed at least one year of study but have not yet graduated. The university will send a questionnaire to an SRS of 100 of these students, drawn from an alphabetized list.

(a) Describe how you will label the students in order to select the sample.

(b) Use Table B, beginning at line 105, to select the *first 5* students in the sample.

3.66 **Sampling college faculty.** A labor organization wants to study the attitudes of college faculty members toward collective bargaining. These attitudes appear to be different depending on the type of college. The American Association of University Professors classifies colleges as follows:

Class I. Offer doctorate degrees and award at least 15 per year.

Class IIA. Award degrees above the bachelor's but are not in Class I.

Class IIB. Award no degrees beyond the bachelor's.

Class III. Two-year colleges.

Discuss the design of a sample of faculty from colleges in your state, with total sample size about 200.

3.67 **Sampling students.** You want to investigate the attitudes of students at your school about the school's policy on sexual harassment. You have a grant that will pay the costs of contacting about 500 students.

(a) Specify the exact population for your study. For example, will you include part-time students?

(b) Describe your sample design. Will you use a stratified sample?

(c) Briefly discuss the practical difficulties that you anticipate. For example, how will you contact the students in your sample?

3.68 **Do antioxidants prevent cancer?** People who eat lots of fruits and vegetables have lower rates of colon cancer than those who eat little of these foods. Fruits and vegetables are rich in "antioxidants" such as vitamins A, C, and E. Will taking antioxidants help prevent colon cancer? A clinical trial studied this question with 864 people who were at risk of colon cancer. The subjects were divided into four groups: daily beta carotene, daily vitamins C and E, all three vitamins every day, and daily placebo. After four years, the researchers were surprised to find no significant difference in colon cancer among the groups.[19]

(a) What are the explanatory and response variables in this experiment?

(b) Outline the design of the experiment. Use your judgment in choosing the group sizes.

(c) Assign labels to the 864 subjects and use Table B starting at line 118 to choose the *first 5* subjects for the beta carotene group.

(d) The study was double-blind. What does this mean?

(e) What does "no significant difference" mean in describing the outcome of the study?

(f) Suggest some lurking variables that could explain why people who eat lots of fruits and vegetables have lower rates of colon cancer. The experiment suggests that these variables, rather than the antioxidants, may be responsible for the observed benefits of fruits and vegetables.

3.69 **Comparing corn varieties.** New varieties of corn with altered amino acid content may have higher nutritional value than standard corn, which is low in the amino acid lysine. An experiment compares two new varieties, called opaque-2 and floury-2, with normal corn. The researchers mix corn-soybean meal diets using each type of corn at each of three protein levels: 12% protein, 16% protein, and 20% protein. They feed each diet to 10 one-day-old male chicks and record their weight gains after 21 days. The weight gain of the chicks is a measure of the nutritional value of their diet.

(a) What are the experimental units and the response variable in this experiment?

(b) How many factors are there? How many treatments? Use a diagram like Figure 3.3 to describe the treatments. How many experimental units does the experiment require?

(c) Use a diagram to describe a completely randomized design for this experiment. (You do not need to actually do the randomization.)

3.70 **An industrial experiment.** A chemical engineer is designing the production process for a new product. The chemical reaction that produces the product may have higher or lower yield, depending on the temperature and the stirring rate in the vessel in which the reaction takes place. The engineer decides to investigate the effects of all combinations of two temperatures (50°C and 60°C) and three stirring rates (60 rpm, 90 rpm, and 120 rpm) on the yield of the process. She will process two batches of the product at each combination of temperature and stirring rate. In Exercise 3.34 (page 188) you identified the treatments.

(a) Outline in graphic form the design of an appropriate experiment.

(b) The randomization in this experiment determines the order in which batches of the product will be processed according to each treatment. Use Table B, starting at line 128, to carry out the randomization and state the result.

3.71 **Speeding the mail?** Is the number of days a letter takes to reach another city affected by the time of day it is mailed and whether or not the zip code is used? Describe briefly the design of a two-factor experiment to investigate this question. Be sure to specify the treatments exactly and to tell how you will handle lurking variables such as the day of the week on which the letter is mailed.

3.72 **McDonald's versus Wendy's.** Do consumers prefer the taste of a cheeseburger from McDonald's or from Wendy's in a blind test in which neither burger is identified? Describe briefly the design of a matched pairs experiment to investigate this question.

3.73 The previous two exercises illustrate the use of statistically designed experiments to answer questions that arise in everyday life. Select a question of interest to you that an experiment might answer and briefly discuss the design of an appropriate experiment.

3.74 **Calcium and blood pressure.** A randomized comparative experiment examines whether a calcium supplement in the diet reduces the blood pressure of healthy men. The subjects receive either a calcium supplement or a placebo for 12 weeks. The researchers conclude that "the blood pressure of the calcium group was significantly lower than that of the placebo group." "Significant" in this conclusion means statistically significant. Explain what statistically significant means in the context of this experiment, as if you were speaking to a doctor who knows no statistics.

3.75 **Reading a medical journal.** The article in the *New England Journal of Medicine* that presents the final results of the Physicians' Health Study begins with these words: "The Physicians' Health Study is a randomized, double-blind, placebo-controlled trial designed to determine whether low-dose aspirin (325 mg every other day) decreases cardiovascular mortality and whether beta carotene reduces the incidence of cancer."[20] Doctors are expected to understand this. Explain to a doctor who knows no statistics what "randomized," "double-blind," and "placebo-controlled" mean.

Understanding Inference

The purpose of statistics is to gain understanding from data. We can seek understanding in different ways, depending on the circumstances. We have studied one approach to data, *exploratory data analysis,* in some detail. In examining the design of sample and experiments, we began to move from data analysis toward *statistical inference.* Both types of reasoning are essential to effective work with data. Here is a brief sketch of the differences between them.

Exploratory data analysis	Statistical inference
Purpose is unrestricted exploration of the data, searching for interesting patterns.	Purpose is to answer specific questions, posed before the data were produced.
Conclusions apply only to the individuals and circumstances for which we have data in hand.	Conclusions apply to a larger group of individuals or a broader class of circumstances.
Conclusions are informal, based on what we see in the data.	Conclusions are formal, backed by a statement of our confidence in them.

These distinctions help us understand how inference differs from data analysis, but in practice the two approaches cooperate. Inference usually requires that the pattern of the data be reasonably regular. Data analysis, especially using graphs, is an essential first step when we want to do

"Oh, sure, he remembers everything. But show me one significant insight he's been able to draw from all that data."

(REPRINTED BY PERMISSION OF HARALD BAKKAN AND MISCHA RICHTER FROM THE CHRONICLE OF HIGHER EDUCATION.)

inference. Data production is also closely linked to inference. A good design for producing data is the best guarantee that inference makes sense.

In Part II of this book you will gain an understanding of the reasoning and basic methods of statistical inference. Chapters 4 and 6 provide the core concepts. Probability is the topic of Chapter 4 and the optional Chapter 5. Chapter 4 stresses sampling distributions, which are the foundation for inference. Chapter 6 presents the reasoning of statistical inference. Chapters 7 and 8 discuss practical examples of inference. The link between inference, data analysis, and data production is clearest when we face real settings in Chapters 7 and 8.

(COURTESY DR. DAVID BLACKWELL, UCA, BERKELEY.)

David Blackwell

Mathematics in the Service of Statistics

> *Statistics has been advanced not only by people concerned with practical problems, but also by people whose first love is mathematics for its own sake.*

Statistical practice rests in part on statistical theory. Statistics has been advanced not only by people concerned with practical problems, from Florence Nightingale to R. A. Fisher and John Tukey, but also by people whose first love is mathematics for its own sake. David Blackwell (1919–) is one of the major contemporary contributors to the mathematical study of statistics.

Blackwell grew up in Illinois, earned a doctorate in mathematics at the age of 22, and in 1944 joined the faculty of Howard University in Washington, D.C. "It was the ambition of every black scholar in those days to get a job at Howard University," he says. "That was the best job you could hope for." Society changed, and in 1954 Blackwell became professor of statistics at the University of California at Berkeley.

Washington, D.C., had an active statistical community, and the young mathematician Blackwell soon began to work on mathematical aspects of statistics. He explored the behavior of statistical procedures which, rather than working with a fixed sample, keep taking observations until there is enough information to reach a firm conclusion. He found insights into statistical inference by thinking of inference as a game in which nature plays against the statistician. Blackwell's work uses probability theory, the mathematics that describes chance behavior. We must travel the same route, though only a short distance. This chapter presents, in a rather informal fashion, the probabilistic ideas needed to understand the reasoning of statistical inference.

Probability and Sampling Distributions

(JEFF GREENBERG/VISUALS UNLIMITED.)

Blackwell thought of statistical problems as playing a game in which "nature" tries to defeat us.

213

Introduction

The reasoning of statistical inference rests on asking, "How often would this method give a correct answer if I used it very many times?" Inference is most secure when we produce data by random sampling or randomized comparative experiments. The reason is that when we use chance to choose respondents or assign subjects, the laws of probability answer the question "What would happen if we did this many times?" The purpose of this chapter is to see what the laws of probability tell us, but without going into the mathematics of probability theory.

4.1 Randomness

What is the mean income of households in the United States? The government's Current Population Survey contacted a sample of 50,000 households in 1997. Their mean income was $\bar{x} = \$49,692$.[1] That $49,692 describes the sample, but we use it to estimate the mean income of all households. We must now take care to keep straight whether a number describes a sample or a population. Here is the vocabulary we use.

> **PARAMETER, STATISTIC**
>
> A **parameter** is a number that describes the population. In statistical practice, the value of a parameter is not known because we cannot examine the entire population.
>
> A **statistic** is a number that can be computed from the sample data without making use of any unknown parameters. In practice, we often use a statistic to estimate an unknown parameter.

EXAMPLE 4.1   Household income

The mean income of the sample of households contacted by the Current Population Survey was $\bar{x} = \$49,692$. The number $49,692 is a *statistic* because it describes this one Current Population Survey sample. The population that the poll wants to draw conclusions about is all 103 million U.S. households. The *parameter* of interest is the mean income of all of these households. We don't know the value of this parameter.

Remember: **s**tatistics come from **s**amples, and **p**arameters come from **p**opulations. As long as we were just doing data analysis, the distinction between population and sample was not important. Now, however, it is essential. The notation we use must reflect this distinction. We write μ (the Greek letter mu) for the **mean of a population**. This is a fixed parameter that is unknown when we use a sample for inference. The **mean of the sample** is the familiar $\bar{x}$, the average of the observations in the sample. This is a statistic that would almost certainly take a different value if we chose another sample from the same population. The sample mean $\bar{x}$ from a sample or an experiment is an estimate of the mean μ of the underlying population.

population mean μ
sample mean $\bar{x}$

APPLY YOUR KNOWLEDGE

State whether each boldface number in Exercises 4.1 to 4.3 is a *parameter* or a *statistic*.

4.1 A carload lot of ball bearings has mean diameter **2.5003** centimeters (cm). This is within the specifications for acceptance of the lot by the purchaser. By chance, an inspector chooses 100 bearings from the lot that have mean diameter **2.5009** cm. Because this is outside the specified limits, the lot is mistakenly rejected.

4.2 A telemarketing firm in Los Angeles uses a device that dials residential telephone numbers in that city at random. Of the first 100 numbers dialed, **48%** are unlisted. This is not surprising because **52%** of all Los Angeles residential phones are unlisted.

4.3 A researcher carries out a randomized comparative experiment with young rats to investigate the effects of a toxic compound in food. She feeds the control group a normal diet. The experimental group receives a diet with 2500 parts per million of the toxic material. After 8 weeks, the mean weight gain is **335** grams for the control group and **289** grams for the experimental group.

The idea of probability

How can $\bar{x}$, based on a sample of only a few of the 100 million American households, be an accurate estimate of μ? After all, a second random sample taken at the same time would choose different households and no doubt produce a different value of $\bar{x}$. This basic fact is called **sampling variability:** the value of a statistic varies in repeated random sampling.

sampling variability

To understand why sampling variability is not fatal, we must look closely at chance behavior. The big fact that emerges is this: **chance behavior is unpredictable in the short run but has a regular and predictable pattern in the long run.**

Toss a coin, or choose an SRS. The result can't be predicted in advance, because the result will vary when you toss the coin or choose the sample repeatedly. But there is still a regular pattern in the results, a pattern that emerges clearly only after many repetitions. This remarkable fact is the basis for the idea of probability.

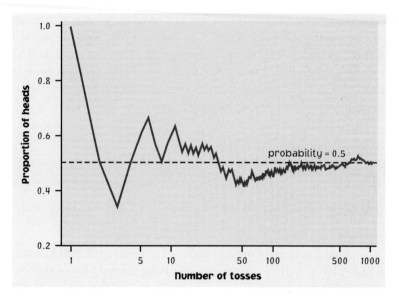

Figure 4.1 *The proportion of tosses of a coin that give a head changes as we make more tosses. Eventually, however, the proportion approaches 0.5, the probability of a head.*

(SUPER STOCK)

EXAMPLE 4.2   Coin tossing

When you toss a coin, there are only two possible outcomes, heads or tails. Figure 4.1 shows the results of tossing a coin 1000 times. For each number of tosses from 1 to 1000, we have plotted the proportion of those tosses that gave a head. The first toss was a head, so the proportion of heads starts at 1. The second toss was a tail, reducing the proportion of heads to 0.5 after two tosses. The next three tosses gave a tail followed by two heads, so the proportion of heads after five tosses is 3/5, or 0.6.

The proportion of tosses that produce heads is quite variable at first, but it settles down as we make more and more tosses. Eventually this proportion gets close to 0.5 and stays there. We say that 0.5 is the *probability* of a head. The probability 0.5 appears as a horizontal line on the graph.

"Random" in statistics is not a synonym for "haphazard" but a description of a kind of order that emerges only in the long run. We often encounter the unpredictable side of randomness in our everyday experience, but we rarely see enough repetitions of the same random phenomenon to observe the long-term regularity that probability describes. You can see that regularity emerging in Figure 4.1. In the very long run, the proportion of tosses that give a head is 0.5. This is the intuitive idea of probability. Probability 0.5 means "occurs half the time in a very large number of trials."

We might suspect that a coin has probability 0.5 of coming up heads just because the coin has two sides. As Exercises 4.4 and 4.5 illustrate, such suspicions are not always correct. The idea of probability is empirical. That is, it is based on observation rather than theorizing. Probability describes what happens in very many trials, and we must actually observe many trials to pin down a probability. In the case of tossing a coin, some diligent people have in fact made thousands of tosses.

EXAMPLE 4.3	**Some coin tossers**

The French naturalist Count Buffon (1707–1788) tossed a coin 4040 times. Result: 2048 heads, or proportion 2048/4040 = 0.5069 for heads.

Around 1900, the English statistician Karl Pearson heroically tossed a coin 24,000 times. Result: 12,012 heads, a proportion of 0.5005.

While imprisoned by the Germans during World War II, the South African mathematician John Kerrich tossed a coin 10,000 times. Result: 5067 heads, a proportion of 0.5067.

RANDOMNESS AND PROBABILITY

We call a phenomenon **random** if individual outcomes are uncertain but there is nonetheless a regular distribution of outcomes in a large number of repetitions.

The **probability** of any outcome of a random phenomenon is the proportion of times the outcome would occur in a very long series of repetitions.

What looks random?

Toss a coin six times and record heads (H) or tails (T) on each toss. Which of these outcomes is more probable: HTHTTH or TTTHHH? Almost everyone says that HTHTTH is more probable, because TTTHHH does not "look random." In fact, both are equally probable. That heads has probability 0.5 says that about half of a very long sequence of tosses will be heads. It doesn't say that heads and tails must come close to alternating in the short run. The coin doesn't know what past outcomes were, and it can't try to create a balanced sequence.

APPLY YOUR KNOWLEDGE

4.4 **Pennies spinning.** Hold a penny upright on its edge under your forefinger on a hard surface, then snap it with your other forefinger so that it spins for some time before falling. Based on 50 spins, estimate the probability of heads.

4.5 **Pennies falling over.** You may feel that it is obvious that the probability of a head in tossing a coin is about 1/2 because the coin has two faces. Such opinions are not always correct. The previous exercise asked you to spin a penny rather than toss it—that changes the probability of a head. Now try another variation. Stand a penny on edge on a hard, flat surface. Pound the surface with your hand so that the penny falls over. What is the probability that it falls with heads upward? Make at least 50 trials to estimate the probability of a head.

Thinking about randomness

That some things are random is an observed fact about the world. The outcome of a coin toss, the time between emissions of particles by a radioactive source, and the sexes of the next litter of lab rats are all random. So is the outcome of a random sample or a randomized experiment. Probability theory is the branch of mathematics that describes random behavior. Of course, we can never observe a probability exactly. We could always continue tossing the coin, for example. Mathematical probability is an idealization based on imagining what would happen in an indefinitely long series of trials.

The best way to understand randomness is to observe random behavior—not only the long-run regularity but the unpredictable results of short runs. You can do this with physical devices, as in Exercises 4.4 to 4.8, but computer simulations (imitations) of random behavior allow faster exploration. Exercises 4.11 and 4.13 suggest some simulations of random behavior. As you explore randomness, remember:

independence
- You must have a long series of **independent** trials. That is, the outcome of one trial must not influence the outcome of any other. Imagine a crooked gambling house where the operator of a roulette wheel can stop it where she chooses—she can prevent the proportion of "red" from settling down to a fixed number. These trials are not independent.

- The idea of probability is empirical. Computer simulations start with given probabilities and imitate random behavior, but we can estimate a real-world probability only by actually observing many trials.

- Nonetheless, computer simulations are very useful because we need long runs of trials. In situations such as coin tossing, the proportion of an outcome often requires several hundred trials to settle down to the probability of that outcome. The kinds of physical random devices suggested in the exercises are too slow for this. Short runs give only rough estimates of a probability.

Section 4.1 Summary

A **random phenomenon** has outcomes that we cannot predict but that nonetheless have a regular distribution in very many repetitions.

The **probability** of an event is the proportion of times the event occurs in many repeated trials of a random phenomenon.

Section 4.1 Exercises

4.6 **Random digits.** The table of random digits (Table B) was produced by a random mechanism that gives each digit probability 0.1 of being a 0. What proportion of the first 200 digits in the table are 0s? This proportion is an estimate, based on 200 repetitions, of the true probability, which in this case is known to be 0.1.

4.7 **How many tosses to get a head?** When we toss a penny, experience shows that the probability (long-term proportion) of a head is close to 1/2. Suppose now that we toss the penny repeatedly until we get a head. What is the probability that the first head comes up in an odd number of tosses (1, 3, 5, and so on)? To find out, repeat this experiment 50 times, and keep a record of the number of tosses needed to get a head on each of your 50 trials.

(a) From your experiment, estimate the probability of a head on the first toss. What value should we expect this probability to have?

(b) Use your results to estimate the probability that the first head appears on an odd-numbered toss.

4.8 **Tossing a thumbtack.** Toss a thumbtack on a hard surface 100 times. How many times did it land with the point up? What is the approximate probability of landing point up?

4.9 **Three of a kind.** You read in a book on poker that the probability of being dealt three of a kind in a five-card poker hand is 1/50. Explain in simple language what this means.

4.10 Probability is a measure of how likely an event is to occur. Match one of the probabilities that follow with each statement of likelihood given. (The probability is usually a more exact measure of likelihood than is the verbal statement.)

$$0 \quad 0.01 \quad 0.3 \quad 0.6 \quad 0.99 \quad 1$$

(a) This event is impossible. It can never occur.

(b) This event is certain. It will occur on every trial.

(c) This event is very unlikely, but it will occur once in a while in a long sequence of trials.

(d) This event will occur more often than not.

4.11 **Shaq's free throws.** The basketball player Shaquille O'Neal makes about half of his free throws over an entire season. Use software to simulate 100 free throws shot independently by a player who has probability 0.5 of making each shot. (In most software, the key phrase to look for is "Bernoulli trials." This is the technical term for independent trials with Yes/No outcomes. Our outcomes here are "Hit" and "Miss.")

(a) What percent of the 100 shots did he hit?

(b) Examine the sequence of hits and misses. How long was the longest run of shots made? Of shots missed? (Sequences of random outcomes often show runs longer than our intuition thinks likely.)

(AP/WIDE WORLD PHOTOS.)

4.12 **Winning a baseball game (EESEE).** A study of the home-field advantage in baseball found that over the period from 1969 to 1989 the league champions won 63% of their home games.[2] The two league champions meet in the baseball World Series. Would you use the study results to assign probability 0.63 to the event that the home team wins in a World Series game? Explain your answer.

4.13 **Simulating an opinion poll.** A recent opinion poll showed that about 73% of married women agree that their husbands do at least their fair share of household chores. Suppose that this is exactly true. Choosing a married woman at random then has probability 0.73 of getting one who agrees that her husband does his share. Use software to simulate choosing many women independently. (In most software, the key phrase to look for is "Bernoulli trials." This is the technical term for independent trials with Yes/No outcomes. Our outcomes here are "Agree" or not.)

(a) Simulate drawing 20 women, then 80 women, then 320 women. What proportion agree in each case? We expect (but because of chance variation we can't be sure) that the proportion will be closer to 0.73 in longer runs of trials.

(b) Simulate drawing 20 women 10 times and record the percents in each trial who agree. Then simulate drawing 320 women 10 times and again record the 10 percents. Which set of 10 results is less variable? We expect the results of 320 trials to be more predictable (less variable) than the results of 20 trials. That is "long-run regularity" showing itself.

4.2 Probability Models

Earlier chapters gave mathematical models for linear relationships (in the form of the equation of a line) and for some distributions of data (in the form of normal density curves). Now we must give a mathematical description or model for randomness. To see how to proceed, think first about a very simple random phenomenon, tossing a coin once. When we toss a coin, we cannot know the outcome in advance. What do we know? We are willing to say that the outcome will be either heads or tails. We believe that each of these outcomes has probability 1/2. This description of coin tossing has two parts:

• A list of possible outcomes.
• A probability for each outcome.

Such a description is the basis for all probability models. Here is the basic vocabulary we use.

> **PROBABILITY MODELS**
>
> The **sample space** S of a random phenomenon is the set of all possible outcomes.
>
> An **event** is any outcome or a set of outcomes of a random phenomenon. That is, an event is a subset of the sample space.
>
> A **probability model** is a mathematical description of a random phenomenon consisting of two parts: a sample space S and a way of assigning probabilities to events.

Figure 4.2 *The 36 possible outcomes in rolling two dice.*

The sample space S can be very simple or very complex. When we toss a coin once, there are only two outcomes, heads and tails. The sample space is $S = \{H, T\}$. If we draw a random sample of 50,000 U.S. households, as the Current Population Survey does, the sample space contains all possible choices of 50,000 of the 103 million households in the country. This S is extremely large. Each member of S is a possible sample, which explains the term *sample space*.

EXAMPLE 4.4   **Rolling dice**

Rolling two dice is a common way to lose money in casinos. There are 36 possible outcomes when we roll two dice and record the up-faces in order (first die, second die). Figure 4.2 displays these outcomes. They make up the sample space S. "Roll a 5" is an event, call it A, that contains four of these 36 outcomes:

$$A = \{ \quad \boxed{\cdot} \;\; \boxed{:\!:} \qquad \boxed{\cdot\cdot} \;\; \boxed{\cdot\!\cdot\!\cdot} \qquad \boxed{\cdot\!\cdot\!\cdot} \;\; \boxed{\cdot\cdot} \qquad \boxed{:\!:} \;\; \boxed{\cdot} \quad \}$$

Gamblers care only about the number of pips on the up-faces of the dice. The sample space for rolling two dice and counting the pips is

$$S = \{2, 3, 4, 5, 6, 7, 8, 9, 10, 11, 12\}$$

Comparing this S with Figure 4.2 reminds us that we can change S by changing the detailed description of the random phenomenon we are describing.

APPLY YOUR KNOWLEDGE

4.14   In each of the following situations, describe a sample space S for the random phenomenon. In some cases, you have some freedom in your choice of S.

   (a) A seed is planted in the ground. It either germinates or fails to grow.

(b) A patient with a usually fatal form of cancer is given a new treatment. The response variable is the length of time that the patient lives after treatment.

(c) A student enrolls in a statistics course and at the end of the semester receives a letter grade.

(d) A basketball player shoots four free throws. You record the sequence of hits and misses.

(e) A basketball player shoots four free throws. You record the number of baskets she makes.

4.15   In each of the following situations, describe a sample space S for the random phenomenon. In some cases you have some freedom in specifying S, especially in setting the largest and the smallest value in S.

(a) Choose a student in your class at random. Ask how much time that student spent studying during the past 24 hours.

(b) The Physicians' Health Study asked 11,000 physicians to take an aspirin every other day and observed how many of them had a heart attack in a five-year period.

(c) In a test of a new package design, you drop a carton of a dozen eggs from a height of 1 foot and count the number of broken eggs.

(d) Choose a student in your class at random. Ask how much cash that student is carrying.

(e) A nutrition researcher feeds a new diet to a young male white rat. The response variable is the weight (in grams) that the rat gains in 8 weeks.

Probability rules

The true probability of any outcome—say, "roll a 5 when we toss two dice"—can be found only by actually tossing two dice many times, and then only approximately. How then can we describe probability mathematically? Rather than try to give "correct" probabilities, we start by laying down facts that must be true for any assignment of probabilities. These facts follow from the idea of probability as "the long-run proportion of repetitions on which an event occurs."

1. **Any probability is a number between 0 and 1**. Any proportion is a number between 0 and 1, so any probability is also a number between 0 and 1. An event with probability 0 never occurs, and an event with probability 1 occurs on every trial. An event with probability 0.5 occurs in half the trials in the long run.

2. **All possible outcomes together must have probability 1**. Because some outcome must occur on every trial, the sum of the probabilities for all possible outcomes must be exactly 1.

3. **The probability that an event does not occur is 1 minus the probability that the event does occur**. If an event occurs in (say) 70% of all trials, it fails to occur in the other 30%. The probability that an event occurs and the probability that it does not occur always add to 100%, or 1.

4. **If two events have no outcomes in common, the probability that one or the other occurs is the sum of their individual probabilities**. If one event occurs in 40% of all trials, a different event occurs in 25% of all trials, and the two can never occur together, then one or the other occurs on 65% of all trials because 40% + 25% = 65%.

We can use mathematical notation to state Facts 1 to 4 more concisely. Capital letters near the beginning of the alphabet denote events. If A is any event, we write its probability as $P(A)$. Here are our probability facts in formal language. As you apply these rules, remember that they are just another form of intuitively true facts about long-run proportions.

PROBABILITY RULES

Rule 1. The probability $P(A)$ of any event A satisfies $0 \leq P(A) \leq 1$.

Rule 2. If S is the sample space in a probability model, then $P(S) = 1$.

Rule 3. For any event A,

$$P(A \text{ does not occur}) = 1 - P(A)$$

Rule 4. Two events A and B are **disjoint** if they have no outcomes in common and so can never occur simultaneously. If A and B are disjoint,

$$P(A \text{ or } B) = P(A) + P(B)$$

This is the **addition rule for disjoint events**.

EXAMPLE 4.5   **Marital status of young women**

Draw a woman aged 25 to 29 years old at random and record her marital status. "At random" means that we give every such woman the same chance to be the one we choose. That is, we choose an SRS of size 1. The probability of any marital status is just the proportion of all women aged 25 to 29 who have that status—if we drew many women, this is the proportion we would get. Here is the probability model:

Marital status	Never married	Married	Widowed	Divorced
Probability	0.353	0.574	0.002	0.071

Each probability is between 0 and 1. The probabilities add to 1, because these outcomes together make up the sample space S.

The probability that the woman we draw is not married is, by Rule 3,

$$P(\text{not married}) = 1 - P(\text{married})$$

$$= 1 - 0.574 = 0.426$$

That is, if 57.4% are married, then the remaining 42.6% are not married.

"Never married" and "divorced" are disjoint events, because no woman can be both never married and divorced. So the addition rule says that

$$P(\text{never married or divorced}) = P(\text{never married}) + P(\text{divorced})$$

$$= 0.353 + 0.071 = 0.424$$

That is, 42.4% of women in this age group are either never married or divorced.

EXAMPLE 4.6 | **Probabilities for rolling dice**

(SUPER STOCK)

Figure 4.2 displays the 36 possible outcomes of rolling two dice. What probabilities should we assign to these outcomes?

Casino dice are carefully made. Their spots are not hollowed out, which would give the faces different weights, but are filled with white plastic of the same density as the colored plastic of the body. For casino dice it is reasonable to assign the same probability to each of the 36 outcomes in Figure 4.2. Because all 36 outcomes together must have probability 1 (Rule 2), each outcome must have probability 1/36.

Gamblers are often interested in the sum of the pips on the up-faces. What is the probability of rolling a 5? Because the event "roll a 5" contains the four outcomes displayed in Example 4.4, the addition rule (Rule 4) says that its probability is

$$P(\text{roll a 5}) = P(\boxed{\cdot}\ \boxed{\vdots}) + P(\boxed{\therefore}\ \boxed{\ddots}) + P(\boxed{\ddots}\ \boxed{\therefore}) + P(\boxed{\vdots}\ \boxed{\cdot})$$

$$= \frac{1}{36} + \frac{1}{36} + \frac{1}{36} + \frac{1}{36}$$

$$= \frac{4}{36} = 0.111$$

What about the probability of rolling a 7? In Figure 4.2 you will find six outcomes for which the sum of the pips is 7. The probability is 6/36, or about 0.167.

APPLY YOUR KNOWLEDGE

4.16 **Moving up.** A sociologist studying social mobility in Denmark finds that the probability that the son of a lower-class father remains in the lower class is 0.46. What is the probability that the son moves to one of the higher classes?

4.17 **Causes of death.** Government data assign a single cause for each death that occurs in the United States. The data show that the probability is 0.45 that a randomly chosen death was due to cardiovascular (mainly heart) disease, and 0.22 that it was due to cancer. What is the probability that a death was due either to cardiovascular disease or to cancer? What is the probability that the death was due to some other cause?

4.18 **Do husbands do their share?** The *New York Times* (August 21, 1989) reported a poll that interviewed a random sample of 1025 women. The married women in the sample were asked whether their husbands did their fair share of household chores. Here are the results:

Outcome	Probability
Does more than his fair share	0.12
Does his fair share	0.61
Does less than his fair share	?

These proportions are probabilities for the random phenomenon of choosing a married woman at random and asking her opinion.

(a) What must be the probability that the woman chosen says that her husband does less than his fair share? Why?

(b) The event "I think my husband does at least his fair share" contains the first two outcomes. What is its probability?

Assigning probabilities: finite number of outcomes

Examples 4.5 and 4.6 illustrate one way to assign probabilities to events: assign a probability to every individual outcome, then add these probabilities to find the probability of any event. If such an assignment is to satisfy the rules of probability, the probabilities of all the individual outcomes must sum to exactly 1.

PROBABILITIES IN A FINITE SAMPLE SPACE

Assign a probability to each individual outcome. These probabilities must be numbers between 0 and 1 and must have sum 1.

The probability of any event is the sum of the probabilities of the outcomes making up the event.

What are the odds?

Gamblers often express chance in terms of *odds* rather than probability. Odds of *A* to *B* against an outcome means that the probability of that outcome is $B/(A + B)$. So "odds of 5 to 1" is another way of saying "probability 1/6." A probability is always between zero and one, but odds range from zero to infinity. Although odds are mainly used in gambling, they give us a way to make very small probabilities clearer. "Odds of 999 to 1" may be easier to understand than "probability 0.001."

Really random digits

For purists, the RAND Corporation long ago published a book titled *One Million Random Digits.* The book lists 1,000,000 digits that were produced by a very elaborate physical randomization and really are random. An employee of RAND once told me that this is not the most boring book that RAND has ever published.

EXAMPLE 4.7 **Random digits**

The table of random digits in the back of the book was produced by software that generates digits between 0 and 9 at random. If we produce one digit, the sample space is

$$S = \{0, 1, 2, 3, 4, 5, 6, 7, 8, 9\}$$

Careful randomization makes each outcome equally likely to be any of the 10 candidates. Because the total probability must be 1, the probability of each of the 10 outcomes must be 1/10. This assignment of probabilities to individual outcomes can be summarized in a table as follows:

Outcome	0	1	2	3	4	5	6	7	8	9
Probability	0.1	0.1	0.1	0.1	0.1	0.1	0.1	0.1	0.1	0.1

The probability of the event that an odd digit is chosen is

$$P(\text{odd outcome}) = P(1) + P(3) + P(5) + P(7) + P(9) = 0.5$$

This assignment of probability satisfies all of our rules. For example, we can find the probability of an even digit from Rule 3:

$$P(\text{even outcome}) = P(\text{outcome is not odd})$$

$$= 1 - P(\text{odd outcome})$$

$$= 1 - 0.5 = 0.5$$

Check that you get the same result by adding the probabilities of all the possible even outcomes.

APPLY YOUR KNOWLEDGE

4.19 **Rolling a die.** Figure 4.3 displays several assignments of probabilities to the six faces of a die. We can learn which assignment is actually *accurate* for a particular die only by rolling the die many times. However, some of the assignments are not *legitimate* assignments of probability. That is, they do not obey the rules. Which are legitimate and which are not? In the case of the illegitimate models, explain what is wrong.

4.20 **High school academic rank.** Select a first-year college student at random and ask what his or her academic rank was in high school. Here are the probabilities, based on proportions from a large sample survey of first-year students:

Outcome	Probability			
	Model 1	Model 2	Model 3	Model 4
⚀	1/7	1/3	1/3	1
⚁	1/7	1/6	1/6	1
⚂	1/7	1/6	1/6	2
⚃	1/7	0	1/6	1
⚄	1/7	1/6	1/6	1
⚅	1/7	1/6	1/6	2

Figure 4.3 *Four assignments of probabilities to the six faces of a die, for Exercise 4.19.*

Rank	Top 20%	Second 20%	Third 20%	Fourth 20%	Lowest 20%
Probability	0.41	0.23	0.29	0.06	0.01

(a) What is the sum of these probabilities? Why do you expect the sum to have this value?

(b) What is the probability that a randomly chosen first-year college student was not in the top 20% of his or her high school class?

(c) What is the probability that a first-year student was in the top 40% in high school?

4.21 **Blood types.** All human blood can be typed as one of O, A, B, or AB, but the distribution of the types varies a bit with race. Here is the probability model for the blood type of a randomly chosen black American:

Blood type	O	A	B	AB
Probability	0.49	0.27	0.20	?

(a) What is the probability of type AB blood? Why?

(b) Maria has type B blood. She can safely receive blood transfusions from people with blood types O and B. What is the probability that a randomly chosen black American can donate blood to Maria?

Assigning probabilities: intervals of outcomes

Suppose that we want to choose a number at random between 0 and 1, allowing *any* number between 0 and 1 as the outcome. A software random number generator will do this. The sample space is now an entire interval of numbers:

$$S = \{\text{all numbers between 0 and 1}\}$$

Call the outcome of the random number generator Y for short. How can we assign probabilities to such events as $\{0.3 \leq Y \leq 0.7\}$? As in the case of selecting a random digit, we would like all possible outcomes to be equally likely. But we cannot assign probabilities to each individual value of Y and then add them, because there are infinitely many possible values.

We use a new way of assigning probabilities directly to events—as *areas under a density curve*. Any density curve has area exactly 1 underneath it, corresponding to total probability 1.

EXAMPLE 4.8 **Random numbers**

The random number generator will spread its output uniformly across the entire interval from 0 to 1 as we allow it to generate a long sequence of numbers. The results of many trials are represented by the uniform density curve shown in Figure 4.4. This density curve has height 1 over the interval from 0 to 1. The area under the curve is 1, and the probability of any event is the area under the curve and above the event in question.

As Figure 4.4(a) illustrates, the probability that the random number generator produces a number between 0.3 and 0.7 is

$$P(0.3 \leq Y \leq 0.7) = 0.4$$

because the area under the density curve and above the interval from 0.3 to 0.7 is 0.4. The height of the curve is 1 and the area of a rectangle is the product of height

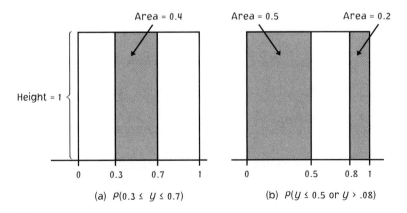

(a) $P(0.3 \leq y \leq 0.7)$ **(b)** $P(y \leq 0.5 \text{ or } y > .08)$

Figure 4.4 *Probability as area under a density curve. This uniform density curve spreads probability evenly between 0 and 1.*

and length, so the probability of any interval of outcomes is just the length of the interval.

Similarly,

$$P(Y \leq 0.5) = 0.5$$

$$P(Y > 0.8) = 0.2$$

$$P(Y \leq 0.5 \text{ or } Y > 0.8) = 0.7$$

Notice that the last event consists of two nonoverlapping intervals, so the total area above the event is found by adding two areas, as illustrated by Figure 4.4(b). This assignment of probabilities obeys all of our rules for probability.

APPLY YOUR KNOWLEDGE

4.22 **Random numbers.** Let X be a random number between 0 and 1 produced by the idealized uniform random number generator described in Example 4.8 and Figure 4.4. Find the following probabilities:

(a) $P(0 \leq X \leq 0.4)$

(b) $P(0.4 \leq X \leq 1)$

(c) $P(0.3 \leq X \leq 0.5)$

(d) $P(0.3 < X < 0.5)$

4.23 **Adding random numbers.** Generate two random numbers between 0 and 1 and take Y to be their sum. The sum Y can take any value between 0 and 2. The density curve of Y is the triangle shown in Figure 4.5.

(a) Verify by geometry that the area under this curve is 1.

(b) What is the probability that Y is less than 1? (Sketch the density curve, shade the area that represents the probability, then find that area. Do this for (c) also.)

(c) What is the probability that Y is less than 0.5?

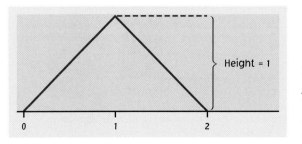

Figure 4.5 *The density curve for the sum of two random numbers. This density curve spreads probability between 0 and 2.*

Normal probability distributions

Any density curve can be used to assign probabilities. The density curves that are most familiar to us are the normal curves. So **normal distributions are probability models**. There is a close connection between a normal distribution as an idealized description for data and a normal probability model. If we look at the heights of all young women, we find that they closely follow the normal distribution with mean $\mu = 64.5$ inches and standard deviation $\sigma = 2.5$ inches. That is a distribution for a large set of data. Now choose one young woman at random. Call her height X. If we repeat the random choice very many times, the distribution of values of X is the same normal distribution.

EXAMPLE 4.9	The height of young women

What is the probability that a randomly chosen young woman has height between 68 and 70 inches?

The height X of the woman we choose has the $N(64.5, 2.5)$ distribution. Find the probability by standardizing and using Table A, the table of standard normal probabilities. We will reserve capital Z for a standard normal variable.

$$P(68 \le X \le 70) = P\left(\frac{68 - 64.5}{2.5} \le \frac{X - 64.5}{2.5} \le \frac{70 - 64.5}{2.5}\right)$$

$$= P(1.4 \le Z \le 2.2)$$

$$= 0.9861 - 0.9192 = 0.0669$$

Figure 4.6 shows the areas under the standard normal curve. The calculation is the same as those we did in Chapter 1. Only the language of probability is new.

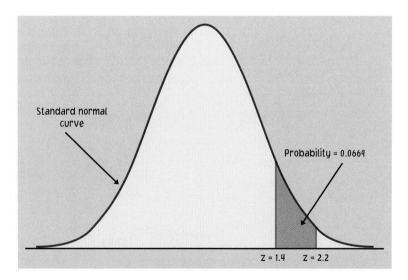

Standard normal curve

Probability = 0.0669

$Z = 1.4$ $Z = 2.2$

Figure 4.6 *The probability in Example 4.9 as an area under the standard normal curve.*

Examples 4.8 and 4.9 use a shorthand notation that is often convenient. We let X stand for the result of choosing a woman at random and measuring her height. We know that X would take a different value if we took another random sample. Because its value changes from one sample to another, we call the height X a *random variable*.

> **RANDOM VARIABLE**
>
> A **random variable** is a variable whose value is a numerical outcome of a random phenomenon.
>
> The **probability distribution** of a random variable X tells us what values X can take and how to assign probabilities to those values.

We usually denote random variables by capital letters near the end of the alphabet, such as X or Y. Of course, the random variables of greatest interest to us are outcomes such as the mean $\bar{x}$ of a random sample, for which we will keep the familiar notation.

APPLY YOUR KNOWLEDGE

4.24 **Iowa Test scores.** The normal distribution with mean $\mu = 6.8$ and standard deviation $\sigma = 1.6$ is a good description of the Iowa Test vocabulary scores of seventh-grade students in Gary, Indiana. Figure 1.13 (page 46) pictures this distribution. Let the random variable X be the Iowa Test score of one Gary seventh grader chosen at random.

    (a) Write the event "the student chosen has a score of 10 or higher" in terms of X.

    (b) Find the probability of this event.

4.25 **Rolling the dice.** Example 4.6 describes the assignment of probabilities to the 36 possible outcomes of tossing two dice. Take the random variable X to be the sum of the values on the two up-faces. Example 4.6 shows that $P(X = 5) = 4/36$ and that $P(X = 7) = 6/36$. Give the complete probability distribution of the random variable X. That is, list the possible values of X and say what the probability of each value is. Start from the 36 outcomes displayed in Figure 4.2.

SECTION 4.2 Summary

A **probability model** for a random phenomenon consists of a sample space S and an assignment of probabilities P.

The **sample space** S is the set of all possible outcomes of the random phenomenon. Sets of outcomes are called **events**. P assigns a number $P(A)$ to an event A as its probability.

Any assignment of probability must obey the rules that state the basic properties of probability:

1. $0 \leq P(A) \leq 1$ for any event A.
2. $P(S) = 1$.
3. For any event A, $P(A \text{ does not occur}) = 1 - P(A)$.
4. **Addition rule:** Events A and B are **disjoint** if they have no outcomes in common. If A and B are disjoint, then $P(A \text{ or } B) = P(A) + P(B)$.

When a sample space S contains finitely many possible values, a probability model assigns each of these values a probability between 0 and 1 such that the sum of all the probabilities is exactly 1. The probability of any event is the sum of the probabilities of all the values that make up the event.

A sample space can contain all values in some interval of numbers. A probability model assigns probabilities as **areas under a density curve**. The probability of any event is the area under the curve above the values that make up the event.

A **random variable** is a variable taking numerical values determined by the outcome of a random phenomenon. The **probability distribution** of a random variable X tells us what the possible values of X are and how probabilities are assigned to those values.

SECTION 4.2 Exercises

4.26 Buy a hot dog and record how many calories it has. Give a reasonable sample space S for your possible results. (Exercise 1.39 on page 42 contains some information to guide you.)

4.27 Choose a student at random and record the number of dollars in bills (ignore change) that he or she is carrying. Give a reasonable sample space S for this random phenomenon. (We don't know the largest amount that a student could reasonably carry, so you will have to make a choice in stating the sample space.)

4.28 **Land in Canada.** Choose an acre of land in Canada at random. The probability is 0.35 that it is forest and 0.03 that it is pasture.

 (a) What is the probability that the acre chosen is not forested?

 (b) What is the probability that it is either forest or pasture?

 (c) What is the probability that a randomly chosen acre in Canada is something other than forest or pasture?

4.29 **Colors of M&M's.** If you draw an M&M candy at random from a bag of the candies, the candy you draw will have one of six colors.

The probability of drawing each color depends on the proportion of each color among all candies made.

(a) Here are the probabilities of each color for a randomly chosen plain M&M:

Color	Brown	Red	Yellow	Green	Orange	Blue
Probability	0.3	0.2	0.2	0.1	0.1	?

What must be the probability of drawing a blue candy?

(b) The probabilities for peanut M&M's are a bit different. Here they are:

Color	Brown	Red	Yellow	Green	Orange	Blue
Probability	0.2	0.1	0.2	0.1	0.1	?

What is the probability that a peanut M&M chosen at random is blue?

(c) What is the probability that a plain M&M is any of red, yellow, or orange? What is the probability that a peanut M&M has one of these colors?

4.30 **Legitimate probabilities?** In each of the following situations, state whether or not the given assignment of probabilities to individual outcomes is legitimate, that is, satisfies the rules of probability. If not, give specific reasons for your answer.

(a) When a coin is spun, $P(H) = 0.55$ and $P(T) = 0.45$.

(b) When two coins are tossed, $P(HH) = 0.4$, $P(HT) = 0.4$, $P(TH) = 0.4$, and $P(TT) = 0.4$.

(c) Plain M&M's have not always had the mixture of colors given in Exercise 4.29. In the past there were no red candies and no blue candies. Tan had probability 0.10 and the other four colors had the same probabilities that are given in Exercise 4.29.

4.31 **Who goes to Paris?** Abby, Deborah, Sam, Tonya, and Roberto work in a firm's public relations office. Their employer must choose two of them to attend a conference in Paris. To avoid unfairness, the choice will be made by drawing two names from a hat. (This is an SRS of size 2.)

(a) Write down all possible choices of two of the five names. This is the sample space.

(b) The random drawing makes all choices equally likely. What is the probability of each choice?

(c) What is the probability that Tonya is chosen?

(d) What is the probability that neither of the two men (Sam and Roberto) is chosen?

4.32 **How big are farms?** Choose an American farm at random and measure its size in acres. Here are the probabilities that the farm chosen falls in several acreage categories:

Acres	< 10	10–49	50–99	100–179	180–499	500–999	1000–1999	≥ 2000
Probability	0.09	0.20	0.15	0.16	0.22	0.09	0.05	0.04

Let A be the event that the farm is less than 50 acres in size, and let B be the event that it is 500 acres or more.

(a) Find $P(A)$ and $P(B)$.

(b) Describe the event "A does not occur" in words and find its probability by Rule 3.

(c) Describe the event "A or B" in words and find its probability by the addition rule.

(FRANK SITEMAN/STOCK, BOSTON.)

4.33 **Roulette.** A roulette wheel has 38 slots, numbered 0, 00, and 1 to 36. The slots 0 and 00 are colored green, 18 of the others are red, and 18 are black. The dealer spins the wheel and at the same time rolls a small ball along the wheel in the opposite direction. The wheel is carefully balanced so that the ball is equally likely to land in any slot when the wheel slows. Gamblers can bet on various combinations of numbers and colors.

(a) What is the probability of any one of the 38 possible outcomes? Explain your answer.

(b) If you bet on "red," you win if the ball lands in a red slot. What is the probability of winning?

(c) The slot numbers are laid out on a board on which gamblers place their bets. One column of numbers on the board contains all multiples of 3, that is, 3, 6, 9, . . ., 36. You place a "column bet" that wins if any of these numbers comes up. What is your probability of winning?

4.34 **Birth order.** A couple plans to have three children. There are 8 possible arrangements of girls and boys. For example, GGB means the first two children are girls and the third child is a boy. All 8 arrangements are (approximately) equally likely.

(a) Write down all 8 arrangements of the sexes of three children. What is the probability of any one of these arrangements?

(b) Let X be the number of girls the couple has. What is the probability that $X = 2$?

(c) Starting from your work in (a), find the distribution of X. That is, what values can X take, and what are the probabilities for each value?

4.35 **Moving up.** A study of social mobility in England looked at the social class reached by the sons of lower-class fathers. Social classes are numbered from 1 (low) to 5 (high). Take the random variable X to be the class of a randomly chosen son of a father in Class 1. The study found that the distribution of X is

Son's class	1	2	3	4	5
Probability	0.48	0.38	0.08	0.05	0.01

(a) What percent of the sons of lower-class fathers reach the highest class, Class 5?

(b) Check that this distribution satisfies the two requirements for a legitimate assignment of probabilities to individual outcomes.

(c) What is $P(X \leq 3)$? (Be careful: the event "$X \leq 3$" includes the value 3.)

(d) What is $P(X < 3)$?

(e) Write the event "a son of a lower-class father reaches one of the two highest classes" in terms of values of X. What is the probability of this event?

4.36 **How large are households?** Choose an American household at random and let the random variable X be the number of persons living in the household. If we ignore the few households with more than seven inhabitants, the probability distribution of X is as follows:

Inhabitants	1	2	3	4	5	6	7
Probability	0.25	0.32	0.17	0.15	0.07	0.03	0.01

(a) Verify that this is a legitimate probability distribution.

(b) What is $P(X \geq 5)$?

(c) What is $P(X > 5)$?

(d) What is $P(2 < X \leq 4)$?

(e) What is $P(X \neq 1)$?

(f) Write the event that a randomly chosen household contains more than two persons in terms of the random variable X. What is the probability of this event?

4.37 **Random numbers.** Many random number generators allow users to specify the range of the random numbers to be produced. Suppose that you specify that the random number Y can take any value between 0 and 2. Then the density curve of the outcomes has constant height between 0 and 2, and height 0 elsewhere.

(a) What is the height of the density curve between 0 and 2? Draw a graph of the density curve.

(b) Use your graph from (a) and the fact that probability is area under the curve to find $P(Y \leq 1)$.

(c) Find $P(0.5 < Y < 1.3)$.

(d) Find $P(Y \geq 0.8)$.

4.3 Sampling Distributions

Statistical inference uses sample data to draw conclusions about the entire population. Because good samples are chosen randomly, statistics such as $\bar{x}$ are random variables. We can describe the behavior of a sample statistic by a probability model that answers the question "What would happen if we did this many times?"

Statistical estimation and the law of large numbers

Here is an example that will lead us toward the probability ideas most important for statistical inference.

EXAMPLE 4.10   Does this wine smell bad?

parameter

Sulfur compounds such as dimethyl sulfide (DMS) are sometimes present in wine. DMS causes "off-odors" in wine, so winemakers want to know the odor threshold, the lowest concentration of DMS that the human nose can detect. Different people have different thresholds, so we start by asking about the mean threshold μ in the population of all adults. The number μ is a **parameter** that describes this population.

To estimate μ, we present tasters with both natural wine and the same wine spiked with DMS at different concentrations to find the lowest concentration at which they identify the spiked wine. Here are the odor thresholds (measured in micrograms of DMS per liter of wine) for 10 randomly chosen subjects:

$$28 \quad 40 \quad 28 \quad 33 \quad 20 \quad 31 \quad 29 \quad 27 \quad 17 \quad 21$$

statistic

The mean threshold for these subjects is $\bar{x} = 27.4$. This sample mean is a **statistic** that we use to estimate the parameter μ, but it is probably not exactly equal to μ. Moreover, we know that a different 10 subjects would give us a different $\bar{x}$.

A parameter, such as the mean odor threshold μ of all adults, is in practice a fixed but unknown number. A statistic, such as the mean threshold $\bar{x}$ of a random sample of 10 adults, is a random variable. It seems reasonable to use

$\bar{x}$ to estimate μ. An SRS should fairly represent the population, so the mean $\bar{x}$ of the sample should be somewhere near the mean μ of the population. Of course, we don't expect $\bar{x}$ to be exactly equal to μ, and we realize that if we choose another SRS, the luck of the draw will probably produce a different $\bar{x}$.

If $\bar{x}$ is rarely exactly right and varies from sample to sample, why is it nonetheless a reasonable estimate of the population mean μ? Here is one answer: if we keep on taking larger and larger samples, the statistic $\bar{x}$ is *guaranteed* to get closer and closer to the parameter μ. We have the comfort of knowing that if we can afford to keep on measuring more subjects, eventually we will estimate the mean odor threshold of all adults very accurately. This remarkable fact is called the *law of large numbers*. It is remarkable because it holds for *any* population, not just for some special class such as normal distributions.

LAW OF LARGE NUMBERS

Draw observations at random from any population with finite mean μ. As the number of observations drawn increases, the mean $\bar{x}$ of the observed values gets closer and closer to the mean μ of the population.

The law of large numbers can be proved mathematically starting from the basic laws of probability. The behavior of $\bar{x}$ is similar to the idea of probability. In the long run, the proportion of outcomes taking any value gets close to the probability of that value, and the average outcome gets close to the population mean. Figure 4.1 (page 216) shows how proportions approach probability in one example. Here is an example of how sample means approach the population mean.

EXAMPLE 4.11 The law of large numbers in action

In fact, the distribution of odor thresholds among all adults has mean 25. The mean $\mu = 25$ is the true value of the parameter we seek to estimate. Figure 4.7 shows how the sample mean $\bar{x}$ of an SRS drawn from this population changes as we add more subjects to our sample.

The first subject in Example 4.10 had threshold 28, so the line in Figure 4.7 starts there. The mean for the first two subjects is

$$\bar{x} = \frac{28 + 40}{2} = 34$$

This is the second point on the graph. At first, the graph shows that the mean of the sample changes as we take more observations. Eventually, however, the mean of the observations gets close to the population mean $\mu = 25$ and settles down at that value.

If we started over, again choosing people at random from the population, we would get a different path from left to right in Figure 4.7. The law of large numbers says that whatever path we get will always settle down at 25 as we draw more and more people.

The probability of dying

The behavior of large groups of people can be as random as coin tosses. Life insurance is based on this fact. We can't predict whether a specific person will die next year. But if we observe millions of people, deaths are random. The proportion of men aged 25 to 34 who will die next year is about 0.0021. This is the probability that a young man will die. For women that age, the probability of death is about 0.0007. An insurance company that sells many policies to people aged 25 to 34 will have to pay off on about 0.21% of the policies sold to men and about 0.07% of the policies sold to women.

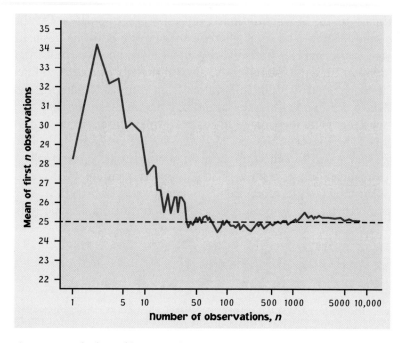

Figure 4.7 *The law of large numbers in action: as we take more observations, the sample mean $\overline{x}$ always approaches the mean μ of the population.*

The law of large numbers is the foundation of such business enterprises as gambling casinos and insurance companies. The winnings (or losses) of a gambler on a few plays are uncertain—that's why gambling is exciting. In Figure 4.7, the mean of even 100 observations is not yet very close to μ. It is only *in the long run* that the mean outcome is predictable. The house plays tens of thousands of times. So the house, unlike individual gamblers, can count on the long-run regularity described by the law of large numbers. The average winnings of the house on tens of thousands of plays will be very close to the mean of the distribution of winnings. Needless to say, this mean guarantees the house a profit. That's why gambling can be a business.

APPLY YOUR KNOWLEDGE ·

4.38   Figure 4.7 shows how the mean of *n* observations behaves as we keep adding more observations to those already in hand. The first 10 observations are given in Example 4.10. Demonstrate that you grasp the idea of Figure 4.7: find the mean of the first one, then two, three, four, and five of these observations and plot the successive means against *n*. Verify that your plot agrees with the first part of the plot in Figure 4.7.

4.39   **Playing the numbers.** The numbers racket is a well-entrenched illegal gambling operation in most large cities. One version works as follows: you choose one of the 1000 three-digit numbers 000 to 999 and pay your local numbers runner a dollar to enter your bet. Each day, one three-digit number is chosen at random and pays off $600. The mean payoff for the population of thousands of bets is $\mu = 60$ cents. Joe makes one bet every day for many years. Explain what the law of large numbers says about Joe's results as he keeps on betting.

Sampling distributions

The law of large numbers assures us that if we measure enough subjects, the statistic $\bar{x}$ will eventually get very close to the unknown parameter μ. But our study in Example 4.10 had just 10 subjects. What can we say about $\bar{x}$ from 10 subjects as an estimate of μ? We ask: "What would happen if we took many samples of 10 subjects from this population?" Here's how to answer this question:

- Take a large number of samples of size 10 from the same population.
- Calculate the sample mean $\bar{x}$ for each sample.
- Make a histogram of the values of $\bar{x}$.
- Examine the distribution displayed in the histogram for shape, center, and spread, as well as outliers or other deviations.

In practice it is too expensive to take many samples from a large population such as all adult U.S. residents. But we can imitate many samples by using software. Using software to imitate chance behavior is called **simulation**.

simulation

EXAMPLE 4.12	**Constructing a sampling distribution**

Extensive studies have found that the DMS odor threshold of adults follows roughly a normal distribution with mean $\mu = 25$ micrograms per liter and standard deviation $\sigma = 7$ micrograms per liter. Figure 4.8 shows this population distribution. With this information, we can simulate many runs of our study with different subjects drawn at random from the population. Figure 4.9 illustrates the process: we take 1000 samples of size 10, find the 1000 sample mean thresholds $\bar{x}$, and make a histogram of these 1000 values.

What can we say about the shape, center, and spread of this distribution?

- **Shape:** It looks normal! Detailed examination confirms that the distribution of $\bar{x}$ from many samples does have a distribution that is very close to normal.
- **Center:** The mean of the 1000 $\bar{x}$'s is 25.073. That is, the distribution is centered very close to the population mean $\mu = 25$.
- **Spread:** The standard deviation of the 1000 $\bar{x}$'s is 2.191, notably smaller than the standard deviation $\sigma = 7$ of the population of individual subjects.

The histogram in Figure 4.9 shows how $\bar{x}$ would behave if we drew many samples. It displays the *sampling distribution* of the statistic $\bar{x}$.

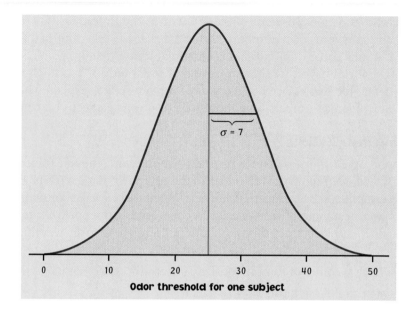

Figure 4.8 *The distribution of odor threshold for DMS in wine in the population of all adults. This is also the distribution of the DMS threshold for one adult chosen at random.*

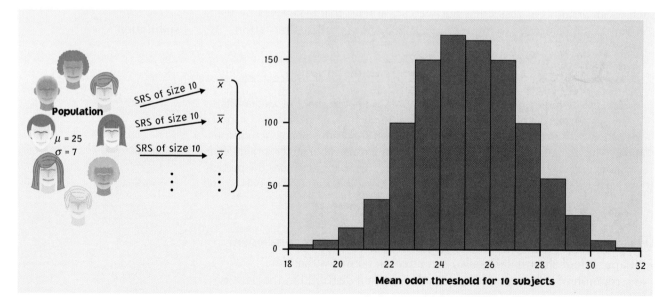

Figure 4.9 *The idea of a sampling distribution: take many samples from the same population, collect the $\bar{x}$'s from all the samples, and display the distribution of the $\bar{x}$'s. The histogram here shows the results of 1000 samples.*

SAMPLING DISTRIBUTION

The **sampling distribution** of a statistic is the distribution of values taken by the statistic in all possible samples of the same size from the same population.

Strictly speaking, the sampling distribution is the ideal pattern that would emerge if we looked at all possible samples of size 10 from our population. A distribution obtained from a fixed number of trials, like the 1000 trials in Figure 4.9, is only an approximation to the sampling distribution. One of the uses of probability theory in statistics is to obtain exact sampling distributions without simulation. The interpretation of a sampling distribution is the same, however, whether we obtain it by simulation or by the mathematics of probability.

APPLY YOUR KNOWLEDGE

4.40 **Generating a sampling distribution.** Let us illustrate the idea of a sampling distribution in the case of a very small sample from a very small population. The population is the scores of 10 students on an exam:

Student	0	1	2	3	4	5	6	7	8	9
Score	82	62	80	58	72	73	65	66	74	62

The parameter of interest is the mean score μ in this population. The sample is an SRS of size $n = 4$ drawn from the population. Because the students are labeled 0 to 9, a single random digit from Table B chooses one student for the sample.

(a) Find the mean of the 10 scores in the population. This is the population mean μ.

(b) Use Table B to draw an SRS of size 4 from this population. Write the four scores in your sample and calculate the mean $\bar{x}$ of the sample scores. This statistic is an estimate of μ.

(c) Repeat this process 10 times using different parts of Table B. Make a histogram of the 10 values of $\bar{x}$. You are constructing the sampling distribution of $\bar{x}$. Is the center of your histogram close to μ?

The mean and the standard deviation of $\bar{x}$

Figure 4.9 suggests that when we choose many SRSs from a population, the sampling distribution of the sample means is centered at the mean of the original population and less spread out than the distribution of individual observations. Here are the full facts.

> ### MEAN AND STANDARD DEVIATION OF A SAMPLE MEAN[3]
>
> Suppose that $\bar{x}$ is the mean of an SRS of size n drawn from a large population with mean μ and standard deviation σ. Then the **mean** of the sampling distribution of $\bar{x}$ is μ and its **standard deviation is** $\sigma/\sqrt{n}$.

Both the mean and the standard deviation of the sampling distribution of $\bar{x}$ have important implications for statistical inference.

- The mean of the statistic $\bar{x}$ is always the same as the mean μ of the population. The sampling distribution of $\bar{x}$ is centered at μ. In repeated sampling, $\bar{x}$ will sometimes fall above the true value of the parameter μ and sometimes below, but there is no systematic tendency to overestimate or underestimate the parameter. This makes the idea of lack of bias in the sense of "no favoritism" more precise. Because the mean of $\bar{x}$ is equal to μ, *unbiased estimator* we say that the statistic $\bar{x}$ is an **unbiased estimator** of the parameter μ.

- An unbiased estimator is "correct on the average" in many samples. How close the estimator falls to the parameter in most samples is determined by the spread of the sampling distribution. If individual observations have standard deviation σ, then sample means $\bar{x}$ from samples of size n have standard deviation $\sigma/\sqrt{n}$. **Averages are less variable than individual**

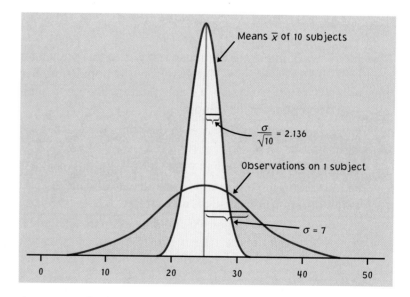

Figure 4.10 *The distribution of single observations compared with the distribution of the means $\bar{x}$ of 10 observations. Averages are less variable than individual observations.*

observations. Figure 4.10 compares the distribution of single observations (the population distribution) for odor thresholds with the sampling distribution of the means $\bar{x}$ of 10 observations.

Not only is the standard deviation of the distribution of $\bar{x}$ smaller than the standard deviation of individual observations, but it gets smaller as we take larger samples. **The results of large samples are less variable than the results of small samples**. If n is large, the standard deviation of $\bar{x}$ is small and almost all samples will give values of $\bar{x}$ that lie very close to the true parameter μ. That is, the sample mean from a large sample can be trusted to estimate the population mean accurately. Notice, however, that the standard deviation of the sampling distribution gets smaller only at the rate $\sqrt{n}$. To cut the standard deviation of $\bar{x}$ in half, we must take four times as many observations, not just twice as many.

APPLY YOUR KNOWLEDGE

4.41   **Measurements in the lab.** Juan makes a measurement in a chemistry laboratory and records the result in his lab report. The standard deviation of students' lab measurements is $\sigma = 10$ milligrams. Juan repeats the measurement 3 times and records the mean $\bar{x}$ of his 3 measurements.

(a) What is the standard deviation of Juan's mean result? (That is, if Juan kept on making 3 measurements and averaging them, what would be the standard deviation of all his $\bar{x}$'s?)

(b) How many times must Juan repeat the measurement to reduce the standard deviation of $\bar{x}$ to 5? Explain to someone who knows no statistics the advantage of reporting the average of several measurements rather than the result of a single measurement.

4.42   **Measuring blood cholesterol.** A study of the health of teenagers plans to measure the blood cholesterol level of an SRS of youth of ages 13 to 16 years. The researchers will report the mean $\bar{x}$ from their sample as an estimate of the mean cholesterol level μ in this population.

(a) Explain to someone who knows no statistics what it means to say that $\bar{x}$ is an "unbiased" estimator of μ.

(b) The sample result $\bar{x}$ is an unbiased estimator of the population truth μ no matter what size SRS the study chooses. Explain to someone who knows no statistics why a large sample gives more trustworthy results than a small sample.

The central limit theorem

We have described the center and spread of the probability distribution of a sample mean $\bar{x}$, but not its shape. The shape of the distribution of $\bar{x}$ depends on the shape of the population distribution. Here is one important case: if the population distribution is normal, then so is the distribution of the sample mean.

SAMPLING DISTRIBUTION OF A SAMPLE MEAN

If a population has the $N(\mu, \sigma)$ distribution, then the sample mean $\bar{x}$ of n independent observations has the $N(\mu, \sigma/\sqrt{n})$ distribution.

We already knew the mean and standard deviation of the sampling distribution. All that we have added now is the normal shape. Figure 4.10 illustrates these facts in the case of measuring odor thresholds. Odor thresholds follow a normal distribution in the population of all adults, so the sampling distribution of the mean threshold for 10 adults also follows a normal distribution.

What happens when the population distribution is not normal? As the sample size increases, the distribution of $\bar{x}$ changes shape: it looks less like that of the population and more like a normal distribution. When the sample is large enough, the distribution of $\bar{x}$ is very close to normal. This is true no matter what shape the population distribution has, as long as the population has a finite standard deviation σ. This famous fact of probability theory is called the *central limit theorem*. It is much more useful than the fact that the distribution of $\bar{x}$ is exactly normal if the population is exactly normal.

CENTRAL LIMIT THEOREM

Draw an SRS of size n from any population with mean μ and finite standard deviation σ. When n is large, the sampling distribution of the sample mean $\bar{x}$ is approximately normal:

$$\bar{x} \text{ is approximately } N(\mu, \sigma/\sqrt{n})$$

More generally, the central limit theorem says that the distribution of a sum or average of many small random quantities is close to normal. This is true even if the quantities are not independent (as long as they are not too highly correlated) and even if they have different distributions (as long as no one random quantity is so large that it dominates the others). The central limit theorem suggests why the normal distributions are common models for observed data. Any variable that is a sum of many small influences will have approximately a normal distribution.

How large a sample size n is needed for $\bar{x}$ to be close to normal depends on the population distribution. More observations are required if the shape of the population distribution is far from normal.

Politically correct

In 1950, the Russian mathematician B. V. Gnedenko (1912–) wrote a text on *The Theory of Probability* that was widely used in the United States and elsewhere. The introduction contains a mystifying paragraph that begins, "We note that the entire development of probability theory shows evidence of how its concepts and ideas were crystallized in a severe struggle between materialistic and idealistic conceptions." It turns out that "materialistic" is jargon for "Marxist-Leninist." It was good for the health of Russian scientists in the Stalin era to add such statements to their books.

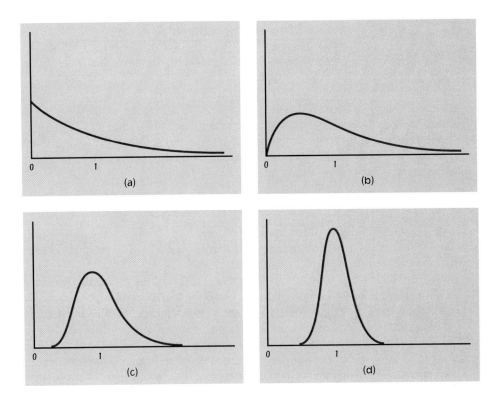

Figure 4.11 *The central limit theorem in action: the distribution of sample means $\bar{x}$ from a strongly nonnormal population becomes more normal as the sample size increases.* **(a)** *The distribution of one observation.* **(b)** *The distribution of $\bar{x}$ for 2 observations.* **(c)** *The distribution of $\bar{x}$ for 10 observations.* **(d)** *The distribution of $\bar{x}$ for 25 observations.*

EXAMPLE 4.13   **The central limit theorem in action**

Figure 4.11 shows how the central limit theorem works for a very nonnormal population. Figure 4.11(a) displays the density curve of a single observation, that is, of the population. The distribution is strongly right-skewed, and the most probable outcomes are near 0. The mean μ of this distribution is 1, and its standard deviation σ is also 1. This particular distribution is called an *exponential distribution*. Exponential distributions are used as models for the lifetime in service of electronic components and for the time required to serve a customer or repair a machine.

Figures 4.11(b), (c), and (d) are the density curves of the sample means of 2, 10, and 25 observations from this population. As n increases, the shape becomes more normal. The mean remains at $\mu = 1$, and the standard deviation decreases, taking the value $1/\sqrt{n}$. The density curve for 10 observations is still somewhat skewed to the right but already resembles a normal curve having $\mu = 1$ and $\sigma = 1/\sqrt{10} = 0.32$. The density curve for $n = 25$ is yet more normal. The contrast between the shapes of the population distribution and of the distribution of the mean of 10 or 25 observations is striking.

The central limit theorem allows us to use normal probability calculations to answer questions about sample means from many observations even when the population distribution is not normal.

EXAMPLE 4.14 | **Maintaining air conditioners**

The time X that a technician requires to perform preventive maintenance on an air-conditioning unit is governed by the exponential distribution whose density curve appears in Figure 4.11(a). The mean time is $\mu = 1$ hour and the standard deviation is $\sigma = 1$ hour. Your company operates 70 of these units. What is the probability that their average maintenance time exceeds 50 minutes?

The central limit theorem says that the sample mean time $\bar{x}$ (in hours) spent working on 70 units has approximately the normal distribution with mean equal to the population mean $\mu = 1$ hour and standard deviation

$$\frac{\sigma}{\sqrt{70}} = \frac{1}{\sqrt{70}} = 0.12 \text{ hour}$$

The distribution of $\bar{x}$ is therefore approximately $N(1, 0.12)$. This normal curve is the solid curve in Figure 4.12.

Because 50 minutes is 50/60 of an hour, or 0.83 hour, the probability we want is $P(\bar{x} > 0.83)$. A normal distribution calculation gives this probability as 0.9222. This is the area to the right of 0.83 under the solid normal curve in Figure 4.12.

Using more mathematics, we could start with the exponential distribution and find the actual density curve of $\bar{x}$ for 70 observations. This is the dashed curve in Figure 4.12. You can see that the solid normal curve is a good approximation. The exactly correct probability is the area under the dashed density curve. It is 0.9294. The central limit theorem normal approximation is off by only about 0.007.

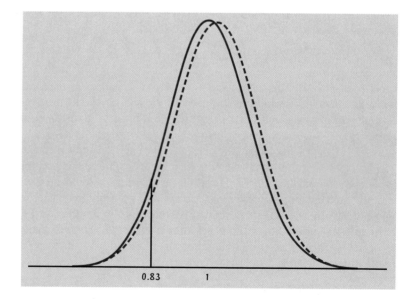

Figure 4.12 *The exact distribution (dashed) and the normal approximation from the central limit theorem (solid) for the average time needed to maintain an air conditioner, for Example 4.14.*

4.43   **ACT scores.** The scores of students on the ACT college entrance examination in a recent year had the normal distribution with mean $\mu = 18.6$ and standard deviation $\sigma = 5.9$.

(a) What is the probability that a single student randomly chosen from all those taking the test scores 21 or higher?

(b) Now take an SRS of 50 students who took the test. What are the mean and standard deviation of the sample mean score $\bar{x}$ of these 50 students?

(c) What is the probability that the mean score $\bar{x}$ of these students is 21 or higher?

4.44   **Flaws in carpets.** The number of flaws per square yard in a type of carpet material varies with mean 1.6 flaws per square yard and standard deviation 1.2 flaws per square yard. The population distribution cannot be normal, because a count takes only whole-number values. An inspector samples 200 square yards of the material, records the number of flaws found in each square yard, and calculates $\bar{x}$, the mean number of flaws per square yard inspected. Use the central limit theorem to find the approximate probability that the mean number of flaws exceeds 2 per square yard.

4.45   **Returns on stocks.** The distribution of annual returns on common stocks is roughly symmetric, but extreme observations are more frequent than in a normal distribution. Because the distribution is not strongly nonnormal, the mean return over even a moderate number of years is close to normal. In the long run, annual real returns on common stocks have varied with mean about 9% and standard deviation about 28%. Andrew plans to retire in 45 years and is considering investing in stocks. What is the probability (assuming that the past pattern of variation continues) that the mean annual return on common stocks over the next 45 years will exceed 15%? What is the probability that the mean return will be less than 5%?

SECTION 4.3   Summary

When we want information about the **population mean** μ for some variable, we often take an SRS and use the **sample mean** $\bar{x}$ to estimate the unknown parameter μ.

The **law of large numbers** states that the actually observed mean outcome $\bar{x}$ must approach the mean μ of the population as the number of observations increases.

The **sampling distribution** of $\bar{x}$ describes how the statistic $\bar{x}$ varies in all possible samples of the same size from the same population.

The **mean** of the sampling distribution is μ, so that $\bar{x}$ is an **unbiased estimator** of μ.

The **standard deviation** of the sampling distribution of $\bar{x}$ is $\sigma/\sqrt{n}$ for an SRS of size n if the population has standard deviation σ. That is, averages are less variable than individual observations.

If the population has a normal distribution, so does $\bar{x}$.

The **central limit theorem** states that for large n the sampling distribution of $\bar{x}$ is approximately normal for any population with finite standard deviation σ. That is, averages are more normal than individual observations. We can use the $N(\mu, \sigma/\sqrt{n})$ distribution to calculate approximate probabilities for events involving $\bar{x}$.

SECTION 4.3 Exercises

4.46　**Roulette.** A roulette wheel has 38 slots, of which 18 are black, 18 are red, and 2 are green. When the wheel is spun, the ball is equally likely to come to rest in any of the slots. One of the simplest wagers chooses red or black. A bet of $1 on red returns $2 if the ball lands in a red slot. Otherwise, the player loses his dollar. When gamblers bet on red or black, the two green slots belong to the house. Because the probability of winning $2 is 18/38, the mean payoff from a $1 bet is twice 18/38, or 94.7 cents. Explain what the law of large numbers tells us about what will happen if a gambler makes a large number of bets on red.

4.47　**Generating a sampling distribution.** Table 1.9 (page 72) gives the survival times of 72 guinea pigs in a medical experiment. Consider these 72 animals to be the population of interest.

(a) Make a histogram of the 72 survival times. This is the population distribution. It is strongly skewed to the right.

(b) Find the mean of the 72 survival times. This is the population mean μ. Mark μ on the x axis of your histogram.

(c) Label the members of the population 01 to 72 and use Table B to choose an SRS of size $n = 12$. What is the mean survival time $\bar{x}$ for your sample? Mark the value of $\bar{x}$ with a point on the axis of your histogram from (a).

(d) Choose four more SRSs of size 12, using different parts of Table B. Find $\bar{x}$ for each sample and mark the values on the axis of your histogram from (a). Would you be surprised if all five $\bar{x}$'s fell on the same side of μ? Why?

(e) If you chose a large number of SRSs of size 12 from this population and made a histogram of the $\bar{x}$-values, where would you expect the center of this sampling distribution to lie?

4.48 **Dust in coal mines.** A laboratory weighs filters from a coal mine to measure the amount of dust in the mine atmosphere. Repeated measurements of the weight of dust on the same filter vary normally with standard deviation $\sigma = 0.08$ milligram (mg) because the weighing is not perfectly precise. The dust on a particular filter actually weighs 123 mg. Repeated weighings will then have the normal distribution with mean 123 mg and standard deviation 0.08 mg.

(a) The laboratory reports the mean of 3 weighings. What is the distribution of this mean?

(b) What is the probability that the laboratory reports a weight of 124 mg or higher for this filter?

4.49 **An opinion poll.** An SRS of 400 American adults is asked, "What do you think is the most serious problem facing our schools?" Suppose that in fact 30% of all adults would answer "drugs" if asked this question. The proportion X of the sample who answer "drugs" will vary in repeated sampling. In fact, we can assign probabilities to values of X using the normal density curve with mean 0.3 and standard deviation 0.023. Use this density curve to find the probabilities of the following events:

(a) At least half of the sample believes that drugs are the schools' most serious problem.

(b) Less than 25% of the sample believes that drugs are the most serious problem.

(c) The sample proportion is between 0.25 and 0.35.

4.50 **Making auto parts.** An automatic grinding machine in an auto parts plant prepares axles with a target diameter $\mu = 40.125$ millimeters (mm). The machine has some variability, so the standard deviation of the diameters is $\sigma = 0.002$ mm. A sample of 4 axles is inspected each hour for process control purposes, and records are kept of the sample mean diameter. What will be the mean and standard deviation of the numbers recorded?

4.51 **Bottling cola.** A bottling company uses a filling machine to fill plastic bottles with cola. The bottles are supposed to contain 300 milliliters (ml). In fact, the contents vary according to a normal distribution with mean $\mu = 298$ ml and standard deviation $\sigma = 3$ ml.

(a) What is the probability that an individual bottle contains less than 295 ml?

(b) What is the probability that the mean contents of the bottles in a six-pack is less than 295 ml?

4.52 **Potassium in the blood.** Judy's doctor is concerned that she may suffer from hypokalemia (low potassium in the blood). There is variation both in the actual potassium level and in the blood test that measures the level. Judy's measured potassium level varies according to the normal distribution with $\mu = 3.8$ and $\sigma = 0.2$. A patient is classified as hypokalemic if the potassium level is below 3.5.

(a) If a single potassium measurement is made, what is the probability that Judy is diagnosed as hypokalemic?

(b) If measurements are made instead on 4 separate days and the mean result is compared with the criterion 3.5, what is the probability that Judy is diagnosed as hypokalemic?

4.53 **Auto accidents.** The number of accidents per week at a hazardous intersection varies with mean 2.2 and standard deviation 1.4. This distribution takes only whole-number values, so it is certainly not normal.

(a) Let $\bar{x}$ be the mean number of accidents per week at the intersection during a year (52 weeks). What is the approximate distribution of $\bar{x}$ according to the central limit theorem?

(b) What is the approximate probability that $\bar{x}$ is less than 2?

(c) What is the approximate probability that there are fewer than 100 accidents at the intersection in a year? (Hint: Restate this event in terms of $\bar{x}$.)

4.54 **Pollutants in auto exhausts.** The level of nitrogen oxides (NOX) in the exhaust of a particular car model varies with mean 0.9 grams per mile (g/mi) and standard deviation 0.15 g/mi. A company has 125 cars of this model in its fleet.

(a) What is the approximate distribution of the mean NOX emission level $\bar{x}$ for these cars?

(b) What is the level L such that the probability that $\bar{x}$ is greater than L is only 0.01? (Hint: This requires a backward normal calculation. See page 61 of Chapter 1 if you need to review.)

4.55 **Testing kindergarten children.** Children in kindergarten are sometimes given the Ravin Progressive Matrices Test (RPMT) to assess their readiness for learning. Experience at Southwark Elementary School suggests that the RPMT scores for its kindergarten pupils have mean 13.6 and standard deviation 3.1. The distribution is close to normal. Mr. Lavin has 22 children in his kindergarten class this year. He suspects that their RPMT scores will be unusually low because the test was interrupted by a fire drill. To check this suspicion, he wants to find the level L such that there is probability only 0.05 that the mean score of 22 children falls below L when the usual Southwark distribution remains true. What is the value of L? (Hint: This requires a backward normal calculation. See page 61 of Chapter 1 if you need to review.)

This chapter lays the foundations for the study of statistical inference. Statistical inference uses data to draw conclusions about the population or process from which the data come. What is special about inference is that the conclusions include a statement, in the language of probability, about how reliable they are. The statement gives a probability that answers the question "What would happen if I used this method very many times?" The probabilities we need come from the sampling distributions of sample statistics.

Sampling distributions are the key to understanding statistical inference. The Statistics in Summary figure on the next page summarizes the facts about the sampling distribution of $\bar{x}$ in a way that reminds us of the big idea of a sampling distribution. Keep taking random samples of size n from a population with mean μ. Find the sample mean $\bar{x}$ for each sample. Collect all the $\bar{x}$'s and display their distribution. That's the sampling distribution of $\bar{x}$. Keep this figure in mind as you go forward.

To think more effectively about sampling distributions, we use the language of probability. Probability, the mathematics that describes randomness, is important in many areas of study. Here, we concentrate on informal probability as the conceptual foundation for statistical inference. Because random samples and randomized comparative experiments use chance, their results vary according to the laws of probability. Here is a review list of the most important things you should be able to do after studying this chapter.

A. PROBABILITY

1. Recognize that some phenomena are random. Probability describes the long-run regularity of random phenomena.

2. Understand that the probability of an event is the proportion of times the event occurs in very many repetitions of a random phenomenon. Use the idea of probability as long-run proportion to think about probability.

3. Use basic probability facts to detect illegitimate assignments of probability: Any probability must be a number between 0 and 1, and the total probability assigned to all possible outcomes must be 1.

4. Use basic probability facts to find the probabilities of events that are formed from other events: The probability that an event does not occur is 1 minus its probability. If two events are disjoint, the probability that one or the other occurs is the sum of their individual probabilities.

5. Find probabilities of events in a finite sample space by adding the probabilities of their outcomes. Find probabilities of events as areas under a density curve.

6. Use the notation of random variables to make compact statements about random outcomes, such as $P(\bar{x} \leq 4) = 0.3$. Be able to read such statements.

STATISTICS IN SUMMARY
Approximate Sampling Distribution of $\bar{x}$

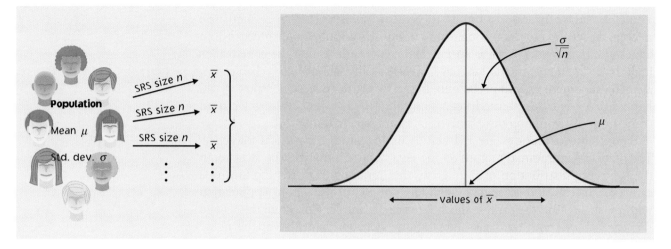

B. SAMPLING DISTRIBUTIONS

1. Identify parameters and statistics in a sample or experiment.
2. Recognize the fact of sampling variability: a statistic will take different values when you repeat a sample or experiment.
3. Interpret a sampling distribution as describing the values taken by a statistic in all possible repetitions of a sample or experiment under the same conditions.
4. Interpret the sampling distribution of a statistic as describing the probabilities of its possible values.

C. THE SAMPLING DISTRIBUTION OF A SAMPLE MEAN

1. Recognize when a problem involves the mean $\bar{x}$ of a sample. Understand that $\bar{x}$ estimates the mean μ of the population from which the sample is drawn.
2. Use the law of large numbers to describe the behavior of $\bar{x}$ as the size of the sample increases.
3. Find the mean and standard deviation of a sample mean $\bar{x}$ from an SRS of size n when the mean μ and standard deviation σ of the population are known.
4. Understand that $\bar{x}$ is an unbiased estimator of μ and that the variability of $\bar{x}$ about its mean μ gets smaller as the sample size increases.
5. Understand that $\bar{x}$ has approximately a normal distribution when the sample is large (central limit theorem). Use this normal distribution to calculate probabilities that concern $\bar{x}$.

CHAPTER 4 Review Exercises

4.56 A random sample of female college students has a mean height of **64.5** inches, which is greater than the **63**-inch mean height of all adult American women. Is each of the bold numbers a parameter or a statistic?

4.57 A sample of students of high academic ability under 13 years of age was given the SAT mathematics examination, which is usually taken by high school seniors. The mean score for the females in the sample was **386**, whereas the mean score of the males was **416**. Is each of the bold numbers a parameter or a statistic?

4.58 If a carefully made die is rolled once, it is reasonable to assign probability 1/6 to each of the six faces. What is the probability of rolling a number less than 3?

4.59 Exactly one of Brown, Chavez, and Williams will be promoted to partner in the law firm that employs them all. Brown thinks that she has probability 0.25 of winning the promotion and that Williams has probability 0.2. What probability does Brown assign to the outcome that Chavez is the one promoted?

4.60 **Predicting the ACC champion.** Las Vegas Zeke, when asked to predict the Atlantic Coast Conference basketball champion, follows the modern practice of giving probabilistic predictions. He says, "North Carolina's probability of winning is twice Duke's. North Carolina State and Virginia each have probability 0.1 of winning, but Duke's probability is three times that. Nobody else has a chance." Has Zeke given a legitimate assignment of probabilities to the eight teams in the conference? Explain your answer.

4.61 **How far do fifth graders go in school?** A study of education followed a large group of fifth-grade children to see how many years of school they eventually completed. Let X be the highest year of school that a randomly chosen fifth grader completes. (Students who go on to college are included in the outcome $X = 12$.) The study found this probability distribution for X:

Years	4	5	6	7	8	9	10	11	12
Probability	0.010	0.007	0.007	0.013	0.032	0.068	0.070	0.041	0.752

(a) What percent of fifth graders eventually finished twelfth grade?

(b) Check that this is a legitimate probability distribution.

(c) Find $P(X \geq 6)$. (Be careful: the event "$X \geq 6$" includes the value 6.)

(d) Find $P(X > 6)$.

(e) What values of X make up the event "the student completed at least one year of high school"? (High school begins with the ninth grade.) What is the probability of this event?

4.62 **Classifying occupations.** Choose an American worker at random and classify his or her occupation into one of the following classes. These classes are used in government employment data.

A Managerial and professional
B Technical, sales, administrative support
C Service occupations
D Precision production, craft, and repair
E Operators, fabricators, and laborers
F Farming, forestry, and fishing

The table below gives the probabilities that a randomly chosen worker falls into each of 12 sex-by-occupation classes:

Class	A	B	C	D	E	F
Male	0.14	0.11	0.06	0.11	0.12	0.03
Female	0.09	0.20	0.08	0.01	0.04	0.01

(a) Verify that this is a legitimate assignment of probabilities to these outcomes.

(b) What is the probability that the worker is female?

(c) What is the probability that the worker is not engaged in farming, forestry, or fishing?

(d) Classes D and E include most mechanical and factory jobs. What is the probability that the worker holds a job in one of these classes?

(e) What is the probability that the worker does not hold a job in Classes D or E?

4.63 **An IQ test.** The Wechsler Adult Intelligence Scale (WAIS) is a common "IQ test" for adults. The distribution of WAIS scores for persons over 16 years of age is approximately normal with mean 100 and standard deviation 15.

(a) What is the probability that a randomly chosen individual has a WAIS score of 105 or higher?

(b) What are the mean and standard deviation of the average WAIS score $\bar{x}$ for an SRS of 60 people?

    (c) What is the probability that the average WAIS score of an SRS of 60 people is 105 or higher?

    (d) Would your answers to any of (a), (b), or (c) be affected if the distribution of WAIS scores in the adult population were distinctly nonnormal?

4.64 **Weights of eggs.** The weight of the eggs produced by a certain breed of hen is normally distributed with mean 65 grams (g) and standard deviation 5 g. Think of cartons of such eggs as SRSs of size 12 from the population of all eggs. What is the probability that the weight of a carton falls between 750 g and 825 g?

4.65 **How many people in a car?** A study of rush-hour traffic in San Francisco counts the number of people in each car entering a freeway at a suburban interchange. Suppose that this count has mean 1.5 and standard deviation 0.75 in the population of all cars that enter at this interchange during rush hours.

    (a) Could the exact distribution of the count be normal? Why or why not?

    (b) Traffic engineers estimate that the capacity of the interchange is 700 cars per hour. According to the central limit theorem, what is the approximate distribution of the mean number of persons $\bar{x}$ in 700 randomly selected cars at this interchange?

    (c) What is the probability that 700 cars will carry more than 1075 people? (Hint: Restate this event in terms of the mean number of people $\bar{x}$ per car.)

A. N. Kolmogorov

(SOVFOTO/EASTFOTO.)

General Laws of Probability

There are national styles in science as well as in cuisine. Statistics, the science of data, was created mainly by British and Americans. Probability, the mathematics of chance, was long led by French and Russians. Andrei Nikolaevich Kolmogorov (1903–1987) was the greatest of the Russian probabilists and one of the most influential mathematicians of the twentieth century. His more than 500 mathematical publications shaped several areas of modern mathematics and apply mathematical ideas to areas as far afield as the rhythms and meters of poetry.

Kolmogorov entered Moscow State University as a student in 1920 and remained there until his death. He was named a Hero of Socialist Labor in 1963, a rare honor for someone whose career was devoted entirely to scholarship.

Kolmogorov's first work in probability concerned the behavior of strings of random observations. The law of large numbers is the starting point for these studies, and Kolmogorov discovered many extensions of that law. Kolmogorov effectively established probability as a field of mathematics in 1933, when he placed it on a firm mathematical foundation by starting with a few general laws from which all else follows. The general laws of probability in this chapter are in the spirit of Kolmogorov.

Statistics, the science of data, was created mainly by British and Americans. Probability, the mathematics of chance, was long led by French and Russians.

Probability Theory

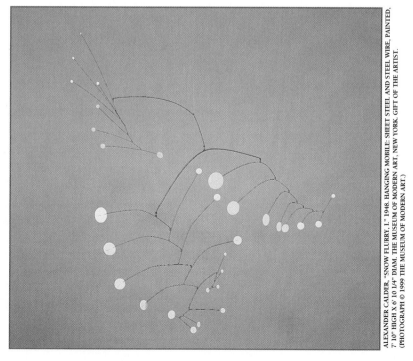

ALEXANDER CALDER, "SNOW FLURRY, 1." 1948. HANGING MOBILE: SHEET STEEL AND STEEL WIRE, PAINTED, 7' 10" HIGH X 6' 10 1/4" DIAM. THE MUSEUM OF MODERN ART, NEW YORK. GIFT OF THE ARTIST. (PHOTOGRAPH © 1999 THE MUSEUM OF MODERN ART.)

Kolmogorov studied linked series of random observations. Might we see a parallel in Alexander Calder's mobiles?

Introduction

The mathematics of probability can provide models to describe the flow of traffic through a highway system, a telephone interchange, or a computer processor; the genetic makeup of populations; the energy states of subatomic particles; the spread of epidemics or rumors; and the rate of return on risky investments. Although we are interested in probability because of its usefulness in statistics, the mathematics of chance is important in many fields of study. This chapter presents a bit more of the theory of probability.

5.1 General Probability Rules

Our study of probability in Chapter 4 concentrated on sampling distributions. Now we return to the general laws of probability. With more probability at our command, we can model more complex random phenomena. We have already met and used four rules.

RULES OF PROBABILITY

Rule 1. $0 \leq P(A) \leq 1$ for any event A

Rule 2. $P(S) = 1$

Rule 3. For any event A,

$$P(A \text{ does not occur}) = 1 - P(A)$$

Rule 4. Addition rule: If A and B are **disjoint** events, then

$$P(A \text{ or } B) = P(A) + P(B)$$

Independence and the multiplication rule

Rule 4, the addition rule for disjoint events, describes the probability that *one or the other* of two events A and B occurs when A and B cannot occur together. Now we will describe the probability that *both* events A and B occur, again only in a special situation.

You may find it helpful to draw a picture to display relations among several events. A picture like Figure 5.1 that shows the sample space S as a rectangular area and events as areas within S is called a **Venn diagram.** The events A and *Venn diagram* B in Figure 5.1 are disjoint because they do not overlap. The Venn diagram in Figure 5.2 illustrates two events that are not disjoint. The event $\{A \text{ and } B\}$ appears as the overlapping area that is common to both A and B.

Suppose that you toss a balanced coin twice. You are counting heads, so two events of interest are

$$A = \text{first toss is a head}$$
$$B = \text{second toss is a head}$$

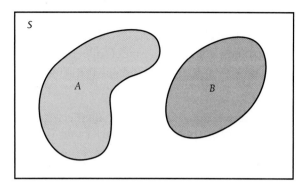

Figure 5.1 *Venn diagram showing disjoint events A and B.*

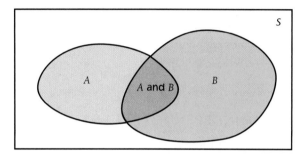

Figure 5.2 *Venn diagram showing events A and B that are not disjoint. The event {A and B} consists of outcomes common to A and B.*

The events *A* and *B* are not disjoint. They occur together whenever both tosses give heads. We want to find the probability of the event {*A* and *B*} that *both* tosses are heads.

The coin tossing of Buffon, Pearson, and Kerrich described at the beginning of Chapter 4 makes us willing to assign probability 1/2 to a head when we toss a coin. So

$$P(A) = 0.5$$

$$P(B) = 0.5$$

What is *P*(*A* and *B*)? Our common sense says that it is 1/4. The first coin will give a head half the time and then the second will give a head on half of those trials, so both coins will give heads on $1/2 \times 1/2 = 1/4$ of all trials in the long run. This reasoning assumes that the second coin still has probability 1/2 of a head after the first has given a head. This is true—we can verify it by tossing two coins many times and observing the proportion of heads on the second toss after the first toss has produced a head. We say that the events "head on the first toss" and "head on the second toss" are **independent.** Independence means that the outcome of the first toss cannot influence the outcome of the second toss.

independence

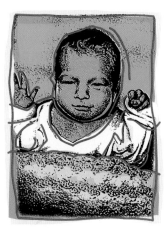

We want a boy

Misunderstanding independence can be disastrous. "Dear Abby" once published a letter from a mother of eight girls. She and her husband had planned a family of four children. When all four were girls, they kept trying. After seven girls, her doctor assured her that "the law of averages was in our favor 100 to one." Unfortunately, having children is like tossing coins. Eight girls in a row is highly unlikely, but once seven girls have been born, it is not at all unlikely that the next child will be a girl—and it was.

EXAMPLE 5.1   Independent or not?

Because a coin has no memory and most coin tossers cannot influence the fall of the coin, it is safe to assume that successive coin tosses are independent. For a balanced coin this means that after we see the outcome of the first toss, we still assign probability 1/2 to heads on the second toss.

On the other hand, the colors of successive cards dealt from the same deck are not independent. A standard 52-card deck contains 26 red and 26 black cards. For the first card dealt from a shuffled deck, the probability of a red card is $26/52 = 0.50$ (equally likely outcomes). Once we see that the first card is red, we know that there are only 25 reds among the remaining 51 cards. The probability that the second card is red is therefore only $25/51 = 0.49$. Knowing the outcome of the first deal changes the probabilities for the second.

If a doctor measures your blood pressure twice, it is reasonable to assume that the two results are independent because the first result does not influence the instrument that makes the second reading. But if you take an IQ test or other mental test twice in succession, the two test scores are not independent. The learning that occurs on the first attempt influences your second attempt.

MULTIPLICATION RULE FOR INDEPENDENT EVENTS

Two events A and B are **independent** if knowing that one occurs does not change the probability that the other occurs. If A and B are independent,

$$P(A \text{ and } B) = P(A)P(B)$$

EXAMPLE 5.2   Mendel's peas

Gregor Mendel used garden peas in some of the experiments that revealed that inheritance operates randomly. The seed color of Mendel's peas can be either green or yellow. Two parent plants are "crossed" (one pollinates the other) to produce seeds. Each parent plant carries two genes for seed color, and each of these genes has probability 1/2 of being passed to a seed. The two genes that the seed receives, one from each parent, determine its color. The parents contribute their genes independently of each other.

Suppose that both parents carry the G (green) and the Y (yellow) genes. The seed will be green if both parents contribute a G gene; otherwise it will be yellow. If M is the event that the male contributes a G gene and F is the event that the female contributes a G gene, then the probability of a green seed is

$$P(M \text{ and } F) = P(M)P(F)$$
$$= (0.5)(0.5) = 0.25$$

In the long run, 1/4 of all seeds produced by crossing these plants will be green.

The multiplication rule $P(A \text{ and } B) = P(A)P(B)$ holds if A and B are independent but not otherwise. The addition rule $P(A \text{ or } B) = P(A) + P(B)$ holds if A and B are disjoint but not otherwise. Resist the temptation to use these simple rules when the circumstances that justify them are not present. You must also be certain not to confuse disjointness and independence. If A and B are disjoint, then the fact that A occurs tells us that B cannot occur—look again at Figure 5.1. So disjoint events are not independent. Unlike disjointness, we cannot picture independence in a Venn diagram, because it involves the probabilities of the events rather than just the outcomes that make up the events.

APPLY YOUR KNOWLEDGE

5.1 **Albinism.** The gene for albinism in humans is recessive. That is, carriers of this gene have probability 1/2 of passing it to a child, and the child is albino only if both parents pass the albinism gene. Parents pass their genes independently of each other. If both parents carry the albinism gene, what is the probability that their first child is albino? If they have two children (who inherit independently of each other), what is the probability that both are albino? That neither is albino?

5.2 **High school rank.** Select a first-year college student at random and ask what his or her academic rank was in high school. Here are the probabilities, based on proportions from a large sample survey of first-year students:

Rank	Top 20%	Second 20%	Third 20%	Fourth 20%	Lowest 20%
Probability	0.41	0.23	0.29	0.06	0.01

(a) Choose two first-year college students at random. Why is it reasonable to assume that their high school ranks are independent?

(b) What is the probability that both were in the top 20% of their high school classes?

(c) What is the probability that the first was in the top 20% and the second was in the lowest 20%?

5.3 **College-educated laborers?** Government data show that 27% of the civilian labor force have at least 4 years of college and that 16% of the labor force work as laborers or operators of machines or vehicles. Can you conclude that because $(0.27)(0.16) = 0.043$ about 4% of the labor force are college-educated laborers or operators? Explain your answer.

Applying the multiplication rule

If two events A and B are independent, the event that A does not occur is also independent of B, and so on. Suppose, for example, that 75% of all registered voters in a suburban district are Republicans. If an opinion poll interviews two voters chosen independently, the probability that the first is a Republican and the second is not a Republican is $(0.75)(0.25) = 0.1875$. The multiplication rule also extends to collections of more than two events, provided that all are independent. Independence of events A, B, and C means that no information about any one or any two can change the probability of the remaining events. Independence is often assumed in setting up a probability model when the events we are describing seem to have no connection. We can then use the multiplication rule freely, as in this example.

EXAMPLE 5.3 **Undersea cables**

A transatlantic data cable contains repeaters at regular intervals to amplify the signal. If a repeater fails, it must be replaced by fishing the cable to the surface at great expense. Each repeater has probability 0.999 of functioning without failure for 10 years. Suppose that repeaters fail independently of each other. (This assumption means that there are no "common causes" such as earthquakes that would affect several repeaters at once.) Denote by A_i the event that the ith repeater operates successfully for 10 years.

The probability that two repeaters both last 10 years is

$$P(A_1 \text{ and } A_2) = P(A_1)P(A_2)$$

$$= 0.999 \times 0.999 = 0.998$$

For a cable with 10 repeaters the probability of no failures in 10 years is

$$P(A_1 \text{ and } A_2 \text{ and } \ldots \text{ and } A_{10}) = P(A_1)P(A_2) \cdots P(A_{10})$$

$$= 0.999 \times 0.999 \times \cdots \times 0.999$$

$$= 0.999^{10} = 0.990$$

Cables with 2 or 10 repeaters are quite reliable. Unfortunately, a transatlantic cable has 300 repeaters. The probability that all 300 work for 10 years is

$$P(A_1 \text{ and } A_2 \text{ and } \ldots \text{ and } A_{300}) = 0.999^{300} = 0.741$$

There is therefore about one chance in four that the cable will have to be fished up for replacement of a repeater sometime during the next 10 years. Repeaters are in fact designed to be much more reliable than 0.999 in 10 years. Some transatlantic cables have served for more than 20 years with no failures.

By combining the rules we have learned, we can compute probabilities for rather complex events. Here is an example.

EXAMPLE 5.4   **False positives in AIDS testing**

A diagnostic test for the presence of the AIDS virus has probability 0.005 of producing a false positive. That is, when a person free of the AIDS virus is tested, the test has probability 0.005 of falsely indicating that the virus is present. If the 140 employees of a medical clinic are tested and all 140 are free of AIDS, what is the probability that at least one false positive will occur?

It is reasonable to assume as part of the probability model that the test results for different individuals are independent. The probability that the test is positive for a single person is 0.005, so the probability of a negative result is $1 - 0.005 = 0.995$ by Rule 3. The probability of at least one false positive among the 140 people tested is therefore

$$P(\text{at least one positive}) = 1 - P(\text{no positives})$$

$$= 1 - P(140 \text{ negatives})$$

$$= 1 - 0.995^{140}$$

$$= 1 - 0.496 = 0.504$$

The probability is greater than 1/2 that at least one of the 140 people will test positive for AIDS, even though no one has the virus.

APPLY YOUR KNOWLEDGE

5.4   **Bright lights?** A string of Christmas lights contains 20 lights. The lights are wired in series, so that if any light fails the whole string will go dark. Each light has probability 0.02 of failing during a 3-year period. The lights fail independently of each other. What is the probability that the string of lights will remain bright for 3 years?

5.5   **Detecting steroids.** An athlete suspected of having used steroids is given two tests that operate independently of each other. Test A has probability 0.9 of being positive if steroids have been used. Test B has probability 0.8 of being positive if steroids have been used. What is the probability that *neither* test is positive if steroids have been used?

5.6   **Psychologists roll the dice.** A six-sided die has four green and two red faces and is balanced so that each face is equally likely to come up. The die will be rolled several times. You must choose one of the following three sequences of colors:

RGRRR

RGRRRG

GRRRR

You will win $25 if the first rolls of the die give the sequence that you have chosen.

(a) What is the probability that one roll gives green? That it gives red?

Figure 5.3 *The addition rule for disjoint events: P(A or B or C) = P(A) + P(B) + P(C) when events A, B, and C are disjoint.*

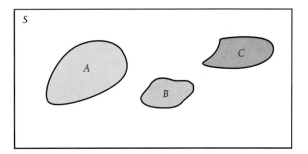

(b) What is the probability of each of the sequences of colors given above? Which sequence would you choose in an attempt to win the $25? Why? (In a psychological experiment, 63% of 260 students who had not studied probability chose the second sequence. This is evidence that our intuitive understanding of probability is not very accurate.[1])

The general addition rule

We know that if A and B are disjoint events, then $P(A \text{ or } B) = P(A) + P(B)$. This addition rule extends to more than two events that are disjoint in the sense that no two have any outcomes in common. The Venn diagram in Figure 5.3 shows three disjoint events A, B, and C. The probability that some one of these events occurs is $P(A) + P(B) + P(C)$.

If events A and B are *not* disjoint, they can occur simultaneously. The probability that one or the other occurs is then *less* than the sum of their probabilities. As Figure 5.4 suggests, the outcomes common to both are counted twice when we add probabilities, so we must subtract this probability once. Here is the addition rule for any two events, disjoint or not.

GENERAL ADDITION RULE FOR ANY TWO EVENTS

For any two events A and B,

$$P(A \text{ or } B) = P(A) + P(B) - P(A \text{ and } B)$$

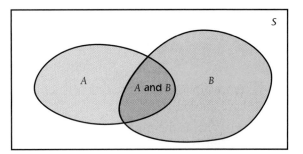

Figure 5.4 *The general addition rule: P(A or B) = P(A) + P(B) − P(A and B) for any events A and B.*

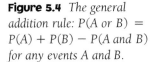

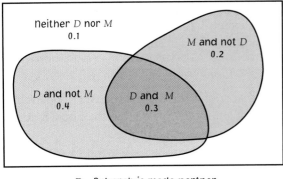

Figure 5.5 *Venn diagram and probabilities for Example 5.5.*

neither *D* nor *M*
0.1

M and not *D*
0.2

D and not *M*
0.4

D and *M*
0.3

D = Deborah is made partner
M = Matthew is made partner

If *A* and *B* are disjoint, the event {*A* and *B*} that both occur contains no outcomes and therefore has probability 0. So the general addition rule includes Rule 4, the addition rule for disjoint events.

EXAMPLE 5.5   Making partner

Deborah and Matthew are anxiously awaiting word on whether they have been made partners of their law firm. Deborah guesses that her probability of making partner is 0.7 and that Matthew's is 0.5. (These are personal probabilities reflecting Deborah's assessment of chance.) This assignment of probabilities does not give us enough information to compute the probability that at least one of the two is promoted. In particular, adding the individual probabilities of promotion gives the impossible result 1.2. If Deborah also guesses that the probability that *both* she and Matthew are made partners is 0.3, then by the general addition rule

$$P(\text{at least one is promoted}) = 0.7 + 0.5 - 0.3 = 0.9$$

The probability that *neither* is promoted is then 0.1 by Rule 3.

Venn diagrams are a great help in finding probabilities because you can just think of adding and subtracting areas. Figure 5.5 shows some events and their probabilities for Example 5.5. What is the probability that Deborah is promoted and Matthew is not? The Venn diagram shows that this is the probability that Deborah is promoted minus the probability that both are promoted, $0.7 - 0.3 = 0.4$. Similarly, the probability that Matthew is promoted and Deborah is not is $0.5 - 0.3 = 0.2$. The four probabilities that appear in the figure add to 1 because they refer to four disjoint events that make up the entire sample space.

Winning the lottery twice

In 1986, Evelyn Marie Adams won the New Jersey lottery for the second time, adding $1.5 million to her previous $3.9 million jackpot. The *New York Times* claimed that the odds of one person winning the big prize twice were 1 in 17 trillion. Nonsense, said two statisticians in a letter to the *Times*. The chance that Evelyn Marie Adams would win twice is indeed tiny, but it is almost certain that *someone* among the millions of lottery players would win two jackpots. Sure enough, Robert Humphries won his second Pennsylvania lottery jackpot ($6.8 million total) in 1988.

APPLY YOUR KNOWLEDGE

5.7   **Prosperity and education.** Call a household prosperous if its income exceeds $75,000. Call the household educated if the householder completed college. Select an American household at random, and let A be the event that the selected household is prosperous and B the event that it is educated. According to the Census Bureau, $P(A) = 0.15$, $P(B) = 0.25$, and the probability that a household is both prosperous and educated is $P(A \text{ and } B) = 0.09$.

(a) Draw a Venn diagram that shows the relation between the events A and B. What is the probability $P(A \text{ or } B)$ that the household selected is either prosperous or educated?

(b) In your diagram, shade the event that the household is educated but not prosperous. What is the probability of this event?

5.8   **Caffeine in the diet.** Common sources of caffeine are coffee, tea, and cola drinks. Suppose that

55% of adults drink coffee

25% of adults drink tea

45% of adults drink cola

and also that

15% drink both coffee and tea

5% drink all three beverages

25% drink both coffee and cola

5% drink only tea

Draw a Venn diagram marked with this information. Use it along with the addition rules to answer the following questions.

(a) What percent of adults drink only cola?

(b) What percent drink none of these beverages?

(JOHN A. RIZZO/PHOTODISC.)

SECTION 5.1 Summary

Events A and B are **disjoint** if they have no outcomes in common. Events A and B are **independent** if knowing that one event occurs does not change the probability we would assign to the other event.

Any assignment of probability obeys these more general rules in addition to those stated in Chapter 4:

Addition rule: If events $A, B, C, \ldots$ are all disjoint in pairs, then

$$P(\text{at least one of these events occurs}) = P(A) + P(B) + P(C) + \cdots$$

Multiplication rule: If events A and B are **independent**, then

$$P(A \text{ and } B) = P(A)P(B)$$

General addition rule: For any two events A and B,

$$P(A \text{ or } B) = P(A) + P(B) - P(A \text{ and } B)$$

SECTION 5.1 Exercises

5.9 **Military strategy.** A general can plan a campaign to fight one major battle or three small battles. He believes that he has probability 0.6 of winning the large battle and probability 0.8 of winning each of the small battles. Victories or defeats in the small battles are independent. The general must win either the large battle or all three small battles to win the campaign. Which strategy should he choose?

5.10 **Playing the lottery.** An instant lottery game gives you probability 0.02 of winning on any one play. Plays are independent of each other. If you play 5 times, what is the probability that you win at least once?

5.11 An automobile manufacturer buys computer chips from a supplier. The supplier sends a shipment containing 5% defective chips. Each chip chosen from this shipment has probability 0.05 of being defective, and each automobile uses 12 chips selected independently. What is the probability that all 12 chips in a car will work properly?

5.12 **A random walk on Wall Street?** The "random walk" theory of securities prices holds that price movements in disjoint time periods are independent of each other. Suppose that we record only whether the price is up or down each year, and that the probability that our portfolio rises in price in any one year is 0.65. (This probability is approximately correct for a portfolio containing equal dollar amounts of all common stocks listed on the New York Stock Exchange.)

(a) What is the probability that our portfolio goes up for three consecutive years?

(b) If you know that the portfolio has risen in price 2 years in a row, what probability do you assign to the event that it will go down next year?

(c) What is the probability that the portfolio's value moves in the same direction in both of the next 2 years?

5.13 **Getting into college.** Ramon has applied to both Princeton and Stanford. He thinks the probability that Princeton will admit him is 0.4, the probability that Stanford will admit him is 0.5, and the probability that both will admit him is 0.2.

(a) Make a Venn diagram with the probabilities given marked.

(b) What is the probability that neither university admits Ramon?

(c) What is the probability that he gets into Stanford but not Princeton?

5.14 **Tastes in music.** Musical styles other than rock and pop are becoming more popular. A survey of college students finds that

(MARK C. BURNETT/STOCK, BOSTON.)

40% like country music, 30% like gospel music, and 10% like both.

(a) Make a Venn diagram with these results.

(b) What percent of college students like country but not gospel?

(c) What percent like neither country nor gospel?

5.15 **Blood types.** The distribution of blood types among white Americans is approximately as follows: 37% type A, 13% type B, 44% type O, and 6% type AB. Suppose that the blood types of married couples are independent and that both the husband and the wife follow this distribution.

(a) An individual with type B blood can safely receive transfusions only from persons with type B or type O blood. What is the probability that the husband of a woman with type B blood is an acceptable blood donor for her?

(b) What is the probability that in a randomly chosen couple the wife has type B blood and the husband has type A?

(c) What is the probability that one of a randomly chosen couple has type A blood and the other has type B?

(d) What is the probability that at least one of a randomly chosen couple has type O blood?

5.16 **Age effects in medical care.** The type of medical care a patient receives may vary with the age of the patient. A large study of women who had a breast lump investigated whether or not each woman received a mammogram and a biopsy when the lump was discovered. Here are some probabilities estimated by the study. The entries in the table are the probabilities that *both* of two events occur; for example, 0.321 is the probability that a patient is under 65 years of age *and* the tests were done. The four probabilities in the table have sum 1 because the table lists all possible outcomes.

	Tests Done?	
	Yes	No
Age under 65	0.321	0.124
Age 65 or over	0.365	0.190

(a) What is the probability that a patient in this study is under 65? That a patient is 65 or over?

(b) What is the probability that the tests were done for a patient? That they were not done?

(c) Are the events A = {the patient was 65 or older} and B = {the tests were done} independent? Were the tests omitted on older patients more or less frequently than would be the case if testing were independent of age?

5.17 **Playing the odds (EESEE)?** A writer on casino games says that the odds against throwing an 11 in the dice game craps are 17 to 1. He then says that the odds against three 11s in a row are $17 \times 17 \times 17$ to 1, or 4913 to 1.[2]

(a) What is the probability that the sum of the up-faces is 11 when you throw a balanced die? (See Figure 4.2 on page 221.) What is the probability of three 11s in three independent throws?

(b) If an event A has probability P, the odds against A are

$$\text{Odds against } A = \frac{1 - P}{P}$$

Gamblers often speak of odds rather than probabilities. The odds against an event that has probability 1/3 are 2 to 1, for example. Find the odds against throwing an 11 and the odds against throwing three straight 11s. Which of the writer's statements are correct?

5.2 The Binomial Distributions

A basketball player shoots 5 free throws. How many does she make? A new treatment for pancreatic cancer is tried on 25 patients. How many survive for five years? You plant 10 dogwood trees. How many live through the winter? In all these situations, we want a probability model for a *count* of successful outcomes.

The binomial setting

The distribution of a count depends on how the data are produced. Here is a common situation.

> **THE BINOMIAL SETTING**
>
> 1. There are a fixed number n of observations.
> 2. The n observations are all **independent.** That is, knowing the result of one observation tells you nothing about the other observations.
> 3. Each observation falls into one of just two categories, which for convenience we call "success" and "failure."
> 4. The probability of a success, call it p, is the same for each observation.

Think of tossing a coin n times as an example of the binomial setting. Each toss gives either heads or tails. Knowing the outcome of one toss doesn't tell us anything about other tosses, so the n tosses are independent. If we call heads a success, then p is the probability of a head and remains the same as long as we toss the same coin. The number of heads we count is a random variable X. The distribution of X is called a *binomial distribution*.

> **BINOMIAL DISTRIBUTION**
>
> The distribution of the count X of successes in the binomial setting is the **binomial distribution** with parameters n and p. The parameter n is the number of observations, and p is the probability of a success on any one observation. The possible values of X are the whole numbers from 0 to n.

The binomial distributions are an important class of probability distributions. Pay attention to the binomial setting, because not all counts have binomial distributions.

EXAMPLE 5.6   Blood types

Genetics says that children receive genes from their parents independently. Each child of a particular pair of parents has probability 0.25 of having type O blood. If these parents have 5 children, the number who have type O blood is the count X of successes in 5 independent trials with probability 0.25 of a success on each trial. So X has the binomial distribution with $n = 5$ and $p = 0.25$.

EXAMPLE 5.7   Dealing cards

Deal 10 cards from a shuffled deck and count the number X of red cards. There are 10 observations, and each gives either a red or a black card. A "success" is a red card. But the observations are *not* independent. If the first card is black, the second is more likely to be red because there are more red cards than black cards left in the deck. The count X does *not* have a binomial distribution.

EXAMPLE 5.8   Choosing an SRS

An engineer chooses an SRS of 10 switches from a shipment of 10,000 switches. Suppose that (unknown to the engineer) 10% of the switches in the shipment are bad. The engineer counts the number X of bad switches in the sample.

This is not quite a binomial setting. Just as removing one card in Example 5.7 changed the makeup of the deck, removing one switch changes the proportion of bad switches remaining in the shipment. So the state of the second switch chosen is not independent of the first. But removing one switch from a shipment of 10,000 changes the makeup of the remaining 9999 switches very little. In practice, the distribution of X is very close to the binomial distribution with $n = 10$ and $p = 0.1$.

Example 5.8 shows how we can use the binomial distributions in the statistical setting of selecting an SRS. When the population is much larger than the sample, a count of successes in an SRS of size n has approximately the binomial distribution with n equal to the sample size and p equal to the proportion of successes in the population.

APPLY YOUR KNOWLEDGE

In each of Exercises 5.18 to 5.20, X is a count. Does X have a binomial distribution? Give your reasons in each case.

5.18 You observe the sex of the next 20 children born at a local hospital; X is the number of girls among them.

5.19 A couple decides to continue to have children until their first girl is born; X is the total number of children the couple has.

5.20 A student studies binomial distributions using computer-assisted instruction. After the lesson, the computer presents 10 problems. The student solves each problem and enters her answer. The computer gives additional instruction between problems if the answer is wrong. The count X is the number of problems that the student gets right.

Binomial probabilities

We can find a formula for the probability that a binomial random variable takes any value by adding probabilities for the different ways of getting exactly that many successes in n observations. Here is the example we will use to show the idea.

EXAMPLE 5.9 Inheriting blood type

Each child born to a particular set of parents has probability 0.25 of having blood type O. If these parents have 5 children, what is the probability that exactly 2 of them have type O blood?

The count of children with type O blood is a binomial random variable X with $n = 5$ tries and probability $p = 0.25$ of a success on each try. We want $P(X = 2)$.

Because the method doesn't depend on the specific example, let's use "S" for success and "F" for failure for short. Do the work in two steps.

Step 1. Find the probability that a specific 2 of the 5 tries, say the first and the third, give successes. This is the outcome SFSFF. Because tries are independent, the multiplication rule for independent events applies. The probability we want is

$$P(SFSFF) = P(S)P(F)P(S)P(F)P(F)$$

$$= (0.25)(0.75)(0.25)(0.75)(0.75)$$

$$= (0.25)^2(0.75)^3$$

Step 2. Observe that the probability of *any one* arrangement of 2 S's and 3 F's has this same probability. This is true because we multiply together 0.25 twice and 0.75 three times whenever we have 2 S's and 3 F's. The probability that $X = 2$ is the probability of getting 2 S's and 3 F's in any arrangement whatsoever. Here are all the possible arrangements:

$$\text{SSFFF} \quad \text{SFSFF} \quad \text{SFFSF} \quad \text{SFFFS} \quad \text{FSSFF}$$
$$\text{FSFSF} \quad \text{FSFFS} \quad \text{FFSSF} \quad \text{FFSFS} \quad \text{FFFSS}$$

There are 10 of them, all with the same probability. The overall probability of 2 successes is therefore

$$P(X = 2) = 10(0.25)^2(0.75)^3 = 0.2637$$

The pattern of this calculation works for any binomial probability. To use it, we must count the number of arrangements of k successes in n observations. We use the following fact to do the counting without actually listing all the arrangements.

BINOMIAL COEFFICIENT

The number of ways of arranging k successes among n observations is given by the **binomial coefficient**

$$\binom{n}{k} = \frac{n!}{k!\,(n-k)!}$$

for $k = 0, 1, 2, \ldots, n$.

---

factorial

The formula for binomial coefficients uses the **factorial** notation. For any positive whole number n, its factorial $n!$ is

$$n! = n \times (n-1) \times (n-2) \times \cdots \times 3 \times 2 \times 1$$

Also, $0! = 1$.

The larger of the two factorials in the denominator of a binomial coefficient will cancel much of the $n!$ in the numerator. For example, the binomial coefficient we need for Example 5.9 is

$$\binom{5}{2} = \frac{5!}{2!\,3!}$$

$$= \frac{(5)(4)(3)(2)(1)}{(2)(1) \times (3)(2)(1)}$$

$$= \frac{(5)(4)}{(2)(1)} = \frac{20}{2} = 10$$

The notation $\binom{n}{k}$ is *not* related to the fraction $\frac{n}{k}$. A helpful way to remember its meaning is to read it as "binomial coefficient n choose k." Binomial coefficients have many uses in mathematics, but we are interested in them only as an aid to finding binomial probabilities. The binomial coefficient $\binom{n}{k}$ counts the number of different ways in which k successes can be arranged among n observations. The binomial probability $P(X = k)$ is this count multiplied by the probability of any specific arrangement of the k successes. Here is the result we seek.

BINOMIAL PROBABILITY

If X has the binomial distribution with n observations and probability p of success on each observation, the possible values of X are 0, 1, 2, ..., n. If k is any one of these values,

$$P(X = k) = \binom{n}{k}p^k(1 - p)^{n-k}$$

EXAMPLE 5.10   Inspecting switches

The number X of switches that fail inspection in Example 5.8 has approximately the binomial distribution with $n = 10$ and $p = 0.1$.

The probability that no more than 1 switch fails is

$$P(X \le 1) = P(X = 1) + P(X = 0)$$

$$= \binom{10}{1}(0.1)^1(0.9)^9 + \binom{10}{0}(0.1)^0(0.9)^{10}$$

$$= \frac{10!}{1!\,9!}(0.1)(0.3874) + \frac{10!}{0!\,10!}(1)(0.3487)$$

$$= (10)(0.1)(0.3874) + (1)(1)(0.3487)$$

$$= 0.3874 + 0.3487 = 0.7361$$

This calculation uses the facts that $0! = 1$ and that $a^0 = 1$ for any number a other than 0. We see that about 74% of all samples will contain no more than 1 bad switch. In fact, 35% of the samples will contain no bad switches. A sample of size 10 cannot be trusted to alert the engineer to the presence of unacceptable items in the shipment.

Was he good or was he lucky?

When a baseball player hits .300, everyone applauds. A .300 hitter gets a hit in 30% of times at bat. Could a .300 year just be luck? Typical major leaguers bat about 500 times a season and hit about .260. A hitter's successive tries seem to be independent. So the number of hits in a season should have the binomial distribution with $n = 500$ and $p = 0.26$. We can calculate that the probability of hitting .300 is about 0.025. Out of 100 run-of-the-mill major league hitters, two or three each year will bat .300 because they were lucky.

APPLY YOUR KNOWLEDGE

5.21   Inheriting blood type. If the parents in Example 5.9 have 5 children, the number who have type O blood is a random variable X that has the binomial distribution with $n = 5$ and $p = 0.25$.

   (a) What are the possible values of X?

(b) Find the probability of each value of X. Draw a histogram to display this distribution. (Because probabilities are long-run proportions, a histogram with the probabilities as the heights of the bars shows what the distribution of X would be in very many repetitions.)

5.22 A factory employs several thousand workers, of whom 30% are Hispanic. If the 15 members of the union executive committee were chosen from the workers at random, the number of Hispanics on the committee would have the binomial distribution with $n = 15$ and $p = 0.3$.

(a) What is the probability that exactly 3 members of the committee are Hispanic?

(b) What is the probability that 3 or fewer members of the committee are Hispanic?

5.23 **Do our athletes graduate?** A university claims that 80% of its basketball players get degrees. An investigation examines the fate of all 20 players who entered the program over a period of several years that ended six years ago. Of these players, 11 graduated and the remaining 9 are no longer in school. If the university's claim is true, the number of players among the 20 who graduate should have the binomial distribution with $n = 20$ and $p = 0.8$. What is the probability that exactly 11 out of 20 players graduate?

Binomial mean and standard deviation

If a count X has the binomial distribution based on n observations with probability p of success, what is its mean μ? That is, in very many repetitions of the binomial setting, what will be the average count of successes? We can guess the answer. If a basketball player makes 80% of her free throws, the mean number made in 10 tries should be 80% of 10, or 8. In general, the mean of a binomial distribution should be $\mu = np$. Here are the facts.

BINOMIAL MEAN AND STANDARD DEVIATION

If a count X has the binomial distribution with number of observations n and probability of success p, the **mean** and **standard deviation** of X are

$$\mu = np$$
$$\sigma = \sqrt{np(1 - p)}$$

Remember that these short formulas are good only for binomial distributions. They can't be used for other distributions.

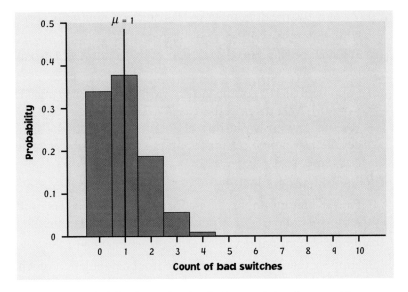

Figure 5.6 *Probability histogram for the binomial distribution with* $n = 10$ *and* $p = 0.1$.

EXAMPLE 5.11　**Inspecting switches**

　　Continuing Example 5.10, the count X of bad switches is binomial with $n = 10$ and $p = 0.1$. The histogram in Figure 5.6 displays this probability distribution. (Because probabilities are long-run proportions, using probabilities as the heights of the bars shows what the distribution of X would be in very many repetitions.) The distribution is strongly skewed. Although X can take any whole-number value from 0 to 10, the probabilities of values larger than 5 are so small that they do not appear in the histogram.

　　The mean and standard deviation of the binomial distribution in Figure 5.6 are

$$\mu = np$$
$$= (10)(0.1) = 1$$
$$\sigma = \sqrt{np(1 - p)}$$
$$= \sqrt{(10)(0.1)(0.9)} = \sqrt{0.9} = 0.9487$$

The mean is marked on the probability histogram in Figure 5.6.

APPLY YOUR KNOWLEDGE

5.24　**Inheriting blood type.** What are the mean and standard deviation of the number of children with type O blood in Exercise 5.21? Mark the location of the mean on the probability histogram you made in that exercise.

5.25   (a) What is the mean number of Hispanics on randomly chosen committees of 15 workers in Exercise 5.22?

(b) What is the standard deviation σ of the count X of Hispanic members?

(c) Suppose that 10% of the factory workers were Hispanic. Then $p = 0.1$. What is σ in this case? What is σ if $p = 0.01$? What does your work show about the behavior of the standard deviation of a binomial distribution as the probability of a success gets closer to 0?

5.26   **Do our athletes graduate?**

(a) Find the mean number of graduates out of 20 players in the setting of Exercise 5.23 if the university's claim is true.

(b) Find the standard deviation σ of the count X.

(c) Suppose that the 20 players came from a population of which $p = 0.9$ graduated. What is the standard deviation σ of the count of graduates? If $p = 0.99$, what is σ? What does your work show about the behavior of the standard deviation of a binomial distribution as the probability p of success gets closer to 1?

The normal approximation to binomial distributions

The formula for binomial probabilities becomes awkward as the number of trials n increases. You can use software or a statistical calculator to handle some problems for which the formula is not practical. Here is another alternative: *as the number of trials n gets larger, the binomial distribution gets close to a normal distribution.* When n is large, we can use normal probability calculations to approximate hard-to-calculate binomial probabilities.

EXAMPLE 5.12   Attitudes toward shopping

Are attitudes toward shopping changing? Sample surveys show that fewer people enjoy shopping than in the past. A recent survey asked a nationwide random sample of 2500 adults if they agreed or disagreed that "I like buying new clothes, but shopping is often frustrating and time-consuming."[3] The population that the poll wants to draw conclusions about is all U.S. residents aged 18 and over. Suppose that in fact 60% of all adult U.S. residents would say "Agree" if asked the same question. What is the probability that 1520 or more of the sample agree?

Because there are more than 195 million adults, we can take the responses of 2500 randomly chosen adults to be independent. So the number in our sample who agree that shopping is frustrating is a random variable X having the binomial distribution with $n = 2500$ and $p = 0.6$. To find the probability that at least 1520 of the people in the sample find shopping frustrating, we must

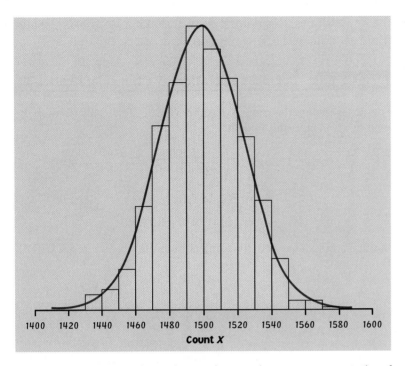

Figure 5.7 *Histogram of 1000 binomial counts (n = 2500, p = 0.6) and the normal density curve that approximates this binomial distribution.*

add the binomial probabilities of all outcomes from $X = 1520$ to $X = 2500$. This isn't practical. Here are three ways to do this problem.

1. Statistical software can do the calculation. The exact result is

$$P(X \geq 1520) = 0.2131$$

2. We can simulate a large number of repetitions of the sample. Figure 5.7 displays a histogram of the counts X from 1000 samples of size 2500 when the truth about the population is $p = 0.6$. Because 221 of these 1000 samples have X at least 1520, the probability estimated from the simulation is

$$P(X \geq 1520) = \frac{221}{1000} = 0.221$$

3. Both of the previous methods require software. Instead, look at the normal curve in Figure 5.7. This is the density curve of the normal distribution with the same mean and standard deviation as the binomial variable X:

$$\mu = np = (2500)(0.6) = 1500$$

$$\sigma = \sqrt{np(1 - p)} = \sqrt{(2500)(0.6)(0.4)} = 24.49$$

As the figure shows, this normal distribution approximates the binomial distribution quite well. So we can do a normal calculation.

EXAMPLE 5.13   **Normal calculation of a binomial probability**

If we act as though the count X has the $N(1500, 24.49)$ distribution, here is the probability we want, using Table A:

$$P(X \geq 1520) = P\left(\frac{X - 1500}{24.49} \geq \frac{1520 - 1500}{24.49}\right)$$

$$= P(Z \geq 0.82)$$

$$= 1 - 0.7939 = 0.2061$$

The normal approximation 0.2061 differs from the software result 0.2131 by only 0.007.

> **NORMAL APPROXIMATION FOR BINOMIAL DISTRIBUTIONS**
>
> Suppose that a count X has the binomial distribution with n trials and success probability p. When n is large, the distribution of X is approximately normal, $N\left(np, \sqrt{np(1 - p)}\right)$.
>
> As a rule of thumb, we will use the normal approximation when n and p satisfy $np \geq 10$ and $n(1 - p) \geq 10$.

The normal approximation is easy to remember because it says that X is normal with its binomial mean and standard deviation. The accuracy of the normal approximation improves as the sample size n increases. It is most accurate for any fixed n when p is close to 1/2, and least accurate when p is near 0 or 1. Whether or not you use the normal approximation should depend on how accurate your calculations need to be. For most statistical purposes great accuracy is not required. Our "rule of thumb" for use of the normal approximation reflects this judgment.

APPLY YOUR KNOWLEDGE

5.27   **A market research survey.** You operate a restaurant. You read that a sample survey by the National Restaurant Association shows that 40% of adults are committed to eating nutritious food when eating away from home. To help plan your menu, you decide to conduct a sample survey in your own area. You will use random digit dialing to contact an SRS of 200 households by telephone.

(a) If the national result holds in your area, it is reasonable to use the binomial distribution with $n = 200$ and $p = 0.4$ to describe the count of X of respondents who seek nutritious food when eating out. Explain why.

(b) What is the mean number of nutrition-conscious people in your sample if $p = 0.4$ is true? What is the standard deviation?

(c) What is the probability that X lies between 75 and 85? (Use the normal approximation.)

5.28 **Are we shipping on time?** Your mail-order company advertises that it ships 90% of its orders within three working days. You select an SRS of 100 of the 5000 orders received in the past week for an audit. The audit reveals that 86 of these orders were shipped on time.

(a) If the company really ships 90% of its orders on time, what is the probability that 86 or fewer in an SRS of 100 orders are shipped on time?

(b) A critic says, "Aha! You claim 90%, but in your sample the on-time percentage is only 86%. So the 90% claim is wrong." Explain in simple language why your probability calculation in (a) shows that the result of the sample does not refute the 90% claim.

5.29 **Checking for survey errors.** One way of checking the effect of undercoverage, nonresponse, and other sources of error in a sample survey is to compare the sample with known facts about the population. About 12% of American adults are black. The number X of blacks in a random sample of 1500 adults should therefore vary with the binomial ($n = 1500$, $p = 0.12$) distribution.

(a) What are the mean and standard deviation of X?

(b) Use the normal approximation to find the probability that the sample will contain 170 or fewer blacks. Be sure to check that you can safely use the approximation.

SECTION 5.2 Summary

A count X of successes has a **binomial distribution** in the **binomial setting:** there are n observations; the observations are independent of each other; each observation results in a success or a failure; and each observation has the same probability p of a success.

The binomial distribution with n observations and probability p of success gives a good approximation to the sampling distribution of the count of successes in an SRS of size n from a large population containing proportion p of successes.

If X has the binomial distribution with parameters n and p, the possible values of X are the whole numbers $0, 1, 2, \ldots, n$. The **binomial probability** that X takes any value is

$$P(X = k) = \binom{n}{k} p^k (1 - p)^{n-k}$$

The **binomial coefficient**

$$\binom{n}{k} = \frac{n!}{k!\,(n - k)!}$$

counts the number of ways k successes can be arranged among n observations. Here the **factorial $n!$** is

$$n! = n \times (n - 1) \times (n - 2) \times \cdots \times 3 \times 2 \times 1$$

for positive whole numbers n, and $0! = 1$.

The **mean** and **standard deviation** of a binomial count X are

$$\mu = np$$

$$\sigma = \sqrt{np(1 - p)}$$

The **normal approximation** to the binomial distribution says that if X is a count having the binomial distribution with parameters n and p, then when n is large, X is approximately $N(np, \sqrt{np(1 - p)})$. We will use this approximation when $np \geq 10$ and $n(1 - p) \geq 10$.

SECTION 5.2 Exercises

5.30 **Binomial setting?** In each situation below, is it reasonable to use a binomial distribution for the random variable X? Give reasons for your answer in each case.

(a) An auto manufacturer chooses one car from each hour's production for a detailed quality inspection. One variable recorded is the count X of finish defects (dimples, ripples, etc.) in the car's paint.

(b) The pool of potential jurors for a murder case contains 100 persons chosen at random from the adult residents of a large city. Each person in the pool is asked whether he or she opposes the death penalty; X is the number who say "Yes."

(c) Joe buys a ticket in his state's "Pick 3" lottery game every week; X is the number of times in a year that he wins a prize.

5.31 **Binomial setting?** In each of the following cases, decide whether or not a binomial distribution is an appropriate model, and give your reasons.

(a) Fifty students are taught about binomial distributions by a television program. After completing their study, all students take the same examination. The number of students who pass is counted.

(b) A student studies binomial distributions using computer-assisted instruction. After the initial instruction is completed, the computer presents 10 problems. The student solves each problem and enters the answer; the computer gives additional instruction between problems if the student's answer is wrong. The number of problems that the student solves correctly is counted.

(c) A chemist repeats a solubility test 10 times on the same substance. Each test is conducted at a temperature 10° higher than the previous test. She counts the number of times that the substance dissolves completely.

5.32　**Random digits.** Each entry in a table of random digits like Table B has probability 0.1 of being a 0, and digits are independent of each other.

(a) What is the probability that a group of five digits from the table will contain at least one 0?

(b) What is the mean number of 0s in lines 40 digits long?

5.33　**Unmarried women.** Among employed women, 25% have never been married. Select 10 employed women at random.

(a) The number in your sample who have never been married has a binomial distribution. What are n and p?

(b) What is the probability that exactly 2 of the 10 women in your sample have never been married?

(c) What is the probability that 2 or fewer have never been married?

(d) What is the mean number of women in such samples who have never been married? What is the standard deviation?

5.34　**Testing ESP.** In a test for ESP (extrasensory perception), a subject is told that cards the experimenter can see but he cannot contain either a star, a circle, a wave, or a square. As the experimenter looks at each of 20 cards in turn, the subject names the shape on the card. A subject who is just guessing has probability 0.25 of guessing correctly on each card.

(a) The count of correct guesses in 20 cards has a binomial distribution. What are n and p?

(b) What is the mean number of correct guesses in many repetitions?

(c) What is the probability of exactly 5 correct guesses?

5.35　**Random stock prices.** A believer in the "random walk" theory of stock markets thinks that an index of stock prices has probability

0.65 of increasing in any year. Moreover, the change in the index in any given year is not influenced by whether it rose or fell in earlier years. Let X be the number of years among the next 5 years in which the index rises.

(a) X has a binomial distribution. What are n and p?

(b) What are the possible values that X can take?

(c) Find the probability of each value of X. Draw a probability histogram for the distribution of X.

(d) What are the mean and standard deviation of this distribution? Mark the location of the mean on your histogram.

5.36 **Lie detectors.** A federal report finds that lie detector tests given to truthful persons have probability about 0.2 of suggesting that the person is deceptive.[4]

(a) A company asks 12 job applicants about thefts from previous employers, using a lie detector to assess their truthfulness. Suppose that all 12 answer truthfully. What is the probability that the lie detector says all 12 are truthful? What is the probability that the lie detector says at least 1 is deceptive?

(b) What is the mean number among 12 truthful persons who will be classified as deceptive? What is the standard deviation of this number?

(c) What is the probability that the number classified as deceptive is less than the mean?

5.37 **Multiple-choice tests.** Here is a simple probability model for multiple-choice tests. Suppose that each student has probability p of correctly answering a question chosen at random from a universe of possible questions. (A strong student has a higher p than a weak student.) Answers to different questions are independent. Jodi is a good student for whom $p = 0.75$.

(a) Use the normal approximation to find the probability that Jodi scores 70% or lower on a 100-question test.

(b) If the test contains 250 questions, what is the probability that Jodi will score 70% or lower?

5.38 **A market research survey.** Return to the restaurant sample described in Exercise 5.27. You find 100 of your 200 respondents concerned about nutrition. Is this reason to believe that the percent in your area is higher than the national 40%? To answer this question, find the probability that X is 100 or larger if $p = 0.4$ is true. If this probability is very small, that is reason to think that p is actually greater than 0.4.

5.39 **Planning a survey.** You are planning a sample survey of small businesses in your area. You will choose an SRS of businesses listed

Because th
to be any
To find

Slim finds
drawn is a
first card i
the *conditi*
says that

Slim will n

Remembe
ability $P(B \mid$
event whose

EXAMPL

About 8
Victorias o
mirror, the

The 85% i
Crown Vic

5.40 Won
A be
that t
data
and p
prob;
holdi

5.41 Buyi
contr

---

in the telephone book's Yellow Pages. Experience shows that only about half the businesses you contact will respond.

(a) If you contact 150 businesses, it is reasonable to use the binomial distribution with $n = 150$ and $p = 0.5$ for the number X who respond. Explain why.

(b) What is the expected number (the mean) who will respond?

(c) What is the probability that 70 or fewer will respond? (Use the normal approximation.)

(d) How large a sample must you take to increase the mean number of respondents to 100?

5.3 Conditional Probability

In Section 2.5 we met the idea of a *conditional distribution*, the distribution of a variable given that a condition is satisfied. Now we will introduce the probability language for this idea.

EXAMPLE 5.14   Suicides

Here is a two-way table of suicides committed in a recent year, classified by the gender of the victim and whether or not a firearm was used:

	Male	Female	Total
Firearm	16,381	2,559	18,940
Other	9,034	3,536	12,570
Total	25,415	6,095	31,510

Choose a suicide at random. What is the probability that a firearm was used? Because "choose at random" gives all 31,510 suicides the same chance, the probability is just the proportion that used a firearm:

$$P(\text{firearm}) = \frac{18,940}{31,510} = 0.60$$

Now we are told that the suicide chosen is female. A glance at the table shows that women are less likely than men to use firearms. The probability that a firearm was used, *given the information that the suicide was female*, is

$$P(\text{firearm} \mid \text{female}) = \frac{2,559}{6,095} = 0.42$$

This is a **conditional probability**.

conditional probability

that there is probability 0.4 that the dollar will fall in value against the Japanese yen in the next month. The treasurer also believes that *if* the dollar falls there is probability 0.8 that the supplier will demand renegotiation of the contract. What probability has the treasurer assigned to the event that the dollar falls and the supplier demands renegotiation?

5.42   **Suicides.** Use the two-way table in Example 5.14 to find these conditional probabilities for suicides.

(a) $P(\text{firearm} \mid \text{male})$

(b) $P(\text{male} \mid \text{firearm})$

Extending the multiplication rule

The multiplication rule extends to the probability that all of several events occur. The key is to condition each event on the occurrence of *all* of the preceding events. For example, we have for three events A, B, and C that

$$P(A \text{ and } B \text{ and } C) = P(A)P(B \mid A)P(C \mid A \text{ and } B)$$

EXAMPLE 5.17   **The future of high school athletes**

Only 5% of male high school basketball, baseball, and football players go on to play at the college level. Of these, only 1.7% enter major league professional sports. About 40% of the athletes who compete in college and then reach the pros have a career of more than 3 years.[5] Define these events:

$$A = \{\text{competes in college}\}$$

$$B = \{\text{competes professionally}\}$$

$$C = \{\text{pro career longer than 3 years}\}$$

What is the probability that a high school athlete competes in college and then goes on to have a pro career of more than 3 years? We know that

$$P(A) = 0.05$$

$$P(B \mid A) = 0.017$$

$$P(C \mid A \text{ and } B) = 0.4$$

The probability we want is therefore

$$P(A \text{ and } B \text{ and } C) = P(A)P(B \mid A)P(C \mid A \text{ and } B)$$

$$= 0.05 \times 0.017 \times 0.40 = 0.00034$$

Only about 3 of every 10,000 high school athletes can expect to compete in college and have a professional career of more than 3 years. High school athletes would be wise to concentrate on studies rather than on unrealistic hopes of fortune from pro sports.

Conditional probability and independence

If we know $P(A)$ and $P(A \text{ and } B)$, we can rearrange the multiplication rule to produce a *definition* of the conditional probability $P(B \mid A)$ in terms of unconditional probabilities.

DEFINITION OF CONDITIONAL PROBABILITY

When $P(A) > 0$, the **conditional probability** of B given A is

$$P(B \mid A) = \frac{P(A \text{ and } B)}{P(A)}$$

The conditional probability $P(B \mid A)$ makes no sense if the event A can never occur, so we require that $P(A) > 0$ whenever we talk about $P(B \mid A)$. The definition of conditional probability reminds us that in principle all probabilities, including conditional probabilities, can be found from the assignment of probabilities to events that describes a random phenomenon. More often, as in Examples 5.15 and 5.17, conditional probabilities are part of the information given to us in a probability model, and the multiplication rule is used to compute $P(A \text{ and } B)$.

The conditional probability $P(B \mid A)$ is generally not equal to the unconditional probability $P(B)$. That is because the occurrence of event A generally gives us some additional information about whether or not event B occurs. If knowing that A occurs gives no additional information about B, then A and B are independent events. The precise definition of independence is expressed in terms of conditional probability.

INDEPENDENT EVENTS

Two events A and B that both have positive probability are **independent** if

$$P(B \mid A) = P(B)$$

This definition makes precise the informal description of independence given in Section 5.1. We now see that the multiplication rule for independent events, $P(A \text{ and } B) = P(A)P(B)$, is a special case of the general multiplication rule, $P(A \text{ and } B) = P(A)P(B \mid A)$, just as the addition rule for disjoint events is a special case of the general addition rule. We will rarely use the definition of independence, because most often independence is part of the information given to us in a probability model.

Gimme the ball

If a basketball player makes several consecutive shots, fans and teammates believe that she has a "hot hand" and is more likely to make the next shot. This is wrong. Careful study has shown that runs of baskets made or missed are no more frequent in basketball than would be expected if each shot is independent of the player's previous shots. Players perform consistently, not in streaks. If a player makes half her shots in the long run, her hits and misses behave just like tosses of a coin.

5.43 **College degrees.** Here are the counts (in thousands) of earned degrees in the United States in a recent year, classified by level and by the sex of the degree recipient:

	Bachelor's	Master's	Professional	Doctorate	Total
Female	616	194	30	16	856
Male	529	171	44	26	770
Total	1145	365	74	42	1626

(a) If you choose a degree recipient at random, what is the probability that the person you choose is a woman?

(b) What is the conditional probability that you choose a woman, given that the person chosen received a professional degree?

(c) Are the events "choose a woman" and "choose a professional degree recipient" independent? How do you know?

5.44 **Prosperity and education.** Call a household prosperous if its income exceeds $75,000. Call the household educated if the householder completed college. Select an American household at random, and let A be the event that the selected household is prosperous and B the event that it is educated. According to the Census Bureau, $P(A) = 0.15$, $P(B) = 0.25$, and the probability that a household is both prosperous and educated is $P(A \text{ and } B) = 0.09$.

(a) Find the conditional probability that a household is educated, given that it is prosperous.

(b) Find the conditional probability that a household is prosperous, given that it is educated.

(c) Are events A and B independent? How do you know?

SECTION 5.3 Summary

The **conditional probability** $P(B \mid A)$ of an event B given an event A is defined by

$$P(B \mid A) = \frac{P(A \text{ and } B)}{P(A)}$$

when $P(A) > 0$. In practice, we most often find conditional probabilities from directly available information rather than from the definition.

Any assignment of probability obeys the **general multiplication rule**
$P(A \text{ and } B) = P(A)P(B \mid A)$.

A and B are **independent** when $P(B \mid A) = P(B)$. The multiplication rule
then becomes $P(A \text{ and } B) = P(A)P(B)$.

SECTION 5.3 Exercises

5.45 **Inspecting switches.** A shipment contains 10,000 switches. Of
these, 1000 are bad. An inspector draws switches at random, so that
each switch has the same chance to be drawn.

   (a) Draw one switch. What is the probability that the switch you
   draw is bad? What is the probability that it is not bad?

   (b) Suppose the first switch drawn is bad. How many switches
   remain? How many of them are bad? Draw a second switch at
   random. What is the conditional probability that this switch is
   bad?

   (c) Answer the questions in (b) again, but now suppose that the first
   switch drawn is not bad.
   Comment: Knowing the result of the first trial changes the
   conditional probability for the second trial, so the trials are not
   independent. But because the shipment is large, the probabilities
   change very little. The trials are almost independent.

5.46 **Tastes in music.** Musical styles other than rock and pop are
becoming more popular. A survey of college students finds that 40%
like country music, 30% like gospel music, and 10% like both.

   (a) What is the conditional probability that a student likes gospel
   music if we know that he or she likes country music?

   (b) What is the conditional probability that a student who does not
   like country music likes gospel music? (A Venn diagram may help
   you.)

5.47 **College degrees.** Exercise 5.43 gives the counts (in thousands) of
earned degrees in the United States in a recent year. Use these data to
answer the following questions.

   (a) What is the probability that a randomly chosen degree recipient is
   a man?

   (b) What is the conditional probability that the person chosen
   received a bachelor's degree, given that he is a man?

   (c) Use the multiplication rule to find the probability of choosing a
   male bachelor's degree recipient. Check your result by finding
   this probability directly from the table of counts.

5.48 **The probability of a flush.** A poker player holds a flush when all
five cards in the hand belong to the same suit. We will find the

probability of a flush when five cards are dealt. Remember that a deck contains 52 cards, 13 of each suit, and that when the deck is well shuffled, each card dealt is equally likely to be any of those that remain in the deck.

(a) We will concentrate on spades. What is the probability that the first card dealt is a spade? What is the conditional probability that the second card is a spade, given that the first is a spade?

(b) Continue to count the remaining cards to find the conditional probabilities of a spade on the third, the fourth, and the fifth card, given in each case that all previous cards are spades.

(c) The probability of being dealt five spades is the product of the five probabilities you have found. Why? What is this probability?

(d) The probability of being dealt five hearts or five diamonds or five clubs is the same as the probability of being dealt five spades. What is the probability of being dealt a flush?

5.49 **Geometric probability.** Choose a point at random in the square with sides $0 \leq x \leq 1$ and $0 \leq y \leq 1$. This means that the probability that the point falls in any region within the square is equal to the area of that region. Let X be the x coordinate and Y the y coordinate of the point chosen. Find the conditional probability $P(Y < 1/2 \mid Y > X)$. (Hint: Draw a diagram of the square and the events $Y < 1/2$ and $Y > X$.)

(TOM MCCARTHY/INDEX STOCK IMAGERY)

5.50 **The probability of a royal flush.** A royal flush is the highest hand possible in poker. It consists of the ace, king, queen, jack, and ten of the same suit. Modify the outline given in Exercise 5.48 to find the probability of being dealt a royal flush in a five-card deal.

5.51 **Income tax returns.** Here is the distribution of the adjusted gross income (in thousands of dollars) reported on individual federal income tax returns in 1994:

Income	< 10	10–29	30–49	50–99	≥ 100
Probability	0.12	0.39	0.24	0.20	0.05

(a) What is the probability that a randomly chosen return shows an adjusted gross income of $50,000 or more?

(b) Given that a return shows an income of at least $50,000, what is the conditional probability that the income is at least $100,000?

5.52 **Classifying occupations.** Exercise 4.62 (page 254) gives the probability distribution of the gender and occupation of a randomly chosen American worker. Use this distribution to answer the following questions:

(a) Given that the worker chosen holds a managerial (Class A) job, what is the conditional probability that the worker is female?

(b) Classes D and E include most mechanical and factory jobs. What is the conditional probability that a worker is female, given that he or she holds a job in one of these classes?

(c) Are gender and job type independent? How do you know?

5.53 **The geometric distributions.** You are tossing a balanced die that has probability 1/6 of coming up 1 on each toss. Tosses are independent. We are interested in how long we must wait to get the first 1.

(a) The probability of a 1 on the first toss is 1/6. What is the probability that the first toss is not a 1 and the second toss is a 1?

(b) What is the probability that the first two tosses are not 1s and the third toss is a 1? This is the probability that the first 1 occurs on the third toss.

(c) Now you see the pattern. What is the probability that the first 1 occurs on the fourth toss? On the fifth toss? Give the general result: what is the probability that the first 1 occurs on the kth toss?

Comment: The distribution of the number of trials to the first success is called a **geometric distribution.** In this problem you have found geometric distribution probabilities when the probability of a success on each trial is $p = 1/6$. The same idea works for any p.

geometric distribution

STATISTICS IN SUMMARY

This chapter concerns some further facts about probability that are useful in modeling but are not needed in our study of statistics. Section 5.1 discusses general rules that all probability models must obey, including the important multiplication rule for independent events. There are many specific probability models for specific situations. Section 5.2 uses the multiplication rule to obtain one of the most important probability models, the binomial distribution for counts. Remember that not all counts have a binomial distribution, just as not all measured variables have a normal distribution. When events are not independent, we need the idea of conditional probability. That is the topic of Section 5.3. At this point, we finally reach the fully general form of the basic rules of probability. Here is a review list of the most important skills you should have acquired from your study of this chapter.

A. PROBABILITY RULES

1. Use Venn diagrams to picture relationships among several events.

2. Use the general addition rule to find probabilities that involve overlapping events.

3. Understand the idea of independence. Judge when it is reasonable to assume independence as part of a probability model.

4. Use the multiplication rule for independent events to find the probability that all of several independent events occur.

5. Use the multiplication rule for independent events in combination with other probability rules to find the probabilities of complex events.

B. BINOMIAL DISTRIBUTIONS

1. Recognize the binomial setting: we are interested in the count X of successes in n independent trials with the same probability p of success on each trial.

2. Use the binomial distribution to find probabilities of events involving the count X in a binomial setting for small values of n.

3. Find the mean and standard deviation of a binomial count X.

4. Recognize when you can use the normal approximation to a binomial distribution. Use the normal approximation to calculate probabilities that concern a binomial count X.

C. CONDITIONAL PROBABILITY

1. Understand the idea of conditional probability. Find conditional probabilities for individuals chosen at random from a table of counts of possible outcomes.

2. Use the general multiplication rule to find $P(A \text{ and } B)$ from $P(A)$ and the conditional probability $P(B \mid A)$.

CHAPTER 5 Review Exercises

5.54 **Leaking gas tanks.** Leakage from underground gasoline tanks at service stations can damage the environment. It is estimated that 25% of these tanks leak. You examine 15 tanks chosen at random, independently of each other.

(a) What is the mean number of leaking tanks in such samples of 15?

(b) What is the probability that 10 or more of the 15 tanks leak?

(c) Now you do a larger study, examining a random sample of 1000 tanks nationally. What is the probability that at least 275 of these tanks are leaking?

5.55 **School vouchers.** An opinion poll asks an SRS of 500 adults whether they favor giving parents of school-age children vouchers that can be exchanged for education at any public or private school of their choice. Each school would be paid by the government on the basis of how many vouchers it collected. Suppose that in fact 45% of the population favor this idea. What is the probability that more than half of the sample are in favor?

5.56 **Reaching dropouts.** High school dropouts make up 14.1% of all Americans aged 18 to 24. A vocational school that wants to attract dropouts mails an advertising flyer to 25,000 persons between the ages of 18 and 24.

(a) If the mailing list can be considered a random sample of the population, what is the mean number of high school dropouts who will receive the flyer?

(b) What is the probability that at least 3500 dropouts will receive the flyer?

5.57 **Is this coin balanced?** While he was a prisoner of the Germans during World War II, John Kerrich tossed a coin 10,000 times. He got 5067 heads. Take Kerrich's tosses to be an SRS from the population of all possible tosses of his coin. If the coin is perfectly balanced, $p = 0.5$. Is there reason to think that Kerrich's coin gave too many heads to be balanced? To answer this question, find the probability that a balanced coin would give 5067 or more heads in 10,000 tosses. What do you conclude?

5.58 **Who is driving?** A sociology professor asks her class to observe cars having a man and a woman in the front seat and record which of the two is the driver.

(a) Explain why it is reasonable to use the binomial distribution for the number of male drivers in n cars if all observations are made in the same location at the same time of day.

(b) Explain why the binomial model may not apply if half the observations are made outside a church on Sunday morning and half are made on campus after a dance.

(c) The professor requires students to observe 10 cars during business hours in a retail district close to campus. Past observations have shown that the man is driving about 85% of cars in this location. What is the probability that the man is driving 8 or fewer of the 10 cars?

(d) The class has 10 students, who will observe 100 cars in all. What is the probability that the man is driving 80 or fewer of these?

5.59 **Income and savings.** A sample survey chooses a sample of households and measures their annual income and their savings. Some events of interest are

$A =$ the household chosen has income at least $100,000

$C =$ the household chosen has at least $50,000 in savings

Based on this sample survey, we estimate that $P(A) = 0.07$ and $P(C) = 0.2$.

(a) We want to find the probability that a household either has income at least $100,000 *or* savings at least $50,000. Explain why we do not have enough information to find this probability. What additional information is needed?

(b) We want to find the probability that a household has income at least $100,000 *and* savings at least $50,000. Explain why we do not have enough information to find this probability. What additional information is needed?

5.60 You have torn a tendon and are facing surgery to repair it. The surgeon explains the risks to you: infection occurs in 3% of such operations, the repair fails in 14%, and both infection and failure occur together in 1%. What percent of these operations succeed and are free from infection?

5.61 Consolidated Builders has bid on two large construction projects. The company president believes that the probability of winning the first contract (event A) is 0.6, that the probability of winning the second (event B) is 0.4, and that the probability of winning both jobs (event $\{A \text{ and } B\}$) is 0.2.

(a) What is the probability of the event $\{A \text{ or } B\}$ that Consolidated will win at least one of the jobs?

(b) You hear that Consolidated won the second job. Given this information, what is the conditional probability that Consolidated won the first job?

5.62 Draw a Venn diagram that illustrates the relation between events A and B in Exercise 5.61. Indicate the events below on your diagram and use the information in Exercise 5.61 to calculate their probabilities.

(a) Consolidated wins both jobs.

(b) Consolidated wins the first job but not the second.

(c) Consolidated does not win the first job but does win the second.

(d) Consolidated does not win either job.

5.63 **Employment data.** In the language of government statistics, the "labor force" includes all civilians at least 16 years of age who are working or looking for work. Select a member of the U.S. labor force at random. Let A be the event that the person selected is white, and B the event that he or she is employed. In 1995, 84.6% of the labor force was white. Of the whites in the labor force, 95.1% were employed. Among nonwhite members of the labor force, 91.9% were employed.

(a) Express each of the percents given as a probability involving the events A and B; for example, $P(A) = 0.846$.

(b) Are the events "an employed person is chosen" and "a white is chosen" independent? How do you know?

(c) Find the probability that the person chosen is an employed white.

(d) Find the probability that an employed nonwhite is chosen.

(e) Find the probability that the person chosen is employed. (Hint: An employed person is either white or nonwhite.)

5.64   Testing for AIDS. ELISA tests are used to screen donated blood for the presence of the AIDS virus. The test actually detects antibodies, substances that the body produces when the virus is present. When antibodies are present, ELISA is positive with probability about 0.997 and negative with probability 0.003. When the blood tested is not contaminated with AIDS antibodies, ELISA gives a positive result with probability about 0.015 and a negative result with probability 0.985.[6] Suppose that 1% of a large population carries the AIDS antibody in their blood.

(a) The information given includes four conditional probabilities and one unconditional probability. Assign letters to the events and restate the information in probability notation ($P(A)$, $P(B \mid A)$, and so on). Use this notation in the rest of this exercise.

(b) What is the probability that the person chosen does not carry the AIDS antibody and tests positive?

(c) What is the probability that the person chosen does carry the AIDS antibody and tests positive?

(d) What is the probability that the person chosen tests positive? (Hint: the person chosen must either carry the AIDS antibody or not. A Venn diagram may help.)

(e) Use the definition of conditional probability and your results in (b) and (d) to find the conditional probability that a person does not carry the AIDS antibody, given that he or she tests positive. (This result illustrates an important fact: when a condition such as AIDS is rare, most positive test results are false.)

Jerzy Neyman

Statistical Confidence

(PHOTO BY G. PAUL BISHOP, COURTESY OF PAUL BISHOP, JR., BERKELEY, CA.)

The most-used methods of statistical inference are confidence intervals and tests of significance. Both are products of the twentieth century. From complex and confusing origins, statistical tests took their current form in the writings of R. A. Fisher, whom we met at the beginning of Chapter 3. Confidence intervals appeared in 1934, the brainchild of Jerzy Neyman (1894–1981).

Age did not slow Neyman's work—he remained active until the end of his long life and almost doubled his list of publications after "retiring."

Neyman was trained in Poland and, like Fisher, worked at an agricultural research institute. He moved to London in 1934 and in 1938 joined the University of California at Berkeley. He founded Berkeley's Statistical Laboratory and remained its head even after his official retirement as a professor in 1961. Age did not slow Neyman's work—he remained active until the end of his long life and almost doubled his list of publications after "retiring." Statistical problems arising from astronomy, biology, and attempts to modify the weather attracted his attention.

Neyman ranks with Fisher as a founder of modern statistical practice. In addition to introducing confidence intervals, he helped systematize the theory of sample surveys and reworked significance tests from a new point of view. Fisher, who was very argumentative, disliked Neyman's approach to tests and said so. Neyman, who wasn't shy, replied vigorously.

Tests and confidence intervals are our topic in this chapter. Like most users of statistics, we will stay close to Fisher's approach to tests. You can find some of Neyman's ideas in the optional final section.

Introduction to Inference

6.1   Estimating with
      Confidence

6.2   Tests of Significance

6.3   Making Sense of
      Statistical Significance

6.4   Error Probabilities and
      Power*

      Statistics in Summary

      Review Exercises

(CORBIS.)

How galaxies are distributed in space is yet another question for statistical study.

Introduction

When we select a sample, we know the responses of the individuals in the sample. Often we are not content with information about the sample. We want to *infer* from the sample data some conclusion about a wider population that the sample represents.

> **STATISTICAL INFERENCE**
>
> **Statistical inference** provides methods for drawing conclusions about a population from sample data.

We cannot be certain that our conclusions are correct—a different sample might lead to different conclusions. Statistical inference uses the language of probability to say how trustworthy its conclusions are.

In this chapter we will meet the two most common types of statistical inference. Section 6.1 concerns *confidence intervals* for estimating the value of a population parameter. Section 6.2 introduces *tests of significance,* which assess the evidence for a claim about a population. Both types of inference are based on the sampling distributions of statistics. That is, both report probabilities that state what would happen if we used the inference method many times.

The reasoning of statistical inference rests on the long-run regular behavior that probability describes. Inference is most reliable when the data are produced by a properly randomized design. **When you use statistical inference, you are acting as if your data are a random sample or come from a randomized experiment.** If this is not true, your conclusions may be open to challenge. Statistical tests and confidence intervals cannot remedy basic flaws in producing the data such as voluntary response samples or uncontrolled experiments. Use the data sense developed in your study of the first three chapters of this book, and apply formal inference only when you are satisfied that the data deserve such analysis.

This chapter introduces the reasoning used in statistical inference. We will illustrate the reasoning by a few specific inference techniques, but these are oversimplified so that they are not very useful in practice. Later chapters will show how to modify these techniques to make them practically useful and will also introduce inference methods for use in most of the settings we met in learning to explore data. There are libraries—both of books and of computer software—full of more elaborate statistical techniques. Informed use of any of these methods requires an understanding of the underlying reasoning. A computer will do the arithmetic, but you must still exercise judgment based on understanding.

6.1 Estimating with Confidence

Young people have a better chance of good jobs and good wages if they are good with numbers. How strong are the quantitative skills of young Americans of working age? One source of data is the National Assessment of Educational Progress (NAEP) Young Adult Literacy Assessment Survey, which is based on a nationwide probability sample of households.

EXAMPLE 6.1   NAEP quantitative scores

The NAEP survey includes a short test of quantitative skills, covering mainly basic arithmetic and the ability to apply it to realistic problems. Scores on the test range from 0 to 500. A person who scores 233 can add the amounts of two checks appearing on a bank deposit slip; someone scoring 325 can determine the price of a meal from a menu; a person scoring 375 can transform a price in cents per ounce into dollars per pound.

In a recent year, 840 men 21 to 25 years of age were in the NAEP sample. Their mean quantitative score was $\bar{x} = 272$. These 840 men are an SRS from the population of all young men. On the basis of this sample, what can we infer about the mean score μ in the population of all 9.5 million young men of these ages?[1]

The law of large numbers tells us that the sample mean $\bar{x}$ from a large SRS will be close to the unknown population mean μ. Because $\bar{x} = 272$, we guess that μ is "somewhere around 272." To make "somewhere around 272" more precise, we ask: *How would the sample mean $\bar{x}$ vary if we took many samples of 840 young men from this same population?*

Recall the essential facts about the sampling distribution of $\bar{x}$:

- The central limit theorem tells us that the mean $\bar{x}$ of 840 scores has a distribution that is close to normal.
- The mean of this normal sampling distribution is the same as the unknown mean μ of the entire population.
- The standard deviation of $\bar{x}$ for an SRS of 840 men is $\sigma/\sqrt{840}$, where σ is the standard deviation of individual NAEP scores among all young men.

Let us suppose that we know that the standard deviation of scores in the population of all young men is $\sigma = 60$. The standard deviation of $\bar{x}$ is then

$$\frac{\sigma}{\sqrt{n}} = \frac{60}{\sqrt{840}} \doteq 2.1$$

(It is not realistic to assume we know σ. We will see in the next chapter how to proceed when σ is not known. For now, we are more interested in statistical reasoning than in details of realistic methods.)

If we choose many samples of size 840 and find the mean NAEP score for each sample, we might get mean $\bar{x} = 272$ from the first sample, $\bar{x} = 268$ from the second, $\bar{x} = 273$ from the third sample, and so on. If we collect all these sample means and display their distribution, we get the normal distribution with mean equal to the unknown μ and standard deviation 2.1. Inference about the unknown μ starts from this sampling distribution. Figure 6.1 displays the distribution. The different values of $\bar{x}$ appear along the axis in the figure, and the normal curve shows how probable these values are.

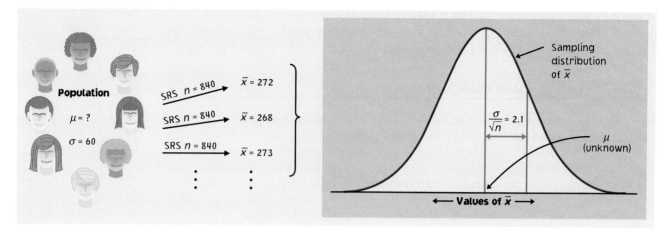

Figure 6.1 *The sampling distribution of the mean score $\bar{x}$ of an SRS of 840 young men on the NAEP quantitative test.*

Statistical confidence

Figure 6.2 is another picture of the same sampling distribution. It illustrates the following line of thought:

- The 68–95–99.7 rule says that in 95% of all samples, the mean score $\bar{x}$ for the sample will be within two standard deviations of the population mean score μ. So the mean $\bar{x}$ of 840 NAEP scores will be within 4.2 points of μ in 95% of all samples.

- Whenever $\bar{x}$ is within 4.2 points of the unknown μ, then μ is within 4.2 points of the observed $\bar{x}$. This happens in 95% of all samples.

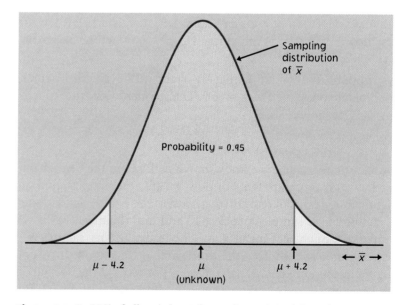

Figure 6.2 *In 95% of all samples, $\bar{x}$ lies within ± 4.2 of the unknown population mean μ. So μ also lies within ± 4.2 of $\bar{x}$ in those samples.*

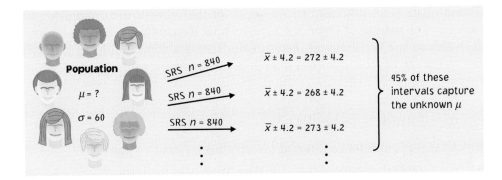

- So in 95% of all samples, the unknown μ lies between $\bar{x} - 4.2$ and $\bar{x} + 4.2$. Figure 6.3 displays this fact in picture form.

This conclusion just restates a fact about the sampling distribution of $\bar{x}$. The language of statistical inference uses this fact about what would happen in the long run to express our confidence in the results of any one sample.

EXAMPLE 6.2   **95% confidence**

Our sample of 840 young men gave $\bar{x} = 272$. We say that we are *95% confident* that the unknown mean NAEP quantitative score for all young men lies between

$$\bar{x} - 4.2 = 272 - 4.2 = 267.8$$

and

$$\bar{x} + 4.2 = 272 + 4.2 = 276.2$$

Be sure you understand the grounds for our confidence. There are only two possibilities:

1. The interval between 267.8 and 276.2 contains the true μ.

2. Our SRS was one of the few samples for which $\bar{x}$ is not within 4.2 points of the true μ. Only 5% of all samples give such inaccurate results.

We cannot know whether our sample is one of the 95% for which the interval $\bar{x} \pm 4.2$ catches μ, or one of the unlucky 5%. The statement that we are 95% confident that the unknown μ lies between 267.8 and 276.2 is shorthand for saying, "We got these numbers by a method that gives correct results 95% of the time."

The interval of numbers between the values $\bar{x} \pm 4.2$ is called a *95% confidence interval* for μ. Like most confidence intervals we will meet, this one has the form

$$\text{estimate} \pm \text{margin of error}$$

margin of error

The estimate ($\bar{x}$ in this case) is our guess for the value of the unknown parameter. The **margin of error** ± 4.2 shows how accurate we believe our guess is, based on the variability of the estimate. This is a *95% confidence interval* because it catches the unknown μ in 95% of all possible samples.

CONFIDENCE INTERVAL

A **level C confidence interval** for a parameter has two parts:

- An interval calculated from the data, usually of the form

  estimate $\pm$ margin of error

- A **confidence level** C, which gives the probability that the interval will capture the true parameter value in repeated samples.

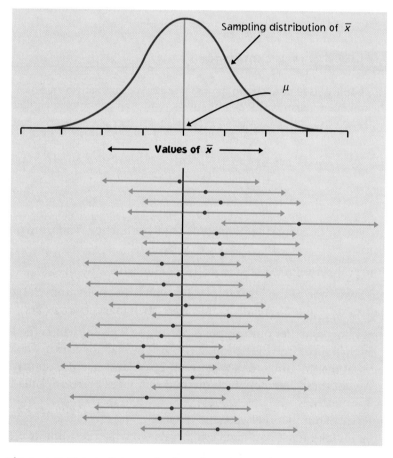

Figure 6.4 *Twenty-five samples from the same population gave these 95% confidence intervals. In the long run, 95% of all samples give an interval that contains the population mean μ.*

Users can choose the confidence level, most often 90% or higher because we most often want to be quite sure of our conclusions. We will use C to stand for the confidence level in decimal form. For example, a 95% confidence level corresponds to $C = 0.95$.

Figure 6.3 is one way to picture the idea of a 95% confidence interval. Figure 6.4 illustrates the idea in a different form. Study these figures carefully. If you understand what they say, you have mastered one of the big ideas of statistics. Figure 6.4 shows the result of drawing many SRSs from the same population and calculating a 95% confidence interval from each sample. The center of each interval is at $\bar{x}$ and therefore varies from sample to sample. The sampling distribution of $\bar{x}$ appears at the top of the figure to show the long-term pattern of this variation. The 95% confidence intervals from 25 SRSs appear below. The center $\bar{x}$ of each interval is marked by a dot. The arrows on either side of the dot span the confidence interval. All except one of these 25 intervals cover the true value of μ. In a very large number of samples, 95% of the confidence intervals would contain μ.

APPLY YOUR KNOWLEDGE

6.1 **Polling women.** A *New York Times* poll on women's issues interviewed 1025 women randomly selected from the United States, excluding Alaska and Hawaii. The poll found that 47% of the women said they do not get enough time for themselves.

(a) The poll announced a margin of error of ±3 percentage points for 95% confidence in its conclusions. What is the 95% confidence interval for the percent of all adult women who think they do not get enough time for themselves?

(b) Explain to someone who knows no statistics why we can't just say that 47% of all adult women do not get enough time for themselves.

(c) Then explain clearly what "95% confidence" means.

6.2 **Explaining confidence.** A student reads that a 95% confidence interval for the mean NAEP quantitative score for men of ages 21 to 25 is 267.8 to 276.2. Asked to explain the meaning of this interval, the student says, "95% of all young men have scores between 267.8 and 276.2." Is the student right? Justify your answer.

6.3 Suppose that you give the NAEP test to an SRS of 1000 people from a large population in which the scores have mean 280 and standard deviation $\sigma = 60$. The mean $\bar{x}$ of the 1000 scores will vary if you take repeated samples.

(a) The sampling distribution of $\bar{x}$ is approximately normal. It has mean $\mu = 280$. What is its standard deviation?

(b) Sketch the normal curve that describes how $\bar{x}$ varies in many samples from this population. Mark the mean $\mu = 280$ and the values one, two, and three standard deviations on either side of the mean.

(c) According to the 68–95–99.7 rule, about 95% of all the values of $\bar{x}$ fall within _____ of the mean of this curve. What is the missing number? Call it m for "margin of error." Shade the region from the mean minus m to the mean plus m on the axis of your sketch, as in Figure 6.2.

(d) Whenever $\bar{x}$ falls in the region you shaded, the true value of the population mean, $\mu = 280$, lies in the confidence interval between $\bar{x} - m$ and $\bar{x} + m$. Draw the confidence interval below your sketch for one value of $\bar{x}$ inside the shaded region and one value of $\bar{x}$ outside the shaded region. (Use Figure 6.4 as a model for the drawing.)

(e) In what percent of all samples will the true mean $\mu = 280$ be covered by the confidence interval $\bar{x} \pm m$?

Confidence intervals for the mean μ

The reasoning we used to find a 95% confidence interval for the population mean μ applies to any confidence level. We start with the sampling distribution of the sample mean $\bar{x}$. If we know μ, we can standardize $\bar{x}$. The result is the *one-sample z statistic*

$$z = \frac{\bar{x} - \mu}{\sigma / \sqrt{n}}$$

The statistic z tells us how far the observed $\bar{x}$ is from μ, in units of the standard deviation of $\bar{x}$. Because $\bar{x}$ has a normal distribution, z has the standard normal distribution $N(0, 1)$.

To find a 95% confidence interval, mark off the central 95% of the area under a normal curve. For confidence level C, mark off central area C. Call z^* the point on the standard normal distribution that catches the central area C of the total area 1 under the density curve.

EXAMPLE 6.3 | **80% confidence**

To find an 80% confidence interval, we must catch the central 80% of the normal sampling distribution of $\bar{x}$. In catching the central 80% we leave out 20%, or 10% in each tail. So z^* is the point with area 0.1 to its right and 0.9 to its left under the standard normal curve. Search the body of Table A to find the point with area 0.9 to its left. The closest entry is $z^* = 1.28$. There is area 0.8 under the standard normal curve between -1.28 and 1.28. Figure 6.5 shows how z^* is related to areas under the curve.

Figure 6.6 shows how z^* and C are related in general: there is area C under the standard normal curve between $-z^*$ and z^*. If we start at the sample

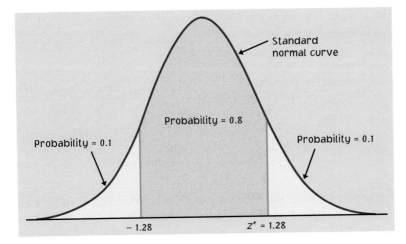

Figure 6.5 *The central probability 0.8 under a standard normal curve lies between −1.28 and 1.28. That is, there is area 0.1 to the right of 1.28 under the curve.*

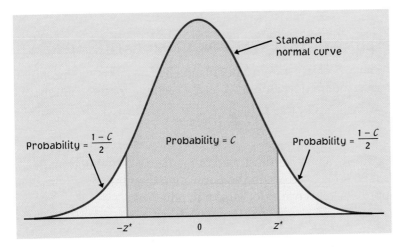

Figure 6.6 *The critical value z^* is the number that catches central probability C under a standard normal curve between $-z^*$ and z^*.*

mean $\bar{x}$ and go out z^* standard deviations, we get an interval that contains the population mean μ in a proportion C of all samples. This interval is

$$\text{from} \quad \bar{x} - z^* \frac{\sigma}{\sqrt{n}} \quad \text{to} \quad \bar{x} + z^* \frac{\sigma}{\sqrt{n}}$$

or

$$\bar{x} \pm z^* \frac{\sigma}{\sqrt{n}}$$

It is a level C confidence interval for μ.

You can find z^* for any C by searching Table A. Here are the results for the most common confidence levels:

Confidence level	Tail area	z^*
90%	0.05	1.645
95%	0.025	1.960
99%	0.005	2.576

Notice that for 95% confidence we use $z^* = 1.960$. This is more exact than the approximate value $z^* = 2$ given by the 68–95–99.7 rule. The bottom row in Table C gives the values z^* for many confidence levels C. This row is labeled z^*. (You can find Table C in the back of the book and on the inside rear cover. We will use the other rows of the table in the next chapter.) Values z^* that mark

critical value   off a specified area under the standard normal curve are often called **critical values** of the distribution.

CONFIDENCE INTERVAL FOR A POPULATION MEAN

Draw an SRS of size n from a population having unknown mean μ and known standard deviation σ. A level C confidence interval for μ is

$$\bar{x} \pm z^* \frac{\sigma}{\sqrt{n}}$$

The critical value z^* is illustrated in Figure 6.6 and found in Table C. This interval is exact when the population distribution is normal and is approximately correct for large n in other cases.

EXAMPLE 6.4   **Analyzing pharmaceuticals**

A manufacturer of pharmaceutical products analyzes a specimen from each batch of a product to verify the concentration of the active ingredient. The chemical analysis is not perfectly precise. Repeated measurements on the same specimen give slightly different results. The results of repeated measurements follow a normal distribution quite closely. The analysis procedure has no bias, so the mean μ of the population of all measurements is the true concentration in the specimen. The standard deviation of this distribution is known to be $\sigma = 0.0068$ grams per liter. The laboratory analyzes each specimen three times and reports the mean result.

Three analyses of one specimen give concentrations

0.8403   0.8363   0.8447

We want a 99% confidence interval for the true concentration μ.

The sample mean of the three readings is

$$\bar{x} = \frac{0.8403 + 0.8363 + 0.8447}{3} = 0.8404$$

For 99% confidence, we see from Table C that $z^* = 2.576$. A 99% confidence interval for μ is therefore

$$\bar{x} \pm z^* \frac{\sigma}{\sqrt{n}} = 0.8404 \pm 2.576\frac{0.0068}{\sqrt{3}}$$

$$= 0.8404 \pm 0.0101$$

$$= 0.8303 \text{ to } 0.8505$$

We are 99% confident that the true concentration lies between 0.8303 and 0.8505 grams per liter.

Suppose that a single measurement gave $x = 0.8404$, the same value that the sample mean took in Example 6.4. Repeating the calculation with $n = 1$ shows that the 99% confidence interval based on a single measurement is

$$\bar{x} \pm z^* \frac{\sigma}{\sqrt{1}} = 0.8404 \pm (2.576)(0.0068)$$

$$= 0.8404 \pm 0.0175$$

$$= 0.8229 \text{ to } 0.8579$$

The mean of three measurements gives a smaller margin of error and therefore a shorter interval than a single measurement. Figure 6.7 illustrates the gain from using three observations.

APPLY YOUR KNOWLEDGE

6.4 **Surveying hotel managers.** A study of the career paths of hotel general managers sent questionnaires to an SRS of 160 hotels belonging to major U.S. hotel chains. There were 114 responses. The average time these 114 general managers had spent with their

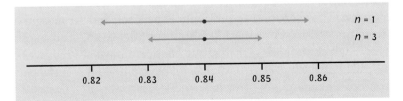

Figure 6.7 *Confidence intervals for n = 3 and n = 1 for Example 6.4. Larger samples give shorter intervals.*

current company was 11.78 years. Give a 99% confidence interval for the mean number of years general managers of major-chain hotels have spent with their current company. (Take it as known that the standard deviation of time with the company for all general managers is 3.2 years.)

6.5   IQ test scores. Here are the IQ test scores of 31 seventh-grade girls in a Midwest school district:[2]

114	100	104	89	102	91	114	114	103	105
108	130	120	132	111	128	118	119	86	72
111	103	74	112	107	103	98	96	112	112	93

(a) We expect the distribution of IQ scores to be close to normal. Make a stemplot of the distribution of these 31 scores (split the stems). Does your plot show outliers, clear skewness, or other nonnormal features?

(b) Treat the 31 girls as an SRS of all seventh-grade girls in the school district. Suppose that the standard deviation of IQ scores in this population is known to be $\sigma = 15$. Give a 99% confidence interval for the mean score in the population.

(c) In fact, the scores are those of all seventh-grade girls in one of the several schools in the district. Explain carefully why your confidence interval from (b) cannot be trusted.

6.6   Blood tests. A test for the level of potassium in the blood is not perfectly precise. Moreover, the actual level of potassium in a person's blood varies slightly from day to day. Suppose that repeated measurements for the same person on different days vary normally with $\sigma = 0.2$.

(a) Julie's potassium level is measured once. The result is $x = 3.2$. Give a 90% confidence interval for her mean potassium level.

(b) If three measurements were taken on different days and the mean result is $\bar{x} = 3.2$, what is a 90% confidence interval for Julie's mean blood potassium level?

How confidence intervals behave

The confidence interval $\bar{x} \pm z^*\sigma/\sqrt{n}$ for the mean of a normal population illustrates several important properties that are shared by all confidence intervals in common use. The user chooses the confidence level, and the margin of error follows from this choice. We would like high confidence and also a small margin of error. High confidence says that our method almost always gives correct answers. A small margin of error says that we have pinned down the parameter quite precisely. The margin of error is

$$\text{margin of error} = z^* \frac{\sigma}{\sqrt{n}}$$

This expression has z^* and σ in the numerator and $\sqrt{n}$ in the denominator. So the margin of error gets smaller when

- z^* gets smaller. Smaller z^* is the same as smaller confidence level C (look at Figure 6.6 again). There is a trade-off between the confidence level and the margin of error. To obtain a smaller margin of error from the same data, you must be willing to accept lower confidence.

- σ gets smaller. The standard deviation σ measures the variation in the population. You can think of the variation among individuals in the population as noise that obscures the average value μ. It is easier to pin down μ when σ is small.

- n gets larger. Increasing the sample size n reduces the margin of error for any fixed confidence level. Because n appears under a square root sign, we must take four times as many observations in order to cut the margin of error in half.

EXAMPLE 6.5   **Changing the margin of error**

Suppose that the pharmaceutical manufacturer in Example 6.4 is content with 90% confidence rather than 99%. Table C gives the critical value for 90% confidence as $z^* = 1.645$. The 90% confidence interval for μ based on three repeated measurements with mean $\bar{x} = 0.8404$ is

$$\bar{x} \pm z^* \frac{\sigma}{\sqrt{n}} = 0.8404 \pm 1.645 \frac{0.0068}{\sqrt{3}}$$

$$= 0.8404 \pm 0.0065$$

$$= 0.8339 \text{ to } 0.8469$$

Settling for 90%, rather than 99%, confidence has reduced the margin of error from ± 0.0101 to ± 0.0065. Figure 6.8 compares these two intervals.

Increasing the number of measurements from 3 to 12 will also reduce the width of the 99% confidence interval in Example 6.4. Check that replacing $\sqrt{3}$ by $\sqrt{12}$ cuts the ± 0.0101 margin of error in half, because we now have four times as many observations.

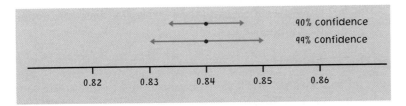

Figure 6.8 *90% and 99% confidence intervals for Example 6.5. Higher confidence requires a wider interval.*

6.7   **Confidence level and interval length.** Examples 6.4 and 6.5 give confidence intervals for the concentration μ based on 3 measurements with $\bar{x} = 0.8404$ and $\sigma = 0.0068$. The 99% confidence interval is 0.8303 to 0.8505 and the 90% confidence interval is 0.8339 to 0.8469.

(a) Find the 80% confidence interval for μ.

(b) Find the 99.9% confidence interval for μ.

(c) Make a sketch like Figure 6.8 to compare all four intervals. How does increasing the confidence level affect the length of the confidence interval?

6.8   **Confidence level and margin of error.** The NAEP test (Example 6.1) was also given to a sample of 1077 women of ages 21 to 25 years. Their mean quantitative score was 275. Take it as known that the standard deviation of all individual scores is $\sigma = 60$.

(a) Give a 95% confidence interval for the mean score μ in the population of all young women.

(b) Give the 90% and 99% confidence intervals for μ.

(c) What are the margins of error for 90%, 95%, and 99% confidence? How does increasing the confidence level affect the margin of error of a confidence interval?

6.9   **Sample size and margin of error.** The NAEP sample of 1077 young women had mean quantitative score $\bar{x} = 275$. Take it as known that the standard deviation of all individual scores is $\sigma = 60$.

(a) Give a 95% confidence interval for the mean score μ in the population of all young women.

(b) Suppose that the same result, $\bar{x} = 275$, had come from a sample of 250 women. Give the 95% confidence interval for the population mean μ in this case.

(c) Then suppose that a sample of 4000 women had produced the sample mean $\bar{x} = 275$, and again give the 95% confidence interval for μ.

(d) What are the margins of error for samples of size 250, 1077, and 4000? How does increasing the sample size affect the margin of error of a confidence interval?

Choosing the sample size

A wise user of statistics never plans data collection without at the same time planning the inference. You can arrange to have both high confidence and a small margin of error by taking enough observations. The margin of error of the confidence interval for the mean of a normally distributed population is

$m = z^*\sigma/\sqrt{n}$. To obtain a desired margin of error m, put in the value of z^* for your desired confidence level, and solve for the sample size n. Here is the result.

SAMPLE SIZE FOR DESIRED MARGIN OF ERROR

The confidence interval for a population mean will have a specified margin of error m when the sample size is

$$n = \left(\frac{z^*\sigma}{m}\right)^2$$

This formula is not the proverbial free lunch. Taking observations costs time and money. The required sample size may be impossibly expensive. Notice that it is the size of the *sample* that determines the margin of error. The size of the *population* does not influence the sample size we need. (This is true as long as the population is much larger than the sample.)

EXAMPLE 6.6   How many observations?

Management asks the laboratory of Example 6.4 to produce results accurate to within ±0.005 with 95% confidence. How many measurements must be averaged to comply with this request?

The desired margin of error is $m = 0.005$. For 95% confidence, Table C gives $z^* = 1.960$. We know that $\sigma = 0.0068$. Therefore,

$$n = \left(\frac{z^*\sigma}{m}\right)^2 = \left(\frac{1.96 \times 0.0068}{0.005}\right)^2 = 7.1$$

Because 7 measurements will give a slightly larger margin of error than desired, and 8 measurements a slightly smaller margin of error, the lab must take 8 measurements on each specimen to meet management's demand. Always round *up* to the next higher whole number when finding n. On learning the cost of this many measurements, management may reconsider its request.

APPLY YOUR KNOWLEDGE

6.10 **Calibrating a scale.** To assess the accuracy of a laboratory scale, a standard weight known to weigh 10 grams is weighed repeatedly. The scale readings are normally distributed with unknown mean (this mean is 10 grams if the scale has no bias). The standard deviation of the scale readings is known to be 0.0002 gram.

(a) The weight is weighed five times. The mean result is 10.0023 grams. Give a 98% confidence interval for the mean of repeated measurements of the weight.

(b) How many measurements must be averaged to get a margin of error of ±0.0001 with 98% confidence?

6.11 How large a sample of the hotel managers in Exercise 6.4 would be needed to estimate the mean μ within ±1 year with 99% confidence?

6.12 How large a sample of schoolgirls in Exercise 6.5 would be needed to estimate the mean μ IQ score within ±5 points with 99% confidence?

Some cautions

Any formula for inference is correct only in specific circumstances. If statistical procedures carried warning labels like those on drugs, most inference methods would have long labels indeed. Our handy formula $\bar{x} \pm z^* \sigma/\sqrt{n}$ for estimating a normal mean comes with the following list of warnings for the user.

- The data must be an SRS from the population. We are completely safe if we actually carried out the random selection of an SRS. We are not in great danger if the data can plausibly be thought of as observations taken at random from a population. That is the case in Examples 6.4 to 6.6, where we have in mind the population resulting from a very large number of repeated analyses of the same specimen.

- The formula is not correct for probability sampling designs more complex than an SRS. Correct methods for other designs are available. We will not discuss confidence intervals based on multistage or stratified samples. If you plan such samples, be sure that you (or your statistical consultant) know how to carry out the inference you desire.

- There is no correct method for inference from data haphazardly collected with bias of unknown size. Fancy formulas cannot rescue badly produced data.

- Because $\bar{x}$ is strongly influenced by a few extreme observations, outliers can have a large effect on the confidence interval. You should search for outliers and try to correct them or justify their removal before computing the interval. If the outliers cannot be removed, ask your statistical consultant about procedures that are not sensitive to outliers.

- If the sample size is small and the population is not normal, the true confidence level will be different from the value C used in computing the interval. Examine your data carefully for skewness and other signs of nonnormality. The interval relies only on the distribution of $\bar{x}$, which even for quite small sample sizes is much closer to normal than the individual observations. When $n \geq 15$, the confidence level is not greatly disturbed by nonnormal populations unless extreme outliers or quite strong skewness are present. We will discuss this issue in more detail in the next chapter.

- You must know the standard deviation σ of the population. This unrealistic requirement renders the interval $\bar{x} \pm z^* \sigma/\sqrt{n}$ of little use in statistical practice. We will learn in the next chapter what to do when σ is unknown. However, if the sample is large, the sample standard deviation s will be close to the unknown σ. Then $\bar{x} \pm z^*s/\sqrt{n}$ is an approximate confidence interval for μ.

The most important caution concerning confidence intervals is a consequence of the first of these warnings. **The margin of error in a confidence interval covers only random sampling errors**. The margin of error is obtained from the sampling distribution and indicates how much error can be expected because of chance variation in randomized data production. Practical difficulties such as undercoverage and nonresponse in a sample survey can cause additional errors that may be larger than the random sampling error. Remember this unpleasant fact when reading the results of an opinion poll or other sample survey. The practical conduct of the survey influences the trustworthiness of its results in ways that are not included in the announced margin of error.

Every inference procedure that we will meet has its own list of warnings. Because many of the warnings are similar to those above, we will not print the full warning label each time. It is easy to state (from the mathematics of probability) conditions under which a method of inference is exactly correct. These conditions are never fully met in practice. For example, no population is exactly normal. Deciding when a statistical procedure should be used often requires judgment assisted by exploratory analysis of the data.

Finally, you should understand what statistical confidence does not say. We are 95% confident that the mean NAEP quantitative score for all men aged 21 to 25 years lies between 267.8 and 276.2. That is, these numbers were calculated by a method that gives correct results in 95% of all possible samples. We *cannot* say that the probability is 95% that the true mean falls between 267.8 and 276.2. No randomness remains after we draw one particular sample and get from it one particular interval. The true mean either is or is not between 267.8 and 276.2. The probability calculations of standard statistical inference describe how often the *method* gives correct answers.

"Hello. I'm taking a survey..."

Most sample surveys are conducted by telephone. Telephone polls use software that dials residential telephone numbers, listed or unlisted, at random. However, about 6% of U.S. households have no telephone. People in the South and people who live alone are less likely to have a phone, and many telephone surveys omit Alaska and Hawaii to save expense. Undercoverage thus causes some bias in telephone surveys. The margin of error announced by the survey does not include this bias.

APPLY YOUR KNOWLEDGE

6.13 **A talk show opinion poll.** A radio talk show invites listeners to enter a dispute about a proposed pay increase for city council members. "What yearly pay do you think council members should get? Call us with your number." In all, 958 people call. The mean pay they suggest is $\bar{x} = \$8740$ per year, and the standard deviation of the responses is $s = \$1125$. For a large sample such as this, s is very close to the unknown population σ. The station calculates the 95% confidence interval for the mean pay μ that all citizens would propose for council members to be $8669 to $8811.

(a) Is the station's calculation correct?

(b) Does their conclusion describe the population of all the city's citizens? Explain your answer.

6.14 **Internet users.** A survey of users of the Internet found that males outnumbered females by nearly 2 to 1. This was a surprise, because earlier surveys had put the ratio of men to women closer to 9 to 1. Later in the article we find this information:

> *Detailed surveys were sent to more than 13,000 organizations on the Internet; 1,468 usable responses were received. According to Mr. Quarterman, the margin of error is 2.8 percent, with a confidence level of 95 percent.*[3]

(a) What was the *response rate* for this survey? (The response rate is the percent of the planned sample that responded.)

(b) Do you think that the small margin of error is a good measure of the accuracy of the survey's results? Explain your answer.

6.15 **Prayer in the schools?** A *New York Times*/CBS News poll recently asked the question "Do you favor an amendment to the Constitution that would permit organized prayer in public schools?" Sixty-six percent of the sample answered "Yes." The article describing the poll says that it "is based on telephone interviews conducted from Sept. 13 to Sept. 18 with 1,664 adults around the United States, excluding Alaska and Hawaii. . . . the telephone numbers were formed by random digits, thus permitting access to both listed and unlisted residential numbers."

(a) The article gives the margin of error as 3 percentage points. Opinion polls customarily announce margins of error for 95% confidence. Make a confidence statement about the percent of all adults who favor a school prayer amendment.

(b) The news article goes on to say: "The theoretical errors do not take into account a margin of additional error resulting from the various practical difficulties in taking any survey of public opinion." List some of the "practical difficulties" that may cause errors in addition to the ±3% margin of error. Pay particular attention to the news article's description of the sampling method.

SECTION 6.1 Summary

A **confidence interval** uses sample data to estimate an unknown population parameter with an indication of how accurate the estimate is and of how confident we are that the result is correct.

Any confidence interval has two parts: an interval computed from the data and a confidence level. The **interval** often has the form

$$\text{estimate} \pm \text{margin of error}$$

The **confidence level** states the probability that the method will give a correct answer. That is, if you use 95% confidence intervals often, in the long run 95% of your intervals will contain the true parameter value. You do not know whether a 95% confidence interval calculated from a particular set of data contains the true parameter value.

A level C **confidence interval for the mean** μ of a normal population with known standard deviation σ, based on an SRS of size n, is given by

$$\bar{x} \pm z^* \frac{\sigma}{\sqrt{n}}$$

The **critical value** z^* is chosen so that the standard normal curve has area C between $-z^*$ and z^*. Because of the central limit theorem, this interval is approximately correct for large samples when the population is not normal.

Other things being equal, the **margin of error** of a confidence interval gets smaller as

- the confidence level C decreases,
- the population standard deviation σ decreases, and
- the sample size n increases.

The sample size required to obtain a confidence interval with specified margin of error m for a normal mean is

$$n = \left(\frac{z^* \sigma}{m}\right)^2$$

where z^* is the critical value for the desired level of confidence. Always round n up when you use this formula.

A specific confidence interval recipe is correct only under specific conditions. The most important conditions concern the method used to produce the data. Other factors such as the form of the population distribution may also be important.

SECTION 6.1   E x e r c i s e s

6.16   **Who should get welfare?**   A news article on a Gallup Poll noted that "28 percent of the 1548 adults questioned felt that those who were able to work should be taken off welfare." The article also said, "The margin of error for a sample size of 1548 is plus or minus three percentage points." Opinion polls usually announce margins of error for 95% confidence. Using this fact, explain to someone who knows no statistics what "margin of error plus or minus three percentage points" means.

6.17   **Hotel computer systems.**   How satisfied are hotel managers with the computer systems their hotels use? A survey was sent to 560 managers in hotels of size 200 to 500 rooms in Chicago and Detroit.[4] In all, 135 managers returned the survey. Two questions concerned their degree of satisfaction with the ease of use of their computer systems and with the level of computer training they had received. The managers responded using a seven-point scale, with 1 meaning "not satisfied," 4 meaning "moderately satisfied," and 7 meaning "very satisfied."

(a) What do you think is the population for this study? There are some major shortcomings in the data production. What are they? These shortcomings reduce the value of the formal inference you are about to do.

(b) The mean response for satisfaction with ease of use was $\bar{x}$ = 5.396. Give a 95% confidence interval for the mean in the entire population. (Assume that the population standard deviation is σ = 1.75.)

(c) For satisfaction with training, the mean response was $\bar{x}$ = 4.398. Taking σ = 1.75, give a 99% confidence interval for the population mean.

(d) The measurements of satisfaction are certainly not normally distributed, because they take only whole-number values from 1 to 7. Nonetheless, the use of confidence intervals based on the normal distribution is justified for this study. Why?

(BLAIR SEITZ/PHOTO RESEARCHERS.)

6.18 **Why do consumers like pharmacies?** Consumers can purchase nonprescription medications at food stores, mass merchandise stores such as Kmart and Wal-Mart, or pharmacies. About 45% of consumers make such purchases at pharmacies. What accounts for the popularity of pharmacies, which often charge higher prices?

A study examined consumers' perceptions of overall performance of the three types of stores, using a long questionnaire that asked about such things as "neat and attractive store," "knowledgeable staff," and "assistance in choosing among various types of nonprescription medication." A performance score was based on 27 such questions. The subjects were 201 people chosen at random from the Indianapolis telephone directory. Here are the means and standard deviations of the performance scores for the sample:[5]

Store type	$\bar{x}$	s
Food stores	18.67	24.95
Mass merchandisers	32.38	33.37
Pharmacies	48.60	35.62

We do not know the population standard deviations, but a sample standard deviation s from so large a sample is usually close to σ. Use s in place of the unknown σ in this exercise.

(a) What population do you think the authors of the study want to draw conclusions about? What population are you certain they can draw conclusions about?

(b) Give 95% confidence intervals for the mean performance for each type of store in the population.

(c) Based on these confidence intervals, are you convinced that consumers think that pharmacies offer higher performance than the other types of stores?

6.19 Healing of skin wounds. Biologists studying the healing of skin wounds measured the rate at which new cells closed a razor cut made in the skin of an anesthetized newt. Here are data from 18 newts, measured in micrometers (millionths of a meter) per hour:[6]

29	27	34	40	22	28	14	35	26
35	12	30	23	18	11	22	23	33

(a) Make a stemplot of the healing rates (split the stems). It is difficult to assess normality from 18 observations, but look for outliers or extreme skewness. What do you find?

(b) Scientists usually assume that animal subjects are SRSs from their species or genetic type. Treat these newts as an SRS and suppose you know that the standard deviation of healing rates for this species of newt is 8 micrometers per hour. Give a 90% confidence interval for the mean healing rate for the species.

(c) A friend who knows almost no statistics follows the formula $\bar{x} \pm z^*\sigma/\sqrt{n}$ in a biology lab manual to get a 95% confidence interval for the mean. Is her interval wider or narrower than yours? Explain to her why it makes sense that higher confidence changes the length of the interval.

6.20 Engine crankshafts. Here are measurements (in millimeters) of a critical dimension on a sample of auto engine crankshafts:

224.120	224.001	224.017	223.982	223.989	223.961
223.960	224.089	223.987	223.976	223.902	223.980
224.098	224.057	223.913	223.999		

The data come from a production process that is known to have standard deviation $\sigma = 0.060$ mm. The process mean is supposed to be $\mu = 224$ mm but can drift away from this target during production.

(a) We expect the distribution of the dimension to be close to normal. Make a stemplot or histogram of these data and describe the shape of the distribution.

(b) Give a 95% confidence interval for the process mean at the time these crankshafts were produced.

6.21 How large a sample would enable you to estimate the mean healing rate of skin wounds in newts (see Exercise 6.19) within a margin of error of 1 micrometer per hour with 90% confidence?

6.22 **A newspaper poll.** A *New York Times* poll on women's issues interviewed 1025 women and 472 men randomly selected from the United States, excluding Alaska and Hawaii. The poll announced a margin of error of ±3 percentage points for 95% confidence in conclusions about women. The margin of error for results concerning men was ±4 percentage points. Why is this larger than the margin of error for women?

6.23 **Calling the election.** The last closely contested presidential election pitted Jimmy Carter against Gerald Ford in 1976. A poll taken immediately before the 1976 election showed that 51% of the sample intended to vote for Carter. The polling organization announced that they were 95% confident that the sample result was within ±2 points of the true percent of all voters who favored Carter.

(a) Explain in plain language to someone who knows no statistics what "95% confident" means in this announcement.

(b) The poll showed Carter leading. Yet the polling organization said the election was too close to call. Explain why.

(c) On hearing of the poll, a nervous politician asked, "What is the probability that over half the voters prefer Carter?" A statistician replied that this question can't be answered from the poll results, and that it doesn't even make sense to talk about such a probability. Explain why.

6.24 **How the poll was conducted.** The *New York Times* includes a box entitled "How the poll was conducted" in news articles about its own opinion polls. Here are quotations from one such box (March 26, 1995). The box also announced a margin of error of plus or minus three percent, with 95% confidence.

> *The latest New York Times/CBS News poll is based on telephone interviews conducted March 9 through 12 with 1,156 adults around the United States, excluding Alaska and Hawaii.* [The box then describes random digit dialing, the method used to select the sample.]
>
> *In addition to sampling error, the practical difficulties of conducting any survey of public opinion may introduce other sources of error into the poll. Variations in question wording or in the order of questions, for instance, can lead to somewhat different results.*

(a) This account mentions several sources of possible errors in the poll's results. List these sources.

(b) Which of the sources of error you listed in (a) are covered by the announced margin of error?

6.2 Tests of Significance

Confidence intervals are one of the two most common types of statistical inference. Use a confidence interval when your goal is to estimate a population parameter. The second common type of inference, called *tests of significance*, has a different goal: to assess the evidence provided by data about some claim concerning a population. Here is the reasoning of statistical tests in a nutshell.

EXAMPLE 6.7   **I'm a great free-throw shooter**

I claim that I make 80% of my basketball free throws. To test my claim, you ask me to shoot 20 free throws. I make only 8 of the 20. "Aha!" you say. "Someone who makes 80% of his free throws would almost never make only 8 out of 20. So I don't believe your claim."

Your reasoning is based on asking what would happen if my claim were true and we repeated the sample of 20 free throws many times—I would almost never make as few as 8. This outcome is so unlikely that it gives strong evidence that my claim is not true.

You can say how strong the evidence against my claim is by giving the probability that I would make as few as 8 out of 20 free throws if I really make 80% in the long run. This probability is 0.0001. I would make as few as 8 of 20 only once in 10,000 tries in the long run if my claim to make 80% is true. The small probability convinces you that my claim is false.

Significance tests use an elaborate vocabulary, but the basic idea is simple: an outcome that would rarely happen if a claim were true is good evidence that the claim is not true.

The reasoning of tests of significance

The reasoning of statistical tests, like that of confidence intervals, is based on asking what would happen if we repeated the sample or experiment many times. We will again start with an unrealistic procedure in order to emphasize the reasoning. Here is an example we will explore.

EXAMPLE 6.8   **Sweetening colas**

Diet colas use artificial sweeteners to avoid sugar. These sweeteners gradually lose their sweetness over time. Manufacturers therefore test new colas for loss of sweetness before marketing them. Trained tasters sip the cola along with drinks of standard sweetness and score the cola on a "sweetness score" of 1 to 10. The cola is then stored for a month at high temperature to imitate the effect of four months' storage at room temperature. Each taster scores the cola again after storage. This is a matched pairs experiment. Our data are the differences (score before storage minus score after storage) in the tasters' scores. The bigger these differences, the bigger the loss of sweetness.

Here are the sweetness losses for a new cola, as measured by 10 trained tasters:

$$2.0 \quad 0.4 \quad 0.7 \quad 2.0 \quad -0.4 \quad 2.2 \quad -1.3 \quad 1.2 \quad 1.1 \quad 2.3$$

Most are positive. That is, most tasters found a loss of sweetness. But the losses are small, and two tasters (the negative scores) thought the cola gained sweetness. *Are these data good evidence that the cola lost sweetness in storage?*

The average sweetness loss for our cola is given by the sample mean,

$$\bar{x} = \frac{2.0 + 0.4 + \cdots + 2.3}{10} = 1.02$$

We will assume (unrealistically) that we know that the standard deviation for all individual tasters is $\sigma = 1$.

The reasoning is as in Example 6.7. We make a claim and ask if the data give evidence *against* it. We seek evidence that there *is* a sweetness loss, so the claim we test is that there *is not* a loss. In that case, the mean loss perceived by the population of all trained testers would be $\mu = 0$.

- If the claim that $\mu = 0$ is true, the sampling distribution of $\bar{x}$ from 10 tasters is normal with mean $\mu = 0$ and standard deviation

$$\frac{\sigma}{\sqrt{n}} = \frac{1}{\sqrt{10}} = 0.316$$

Figure 6.9 shows this sampling distribution. We can judge whether any observed $\bar{x}$ is surprising by locating it on this distribution.

- Suppose that our 10 tasters had mean loss $\bar{x} = 0.3$. It is clear from Figure 6.9 that an $\bar{x}$ this large could easily occur just by chance when the population mean is $\mu = 0$. That 10 tasters find $\bar{x} = 0.3$ is not evidence of a sweetness loss.

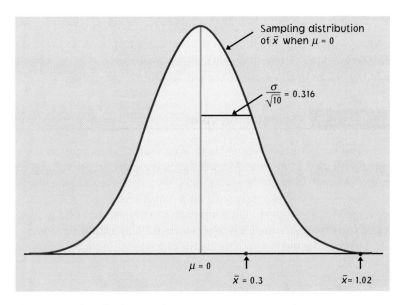

Figure 6.9 *If a cola does not lose sweetness in storage, the mean score $\bar{x}$ for 10 tasters will have this sampling distribution. The actual result for one cola was $\bar{x} = 0.3$. That could easily happen just by chance. Another cola had $\bar{x} = 1.02$. That's so far out on the normal curve that it is good evidence that this cola did lose sweetness.*

- In fact, our taste test produced $\bar{x} = 1.02$. That's way out on the normal curve in Figure 6.9—so far out that an observed value this large would rarely occur just by chance if the true μ were 0. This observed value is good evidence that in fact the true μ is greater than 0, that is, that the cola lost sweetness. The manufacturer must reformulate the cola and try again.

The vocabulary of significance tests

A statistical test starts with a careful statement of the claims we want to compare. The claims concern a population, so we express them in terms of a population parameter. In Example 6.8, the parameter is the population mean μ, the average loss in sweetness that a very large number of tasters would detect in the cola. Because the reasoning we have outlined looks for evidence *against* a claim, we start with the claim we seek evidence against, such as "no loss of sweetness." This claim is our *null hypothesis*.

> #### NULL HYPOTHESIS H_0
>
> The statement being tested in a statistical test is called the **null hypothesis**. The test is designed to assess the strength of the evidence against the null hypothesis. Usually the null hypothesis is a statement of "no effect" or "no difference."

The claim about the population that we are trying to find evidence *for* is the **alternative hypothesis**, written H_a. In Example 6.8, we are seeking evidence of a loss in sweetness. The null hypothesis says "no loss" on the average in a large population of tasters. The alternative hypothesis says "there is a loss." So the hypotheses are

alternative hypothesis

$$H_0 : \mu = 0$$

$$H_a : \mu > 0$$

The null and alternative hypotheses are precise statements of just what claims we are testing. If we get an outcome that would be unlikely if H_0 were true and is in the direction suggested by H_a, we have evidence against H_0 in favor of H_a. We make "unlikely" precise by calculating a probability.

> #### *P*-VALUE
>
> The probability, computed assuming that H_0 is true, that the observed outcome would take a value as extreme or more extreme than that actually observed is called the **P-value** of the test. The smaller the P-value is, the stronger is the evidence against H_0 provided by the data.

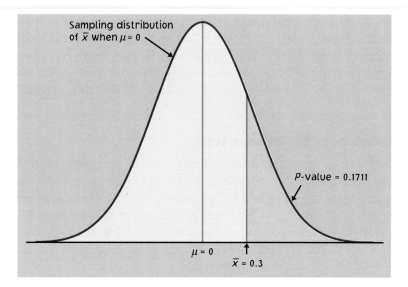

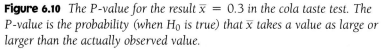

Figure 6.10 *The P-value for the result* $\bar{x} = 0.3$ *in the cola taste test. The P-value is the probability (when H_0 is true) that $\bar{x}$ takes a value as large or larger than the actually observed value.*

EXAMPLE 6.9   What is a P-value?

Figure 6.10 shows the P-value when 10 tasters give mean sweetness loss $\bar{x} = 0.3$. It is the probability that, *if $\mu = 0$ is true*, we observe a sample mean at least as large as 0.3. This probability is $P = 0.1711$. That is, we would observe a sweetness loss this large or larger about 17% of the time, just by the luck of the draw in choosing 10 tasters, even if the entire population of tasters would find no loss on the average. A result that would occur this often when H_0 is true is not good evidence against H_0.

In fact, the 10 tasters found $\bar{x} = 1.02$. The P-value is again the probability of observing an $\bar{x}$ this large or larger if in fact $\mu = 0$. This probability is $P = 0.0006$. We would very rarely observe a sample sweetness loss this large if H_0 were true. The small P-value provides strong evidence against H_0 and in favor of the alternative $H_a : \mu > 0$.

Small P-values are evidence against H_0, because they say that the observed result is unlikely to occur just by chance. Large P-values fail to give evidence against H_0.

Recipes for significance tests hide the underlying reasoning. In fact, statistical software often just gives a P-value. Look again at Figure 6.9, with its $\bar{x}$-values from taste tests of two colas. We can see that one result is not surprising if the true mean score in the population is 0, and that the other is surprising. A significance test simply says that more precisely.

6.25 **Students' attitudes.** The Survey of Study Habits and Attitudes (SSHA) is a psychological test that measures students' study habits and attitude toward school. Scores range from 0 to 200. The mean score for U.S. college students is about 115, and the standard deviation is about 30. A teacher suspects that older students have better attitudes toward school. She gives the SSHA to 25 students who are at least 30 years old. Assume that scores in the population of older students are normally distributed with standard deviation $\sigma = 30$. The teacher wants to test the hypotheses

$$H_0 : \mu = 115$$

$$H_a : \mu > 115$$

(a) What is the sampling distribution of the mean score $\bar{x}$ of a sample of 25 older students if the null hypothesis is true? Sketch the density curve of this distribution. (Hint: Sketch a normal curve first, then mark the axis using what you know about locating μ and σ on a normal curve.)

(b) Suppose that the sample data give $\bar{x} = 118.6$. Mark this point on the axis of your sketch. In fact, the result was $\bar{x} = 125.7$. Mark this point on your sketch. Using your sketch, explain in simple language why one result is good evidence that the mean score of all older students is greater than 115 and why the other outcome is not.

(c) Shade the area under the curve that is the P-value for the sample result $\bar{x} = 118.6$.

6.26 **Spending on housing.** The Census Bureau reports that households spend an average of 31% of their total spending on housing. A homebuilders association in Cleveland believes that this average is lower in their area. They interview a sample of 40 households in the Cleveland metropolitan area to learn what percent of their spending goes toward housing. Take μ to be the mean percent of spending devoted to housing among all Cleveland households. We want to test the hypotheses

$$H_0 : \mu = 31\%$$

$$H_a : \mu < 31\%$$

The population standard deviation is $\sigma = 9.6\%$.

(a) What is the sampling distribution of the mean percent $\bar{x}$ that the sample spends on housing if the null hypothesis is true? Sketch the density curve of the sampling distribution. (Hint: Sketch a normal curve first, then mark the axis using what you know about locating μ and σ on a normal curve.)

(b) Suppose that the study finds $\bar{x} = 30.2\%$ for the 40 households in the sample. Mark this point on the axis in your sketch. Then suppose that the study result is $\bar{x} = 27.6\%$. Mark this point on your sketch. Referring to your sketch, explain in simple language why one result is good evidence that average Cleveland spending on housing is less than 31% and the other result is not.

(c) Shade the area under the curve that gives the *P*-value for the result $\bar{x} = 30.2\%$. (Note that we are looking for evidence that spending is *less* than the null hypothesis states.)

More detail: stating hypotheses

The first step in a test of significance is to state the hypotheses. The null hypothesis is a claim that we will try to find evidence *against*. The alternative hypothesis H_a is the claim about the population that we are trying to find evidence *for*. In Example 6.8, we were seeking evidence of a loss in sweetness. The null hypothesis says "no loss" on the average in a large population of tasters. The alternative hypothesis says "there is a loss." So the hypotheses are

$$H_0 : \mu = 0$$
$$H_a : \mu > 0$$

one-sided alternative   This alternative hypothesis is **one-sided** because we are interested only in deviations from the null hypothesis in one direction. Here is another example.

| EXAMPLE 6.10 | **Studying job satisfaction** |

Does the job satisfaction of assembly workers differ when their work is machine-paced rather than self-paced? One study chose 28 subjects at random from a group of women who worked at assembling electronic devices. Half of the subjects were assigned at random to each of two groups. Both groups did similar assembly work, but one work setup allowed workers to pace themselves and the other featured an assembly line that moved at fixed time intervals so that the workers were paced by machine. After two weeks, all subjects took the Job Diagnosis Survey (JDS), a test of job satisfaction. Then they switched work setups and took the JDS again after two more weeks. This is another matched pairs design. The response variable is the difference in JDS scores, self-paced minus machine-paced.[7]

The parameter of interest is the mean μ of the differences in JDS scores in the population of all female assembly workers. The null hypothesis says that there is no difference between self-paced and machine-paced work, that is,

$$H_0 : \mu = 0$$

The authors of the study wanted to know if the two work conditions have different levels of job satisfaction. They did not specify the direction of the difference. The

two-sided alternative   alternative hypothesis is therefore **two-sided**,

$$H_a : \mu \neq 0$$

Hypotheses always refer to some population, not to a particular outcome. So be sure to state H_0 and H_a in terms of population parameters. Because H_a expresses the effect that we hope to find evidence *for*, it is often easier to begin by stating H_a and then set up H_0 as the statement that the hoped-for effect is not present.

It is not always easy to decide whether H_a should be one-sided or two-sided. In Example 6.10, the alternative $H_a : \mu \neq 0$ is two-sided. It simply says there is a difference in job satisfaction without specifying the direction of the difference. The alternative $H_a : \mu > 0$ in the taste test example is one-sided. Because colas can only lose sweetness in storage, we are interested only in detecting an upward shift in μ. The alternative hypothesis should express the hopes or suspicions we bring to the data. It is cheating to first look at the data and then frame H_a to fit what the data show. Thus the fact that the workers in the study of Example 6.10 were more satisfied with self-paced work should not influence our choice of H_a. If you do not have a specific direction firmly in mind in advance, use a two-sided alternative.

APPLY YOUR KNOWLEDGE

Each of the following situations calls for a significance test for a population mean μ. State the null hypothesis H_0 and the alternative hypothesis H_a in each case.

6.27 The diameter of a spindle in a small motor is supposed to be 5 mm. If the spindle is either too small or too large, the motor will not work properly. The manufacturer measures the diameter in a sample of motors to determine whether the mean diameter has moved away from the target.

6.28 Census Bureau data show that the mean household income in the area served by a shopping mall is $52,500 per year. A market research firm questions shoppers at the mall. The researchers suspect the mean household income of mall shoppers is higher than that of the general population.

6.29 The examinations in a large accounting class are scaled after grading so that the mean score is 50. The professor thinks that one teaching assistant is a poor teacher and suspects that his students have a lower mean score than the class as a whole. The TA's students this semester can be considered a sample from the population of all students in the course, so the professor compares their mean score with 50.

6.30 Last year, your company's service technicians took an average of 2.6 hours to respond to trouble calls from business customers who had purchased service contracts. Do this year's data show a different average response time?

Computer-assisted interviewing

The days of the interviewer with a clipboard are past. Contemporary interviewers read questions from a computer screen and use the keyboard to enter responses. The computer skips irrelevant items—once a woman says that she has no children, further questions about her children never appear. The computer can even present questions in random order to avoid bias due to always following the same order. Computer software keeps records of who has responded and prepares a file of data from the responses. The tedious process of transferring responses from paper to computer, once a source of errors, has disappeared.

More detail: *P*-values and statistical significance

test statistic   A significance test uses data in the form of a **test statistic**. The test statistic is usually based on a statistic that estimates the parameter that appears in the hypotheses. In our examples, the parameter is μ and the test statistic is the sample mean $\bar{x}$.

A test of significance assesses the evidence against the null hypothesis by giving a probability, the *P*-value. If the test statistic falls far from the value suggested by the null hypothesis in the direction specified by the alternative hypothesis, it is good evidence against H_0 and in favor of H_a. The *P*-value describes how strong the evidence is because it is the probability of getting an outcome *as extreme or more extreme than the actually observed outcome.* "Extreme" means "far from what we would expect if H_0 were true." The direction or directions that count as "far from what we would expect" are determined by the alternative hypothesis H_a.

EXAMPLE 6.11   **Calculating a one-sided *P*-value**

In Example 6.8 the observations are an SRS of size $n = 10$ from a normal population with $\sigma = 1$. The observed mean sweetness loss for one cola was $\bar{x} = 0.3$. The *P*-value for testing

$$H_0 : \mu = 0$$

$$H_a : \mu > 0$$

is therefore

$$P(\bar{x} \geq 0.3)$$

calculated assuming that H_0 is true. When H_0 is true, $\bar{x}$ has the normal distribution with mean 0 and standard deviation

$$\frac{\sigma}{\sqrt{n}} = \frac{1}{\sqrt{10}} = 0.316$$

Find the *P*-value by a normal probability calculation. Start by drawing a picture that shows the *P*-value as an area under a normal curve. Figure 6.11 is the picture for this example. Then standardize $\bar{x}$ to get a standard normal Z and use Table A:

$$P(\bar{x} \geq 0.3) = P\left(\frac{\bar{x} - 0}{0.316} \geq \frac{0.3 - 0}{0.316}\right)$$

$$= P(Z \geq 0.95)$$

$$= 1 - 0.8289 = 0.1711$$

This is the value that was reported on page 322.

We sometimes take one final step to assess the evidence against H_0. We can compare the *P*-value with a fixed value that we regard as decisive. This amounts to announcing in advance how much evidence against H_0 we will *significance level*   insist on. The decisive value of *P* is called the **significance level**. We write it as α, the Greek letter alpha. If we choose $\alpha = 0.05$, we are requiring that the

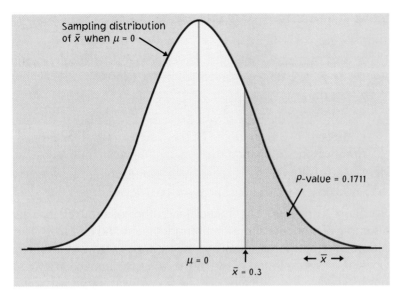

Figure 6.11 *The P-value for the one-sided test in Example 6.11.*

data give evidence against H_0 so strong that it would happen no more than 5% of the time (1 time in 20 samples in the long run) when H_0 is true. If we choose $\alpha = 0.01$, we are insisting on stronger evidence against H_0, evidence so strong that it would appear only 1% of the time (1 time in 100 samples) if H_0 is in fact true.

STATISTICAL SIGNIFICANCE

If the P-value is as small or smaller than α, we say that the data are **statistically significant at level α**.

"Significant" in the statistical sense does not mean "important." It means simply "not likely to happen just by chance." The significance level α makes "not likely" more exact. Significance at level 0.01 is often expressed by the statement "The results were significant ($P < 0.01$)." Here P stands for the P-value. The P-value is more informative than a statement of significance, because it allows us to assess significance at any level we choose. For example, a result with $P = 0.03$ is significant at the $\alpha = 0.05$ level but is not significant at the $\alpha = 0.01$ level.

APPLY YOUR KNOWLEDGE

6.31 Return to Exercise 6.25 (page 323).

(a) Starting from the picture you drew there, calculate the P-values for both $\bar{x} = 118.6$ and $\bar{x} = 125.7$. The two P-values express in numbers the comparison you made informally in Exercise 6.25.

(b) Which of the two observed values of $\bar{x}$ are statistically significant at the $\alpha = 0.05$ level? At the $\alpha = 0.01$ level?

6.32 Return to Exercise 6.26 (page 323).

(a) Starting from the picture you drew there, calculate the P-values for both $\bar{x} = 30.2\%$ and $\bar{x} = 27.6\%$. The two P-values express in numbers the comparison you made informally in Exercise 6.26.

(b) Is the result $\bar{x} = 27.6$ statistically significant at the $\alpha = 0.05$ level? Is it significant at the $\alpha = 0.01$ level?

6.33 **Students' earnings.** The financial aid office of a university asks a sample of students about their employment and earnings. The report says that "for academic year earnings, a significant difference $(P = 0.038)$ was found between the sexes, with men earning more on the average. No difference $(P = 0.476)$ was found between the earnings of black and white students." Explain both of these conclusions, for the effects of sex and of race on mean earnings, in language understandable to someone who knows no statistics.[8]

Tests for a population mean

There are three steps in carrying out a significance test:

1. State the hypotheses.

2. Calculate the test statistic.

3. Find the P-value.

Once you have stated your hypotheses and identified the proper test, you or your computer can do Steps 2 and 3 by following a recipe. We now develop the recipe for the test we have used in our examples.

We have an SRS of size n drawn from a normal population with unknown mean μ. We want to test the hypothesis that μ has a specified value. Call the specified value μ_0. The null hypothesis is

$$H_0 : \mu = \mu_0$$

The test is based on the sample mean $\bar{x}$. Because normal calculations require standardized variables, we will use as our test statistic the *standardized* sample mean

$$z = \frac{\bar{x} - \mu_0}{\sigma/\sqrt{n}}$$

one-sample z statistic   This **one-sample z statistic** has the standard normal distribution when H_0 is true. If the alternative is one-sided on the high side

$$H_a : \mu > \mu_0$$

then the P-value is the probability that a standard normal variable Z takes a value at least as large as the observed z. That is,

$$P = P(Z \geq z)$$

Example 6.11 calculates this P-value for the cola taste test. There, $\mu_0 = 0$, the standardized sample mean was $z = 0.95$, and the P-value was $P(Z \geq 0.95) = 0.1711$. Similar reasoning applies whenever the alternative hypothesis is one-sided.

When H_a states that μ is simply unequal to μ_0 (two-sided), values of z away from zero in either direction count against the null hypothesis. The P-value is the probability that a standard normal Z is at least as far from zero *in either direction* as the observed z.

EXAMPLE 6.12   Calculating a two-sided *P*-value

Suppose that the z statistic for a two-sided test is $z = 1.7$. The two-sided P-value is the probability that $Z \leq -1.7$ or $Z \geq 1.7$. Figure 6.12 shows this probability as areas under the standard normal curve. Because the standard normal distribution is symmetric, we can calculate this probability by finding $P(Z \geq 1.7)$ and *doubling* it.

$$P(Z \leq -1.7 \text{ or } Z \geq 1.7) = 2P(Z \geq 1.7)$$
$$= 2(1 - 0.9554) = 0.0892$$

We would make exactly the same calculation if we observed $z = -1.7$. It is the absolute value $|z|$ that matters, not whether z is positive or negative.

z TEST FOR A POPULATION MEAN

To test the hypothesis $H_0 : \mu = \mu_0$ based on an SRS of size n from a population with unknown mean μ and known standard deviation σ, compute the **one-sample z statistic**

$$z = \frac{\bar{x} - \mu_0}{\sigma/\sqrt{n}}$$

In terms of a variable Z having the standard normal distribution, the P-value for a test of H_0 against

$H_a : \mu > \mu_0$ is $P(Z \geq z)$

$H_a : \mu < \mu_0$ is $P(Z \leq z)$

$H_a : \mu \neq \mu_0$ is $2P(Z \geq |z|)$

These P-values are exact if the population distribution is normal and are approximately correct for large n in other cases.

Figure 6.12 *The P-value for the two-sided test in Example 6.12.*

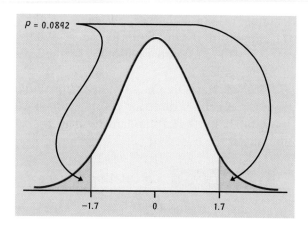

$P = 0.0892$

−1.7   0   1.7

EXAMPLE 6.13   **Executives' blood pressures**

The National Center for Health Statistics reports that the mean systolic blood pressure for males 35 to 44 years of age is 128 and the standard deviation in this population is 15. The medical director of a large company looks at the medical records of 72 executives in this age group and finds that the mean systolic blood pressure in this sample is $\bar{x} = 126.07$. Is this evidence that the company's executives have a different mean blood pressure from the general population? As usual in this chapter, we make the unrealistic assumption that we know the population standard deviation. Assume that executives have the same $\sigma = 15$ as the general population of middle-aged males.

Step 1. Hypotheses. The null hypothesis is "no difference" from the national mean $\mu_0 = 128$. The alternative is two-sided, because the medical director did not have a particular direction in mind before examining the data. So the hypotheses about the unknown mean μ of the executive population are

$$H_0 : \mu = 128$$

$$H_a : \mu \neq 128$$

Step 2. Test statistic. The one-sample z statistic is

$$z = \frac{\bar{x} - \mu_0}{\sigma/\sqrt{n}} = \frac{126.07 - 128}{15/\sqrt{72}}$$

$$= -1.09$$

Step 3. P-value. You should still draw a picture to help find the P-value, but now you can sketch the standard normal curve with the observed value of z. Figure 6.13 shows that the P-value is the probability that a standard normal variable Z takes a value at least 1.09 away from zero. From Table A we find that this probability is

$$P = 2P(Z \geq 1.09) = 2(1 - 0.8621) = 0.2758$$

Conclusion: More than 27% of the time, an SRS of size 72 from the general male population would have a mean blood pressure at least as

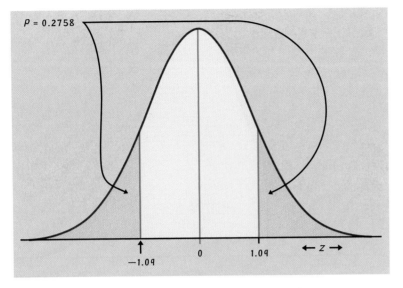

$P = 0.2758$

-1.09 0   1.09 $\leftarrow z \rightarrow$

Figure 6.13 *The P-value for the two-sided test in Example 6.13.*

far from 128 as that of the executive sample. The observed $\bar{x} = 126.07$ is therefore not good evidence that executives differ from other men.

The z test assumes that the 72 executives in the sample are an SRS from the population of all middle-aged male executives in the company. We should check this assumption by asking how the data were produced. If medical records are available only for executives with recent medical problems, for example, the data are of little value for our purpose. It turns out that all executives are given a free annual medical exam, and that the medical director selected 72 exam results at random.

The data in Example 6.13 do *not* establish that the mean blood pressure μ for this company's executives is 128. We sought evidence that μ differed from 128 and failed to find convincing evidence. That is all we can say. No doubt the mean blood pressure of the entire executive population is not exactly equal to 128. A large enough sample would give evidence of the difference, even if it is very small. Tests of significance assess the evidence *against* H_0. If the evidence is strong, we can confidently reject H_0 in favor of the alternative. Failing to find evidence against H_0 means only that the data are consistent with H_0, not that we have clear evidence that H_0 is true.

EXAMPLE 6.14   **Can you balance your checkbook?**

In a discussion of the education level of the American workforce, someone says, "The average young person can't even balance a checkbook." The NAEP survey says that a score of 275 or higher on its quantitative test (see Example 6.1 on page 299) reflects the skill needed to balance a checkbook. The NAEP random sample of 840 young men had a mean score of $\bar{x} = 272$, a bit below the checkbook-balancing

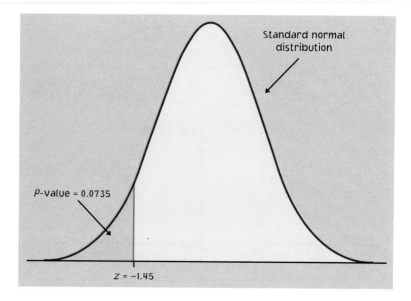

Figure 6.14 *The P-value for the one-sided test in Example 6.14.*

level. Is this sample result good evidence that the mean for *all* young men is less than 275? As in Example 6.1, assume that $\sigma = 60$.

Step 1. Hypotheses. The hypotheses are

$$H_0 : \mu = 275$$

$$H_a : \mu < 275$$

Step 2. Test statistic. The z statistic is

$$z = \frac{\bar{x} - \mu_0}{\sigma/\sqrt{n}} = \frac{272 - 275}{60/\sqrt{840}}$$

$$= -1.45$$

Step 3. *P*-value. Because H_a is one-sided on the low side, small values of z count against H_0. Figure 6.14 illustrates the *P*-value. Using Table A, we find that

$$P = P(Z \leq -1.45) = 0.0735$$

Conclusion: A mean score as low as 272 would occur about 7 times in 100 samples if the population mean were 275. This is modest evidence that the mean NAEP score for all young men is less than 275, but it is not significant at the $\alpha = 0.05$ level.

APPLY YOUR KNOWLEDGE · · · · · · · · · · · · · · · · · ·

6.34 **Sales of coffee.** Weekly sales of regular ground coffee at a supermarket have in the recent past varied according to a normal distribution with mean $\mu = 354$ units per week and standard deviation $\sigma = 33$ units. The store reduces the price by 5%. Sales in the next three weeks are 405, 378, and 411 units. Is this good evidence that average sales are now higher? The hypotheses are

$$H_0 : \mu = 354$$

$$H_a : \mu > 354$$

Assume that the standard deviation of the population of weekly sales remains $\sigma = 33$.

(a) Find the sample mean $\bar{x}$ and the value of the one-sample z test statistic.

(b) Calculate the P-value. Sketch a standard normal curve with the area corresponding to the P-value shaded.

(c) Is the result statistically significant at the $\alpha = 0.05$ level? Is it significant at the $\alpha = 0.01$ level? Do you think there is convincing evidence that mean sales are higher?

6.35 **Engine crankshafts.** Here are measurements (in millimeters) of a critical dimension on a sample of automobile engine crankshafts:

224.120	224.001	224.017	223.982	223.989	223.961
223.960	224.089	223.987	223.976	223.902	223.980
224.098	224.057	223.913	223.999		

The manufacturing process is known to vary normally with standard deviation $\sigma = 0.060$ mm. The process mean is supposed to be 224 mm. Do these data give evidence that the process mean is not equal to the target value 224 mm?

(a) State H_0 and H_a.

(b) Calculate the test statistic z.

(c) Give the P-value of the test. Are you convinced that the process mean is not 224 mm?

6.36 **Filling cola bottles.** Bottles of a popular cola are supposed to contain 300 milliliters (ml) of cola. There is some variation from bottle to bottle because the filling machinery is not perfectly precise. The distribution of the contents is normal with standard deviation $\sigma = 3$ ml. An inspector who suspects that the bottler is underfilling measures the contents of six bottles. The results are

299.4	297.7	301.0	298.9	300.2	297.0

Is this convincing evidence that the mean content of cola bottles is less than the advertised 300 ml?

(a) State the hypotheses that you will test.

(b) Calculate the test statistic.

(c) Find the P-value and state your conclusion.

Tests with fixed significance level

Sometimes we demand a specific degree of evidence in order to reject the null hypothesis. A level of significance α says how much evidence we require. In terms of the P-value, the outcome of a test is significant at level α if $P \leq \alpha$. Significance at any level is easy to assess once you have the P-value. When you do not use statistical software, the P-value can be difficult to calculate. Fortunately, you can decide whether a result is statistically significant without calculating P. The following example illustrates how to assess significance at a fixed level α by using a table of critical values, the same table we used for confidence intervals. First, here is a more complete description of critical values.

CRITICAL VALUES

The number z^* with probability p lying to its right under the standard normal curve is called the **upper p critical value** of the standard normal distribution.

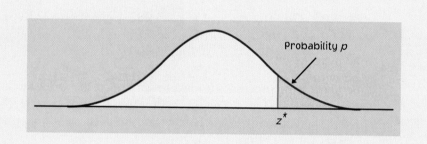

The z^* row in Table C gives critical values for the probabilities p in the top row.

EXAMPLE 6.15   Is it significant?

In Example 6.14, we examined whether the mean NAEP quantitative score of young men is less than 275. The hypotheses are

$$H_0 : \mu = 275$$

$$H_a : \mu < 275$$

The z statistic takes the value $z = -1.45$. Is the evidence against H_0 statistically significant at the 5% level?

To determine significance, just compare the observed $z = -1.45$ with the 0.05 critical value $z^* = 1.645$ from Table C. Because $z = -1.45$ is *not* farther from 0 than -1.645, it is *not* significant at level $\alpha = 0.05$.

Figure 6.15 shows how -1.645 separates values of z that are significant at the $\alpha = 0.05$ level from those that are not.

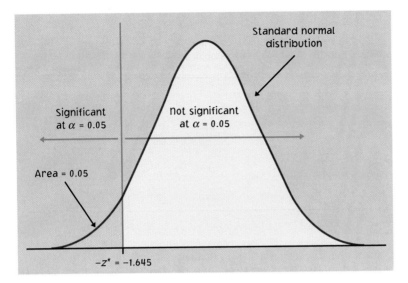

Figure 6.15 *Deciding whether a z statistic is significant at the $\alpha = 0.05$ level in the one-sided test of Example 6.15.*

TESTS WITH FIXED SIGNIFICANCE LEVEL

To test the hypothesis $H_0 : \mu = \mu_0$ based on an SRS of size n from a population with unknown mean μ and known standard deviation σ, compute the z test statistic

$$z = \frac{\bar{x} - \mu_0}{\sigma/\sqrt{n}}$$

Reject H_0 at significance level α against a one-sided alternative

$$H_a : \mu > \mu_0 \quad \text{if} \quad z \geq z^*$$
$$H_a : \mu < \mu_0 \quad \text{if} \quad z \leq -z^*$$

where z^* is the upper α critical value from Table C. Reject H_0 at significance level α against a two-sided alternative

$$H_a : \mu \neq \mu_0 \text{ if } |z| \geq z^*$$

where z^* is the upper $\alpha/2$ critical value from Table C.

EXAMPLE 6.16   **Is the concentration OK?**

The analytical laboratory of Example 6.4 (page 306) is asked to evaluate the claim that the concentration of the active ingredient in a specimen is 0.86%. The lab makes 3 repeated analyses of the specimen. The mean result is $\bar{x} = 0.8404$. The true concentration is the mean μ of the population of all analyses of the specimen. The standard deviation of the analysis process is known to be $\sigma = 0.0068$. Is there significant evidence at the 1% level that $\mu \neq 0.86$?

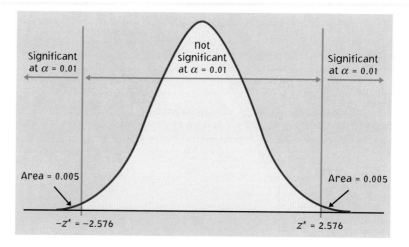

Figure 6.16 *Deciding whether a z statistic is significant at the* $\alpha = 0.01$ *level in the two-sided test of Example 6.16.*

Step 1. Hypotheses. The hypotheses are

$$H_0 : \mu = 0.86$$

$$H_a : \mu \neq 0.86$$

Step 2. Test statistic. The z statistic is

$$z = \frac{0.8404 - 0.86}{0.0068/\sqrt{3}} = -4.99$$

Step 3. Significance. Because the alternative is two-sided, we compare $|z| = 4.99$ with the $\alpha/2 = 0.005$ critical value from Table C. This critical value is $z^* = 2.576$. Figure 6.16 shows how this critical value separates values of z that are statistically significant from those that are not significant. Because $|z| > 2.576$, we reject the null hypothesis and conclude (at the 1% significance level) that the concentration is not as claimed.

The observed result in Example 6.16 was $z = -4.99$. The conclusion that this result is significant at the 1% level does not tell the whole story. The observed z is far beyond the 1% critical value, and the evidence against H_0 is much stronger than 1% significance suggests. The P-value

$$P = 2P(Z \geq 4.99) = 0.0000006$$

gives a better sense of how strong the evidence is. The P-value is the smallest level α at which the data are significant. Knowing the P-value allows us to assess significance at any level.

In addition to assessing significance at fixed levels, a table of critical values allows us to estimate P-values without a calculation. In Example 6.16, compare the observed $z = -4.99$ with the normal critical values in the bottom row of Table C. It is beyond even 3.291, the critical value for $P = 0.0005$. So we know that for the two-sided test, $P < 0.001$. In Example 6.15, $z = -1.45$

lies between the 0.05 and 0.10 entries in the table. So the P-value for the one-sided test lies between 0.05 and 0.10. This approximation is accurate enough for most purposes.

Because the practice of statistics almost always employs software that calculates P-values automatically, tables of critical values are becoming outdated. Tables of critical values such as Table C appear in this book for learning purposes and to rescue students without good computing facilities.

APPLY YOUR KNOWLEDGE

6.37 **Testing a random number generator.** A random number generator is supposed to produce random numbers that are uniformly distributed on the interval from 0 to 1. If this is true, the numbers generated come from a population with $\mu = 0.5$ and $\sigma = 0.2887$. A command to generate 100 random numbers gives outcomes with mean $\bar{x} = 0.4365$. Assume that the population σ remains fixed. We want to test

$$H_0 : \mu = 0.5$$

$$H_a : \mu \neq 0.5$$

(a) Calculate the value of the z test statistic.

(b) Is the result significant at the 5% level ($\alpha = 0.05$)?

(c) Is the result significant at the 1% level ($\alpha = 0.01$)?

(d) Between which two normal critical values in the bottom row of Table C does z lie? Between what two numbers does the P-value lie?

6.38 **Nicotine in cigarettes.** To determine whether the mean nicotine content of a brand of cigarettes is greater than the advertised value of 1.4 milligrams, a health advocacy group tests

$$H_0 : \mu = 1.4$$

$$H_a : \mu > 1.4$$

The calculated value of the test statistic is $z = 2.42$.

(a) Is the result significant at the 5% level?

(b) Is the result significant at the 1% level?

(c) By comparing z with the critical values in the bottom row of Table C, give two numbers that catch the P-value between them.

Tests from confidence intervals

The calculation in Example 6.16 for a 1% significance test is very similar to that in Example 6.4 for a 99% confidence interval. In fact, a two-sided test at significance level α can be carried out directly from a confidence interval with confidence level $C = 1 - \alpha$.

CONFIDENCE INTERVALS AND TWO-SIDED TESTS

A level α two-sided significance test rejects a hypothesis $H_0 : \mu = \mu_0$ exactly when the value μ_0 falls outside a level $1 - \alpha$ confidence interval for μ.

EXAMPLE 6.17   **Tests from a confidence interval**

The 99% confidence interval for μ in Example 6.4 is

$$\bar{x} \pm z^* \frac{\sigma}{\sqrt{n}} = 0.8404 \pm 0.0101$$

$$= 0.8303 \text{ to } 0.8505$$

The hypothesized value $\mu_0 = 0.86$ in Example 6.16 falls outside this confidence interval, so we reject

$$H_0 : \mu = 0.86$$

at the 1% significance level. On the other hand, we cannot reject

$$H_0 : \mu = 0.85$$

at the 1% level in favor of the two-sided alternative $H_a : \mu \neq 0.85$, because 0.85 lies inside the 99% confidence interval for μ. Figure 6.17 illustrates both cases.

Figure 6.17 *Values of μ falling outside a 99% confidence interval can be rejected at the 1% significance level. Values falling inside the interval cannot be rejected.*

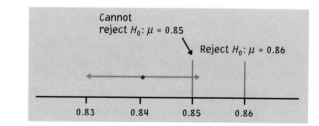

APPLY YOUR KNOWLEDGE

6.39   **IQ test scores.** Here are the IQ test scores of 31 seventh-grade girls in a Midwest school district:[9]

114	100	104	89	102	91	114	114	103	105
108	130	120	132	111	128	118	119	86	72
111	103	74	112	107	103	98	96	112	112	93

Treat the 31 girls as an SRS of all seventh-grade girls in the school district. Suppose that the standard deviation of IQ scores in this population is known to be $\sigma = 15$.

(a) Give a 95% confidence interval for the mean IQ score μ in the population.

(b) Is there significant evidence at the 5% level that the mean IQ score in the population differs from 100? State hypotheses and use your confidence interval to answer the question without more calculations.

SECTION 6.2 Summary

A **test of significance** assesses the evidence provided by data against a **null hypothesis** H_0 in favor of an **alternative hypothesis** H_a.

The hypotheses are stated in terms of population parameters. Usually H_0 is a statement that no effect is present, and H_a says that a parameter differs from its null value in a specific direction (**one-sided alternative**) or in either direction (**two-sided alternative**).

The essential reasoning of a significance test is as follows. Suppose for the sake of argument that the null hypothesis is true. If we repeated our data production many times, would we often get data as inconsistent with H_0 as the data we actually have? If the data are unlikely when H_0 is true, they provide evidence against H_0.

A test is based on a **test statistic**. The **P-value** is the probability, computed supposing H_0 to be true, that the test statistic will take a value at least as extreme as that actually observed. Small P-values indicate strong evidence against H_0. Calculating P-values requires knowledge of the sampling distribution of the test statistic when H_0 is true.

If the P-value is as small or smaller than a specified value α, the data are **statistically significant** at significance level α.

Significance tests for the hypothesis $H_0 : \mu = \mu_0$ concerning the unknown mean μ of a population are based on the **one-sample** z statistic

$$z = \frac{\bar{x} - \mu_0}{\sigma/\sqrt{n}}$$

The z test assumes an SRS of size n, known population standard deviation σ, and either a normal population or a large sample. P-values are computed from the normal distribution (Table A). Fixed α tests use the table of standard normal **critical values** (bottom row of Table C).

SECTION 6.2 Exercises

6.40 **Studying job satisfaction.** The job satisfaction study of Example 6.10 (page 324) measured the JDS job satisfaction score of 28 female assemblers doing both self-paced and machine-paced work. The parameter μ is the mean amount by which the self-paced score exceeds the machine-paced score in the population of all such workers. Scores are normally distributed. The population standard deviation is $\sigma = 0.60$. The hypotheses are

$$H_0 : \mu = 0$$

$$H_a : \mu \neq 0$$

(a) What is the sampling distribution of the mean JDS score $\bar{x}$ for 28 workers if the null hypothesis is true? Sketch the density curve of this distribution. (Hint: Sketch a normal curve first, then mark the axis using what you know about locating μ and σ on a normal curve.)

(b) Suppose that the study had found $\bar{x} = 0.09$. Mark this point on the axis in your sketch. In fact, the study found $\bar{x} = 0.27$ for these 28 workers. Mark this point on your sketch. Referring to your sketch, explain in simple language why one result is good evidence that H_0 is not true, and why the other is not.

(c) Make another copy of your sketch. Shade the area under the curve that gives the P-value for the result $\bar{x} = 0.09$. Then calculate this P-value. (Note that H_a is two-sided.)

(d) Calculate the P-value for the result $\bar{x} = 0.27$ also. The two P-values express your explanation in (b) in numbers.

6.41 The mean area of the several thousand apartments in a new development is advertised to be 1250 square feet. A tenant group thinks that the apartments are smaller than advertised. They hire an engineer to measure a sample of apartments to test their suspicion. What are the null hypothesis H_0 and alternative hypothesis H_a?

6.42 Experiments on learning in animals sometimes measure how long it takes mice to find their way through a maze. The mean time is 18 seconds for one particular maze. A researcher thinks that a loud noise will cause the mice to complete the maze faster. She measures how long each of 10 mice takes with a noise as stimulus. What are the null hypothesis H_0 and alternative hypothesis H_a?

6.43 **Is this milk watered down?** Cobra Cheese Company buys milk from several suppliers. Cobra suspects that some producers are adding water to their milk to increase their profits. Excess water can be detected by measuring the freezing point of the milk. The freezing temperature of natural milk varies normally, with mean $\mu = -0.545°$ Celsius (C) and standard deviation $\sigma = 0.008°C$. Added water raises the freezing temperature toward 0°C, the freezing point of water. Cobra's laboratory manager measures the freezing temperature of five consecutive lots of milk from one producer. The mean measurement is $\bar{x} = -0.538°C$. Is this good evidence that the producer is adding water to the milk? State hypotheses, carry out the test, give the P-value, and state your conclusion.

6.44   **CEO pay.** A study of the pay of corporate chief executive officers (CEOs) examined the increase in cash compensation of the CEOs of 104 companies, adjusted for inflation, in a recent year. The mean increase in real compensation was $\bar{x} = 6.9\%$ and the standard deviation of the increases was $s = 55\%$. Is this good evidence that the mean real compensation μ of all CEOs increased that year? The hypotheses are

$$H_0 : \mu = 0 (\text{no increase})$$

$$H_a : \mu > 0 (\text{an increase})$$

Because the sample size is large, the sample s is close to the population σ, so take $\sigma = 55\%$.

(a) Sketch the normal curve for the sampling distribution of $\bar{x}$ when H_0 is true. Shade the area that represents the P-value for the observed outcome $\bar{x} = 6.9\%$.

(b) Calculate the P-value.

(c) Is the result significant at the $\alpha = 0.05$ level? Do you think the study gives strong evidence that the mean compensation of all CEOs went up?

6.45   A social psychologist reports: "In our sample, ethnocentrism was significantly higher ($P < 0.05$) among church attenders than among nonattenders." Explain what this means in language understandable to someone who knows no statistics. Do not use the word "significance" in your answer.

6.46   There are other z statistics that we have not yet studied. You can use Table C to assess the significance of any z statistic. A study compares American-Japanese joint ventures in which the U.S. company is larger than its Japanese partner with joint ventures in which the U.S. company is smaller. One variable measured is the excess returns earned by shareholders in the American company. The null hypothesis is "no difference" between the means for the two populations. The alternative hypothesis is two-sided. The value of the test statistic is $z = -1.37$.

(a) Is this result significant at the 5% level?

(b) Is the result significant at the 10% level?

6.47   Use Table C to find the approximate P-value for both values of $\bar{x}$ in Exercise 6.40 without calculation. That is, between what two numbers obtained from the table must each P-value lie?

6.48   Use Table C to approximate the P-value in Exercise 6.44 without a calculation. That is, give two values from the table between which P must lie.

6.49   Between what values from Table C does the P-value for the outcome $z = -1.37$ in Exercise 6.46 lie? (Remember that H_a is two-sided.)

Calculate the *P*-value using Table A, and verify that it lies between the values you found from Table C.

6.50 **Benefits of patent protection?** Market pioneers, companies that are among the first to develop a new product or service, tend to have higher market shares than latecomers to the market. What accounts for this advantage? Here is an excerpt from the conclusions of a study of a sample of 1209 manufacturers of industrial goods:

> *Can patent protection explain pioneer share advantages? Only 21% of the pioneers claim a significant benefit from either a product patent or a trade secret. Though their average share is two points higher than that of pioneers without this benefit, the increase is not statistically significant ($z = 1.13$). Thus, at least in mature industrial markets, product patents and trade secrets have little connection to pioneer share advantages.*[10]

Find the *P*-value for the given *z*. Then explain to someone who knows no statistics what "not statistically significant" in the study's conclusion means. Why does the author conclude that patents and trade secrets don't help, even though they contributed 2 percentage points to average market share?

6.51 **Does she look 25?** The cigarette industry has adopted a voluntary code requiring that models appearing in its advertising must appear to be at least 25 years old. Studies have shown, however, that consumers think many of the models are younger. Here is a quote from a study that asked whether different brands of cigarettes use models that appear to be of different ages.

(I. BERTRAND/EXPLORER/PHOTO RESEARCHERS.)

> *The ANCOVA revealed that the brand variable is highly significant ($P < 0.001$), indicating that the average perceived age of the models is not equal across the 12 brands. As discussed previously, certain brands such as Lucky Strike Lights, Kool Milds, and Virginia Slims tended to have younger models . . .*[11]

ANCOVA is an advanced statistical technique, but significance and *P*-values have their usual meaning. Explain to someone who knows no statistics what "highly significant ($P < 0.001$)" means and why this is good evidence of differences among all advertisements of these brands even though the subjects saw only a sample of ads.

6.52 Explain in plain language why a significance test that is significant at the 1% level must always be significant at the 5% level. If a test is significant at the 5% level, what can you say about its significance at the 1% level?

6.53 Asked to explain the meaning of "statistically significant at the $\alpha = 0.05$ level," a student says: "This means that the probability that the null hypothesis is true is less than 0.05." Is this explanation correct? Why or why not?

6.3 Making Sense of Statistical Significance

Significance tests are widely used in reporting the results of research in many fields of applied science and in industry. New pharmaceutical products require significant evidence of effectiveness and safety. Courts inquire about statistical significance in hearing class action discrimination cases. Marketers want to know whether a new ad campaign significantly outperforms the old one, and medical researchers want to know whether a new therapy performs significantly better. In all these uses, statistical significance is valued because it points to an effect that is unlikely to occur simply by chance.

Carrying out a test of significance is often quite simple, especially if you get a P-value effortlessly from a calculator or computer. Using tests wisely is not so simple. Here are some points to keep in mind when using or interpreting significance tests.

How small a P is convincing?

The purpose of a test of significance is to describe the degree of evidence provided by the sample against the null hypothesis. The P-value does this. But how small a P-value is convincing evidence against the null hypothesis? This depends mainly on two circumstances:

- *How plausible is H_0?* If H_0 represents an assumption that the people you must convince have believed for years, strong evidence (small P) will be needed to persuade them.
- *What are the consequences of rejecting H_0?* If rejecting H_0 in favor of H_a means making an expensive changeover from one type of product packaging to another, you need strong evidence that the new packaging will boost sales.

These criteria are a bit subjective. Different people will often insist on different levels of significance. Giving the P-value allows each of us to decide individually if the evidence is sufficiently strong.

Users of statistics have often emphasized standard levels of significance such as 10%, 5%, and 1%. For example, courts have tended to accept 5% as a standard in discrimination cases.[12] This emphasis reflects the time when tables of critical values rather than computer software dominated statistical practice. The 5% level ($\alpha = 0.05$) is particularly common. **There is no sharp border between "significant" and "insignificant," only increasingly strong evidence as the P-value decreases.** There is no practical distinction between the P-values 0.049 and 0.051. It makes no sense to treat $P \leq 0.05$ as a universal rule for what is significant.

APPLY YOUR KNOWLEDGE

6.54 **Is it significant?** Suppose that in the absence of special preparation Scholastic Assessment Test mathematics (SATM) scores vary normally with mean $\mu = 475$ and $\sigma = 100$. One hundred students go

through a rigorous training program designed to raise their SATM scores by improving their mathematics skills. Carry out a test of

$$H_0 : \mu = 475$$
$$H_a : \mu > 475$$

in each of the following situations:

(a) The students' average score is $\overline{x} = 491.4$. Is this result significant at the 5% level?

(b) The average score is $\overline{x} = 491.5$. Is this result significant at the 5% level?

The difference between the two outcomes in (a) and (b) is of no importance. Beware attempts to treat $\alpha = 0.05$ as sacred.

Statistical significance and practical significance

When a null hypothesis ("no effect" or "no difference") can be rejected at the usual levels, $\alpha = 0.05$ or $\alpha = 0.01$, there is good evidence that an effect is present. But that effect may be very small. When large samples are available, even tiny deviations from the null hypothesis will be significant.

Should tests be banned?

Significance tests don't tell us how large or how important an effect is. Research psychologists have emphasized tests, so much so that some think their weaknesses should ban them from use. The American Psychological Association asked a group of experts. They said: use anything that sheds light on your study. Use more data analysis and confidence intervals. But: "The task force does not support any action that could be interpreted as banning the use of null hypothesis significance testing or *P*-values in psychological research and publication."

> **EXAMPLE 6.18   It's significant. So what?**
>
> We are testing the hypothesis of no correlation between two variables. With 1000 observations, an observed correlation of only $r = 0.08$ is significant evidence at the $\alpha = 0.01$ level that the correlation in the population is not zero but positive. The low significance level does not mean there is a strong association, only that there is strong evidence of some association. The true population correlation is probably quite close to the observed sample value, $r = 0.08$. We might well conclude that for practical purposes we can ignore the association between these variables, even though we are confident (at the 1% level) that the correlation is positive.

Exercise 6.55 demonstrates in detail how increasing the sample size drives down *P*. Remember the wise saying: **Statistical significance is not the same thing as practical significance**.

The remedy for attaching too much importance to statistical significance is to pay attention to the actual data as well as to the *P*-value. Plot your data and examine them carefully. Are there outliers or other deviations from a consistent pattern? A few outlying observations can produce highly significant results if you blindly apply common tests of significance. Outliers can also destroy the significance of otherwise convincing data. The foolish user of statistics who feeds the data to a computer without exploratory analysis will often be embarrassed. Is the effect you are seeking visible in your plots? If not, ask yourself if the effect is large enough to be practically important. Give a confidence interval for the parameter in which you are interested. A confidence interval actually estimates the size of an effect, rather than simply asking if it is too large to reasonably occur by chance alone. Confidence intervals are not used as often as they should be, while tests of significance are perhaps overused.

6.55   **Coaching and the SAT.** Suppose that SATM scores in the absence of coaching vary normally with mean $\mu = 475$ and $\sigma = 100$. Suppose also that coaching may change μ but does not change σ. An increase in the SATM score from 475 to 478 is of no importance in seeking admission to college, but this unimportant change can be statistically very significant. To see this, calculate the P-value for the test of

$$H_0 : \mu = 475$$
$$H_a : \mu > 475$$

in each of the following situations:

(a) A coaching service coaches 100 students. Their SATM scores average $\overline{x} = 478$.

(b) By the next year, the service has coached 1000 students. Their SATM scores average $\overline{x} = 478$.

(c) An advertising campaign brings the number of students coached to 10,000. Their average score is still $\overline{x} = 478$.

6.56   Give a 99% confidence interval for the mean SATM score μ after coaching in each part of the previous exercise. For large samples, the confidence interval tells us, "Yes, the mean score is higher than 475 after coaching, but only by a small amount."

Statistical inference is not valid for all sets of data

We know that badly designed surveys or experiments often produce useless results. Formal statistical inference cannot correct basic flaws in the design. A statistical test is valid only in certain circumstances, with properly produced data being particularly important. The z test, for example, should bear the same warning label that we attached on page 312 to the z confidence interval. Similar warnings accompany the other tests that we will learn.

EXAMPLE 6.19   Mammary artery ligation

Angina is the severe pain caused by inadequate blood supply to the heart. Perhaps we can relieve angina by tying off the mammary arteries to force the body to develop other routes to supply blood to the heart. Surgeons tried this procedure, called "mammary artery ligation." Patients reported a statistically significant reduction in angina pain.

Statistical significance says that something other than chance is at work, but it does not say what that something is. The mammary artery ligation experiment was uncontrolled, so that the reduction in pain might be nothing more than the placebo effect. Sure enough, a randomized comparative experiment showed that ligation was no more effective than a placebo. Surgeons abandoned the operation at once.

Tests of significance and confidence intervals are based on the laws of probability. Randomization in sampling or experimentation ensures that these laws apply. Yet we must often analyze data that do not arise from randomized samples or experiments. To apply statistical inference to such data, we must have confidence in the use of probability to describe the data. The diameters of successive holes bored in auto engine blocks during production, for example, may behave like a random sample from a normal distribution. We can check this probability model by examining the data. If the model appears correct, we can apply the recipes of this chapter to do inference about the process mean diameter μ. Always ask how the data were produced, and don't be too impressed by P-values on a printout until you are confident that the data deserve a formal analysis.

APPLY YOUR KNOWLEDGE

6.57 **A call-in opinion poll.** A local television station announces a question for a call-in opinion poll on the six o'clock news, then gives the response on the eleven o'clock news. Today's question concerns a proposed gun-control ordinance. Of the 2372 calls received, 1921 oppose the new law. The station, following standard statistical practice, makes a confidence statement: "81% of the Channel 13 Pulse Poll sample oppose gun control. We can be 95% confident that the proportion of all viewers who oppose the law is within 1.6% of the sample result." Is the station's conclusion justified? Explain your answer.

Beware of multiple analyses

Statistical significance ought to mean that you have found an effect that you were looking for. The reasoning behind statistical significance works well if you decide what effect you are seeking, design a study to search for it, and use a test of significance to weigh the evidence you get. In other settings, significance may have little meaning.

EXAMPLE 6.20 Predicting success of trainees

You want to learn what distinguishes managerial trainees who eventually become executives from those who, after expensive training, don't succeed and leave the company. You have abundant data on past trainees—data on their personalities and goals, their college preparation and performance, even their family backgrounds and their hobbies. Statistical software makes it easy to perform dozens of significance tests on these dozens of variables to see which ones best predict later success. Aha! You find that future executives are significantly more likely than washouts to have an urban or suburban upbringing and an undergraduate degree in a technical field.

Before you base future recruiting on these findings, recall that results significant at the 5% level occur 5 times in 100 in the long run even when H_0 is true. When you make dozens of tests at the 5% level, you expect a few of them to be significant by chance alone. Running one test and reaching the $\alpha = 0.05$ level is reasonably good evidence that you have found something. Running several dozen tests and reaching that level once or twice is not.

Rather than testing all of the variables in the trainee data, you might search for the one variable with the biggest difference between future washouts and future executives, then test whether that difference is significant. This is also bad statistics. The P-value assumes you had one specific variable in mind before you looked at the data. It is cheating to choose the biggest of many differences and then act as if you had this one variable in mind all along.

Searching data for suggestive patterns is certainly legitimate. Exploratory data analysis is an important part of statistics. But the reasoning of formal inference does not apply when your search for a striking effect in the data is successful. The remedy is clear. Once you have a hypothesis, design a study to search specifically for the effect you now think is there. If the result of this study is statistically significant, you have real evidence.

APPLY YOUR KNOWLEDGE

6.58 **Searching for ESP.** A researcher looking for evidence of extrasensory perception (ESP) tests 500 subjects. Four of these subjects do significantly better ($P < 0.01$) than random guessing.

(a) Is it proper to conclude that these four people have ESP? Explain your answer.

(b) What should the researcher now do to test whether any of these four subjects have ESP?

SECTION 6.3 Summary

There is no universal rule for how small a P-value is convincing. Beware of placing too much weight on traditional significance levels such as $\alpha = 0.05$.

Very small effects can be highly significant (small P) when a test is based on a large sample. A statistically significant effect need not be practically important. Plot the data to display the effect you are seeking, and use confidence intervals to estimate the actual values of parameters.

On the other hand, lack of significance does not imply that H_0 is true, especially when the test is based on just a few observations.

Significance tests are not always valid. Faulty data collection, outliers in the data, and testing a hypothesis on the same data that suggested the hypothesis can invalidate a test.

Many tests run at once will probably produce some significant results by chance alone, even if all the null hypotheses are true.

SECTION 6.3 Exercises

6.59 **What is significance good for?** Which of the following questions does a test of significance answer?

(a) Is the sample or experiment properly designed?

(b) Is the observed effect due to chance?

(c) Is the observed effect important?

6.60 **Radar detectors and speeding (EESEE).** Researchers observed the speed of cars on a rural highway (speed limit 55 miles per hour) before and after police radar was directed at them. They compared the speed of cars that had radar detectors with the speed of cars without detectors. Here are the mean speeds (miles per hour) observed for 22 cars with radar detectors and 46 cars without:[13]

	Radar Detector?	
	Yes	No
Before radar	70	68
At radar	59	67

The study report says: "Those vehicles with radar detectors were significantly faster ($P < 0.01$) than those without them before radar exposure . . . and significantly slower immediately after ($P < 0.0001$)."

(a) Explain in simple language why these P-values are good evidence that drivers with radar detectors do behave differently.

(b) Despite the fact that $P < 0.01$, there is little difference in the mean speeds of drivers with and without radar detectors before the radar is turned on. Explain in simple language how so small a difference can be statistically significant.

6.61 A company compares two package designs for a laundry detergent by placing bottles with both designs on the shelves of several markets. Checkout scanner data on more than 5000 bottles bought show that more shoppers bought Design A than Design B. The difference is statistically significant ($P = 0.02$). Can we conclude that consumers strongly prefer Design A? Explain your answer.

6.62 **What distinguishes schizophrenics?** A group of psychologists once measured 77 variables on a sample of schizophrenic people and a sample of people who were not schizophrenic. They compared the two samples using 77 separate significance tests. Two of these tests were significant at the 5% level. Suppose that there is in fact no difference on any of the 77 variables between people who are and people who are not schizophrenic in the adult population. Then all 77 null hypotheses are true.

(a) What is the probability that one specific test shows a difference significant at the 5% level?

(b) Why is it not surprising that 2 of the 77 tests were significant at the 5% level?

6.4 Error Probabilities and Power*

Tests of significance assess the strength of evidence against the null hypothesis. We measure evidence by the P-value, which is a probability computed under the assumption that H_0 is true. The alternative hypothesis (the statement we seek evidence for) enters the test only to help us see what outcomes count against the null hypothesis.

Using significance tests with fixed level α, however, suggests another way of thinking. A level of significance α chosen in advance points to the outcome of the test as a *decision*. If our result is significant at level α, we reject H_0 in favor of H_a. Otherwise, we fail to reject H_0. The transition from measuring the strength of evidence to making a decision is not a small step. Many statisticians feel that making decisions should be left to the user rather than built into the statistical test. A test result is only one among many factors that influence a decision.

Yet there are circumstances that call for a decision or action as the end result of inference. **Acceptance sampling** is one such circumstance. A producer of bearings and the consumer of the bearings agree that each carload lot must meet certain quality standards. When a carload arrives, the consumer inspects a sample of the bearings. On the basis of the sample outcome, the consumer either accepts or rejects the carload. We will use acceptance sampling to introduce new ways to describe how a test performs.

acceptance sampling

Type I and Type II errors

In the acceptance sampling problem, we must decide between

H_0: the lot of bearings meets standards

H_a: the lot does not meet standards

on the basis of a sample of bearings. We hope that our decision will be correct, but sometimes it will be wrong. There are two types of incorrect decisions. We can accept a bad lot of bearings, or we can reject a good lot. Accepting a

---

*This more advanced section introduces additional ideas used to describe statistical tests. It is not needed to read the rest of the book.

Figure 6.18 *The two types of error in testing hypotheses.*

		H_0 true	H_a true
Decision based on sample	Reject H_0	Type I error	Correct decision
	Accept H_0	Correct decision	Type II error

Truth about the population

bad lot injures the consumer, while rejecting a good lot hurts the producer. To distinguish these two types of error, we give them specific names.

> **TYPE I AND TYPE II ERRORS**
>
> If we reject H_0 when in fact H_0 is true, this is a **Type I error**.
> If we fail to reject H_0 when in fact H_a is true, this is a **Type II error**.

The possibilities are summed up in Figure 6.18. If H_0 is true, our decision is correct if we accept H_0 and is a Type I error if we reject H_0. If H_a is true, our decision is either correct or a Type II error. Only one error is possible at one time.

Error probabilities

Significance tests with fixed level α either reject H_0 or fail to reject it. If we think of the test as making a decision, failing to reject H_0 means deciding that H_0 is true. We can then describe the performance of a test by the probabilities of Type I and Type II errors. This is in keeping with the idea that statistical inference is based on asking, "What would happen if I used this procedure many times?"

EXAMPLE 6.21 Are these bearings OK?

The mean diameter of a type of bearing is supposed to be 2.000 centimeters (cm). The bearing diameters vary normally with standard deviation $\sigma = 0.010$ cm. When a lot of the bearings arrives, the consumer takes an SRS of 5 bearings from the lot and measures their diameters. The consumer rejects the bearings if the sample mean diameter is significantly different from 2 at the 5% significance level.

This is a test of the hypotheses

$$H_0 : \mu = 2$$

$$H_a : \mu \neq 2$$

To carry out the test, the consumer computes the z statistic

$$z = \frac{\bar{x} - 2}{0.01/\sqrt{5}}$$

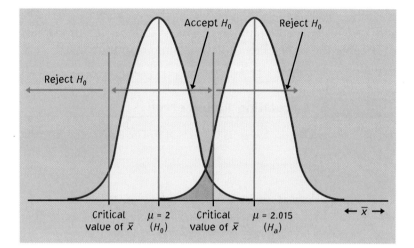

Figure 6.19 *The two error probabilities for Example 6.21. The probability of a Type I error (light shaded area) is the probability of rejecting $H_0 : \mu = 2$ when in fact $\mu = 2$. The probability of a Type II error (dark shaded area) is the probability of accepting H_0 when in fact $\mu = 2.015$.*

and rejects H_0 if $z < -1.96$ or $z > 1.96$. A Type I error is to reject H_0 when in fact $\mu = 2$.

What about Type II errors? Because there are many values of μ in H_a, we will concentrate on one value. The producer and the consumer agree that a lot of bearings with mean diameter 2.015 cm should be rejected. So a particular Type II error is to accept H_0 when in fact $\mu = 2.015$.

Figure 6.19 shows how the two probabilities of error are obtained from the *two* sampling distributions of $\bar{x}$, for $\mu = 2$ and for $\mu = 2.015$. When $\mu = 2$, H_0 is true and to reject H_0 is a Type I error. When $\mu = 2.015$, H_a is true and to accept H_0 is a Type II error. We will now calculate these error probabilities.

The probability of a Type I error is the probability of rejecting H_0 when it is really true. This is the probability that $|z| \geq 1.96$ when $\mu = 2$. But this is exactly the significance level of the test. The critical value 1.96 was chosen to make this probability 0.05, so we do not have to compute it again. The definition of "significance level 0.05" is that values of z this extreme will occur with probability 0.05 when H_0 is true.

SIGNIFICANCE AND TYPE I ERROR

The significance level α of any fixed level test is the probability of a Type I error. That is, α is the probability that the test will reject the null hypothesis H_0 when H_0 is in fact true.

The probability of a Type II error for the particular alternative $\mu = 2.015$ in Example 6.21 is the probability that the test will accept H_0 when μ has this alternative value. This is the probability that the test statistic z falls between -1.96 and 1.96, calculated assuming that $\mu = 2.015$. This probability is *not* $1 - 0.05$, because the probability 0.05 was found assuming that $\mu = 2$. Here is the calculation for Type II error.

EXAMPLE 6.22 | **Calculating Type II error**

Step 1. *Write the rule for accepting H_0 in terms of $\bar{x}$.* The test accepts H_0 when

$$-1.96 \leq \frac{\bar{x} - 2}{0.01/\sqrt{5}} \leq 1.96$$

This is the same as

$$2 - 1.96\left(\frac{0.01}{\sqrt{5}}\right) \leq \bar{x} \leq 2 + 1.96\left(\frac{0.01}{\sqrt{5}}\right)$$

or, doing the arithmetic,

$$1.9912 \leq \bar{x} \leq 2.0088$$

This step does not involve the particular alternative $\mu = 2.015$.

Step 2. *Find the probability of accepting H_0 assuming that the alternative is true.* Take $\mu = 2.015$ and standardize to find the probability.

$$P(\text{Type II error}) = P(1.9912 \leq \bar{x} \leq 2.0088)$$

$$= P\left(\frac{1.9912 - 2.015}{0.01/\sqrt{5}} \leq \frac{\bar{x} - 2.015}{0.01/\sqrt{5}} \leq \frac{2.0088 - 2.015}{0.01/\sqrt{5}}\right)$$

$$= P(-5.32 \leq Z \leq -1.39)$$

$$= 0.0823$$

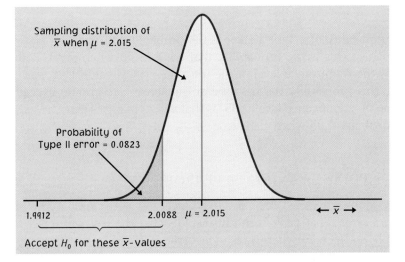

Figure 6.20 *The probability of a Type II error for Example 6.22. This is the probability that the test accepts H_0 when the alternative hypothesis is true.*

Figure 6.20 illustrates this error probability in terms of the sampling distribution of $\bar{x}$ when $\mu = 2.015$. The test will wrongly accept the hypothesis that $\mu = 2$ in about 8% of all samples when in fact $\mu = 2.015$.

This test will reject 5% of all good lots of bearings (for which $\mu = 2$). It will accept 8% of lots so bad that $\mu = 2.015$. Calculations of error probabilities help the producer and consumer decide whether the test is satisfactory.

APPLY YOUR KNOWLEDGE

6.63 Your company markets a computerized medical diagnostic program. The program scans the results of routine medical tests (pulse rate, blood tests, etc.) and either clears the patient or refers the case to a doctor. The program is used to screen thousands of people who do not have specific medical complaints. The program makes a decision about each person.

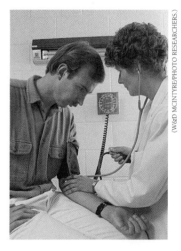

(W&D MCINTYRE/PHOTO RESEARCHERS.)

(a) What are the two hypotheses and the two types of error that the program can make? Describe the two types of error in terms of "false positive" and "false negative" test results.

(b) The program can be adjusted to decrease one error probability, at the cost of an increase in the other error probability. Which error probability would you choose to make smaller, and why? (This is a matter of judgment. There is no single correct answer.)

6.64 You have the NAEP quantitative scores for an SRS of 840 young men. You plan to test hypotheses about the population mean score,

$$H_0 : \mu = 275$$

$$H_a : \mu < 275$$

at the 1% level of significance. The population standard deviation is known to be $\sigma = 60$. The z test statistic is

$$z = \frac{\bar{x} - 275}{60/\sqrt{840}}$$

(a) What is the rule for rejecting H_0 in terms of z?

(b) What is the probability of a Type I error?

(c) You want to know whether this test will usually reject H_0 when the true population mean is 270, 5 points lower than the null hypothesis claims. Answer this question by calculating the probability of a Type II error when $\mu = 270$.

6.65 You have an SRS of size $n = 9$ from a normal distribution with $\sigma = 1$. You wish to test

$$H_0 : \mu = 0$$

$$H_a : \mu > 0$$

You decide to reject H_0 if $\bar{x} > 0$ and to accept H_0 otherwise.

(a) Find the probability of a Type I error. That is, find the probability that the test rejects H_0 when in fact $\mu = 0$.

(b) Find the probability of a Type II error when $\mu = 0.3$. This is the probability that the test accepts H_0 when in fact $\mu = 0.3$.

(c) Find the probability of a Type II error when $\mu = 1$.

Power

A test makes a Type II error when it fails to reject a null hypothesis that really is false. A high probability of a Type II error for a particular alternative means that the test is not sensitive enough to usually detect that alternative. Calculations of the probability of Type II errors are therefore useful even if you don't think that a statistical test should be viewed as making a decision. It is usual to report the probability that a test *does* reject H_0 when an alternative is true. The higher this probability is, the more sensitive the test is.

Fish, fishermen, and Type II errors

Are the stocks of cod in the ocean off eastern Canada declining? Studies over many years failed to find significant evidence of a decline. These studies had low power— that is, they might fail to find a decline even if one was present. When it became clear that the cod were vanishing, quotas on fishing ravaged the economy in parts of Canada. It appears that the earlier studies had made Type II errors. If they had seen the decline, early action might have reduced the economic and environmental costs.

> **POWER**
>
> The probability that a fixed level α significance test will reject H_0 when a particular alternative value of the parameter is true is called the **power** of the test against that alternative.
>
> The power of a test against any alternative is 1 minus the probability of a Type II error for that alternative.

Calculations of power are essentially the same as calculations of the probability of a Type II error. In Example 6.22, the power is the probability of *rejecting* H_0 in Step 2 of the calculation. It is $1 - 0.0823$, or 0.9177.

Calculations of P-values and calculations of power both say what would happen if we repeated the test many times. A P-value describes what would happen supposing that the null hypothesis is true. Power describes what would happen supposing that a particular alternative is true.

In planning an investigation that will include a test of significance, a careful user of statistics decides what alternatives the test should detect and checks that the power is adequate. If the power is too low, a larger sample size will increase the power for the same significance level α. In order to calculate power, we must fix an α so that there is a fixed rule for rejecting H_0. We prefer to report P-values rather than to use a fixed significance level. The usual practice is to calculate the power at a common significance level such as $\alpha = 0.05$ even though you intend to report a P-value.

6.66 The cola maker of Example 6.8 (page 319) determines that a sweetness loss is too large to accept if the mean response for all tasters is $\mu = 1.1$. Will a 5% significance test of the hypotheses

$$H_0 : \mu = 0$$
$$H_a : \mu > 0$$

based on a sample of 10 tasters usually detect a change this great?
We want the power of the test against the alternative $\mu = 1.1$. This is the probability that the test rejects H_0 when $\mu = 1.1$ is true. The calculation method is similar to that for Type II error.

(a) **Step 1.** *Write the rule for rejecting H_0 in terms of $\bar{x}$.* We know that $\sigma = 1$, so the test rejects H_0 at the $\alpha = 0.05$ level when

$$z = \frac{\bar{x} - 0}{1/\sqrt{10}} \geq 1.645$$

Restate this in terms of $\bar{x}$.

(b) **Step 2.** *The power is the probability of this event supposing that the alternative is true.* Standardize using $\mu = 1.1$ to find the probability that $\bar{x}$ takes a value that leads to rejection of H_0.

6.67 Exercise 6.36 (page 333) concerns a test about the mean contents of cola bottles. The hypotheses are

$$H_0 : \mu = 300$$
$$H_a : \mu < 300$$

The sample size is $n = 6$, and the population is assumed to have a normal distribution with $\sigma = 3$. A 5% significance test rejects H_0 if $z \leq -1.645$, where the test statistic z is

$$z = \frac{\bar{x} - 300}{3/\sqrt{6}}$$

Power calculations help us see how large a shortfall in the bottle contents the test can be expected to detect.

(a) Find the power of this test against the alternative $\mu = 299$.

(b) Find the power against the alternative $\mu = 295$.

(c) Is the power against $\mu = 290$ higher or lower than the value you found in (b)? (Don't actually calculate that power.) Explain your answer.

6.68 Increasing the sample size increases the power of a test when the level α is unchanged. Suppose that in the previous exercise a sample of n bottles had been measured. In that exercise, $n = 6$. The 5% significance test still rejects H_0 when $z \leq -1.645$, but the z statistic is now

$$z = \frac{\bar{x} - 300}{3/\sqrt{n}}$$

(a) Find the power of this test against the alternative $\mu = 299$ when $n = 25$.

(b) Find the power against $\mu = 299$ when $n = 100$.

SECTION 6.4 Summary

Significance testing with fixed level α is sometimes used to decide which of H_0 and H_a should be accepted.

We describe the performance of a test at fixed level α by giving the probabilities of two types of error. A **Type I error** occurs if we reject H_0 when it is in fact true. A **Type II error** occurs if we fail to reject H_0 when in fact H_a is true.

The **power** of a significance test measures its ability to detect an alternative hypothesis. The power against a specific alternative is the probability that the test will reject H_0 when the alternative is true.

In a fixed level α significance test, the significance level α is the probability of a Type I error, and the power against a specific alternative is 1 minus the probability of a Type II error for that alternative.

Increasing the size of the sample increases the power (reduces the probability of a Type II error) when the significance level remains fixed.

SECTION 6.4 Exercises

6.69 Power calculations for two-sided tests follow the same outline as for one-sided tests. Example 6.16 (page 335) presents a test of

$$H_0 : \mu = 0.86$$
$$H_a : \mu \neq 0.86$$

at the 1% level of significance. The sample size is $n = 3$ and $\sigma = 0.0068$. We will find the power of this test against the alternative $\mu = 0.845$.

(a) The test in Example 6.16 rejects H_0 when $|z| \geq 2.576$. The test statistic z is

$$z = \frac{\bar{x} - 0.86}{0.0068/\sqrt{3}}$$

Write the rule for rejecting H_0 in terms of the values of $\bar{x}$. (Because the test is two-sided, it rejects when $\bar{x}$ is either too large or too small.)

(b) Now find the probability that $\bar{x}$ takes values that lead to rejecting H_0 if the true mean is $\mu = 0.845$. This probability is the power.

(c) What is the probability that this test makes a Type II error when $\mu = 0.845$?

6.70 In Example 6.13 (page 330), a company medical director failed to find significant evidence that the mean blood pressure of a population of executives differed from the national mean $\mu = 128$. The medical director now wonders if the test used would detect an important difference if one were present. For the SRS of size 72 from a population with standard deviation $\sigma = 15$, the z statistic is

$$z = \frac{\bar{x} - 128}{15/\sqrt{72}}$$

The two-sided test rejects $H_0 : \mu = 128$ at the 5% level of significance when $|z| \geq 1.96$.

(a) Find the power of the test against the alternative $\mu = 134$.

(b) Find the power of the test against $\mu = 122$. Can the test be relied on to detect a mean that differs from 128 by 6?

(c) If the alternative were farther from H_0, say $\mu = 136$, would the power be higher or lower than the values calculated in (a) and (b)?

6.71 In Exercise 6.67 you found the power of a test against the alternative $\mu = 295$. Use the result of that exercise to find the probabilities of Type I and Type II errors for that test and that alternative.

6.72 In Exercise 6.64 you found the probabilities of the two types of error for a test of $H_0 : \mu = 275$, with the specific alternative that $\mu = 270$. Use the result of that exercise to give the power of the test against the alternative $\mu = 270$.

6.73 **Is the stock market efficient?** You are reading an article in a business journal that discusses the "efficient market hypothesis" for the behavior of securities prices. The author admits that most tests of this hypothesis have failed to find significant evidence against it. But he says this failure is a result of the fact that the tests used have low power. "The widespread impression that there is strong evidence for market efficiency may be due just to a lack of appreciation of the low power of many statistical tests."[14]

Explain in simple language why tests having low power often fail to give evidence against a hypothesis even when the hypothesis is really false.

STATISTICS in SUMMARY

Statistical inference draws conclusions about a population on the basis of sample data and uses probability to indicate how reliable the conclusions are. A confidence interval estimates an unknown parameter. A significance test shows how strong the evidence is for some claim about a parameter.

The probabilities in both confidence intervals and tests tell us what would happen if we used the recipe for the interval or test very many times. A confidence level is the probability that the recipe for a confidence interval actually produces an interval that contains the unknown parameter. A 95% confidence interval gives a correct result 95% of the time when we use it repeatedly. A P-value is the probability that the test would produce a result at least as extreme as the observed result if the null hypothesis really were true. That is, a P-value tells us how surprising the observed outcome is. Very surprising outcomes (small P-values) are good evidence that the null hypothesis is not true.

The Statistics in Summary figures on the next page present the ideas of confidence intervals and tests in picture form. These ideas are the foundation for the rest of this book. We will have much to say about many statistical methods and their use in practice. In every case, the basic reasoning of confidence intervals and significance tests remains the same. Here are the most important things you should be able to do after studying this chapter.

A. CONFIDENCE INTERVALS

1. State in nontechnical language what is meant by "95% confidence" or other statements of confidence in statistical reports.

2. Calculate a confidence interval for the mean μ of a normal population with known standard deviation σ, using the recipe $\bar{x} \pm z^* \sigma / \sqrt{n}$.

3. Recognize when you can safely use this confidence interval recipe and when the data collection design or a small sample from a skewed population makes it inaccurate. Understand that the margin of error does not include the effects of undercoverage, nonresponse, or other practical difficulties.

4. Understand how the margin of error of a confidence interval changes with the sample size and the level of confidence C.

5. Find the sample size required to obtain a confidence interval of specified margin of error m when the confidence level and other information are given.

B. SIGNIFICANCE TESTS

1. State the null and alternative hypotheses in a testing situation when the parameter in question is a population mean μ.

STATISTICS IN SUMMARY
The Idea of a Confidence Interval

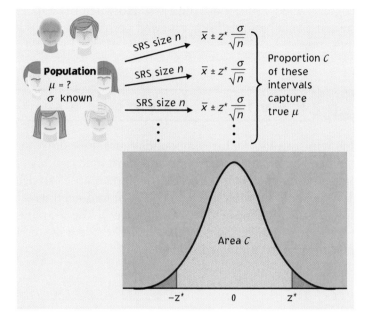

STATISTICS IN SUMMARY
The Idea of a Significance Test

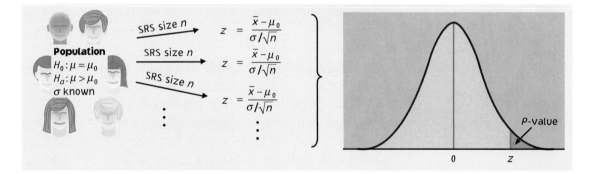

2. Explain in nontechnical language the meaning of the P-value when you are given the numerical value of P for a test.

3. Calculate the one-sample z statistic and the P-value for both one-sided and two-sided tests about the mean μ of a normal population.

4. Assess statistical significance at standard levels α, either by comparing P to α or by comparing z to standard normal critical values.

5. Recognize that significance testing does not measure the size or importance of an effect.

6. Recognize when you can use the z test and when the data collection design or a small sample from a skewed population makes it inappropriate.

CHAPTER 6 Review Exercises

6.74 **This wine stinks.** Sulfur compounds cause "off-odors" in wine, so winemakers want to know the odor threshold, the lowest concentration of a compound that the human nose can detect. The odor threshold for dimethyl sulfide (DMS) in trained wine tasters is about 25 micrograms per liter of wine (μg/l). The untrained noses of consumers may be less sensitive, however. Here are the DMS odor thresholds for 10 untrained students:

31   31   43   36   23   34   32   30   20   24

Assume that the standard deviation of the odor threshold for untrained noses is known to be $\sigma = 7\,\mu$g/l.

(a) Make a stemplot to verify that the distribution is roughly symmetric with no outliers. (More data confirm that there are no systematic departures from normality.)

(b) Give a 95% confidence interval for the mean DMS odor threshold among all students.

(c) Are you convinced that the mean odor threshold for students is higher than the published threshold, 25 μg/l? Carry out a significance test to justify your answer.

6.75 **Cellulose in hay.** An agronomist examines the cellulose content of alfalfa hay. Suppose that the cellulose content in the population has standard deviation $\sigma = 8$ mg/g. A sample of 15 cuttings has mean cellulose content $\bar{x} = 145$ mg/g.

(a) Give a 90% confidence interval for the mean cellulose content in the population.

(b) A previous study claimed that the mean cellulose content was $\mu = 140$ mg/g, but the agronomist believes that the mean is higher than that figure. State H_0 and H_a and carry out a significance test to see if the new data support this belief.

(c) The statistical procedures used in (a) and (b) are valid when several assumptions are met. What are these assumptions?

6.76 **Iron deficiency in infants.** Researchers studying iron deficiency in infants examined infants who were following different feeding patterns. One group of 26 infants was being breast-fed. At 6 months of age, these children had mean hemoglobin level $\bar{x} = 12.9$ grams per 100 milliliters of blood. Assume that the population standard deviation is $\sigma = 1.6$. Give a 95% confidence interval for the mean hemoglobin level of breast-fed infants. What assumptions (other than the unrealistic assumption that we know σ) does the method you used to get the confidence interval require?

6.77 **Investing in Treasury bills.** U.S. Treasury bills are safe investments, but how much do they pay investors? Here are data on the total returns (in percent) on Treasury bills for the years 1970 to 1996:

Year	1970	1971	1972	1973	1974	1975	1976	1977	1978	1979
Return	6.45	4.37	4.17	7.20	8.00	5.89	5.06	5.43	7.46	10.56

Year	1980	1981	1982	1983	1984	1985	1986	1987	1988	1989
Return	12.18	14.71	10.84	8.98	9.89	7.65	6.10	5.89	6.95	8.43

Year	1990	1991	1992	1993	1994	1995	1996
Return	7.72	5.46	3.50	3.04	4.37	5.60	5.13

(a) Make a histogram of these data, using bars 2 percentage points wide. What kind of deviation from normality do you see? Thanks to the central limit theorem, we can nonetheless treat $\bar{x}$ as approximately normal.

(b) Suppose that we can regard these 27 years' results as a random sample of returns on Treasury bills. Give a 90% confidence interval for the long-term mean return. (Assume you know that the standard deviation of all returns is $\sigma = 2.75\%$.)

(c) The rate of inflation during these years averaged about 5.5%. Are you convinced that Treasury bills have a mean return higher than 5.5%? State hypotheses and give a test statistic and a P-value.

(d) Make a time plot of the data. There are strong up-and-down cycles in the returns, which follow cycles of interest rates in the economy. The time plot makes it clear that returns in successive years are strongly correlated, so it is not proper to treat these data as an SRS. You should always check for such *serial correlation* in data collected over time.

6.78 **Workers' income.** The Bureau of Labor Statistics generally uses 90% confidence in its reports. One report gives a 90% confidence interval for the median income of full-time male workers in 1997 as $33,148 to $34,200. This result was calculated by advanced methods from the Current Population Survey, a multistage random sample of about 50,000 households.

(a) Would a 95% confidence interval be wider or narrower? Explain your answer.

(b) Would the null hypothesis that the 1997 median income of all full-time male workers was $33,000 be rejected at the 10% significance level in favor of the two-sided alternative? What about the null hypothesis that the median was $34,000?

6.79 **Why are larger samples better?** Statisticians prefer large samples. Describe briefly the effect of increasing the size of a sample (or the number of subjects in an experiment) on each of the following:

(a) The margin of error of a 95% confidence interval.

(b) The P-value of a test, when H_0 is false and all facts about the population remain unchanged as n increases.

(c) **(Optional)** The power of a fixed level α test, when α, the alternative hypothesis, and all facts about the population remain unchanged.

6.80 A roulette wheel has 18 red slots among its 38 slots. You observe many spins and record the number of times that red occurs. Now you want to use these data to test whether the probability p of a red has the value that is correct for a fair roulette wheel. State the hypotheses H_0 and H_a that you will test. (We will describe the test for this situation in Chapter 8.)

6.81 When asked to explain the meaning of "the P-value was $P = 0.03$," a student says, "This means there is only probability 0.03 that the null hypothesis is true." Is this an essentially correct explanation? Explain your answer.

6.82 Another student, when asked why statistical significance appears so often in research reports, says, "Because saying that results are significant tells us that they cannot easily be explained by chance variation alone." Do you think that this statement is essentially correct? Explain your answer.

6.83 **Helping welfare mothers.** A study compares two groups of mothers with young children who were on welfare two years ago. One group attended a voluntary training program that was offered free of charge at a local vocational school and was advertised in the local news media. The other group did not choose to attend the training program. The study finds a significant difference ($P < 0.01$) between the proportions of the mothers in the two groups who are still on welfare. The difference is not only significant but quite large. The report says that with 95% confidence the percent of the nonattending group still on welfare is 21% $\pm$ 4% higher than that of the group who attended the program. You are on the staff of a member of Congress who is interested in the plight of welfare mothers, and who asks you about the report.

(a) Explain in simple language what "a significant difference ($P <$ 0.01)" means.

(b) Explain clearly and briefly what "95% confidence" means.

(c) Is this study good evidence that requiring job training of all welfare mothers would greatly reduce the percent who remain on welfare for several years?

William S. Gosset

(UNIVERSITY COLLEGE LIBRARY, LONDON.)

Brewing Better Beer,
Brewing New Statistics

What would cause the head brewer of the famous Guinness brewery in Dublin, Ireland, not only to use statistics but to invent new statistical methods? The search for better beer, of course. William S. Gosset (1876–1937), fresh from Oxford University, joined Guinness as a brewer in 1899. He soon became involved in experiments and in statistics to understand the data from these experiments. What are the best varieties of barley and hops for brewing? How should they be grown, dried, and stored? The results of the field experiments, as you can guess, varied. Statistical inference can uncover the pattern behind the variation. The statistical methods available at the turn of the century ended with a version of the z test for means — even confidence intervals were not yet available.

The new t test identified the best barley variety, and Guinness promptly bought up all the available seed.

Gosset faced the problem we noted in using the z test to introduce the reasoning of statistical tests: he didn't know the population standard deviation σ. What is more, field experiments give only small numbers of observations. Just replacing σ by s in the z statistic and calling the result roughly normal wasn't accurate enough. So Gosset asked the key question: what is the exact sampling distribution of the statistic $(\bar{x} - \mu)/s$?

By 1907 Gosset was brewer-in-charge of Guinness's experimental brewery. He also had the answer to his question and had calculated a table of critical values for his new distribution. We call it the t distribution. The new t test identified the best barley variety, and Guinness promptly bought up all the available seed. Guinness allowed Gosset to publish his discoveries, but not under his own name. He used the name "Student," and the t test is sometimes called "Student's t" in his honor. Gosset's statistical work helped him become head brewer, a more interesting title than professor of statistics.

Inference for Distributions

(BILL BACHMAN/PHOTO RESEARCHERS.)

Barley, the main ingredient in beer and Gosset's motivation for discovering the t test.

Introduction

With the principles in hand, we proceed to practice. This chapter describes confidence intervals and significance tests for the mean of a single population and for comparing the means of two populations. An optional section discusses a test for comparing the standard deviations of two populations. Later chapters present procedures for inference about population proportions, for comparing the means of more than two populations, and for studying relationships among variables.

7.1 Inference for the Mean of a Population

Confidence intervals and tests of significance for the mean μ of a normal population are based on the sample mean $\bar{x}$. The sampling distribution of $\bar{x}$ has μ as its mean. That is, $\bar{x}$ is an unbiased estimator of the unknown μ. The spread of $\bar{x}$ depends on the sample size and also on the population standard deviation σ. In the previous chapter we made the unrealistic assumption that we knew the value of σ. In practice, σ is unknown. We must estimate σ from the data even though we are primarily interested in μ. The need to estimate σ changes some details of tests and confidence intervals for μ, but not their interpretation.

Here are the assumptions we make in order to do inference about a population mean.

ASSUMPTIONS FOR INFERENCE ABOUT A MEAN

- Our data are a **simple random sample** (SRS) of size n from the population. This assumption is very important.

- Observations from the population have a **normal distribution** with mean μ and standard deviation σ. In practice, it is enough that the distribution be symmetric and single-peaked unless the sample is very small. Both μ and σ are unknown parameters.

In this setting, the sample mean $\bar{x}$ has the normal distribution with mean μ and standard deviation $\sigma/\sqrt{n}$. Because we don't know σ, we estimate it by the sample standard deviation s. We then estimate the standard deviation of $\bar{x}$ by $s/\sqrt{n}$. This quantity is called the *standard error* of the sample mean $\bar{x}$.

STANDARD ERROR

When the standard deviation of a statistic is estimated from the data, the result is called the **standard error** of the statistic. The standard error of the sample mean $\bar{x}$ is $s/\sqrt{n}$.

The *t* distributions

When we know the value of σ, we base confidence intervals and tests for μ on the one-sample z statistic

$$z = \frac{\bar{x} - \mu}{\sigma/\sqrt{n}}$$

This z statistic has the standard normal distribution $N(0, 1)$. When we do not know σ, we substitute the standard error $s/\sqrt{n}$ of $\bar{x}$ for its standard deviation $\sigma/\sqrt{n}$. The statistic that results does not have a normal distribution. It has a distribution that is new to us, called a t *distribution*.

THE ONE-SAMPLE *t* STATISTIC AND THE *t* DISTRIBUTIONS

Draw an SRS of size n from a population that has the normal distribution with mean μ and standard deviation σ. The **one-sample t statistic**

$$t = \frac{\bar{x} - \mu}{s/\sqrt{n}}$$

has the **t distribution** with $n - 1$ degrees of freedom.

The t statistic has the same interpretation as any standardized statistic: it says how far $\bar{x}$ is from its mean μ in standard deviation units. There is a different t distribution for each sample size. We specify a particular t distribution by giving its **degrees of freedom**. The degrees of freedom for the one-sample t statistic come from the sample standard deviation s in the denominator of t. We saw in Chapter 1 (page 39) that s has $n - 1$ degrees of freedom. There are other t statistics with different degrees of freedom, some of which we will meet later. We will write the t distribution with k degrees of freedom as $t(k)$ for short.

degrees of freedom

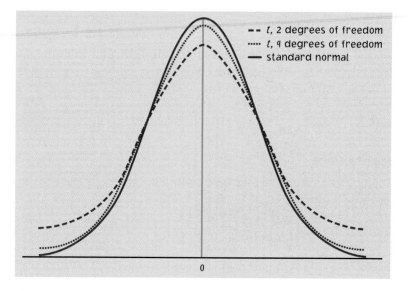

Figure 7.1 *Density curves for the t distributions with 2 and 9 degrees of freedom and the standard normal distribution. All are symmetric with center 0. The t distributions have more probability in the tails than does the standard normal.*

Figure 7.1 compares the density curves of the standard normal distribution and the *t* distributions with 2 and 9 degrees of freedom. The figure illustrates these facts about the *t* distributions:

- The density curves of the *t* distributions are similar in shape to the standard normal curve. They are symmetric about zero, single-peaked, and bell-shaped.

- The spread of the *t* distributions is a bit greater than that of the standard normal distribution. The *t* distributions in Figure 7.1 have more probability in the tails and less in the center than does the standard normal. This is true because substituting the estimate *s* for the fixed parameter σ introduces more variation into the statistic.

- As the degrees of freedom *k* increase, the $t(k)$ density curve approaches the $N(0, 1)$ curve ever more closely. This happens because *s* estimates σ more accurately as the sample size increases. So using *s* in place of σ causes little extra variation when the sample is large.

Table C in the back of the book gives critical values for the *t* distributions. Each row in the table contains critical values for one of the *t* distributions; the degrees of freedom appear at the left of the row. For convenience, we label the table entries both by *p*, the upper tail probability needed for significance tests, and by the confidence level *C* (in percent) required for confidence intervals. You have already used the standard normal critical values z^* in the bottom row of Table C. By looking down any column, you can check that the t critical values approach the normal values as the degrees of freedom increase. As in the case of the normal table, statistical software often makes Table C unnecessary.

APPLY YOUR KNOWLEDGE

7.1   Writers in some fields often summarize data by giving $\bar{x}$ and its standard error rather than $\bar{x}$ and s. The standard error of the mean $\bar{x}$ is often abbreviated as SEM.

   (a) A medical study finds that $\bar{x} = 114.9$ and $s = 9.3$ for the seated systolic blood pressure of the 27 members of one treatment group. What is the standard error of the mean?

   (b) Biologists studying the levels of several compounds in shrimp embryos reported their results in a table, with the note, "Values are means $\pm$ SEM for three independent samples." The table entry for the compound ATP was 0.84 ± 0.01. The researchers made three measurements of ATP, which had $\bar{x} = 0.84$. What was the sample standard deviation s for these measurements?

7.2   What critical value t^* from Table C satisfies each of the following conditions?

   (a) The t distribution with 5 degrees of freedom has probability 0.05 to the right of t^*.

   (b) The t distribution with 21 degrees of freedom has probability 0.99 to the left of t^*.

7.3   What critical value t^* from Table C satisfies each of the following conditions?

   (a) The one-sample t statistic from a sample of 15 observations has probability 0.025 to the right of t^*.

   (b) The one-sample t statistic from an SRS of 20 observations has probability 0.75 to the left of t^*.

The t confidence intervals and tests

To analyze samples from normal populations with unknown σ, just replace the standard deviation $\sigma/\sqrt{n}$ of $\bar{x}$ by its standard error $s/\sqrt{n}$ in the z procedures of Chapter 6. The z procedures then become *one-sample t procedures*. Use P-values or critical values from the t distribution with $n - 1$ degrees of freedom in place of the normal values. The one-sample t procedures are similar in both reasoning and computational detail to the z procedures of Chapter 6. So we will now pay more attention to questions about using these methods in practice.

THE ONE-SAMPLE t PROCEDURES

Draw an SRS of size n from a population having unknown mean μ. A level C confidence interval for μ is

$$\bar{x} \pm t^* \frac{s}{\sqrt{n}}$$

where t^* is the upper $(1 - C)/2$ critical value for the $t(n - 1)$ distribution. This interval is exact when the population distribution is normal and is approximately correct for large n in other cases.

To test the hypothesis $H_0 : \mu = \mu_0$ based on an SRS of size n, compute the one-sample t statistic

$$t = \frac{\bar{x} - \mu_0}{s/\sqrt{n}}$$

In terms of a variable T having the $t(n - 1)$ distribution, the P-value for a test of H_0 against

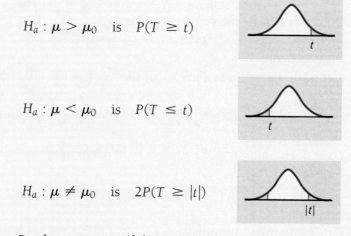

$H_a : \mu > \mu_0$ is $P(T \geq t)$

$H_a : \mu < \mu_0$ is $P(T \leq t)$

$H_a : \mu \neq \mu_0$ is $2P(T \geq |t|)$

These P-values are exact if the population distribution is normal and are approximately correct for large n in other cases.

EXAMPLE 7.1 Cockroach metabolism

To study the metabolism of insects, researchers fed cockroaches measured amounts of a sugar solution. After 2, 5, and 10 hours, they dissected some of the cockroaches and measured the amount of sugar in various tissues.[1] Five roaches fed the sugar D-glucose and dissected after 10 hours had the following amounts (in micrograms) of D-glucose in their hindguts:

55.95   68.24   52.73   21.50   23.78

The researchers gave a 95% confidence interval for the mean amount of D-glucose in cockroach hindguts under these conditions.

(TOM McHUGH/PHOTO RESEARCHERS.)

First calculate that

$$\bar{x} = 44.44 \quad \text{and} \quad s = 20.741$$

The degrees of freedom are $n - 1 = 4$. From Table C we find that for 95% confidence $t^* = 2.776$. The confidence interval is

$$\bar{x} \pm t^* \frac{s}{\sqrt{n}} = 44.44 \pm 2.776 \frac{20.741}{\sqrt{5}}$$

$$= 44.44 \pm 25.75$$

$$= 18.69 \text{ to } 70.19$$

Comparing this estimate with those for other body tissues and different times before dissection led to new insight into cockroach metabolism and to new ways of eliminating roaches from homes and restaurants. The large margin of error is due to the small sample size and the rather large variation among the cockroaches, reflected in the large value of s.

The one-sample t confidence interval has the form

$$\text{estimate} \pm t^* \text{SE}_{\text{estimate}}$$

where "SE" stands for "standard error." We will meet a number of confidence intervals that have this common form. Like the confidence interval, t tests are close in form to the z tests we met earlier. Here is an example. In Chapter 6 we used the z test on these data. That required the unrealistic assumption that we knew the population standard deviation σ. Now we can do a realistic analysis.

EXAMPLE 7.2 **Sweetening colas**

Cola makers test new recipes for loss of sweetness during storage. Trained tasters rate the sweetness before and after storage. Here are the sweetness losses (sweetness before storage minus sweetness after storage) found by 10 tasters for one new cola recipe:

$$2.0 \quad 0.4 \quad 0.7 \quad 2.0 \quad -0.4 \quad 2.2 \quad -1.3 \quad 1.2 \quad 1.1 \quad 2.3$$

Are these data good evidence that the cola lost sweetness?

Step 1: Hypotheses. Tasters vary in their perception of sweetness loss. So we ask the question in terms of the mean loss μ for a large population of tasters. The null hypothesis is "no loss," and the alternative hypothesis says "there is a loss."

$$H_0 : \mu = 0$$

$$H_a : \mu > 0$$

Step 2: Test statistic. The basic statistics are

$$\bar{x} = 1.02 \quad \text{and} \quad s = 1.196$$

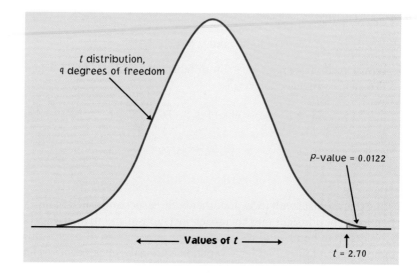

Figure 7.2 *The P-value for the one-sided t test in Example 7.2.*

The one-sample t test statistic is

$$t = \frac{\bar{x} - \mu_0}{s/\sqrt{n}} = \frac{1.02 - 0}{1.196/\sqrt{10}}$$

$$= 2.70$$

df = 9

p	.02	.01
t^*	2.398	2.821

Step 3: P-value. The P-value for $t = 2.70$ is the area to the right of 2.70 under the t distribution curve with degrees of freedom $n - 1 = 9$. Figure 7.2 shows this area. We can't find the exact value of P without software. But we can pin P between two values by using Table C. Search the df = 9 row of Table C for entries that bracket $t = 2.70$. Because the observed t lies between the critical values for 0.02 and 0.01, the P-value lies between 0.01 and 0.02. Computer software gives the more exact result $P = 0.0122$. There is quite strong evidence for a loss of sweetness.

Using the one-sample *t* procedures

Are the conclusions we drew in Examples 7.1 and 7.2 trustworthy? Both studies are experiments. In both cases the researchers took pains to avoid bias. The cockroaches were assigned at random to different sugars and different times before dissection and treated identically in every other way. The taste testers worked in isolation booths to avoid influence by other tasters. So we trust the results for these particular cockroaches and tasters.

We must be willing to treat the cockroaches and the tasters as SRSs from larger populations if we want to draw conclusions about cockroaches in general or tasters in general. The roaches were chosen at random from a population grown in the laboratory for research purposes. The tasters all have the same training. Even though we don't actually have SRSs from the populations we are interested in, we are willing to act as if we did. This is a matter of judgment.

```
2 | 2 4        -1 | 3
3 |            -0 | 4
4 |             0 | 4 7
5 | 3 6         1 | 1 2
6 | 8           2 | 0 0 2 3
7 |             3 |
8 |
      (a)           (b)
```

Figure 7.3 *Stemplots of the data in* (a) *Example 7.1 and* (b) *Example 7.2.*

We attract really good students

The guidebooks that students consult when choosing a college print information supplied by the colleges themselves. Surely no college would simply lie about, say, the average SAT score of its entering students. But we do want our scores to look good. How about leaving out the scores of international and remedial students? Northeastern University did this, raising the mean SAT score of its freshman class 50 points. And if we admit disadvantaged students sponsored by the state, surely we can leave out their SAT scores? New York University did this.

The *t* procedures also assume that the population of all roaches or all tasters has a normal distribution. We cannot effectively check normality with only 5 or 10 observations. In part, the researchers rely on experience with similar variables. They also look at the data. Stemplots of both distributions appear in Figure 7.3 (we rounded the cockroach data). The distribution of the 10 taste scores doesn't have a regular shape, but there are no gaps or outliers or other signs of nonnormal behavior. The cockroach data have a wide gap between the two smallest and the three largest observations. In observational data, this might suggest two different species of roaches. In this case we know that all five cockroaches came from a single population grown in the laboratory. The gap is just chance variation in a very small sample. Fortunately, we will see that confidence levels and *P*-values from *t* procedures are not very sensitive to lack of normality.

APPLY YOUR KNOWLEDGE

7.4 What critical value t^* from Table C would you use for a confidence interval for the mean of the population in each of the following situations?

(a) A 95% confidence interval based on $n = 10$ observations.

(b) A 99% confidence interval from an SRS of 20 observations.

(c) An 80% confidence interval from a sample of size 7.

7.5 The one-sample t statistic for testing

$$H_0 : \mu = 0$$

$$H_a : \mu > 0$$

from a sample of $n = 15$ observations has the value $t = 1.82$.

(a) What are the degrees of freedom for this statistic?

(b) Give the two critical values t^* from Table C that bracket t. What are the right-tail probabilities p for these two entries?

(c) Between what two values does the *P*-value of the test fall?

(d) Is the value $t = 1.82$ significant at the 5% level? Is it significant at the 1% level?

7.6   The one-sample t statistic from a sample of $n = 25$ observations for the two-sided test of

$$H_0: \mu = 64$$
$$H_a: \mu \neq 64$$

has the value $t = 1.12$.

(a) What are the degrees of freedom for t?

(b) Locate the two critical values t^* from Table C that bracket t. What are the right-tail probabilities p for these two values?

(c) Between what two values does the P-value of the test fall? (Note that H_a is two-sided.)

(d) Is the value $t = 1.12$ statistically significant at the 10% level? At the 5% level?

7.7   **DDT poisoning.** Poisoning by the pesticide DDT causes tremors and convulsions. In a study of DDT poisoning, researchers fed several rats a measured amount of DDT. They then made measurements on the rats' nervous systems that might explain how DDT poisoning causes tremors. One important variable was the "absolutely refractory period," the time required for a nerve to recover after a stimulus. This period varies normally. Measurements on four rats gave the data below (in milliseconds):[2]

$$1.6 \quad 1.7 \quad 1.8 \quad 1.9$$

(a) Find the mean refractory period $\bar{x}$ and the standard error of the mean.

(b) Give a 90% confidence interval for the mean absolutely refractory period for all rats of this strain when subjected to the same treatment.

(c) Suppose that the mean absolutely refractory period for unpoisoned rats is known to be 1.3 milliseconds. DDT poisoning should slow nerve recovery and so increase this period. Do the data give good evidence for this claim? State H_0 and H_a and do a t test. Between what levels from Table C does the P-value lie? What do you conclude from the test?

Matched pairs t procedures

The cockroach study in Example 7.1 estimated the mean amount of sugar in the hindgut, but the researchers then compared results for several body tissues and several times before dissection to get a bigger picture. The taste test in Example 7.2 was a matched pairs study in which the same 10 tasters rated before-and-after sweetness. Comparative studies are more convincing than single-sample investigations. For that reason, one-sample inference is less common than comparative inference. One common design to compare two treatments makes use of one-sample procedures. In a **matched pairs design**, subjects are matched in pairs and each treatment is given to one subject in each pair. The experimenter can toss a coin to assign two treatments to the two subjects in each pair. Another situation calling for matched pairs is before-and-after observations on the same subjects, as in the taste test of Example 7.2.

matched pairs design

MATCHED PAIRS t PROCEDURES

To compare the responses to the two treatments in a matched pairs design, apply the one-sample t procedures to the observed differences.

The parameter μ in a matched pairs t procedure is the mean difference in the responses to the two treatments within matched pairs of subjects in the entire population.

EXAMPLE 7.3   Floral scents and learning (EESEE)

We hear that listening to Mozart improves students' performance on tests. Perhaps pleasant odors have a similar effect. To test this idea, 21 subjects worked a paper-and-pencil maze while wearing a mask. The mask was either unscented or carried a floral scent. The response variable is their average time on three trials. Each subject worked the maze with both masks, in a random order. The randomization is important because subjects tend to improve their times as they work a maze repeatedly. Table 7.1 gives the subjects' average times with both masks.[3]

Table 7.1 Average time to complete a maze

Subject	Unscented (seconds)	Scented (seconds)	Difference	Subject	Unscented (seconds)	Scented (seconds)	Difference
1	30.60	37.97	−7.37	12	58.93	83.50	−24.57
2	48.43	51.57	−3.14	13	54.47	38.30	16.17
3	60.77	56.67	4.10	14	43.53	51.37	−7.84
4	36.07	40.47	−4.40	15	37.93	29.33	8.60
5	68.47	49.00	19.47	16	43.50	54.27	−10.77
6	32.43	43.23	−10.80	17	87.70	62.73	24.97
7	43.70	44.57	−0.87	18	53.53	58.00	−4.47
8	37.10	28.40	8.70	19	64.30	52.40	11.90
9	31.17	28.23	2.94	20	47.37	53.63	−6.26
10	51.23	68.47	−17.24	21	53.67	47.00	6.67
11	65.40	51.10	14.30				

```
-2 | 5
-1 | 7 1 1
-0 | 8 7 6 4 4 3 1
 0 | 3 4 7 9 9
 1 | 2 4 6 9
 2 | 5
```

Figure 7.4 *Stemplot of the differences in time to complete a maze for the 21 subjects in Example 7.3. The data are rounded to the nearest whole second. Notice that the stem 0 must appear twice, to display differences between −9 and 0 and between 0 and 9.*

To analyze these data, subtract the scented time from the unscented time for each subject. The 21 differences form a single sample. They appear in the "Difference" column in Table 7.1. The first subject, for example, was 7.37 seconds slower wearing the scented mask, so the difference is negative. Because shorter times represent better performance, positive differences show that the subject did better when wearing the scented mask. A stemplot of the differences (Figure 7.4) shows that their distribution is symmetric and appears reasonably normal in shape.

Step 1: Hypotheses. To assess whether the floral scent significantly improved performance, we test

$$H_0 : \mu = 0$$

$$H_a : \mu > 0$$

Here μ is the mean difference in the population from which the subjects were drawn. The null hypothesis says that no improvement occurs, and H_a says that unscented times are longer than scented times on the average.

Step 2: Test statistic. The 21 differences have

$$\bar{x} = 0.9567 \quad \text{and} \quad s = 12.5479$$

The one-sample t statistic is therefore

$$t = \frac{\bar{x} - 0}{s/\sqrt{n}} = \frac{0.9567 - 0}{12.5479/\sqrt{21}}$$

$$= 0.349$$

df = 20		
p	.25	.20
t^*	0.687	0.860

Step 3: P-value. Find the P-value from the $t(20)$ distribution. (Remember that the degrees of freedom are 1 less than the sample size.) Table C shows that 0.349 is less than the 0.25 critical value of the $t(20)$ distribution. The P-value is therefore greater than 0.25. Statistical software gives the value $P = 0.3652$.

Conclusion: The data do not support the claim that floral scents improve performance. The average improvement is small, just 0.96 seconds over the 50 seconds that the average subject took when wearing the unscented mask. This small improvement is not statistically significant at even the 25% level.

Example 7.3 illustrates how to turn matched pairs data into single-sample data by taking differences within each pair. We are making inferences about a single population, the population of all differences within matched pairs. It is incorrect to ignore the pairs and analyze the data as if we had two samples, one from subjects who wore unscented masks and a second from subjects who wore scented masks. Inference procedures for comparing two samples assume that the samples are selected independently of each other. This assumption does not hold when the same subjects are measured twice. The proper analysis depends on the design used to produce the data.

Because the t procedures are so common, all statistical software will do the calculations for you. If you are familiar with the t procedures, you can understand the output from any software. Figure 7.5 shows the output for Example

Paired T-Test and Confidence Level

```
Paired T for Unscented - Scented

                  N        Mean        StDev       SE Mean
Unscented        21        50.01       14.36          3.13
Scented          21        49.06       13.39          2.92
Difference       21         0.96       12.55          2.74

95% CI for mean difference:  (-4.76, 6.67)
T-Test of mean difference = 0 (vs > 0): T-Value = 0.35  P-Value = 0.365
```

(a) Minitab

Unscented – Scented:
```
Test Ho:μ(Unscented–Scented) = 0 vs Ha:μ(Unscented–Scented) > 0
Mean of Paired Differences = 0.956667 t-Statistic = 0.349 w/20 df
Fail to reject Ho at Alpha = 0.0500
p = 0.3652
```

(b) Data Desk

t-Test: Paired Two Sample for Means

	Variable 1	Variable 2
Mean	50.01429	49.05762
Variance	206.3097	179.1748
Observations	21	21
Pearson Correlation	0.593026	
Hypothesized Mean Difference	0	
df	20	
t Stat	0.349381	
P(T<=t) one-tail	0.365227	
t Critical one-tail	1.724718	
P(T<=t) two-tail	0.730455	
t Critical two-tail	2.085962	

(c) Excel

Figure 7.5 *Output for the matched pairs t test of Example 7.3 from three statistical software packages. You can easily locate the basic results in output from any statistical software.*

7.3 from Minitab, Data Desk, and Excel. In each case, we entered the data and asked for the one-sided matched pairs t test. The three outputs report slightly different information, but all include the basic facts: $t = 0.349$, $P = 0.365$.

(LARRY MULVEHILL/THE IMAGE WORKS.)

APPLY YOUR KNOWLEDGE

Many exercises from this point on ask you to give the P-value of a t test. If you have a suitable calculator or computer software, give the exact P-value. Otherwise, use Table C to give two values between which P lies.

7.8 **Growing tomatoes.** An agricultural field trial compares the yield of two varieties of tomatoes for commercial use. The researchers divide in half each of 10 small plots of land in different locations and plant each tomato variety on one half of each plot. After harvest, they compare the yields in pounds per plant at each location. The 10 differences (Variety A − Variety B) give $\bar{x} = 0.34$ and $s = 0.83$. Is there convincing evidence that Variety A has the higher mean yield?

    (a) Describe in words what the parameter μ is in this setting.

    (b) State H_0 and H_a.

    (c) Find the t statistic and give a P-value. What do you conclude?

7.9 **Right versus left.** The design of controls and instruments affects how easily people can use them. A student project investigated this effect by asking 25 right-handed students to turn a knob (with their right hands) that moved an indicator by screw action. There were two identical instruments, one with a right-hand thread (the knob turns clockwise) and the other with a left-hand thread (the knob must be turned counterclockwise). Table 7.2 gives the times in seconds each subject took to move the indicator a fixed distance.[4]

    (a) Each of the 25 students used both instruments. Discuss briefly how you would use randomization in arranging the experiment.

    (b) The project hoped to show that right-handed people find right-hand threads easier to use. What is the parameter μ for a matched pairs t test? State H_0 and H_a in terms of μ.

    (c) Carry out a test of your hypotheses. Give the P-value and report your conclusions.

7.10 Give a 90% confidence interval for the mean time advantage of right-hand over left-hand threads in the setting of Exercise 7.9. Do you think that the time saved would be of practical importance if the task were performed many times—for example, by an assembly line worker? To help answer this question, find the mean time for right-hand threads as a percent of the mean time for left-hand threads.

Table 7.2 Performance times (seconds) using right-hand and left-hand threads

Subject	Right thread	Left thread	Subject	Right thread	Left thread
1	113	137	14	107	87
2	105	105	15	118	166
3	130	133	16	103	146
4	101	108	17	111	123
5	138	115	18	104	135
6	118	170	19	111	112
7	87	103	20	89	93
8	116	145	21	78	76
9	75	78	22	100	116
10	96	107	23	89	78
11	122	84	24	85	101
12	103	148	25	88	123
13	116	147			

Robustness of *t* procedures

The *t* confidence interval and test are exactly correct when the distribution of the population is exactly normal. No real data are exactly normal. The usefulness of the *t* procedures in practice therefore depends on how strongly they are affected by lack of normality.

> **ROBUST PROCEDURES**
>
> A confidence interval or significance test is called **robust** if the confidence level or *P*-value does not change very much when the assumptions of the procedure are violated.

Because the tails of normal curves drop off quickly, samples from normal distributions will have very few outliers. Outliers suggest that your data are not a sample from a normal population. **Like $\bar{x}$ and s, the t procedures are strongly influenced by outliers.**

EXAMPLE 7.4   **The effect of outliers**

The cockroach data in Example 7.1 were

$$55.95 \quad 68.24 \quad 52.73 \quad 21.50 \quad 23.78$$

If the two very low observations moved up by 20 micrograms, to 41.5 and 43.78, the mean would rise from $\bar{x} = 44.44$ to $\bar{x} = 52.44$, and the standard deviation would drop from $s = 20.74$ to $s = 10.69$. The t confidence interval for μ would be only about half as wide as in Example 7.1.

Fortunately, the t procedures are quite robust against nonnormality of the population when there are no outliers, especially when the distribution is roughly symmetric. Larger samples improve the accuracy of P-values and critical values from the t distributions when the population is not normal. The main reason for this is the central limit theorem. The t statistic is based on the sample mean $\bar{x}$, which becomes more nearly normal as the sample size gets larger even when the population does not have a normal distribution.

Always make a plot to check for skewness and outliers before you use the t procedures for small samples. For most purposes, you can safely use the one-sample t procedures when $n \geq 15$ unless an outlier or quite strong skewness is present. Here are practical guidelines for inference on a single mean.[5]

USING THE t PROCEDURES

- Except in the case of small samples, the assumption that the data are an SRS from the population of interest is more important than the assumption that the population distribution is normal.

- *Sample size less than 15.* Use t procedures if the data are close to normal. If the data are clearly nonnormal or if outliers are present, do not use t.

- *Sample size at least 15.* The t procedures can be used except in the presence of outliers or strong skewness.

- *Large samples.* The t procedures can be used even for clearly skewed distributions when the sample is large, roughly $n \geq 40$.

EXAMPLE 7.5   Can we use t?

Consider several of the data sets we graphed in Chapter 1. Figure 7.6 shows the histograms.

- Figure 7.6(a) is a histogram of the percent of each state's residents who are at least 65 years of age. *We have data on the entire population of 50 states, so formal inference makes no sense.* We can calculate the exact mean for the population. There is no uncertainty due to having only a sample from the population, and no need for a confidence interval or test.

- Figure 7.6(b) shows the time of the first lightning strike each day in a mountain region in Colorado. The data contain more than 70 observations that have a symmetric distribution. You can use the t procedures to draw conclusions about the mean time of a day's first lightning strike with complete confidence.

- Figure 7.6(c) shows that the distribution of word lengths in Shakespeare's plays is skewed to the right. We aren't told how large the sample is. You can use the t procedures for a distribution like this if the sample size is roughly 40 or larger.

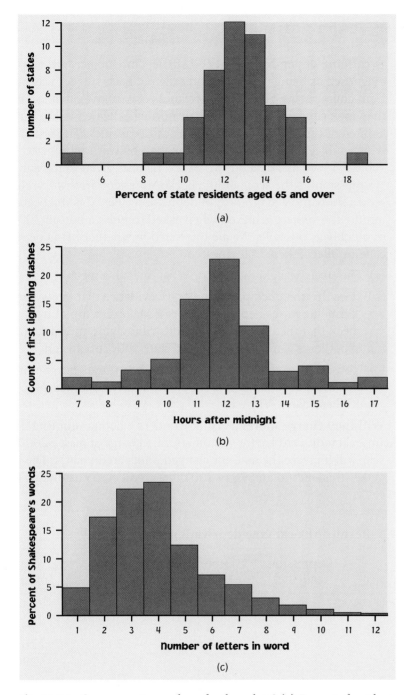

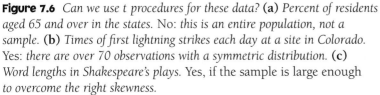

Figure 7.6 *Can we use t procedures for these data?* (**a**) *Percent of residents aged 65 and over in the states.* No: this is an entire population, not a sample. (**b**) *Times of first lightning strikes each day at a site in Colorado.* Yes: *there are over 70 observations with a symmetric distribution.* (**c**) *Word lengths in Shakespeare's plays.* Yes, if the sample is large enough to overcome the right skewness.

(PHOTOLINK/PHOTODISC.)

7.11   Is caffeine dependence real (EESEE)? Our subjects are 11 people diagnosed as being dependent on caffeine. Each subject was barred from coffee, colas, and other substances containing caffeine. Instead, they took capsules containing their normal caffeine intake. During a different time period, they took placebo capsules. The order in which subjects took caffeine and the placebo was randomized. Table 7.3 contains data on two of several tests given to the subjects. "Depression" is the score on the Beck Depression Inventory. Higher scores show more symptoms of depression. "Beats" is the beats per minute the subject achieved when asked to press a button 200 times as quickly as possible. We are interested in whether being deprived of caffeine affects these outcomes.[6]

(a) The study was double-blind. What does this mean?

(b) Does this matched pairs study give evidence that being deprived of caffeine raises depression scores? Make a stemplot to check that the differences are not strikingly nonnormal, state hypotheses, carry out a test, and state your conclusion.

(c) Now make a stemplot of the differences in beats per minute with and without caffeine. You should hesitate to use the t procedures on these data. Why?

7.12   Will they charge more? A bank wonders whether omitting the annual credit card fee for customers who charge at least $2400 in a year will increase the amount charged on its credit cards. The bank makes this offer to an SRS of 200 of its credit card customers. It

Table 7.3 Results of a caffeine-deprivation study

Subject	Depression (caffeine)	Depression (placebo)	Beats (caffeine)	Beats (placebo)
1	5	16	281	201
2	5	23	284	262
3	4	5	300	283
4	3	7	421	290
5	8	14	240	259
6	5	24	294	291
7	0	6	377	354
8	0	3	345	346
9	2	15	303	283
10	11	12	340	391
11	1	0	408	411

then compares how much these customers charge this year with the amount that they charged last year. The mean increase in the sample is $332, and the standard deviation is $108.

(a) Is there significant evidence at the 1% level that the mean amount charged increases under the no-fee offer? State H_0 and H_a and carry out a t test.

(b) Give a 99% confidence interval for the mean amount of the increase.

(c) The distribution of the amount charged is skewed to the right, but outliers are prevented by the credit limit that the bank enforces on each card. Use of the t procedures is justified in this case even though the population distribution is not normal. Explain why.

(d) A critic points out that the customers would probably have charged more this year than last even without the new offer, because the economy is more prosperous and interest rates are lower. Briefly describe the design of an experiment to study the effect of the no-fee offer that would avoid this criticism.

7.13 **Auto crankshafts.** Here are measurements (in millimeters) of a critical dimension for 16 auto engine crankshafts:

224.120	224.001	224.017	223.982	223.989	223.961
223.960	224.089	223.987	223.976	223.902	223.980
224.098	224.057	223.913	223.999		

The dimension is supposed to be 224 mm and the variability of the manufacturing process is unknown. Is there evidence that the mean dimension is not 224 mm?

(a) Check the data graphically for outliers or strong skewness that might threaten the validity of the t procedures. What do you conclude?

(b) State H_0 and H_a and carry out a t test. Give the P-value (from Table C or software). What do you conclude?

SECTION 7.1 S u m m a r y

Tests and confidence intervals for the mean μ of a normal population are based on the sample mean $\bar{x}$ of an SRS. Because of the central limit theorem, the resulting procedures are approximately correct for other population distributions when the sample is large.

The standardized sample mean is the **one-sample z statistic**,

$$z = \frac{\bar{x} - \mu}{\sigma / \sqrt{n}}$$

When we know σ, we use the z statistic and the standard normal distribution.

In practice, we do not know σ. Replace the standard deviation $\sigma/\sqrt{n}$ of $\bar{x}$ by the **standard error** $s/\sqrt{n}$ to get the **one-sample t statistic**,

$$t = \frac{\bar{x} - \mu}{s/\sqrt{n}}$$

The t statistic has the **t distribution** with $n - 1$ degrees of freedom.

There is a t distribution for every positive **degrees of freedom** k. All are symmetric distributions similar in shape to the standard normal distribution. The $t(k)$ distribution approaches the $N(0, 1)$ distribution as k increases.

An exact level C **confidence interval** for the mean μ of a normal population is

$$\bar{x} \pm t^* \frac{s}{\sqrt{n}}$$

where t^* is the upper $(1 - C)/2$ critical value of the $t(n - 1)$ distribution.

Significance tests for $H_0: \mu = \mu_0$ are based on the t statistic. Use P-values or fixed significance levels from the $t(n - 1)$ distribution.

Use these one-sample procedures to analyze **matched pairs** data by first taking the difference within each matched pair to produce a single sample.

The t procedures are relatively **robust** when the population is nonnormal, especially for larger sample sizes. The t procedures are useful for nonnormal data when $n \geq 15$ unless the data show outliers or strong skewness.

SECTION 7.1 Exercises

When an exercise asks for a P-value, give P exactly if you have a suitable calculator or computer software. Otherwise, use Table C to give two values between which P lies.

7.14 The one-sample t statistic for a test of

$$H_0 : \mu = 10$$
$$H_a : \mu < 10$$

based on $n = 10$ observations has the value $t = -2.25$.

(a) What are the degrees of freedom for this statistic?

(b) Between what two probabilities p from Table C does the P-value of the test fall?

7.15 **Market research.** A manufacturer of small appliances employs a market research firm to estimate retail sales of its products by gathering information from a sample of retail stores. This month an SRS of 75 stores in the Midwest sales region finds that these stores sold an average of 24 of the manufacturer's hand mixers, with standard deviation 11.

(a) Give a 95% confidence interval for the mean number of mixers sold by all stores in the region.

(b) The distribution of sales is strongly right-skewed, because there are many smaller stores and a few very large stores. The use of t in (a) is reasonably safe despite this violation of the normality assumption. Why?

7.16 **Sharks.** Great white sharks are big and hungry. Here are the lengths in feet of 44 great whites:[7]

18.7	12.3	18.6	16.4	15.7	18.3	14.6	15.8	14.9	17.6	12.1
16.4	16.7	17.8	16.2	12.6	17.8	13.8	12.2	15.2	14.7	12.4
13.2	15.8	14.3	16.6	9.4	18.2	13.2	13.6	15.3	16.1	13.5
19.1	16.2	22.8	16.8	13.6	13.2	15.7	19.7	18.7	13.2	16.8

(KELVIN AITKEN/PETER ARNOLD.)

(a) Examine these data for shape, center, spread, and outliers. The distribution is reasonably normal except for one outlier in each direction. Because these are not extreme and preserve the symmetry of the distribution, use of the t procedures is safe with 44 observations.

(b) Give a 95% confidence interval for the mean length of great white sharks. Based on this interval, is there significant evidence at the 5% level to reject the claim "Great white sharks average 20 feet in length"?

(c) It isn't clear exactly what parameter μ you estimated in (b). What information do you need to say what μ is?

7.17 **Calcium and blood pressure.** In a randomized comparative experiment on the effect of calcium in the diet on blood pressure, researchers divided 54 healthy white males at random into two groups. One group received calcium; the other, a placebo. At the beginning of the study, the researchers measured many variables on the subjects. The paper reporting the study gives $\bar{x} = 114.9$ and $s = 9.3$ for the seated systolic blood pressure of the 27 members of the placebo group.

(a) Give a 95% confidence interval for the mean blood pressure in the population from which the subjects were recruited.

(b) What assumptions about the population and the study design are required by the procedure you used in (a)? Which of these assumptions are important for the validity of the procedure in this case?

7.18 **Calibrating an instrument.** Gas chromatography is a sensitive technique used to measure small amounts of compounds. The response of a gas chromatograph is calibrated by repeatedly testing specimens containing a known amount of the compound to be measured. A calibration study for a specimen containing 1 nanogram (that's 10^{-9} gram) of a compound gave the following response readings:[8]

21.6   20.0   25.0   21.9

The response is known from experience to vary according to a normal distribution unless an outlier indicates an error in the analysis. Estimate the mean response to 1 nanogram of this substance, and give the margin of error for your choice of confidence level. Then explain to a chemist who knows no statistics what your margin of error means.

7.19 **Shrimp embryos.** The embryos of brine shrimp can enter a dormant phase in which metabolic activity drops to a low level. Researchers studying this dormant phase measured the level of several compounds important to normal metabolism. They reported their results in a table, with the note, "Values are means ± SEM for three independent samples." The table entry for the compound ATP was 0.84 ± 0.01. Biologists reading the article must be able to decipher this.[9]

(a) What does the abbreviation SEM stand for?

(b) The researchers made three measurements of ATP, which had $\bar{x} = 0.84$. Give a 90% confidence interval for the mean ATP level in dormant brine shrimp embryos.

7.20 **Measuring acculturation.** The Acculturation Rating Scale for Mexican Americans (ARSMA) measures the extent to which Mexican Americans have adopted Anglo/English culture. During the development of ARSMA, the test was given to a group of 17 Mexicans. Their scores, from a possible range of 1.00 to 5.00, had a symmetric distribution with $\bar{x} = 1.67$ and $s = 0.25$. Because low scores should indicate a Mexican cultural orientation, these results helped to establish the validity of the test.[10]

(a) Give a 95% confidence interval for the mean ARSMA score of Mexicans.

(b) What assumptions does your confidence interval require? Which of these assumptions is most important in this case?

7.21 **Does nature heal best?** Differences of electric potential occur naturally from point to point on a body's skin. Is the natural electric field strength best for helping wounds to heal? If so, changing the field will slow healing. The research subjects are anesthetized newts.

Table 7.4 Healing rates (micrometers per hour) for newts

Newt	Experimental limb	Control limb	Difference in healing
13	24	25	−1
14	23	13	10
15	47	44	3
16	42	45	−3
17	26	57	−31
18	46	42	4
19	38	50	−12
20	33	36	−3
21	28	35	−7
22	28	38	−10
23	21	43	−22
24	27	31	−4
25	25	26	−1
26	45	48	−3

Make a razor cut in both hind limbs. Let one heal naturally (the control). Use an electrode to change the electric field in the other to half its normal value. After two hours, measure the healing rate. Table 7.4 gives the healing rates (in micrometers per hour) for 14 newts.[11]

(a) As is usual, the paper did not report these raw data. Readers are expected to be able to interpret the summaries that the paper did report. The paper summarized the differences in the table above as "−5.71 ± 2.82" and said, "All values are expressed as means ± standard error of the mean." Show carefully where the numbers −5.71 and 2.82 come from.

(b) The researchers want to know if changing the electric field reduces the mean healing rate for all newts. State hypotheses, carry out a test, and give your conclusion. Is the result statistically significant at the 5% level? At the 1% level? (The researchers compared several field strengths and concluded that the natural strength is about right for fastest healing.)

(c) Give a 90% confidence interval for the amount by which changing the field changes the rate of healing. Then explain in a sentence what it means to say that you are "90% confident" of your result.

7.22 **ARSMA versus BI.** The ARSMA test (Exercise 7.20) was compared with a similar test, the Bicultural Inventory (BI), by administering both tests to 22 Mexican Americans. Both tests have the same range of scores (1.00 to 5.00) and are scaled to have similar means for the groups used to develop them. There was a high correlation between the two scores, giving evidence that both are measuring the same characteristics. The researchers wanted to know whether the population mean scores for the two tests are the same. The differences in scores (ARSMA − BI) for the 22 subjects had $\bar{x} = 0.2519$ and $s = 0.2767$.

(a) Describe briefly how to arrange the administration of the two tests to the subjects, including randomization.

(b) Carry out a significance test for the hypothesis that the two tests have the same population mean. Give the P-value and state your conclusion.

(c) Give a 95% confidence interval for the difference between the two population mean scores.

7.23 **CEO pay.** A study of the pay of corporate CEOs (chief executive officers) examined the cash compensation, adjusted for inflation, of the CEOs of 104 corporations over the period 1977 to 1988. Among the data are the average annual pay increases for each of the 104 CEOs. The mean percent increase in pay was 6.9%. The data showed great variation, with a standard deviation of 17.4%. The distribution was strongly skewed to the right.[12]

(a) Despite the skewness of the distribution, there were no extreme outliers. Explain why we can use t procedures for these data.

(b) What are the degrees of freedom? When the exact degrees of freedom do not appear in Table C, use the next lower degrees of freedom in the table.

(c) Give a 99% confidence interval for the mean increase in pay for all corporate CEOs. What essential condition must the data satisfy if we are to trust your result?

7.24 **Comparing two drugs.** Makers of generic drugs must show that they do not differ significantly from the "reference" drug that they imitate. One aspect in which drugs might differ is their extent of absorption in the blood. Table 7.5 gives data taken from 20 healthy nonsmoking male subjects for one pair of drugs.[13] This is a matched pairs design. Subjects 1 to 10 received the generic drug first, and Subjects 11 to 20 received the reference drug first. In all cases, a washout period separated the two drugs so that the first had disappeared from the blood before the subject took the second. The subject numbers in the table were assigned at random to decide the order of the drugs for each subject.

(a) Do a data analysis of the differences between the absorption measures for the generic and reference drugs. Is there any reason not to apply t procedures?

Table 7.5 Absorption extent for two versions of a drug

Subject	Reference drug	Generic drug	Subject	Reference drug	Generic drug
15	4108	1755	4	2344	2738
3	2526	1138	16	1864	2302
9	2779	1613	6	1022	1284
13	3852	2254	10	2256	3052
12	1833	1310	5	938	1287
8	2463	2120	7	1339	1930
18	2059	1851	14	1262	1964
20	1709	1878	11	1438	2549
17	1829	1682	1	1735	3340
2	2594	2613	19	1020	3050

(b) Use a t test to answer the key question: do the drugs differ significantly in absorption?

7.25 Ages of the presidents. Table 1.7 (page 37) gives the ages of U.S. presidents when they took office. It does not make sense to use the t procedures (or any other statistical procedures) to give a 95% confidence interval for the mean age of the presidents. Explain why not.

7.26 Optional: The power of a t test. The bank in Exercise 7.12 tested a new idea on a sample of 200 customers. The bank wants to be quite certain of detecting a mean increase of $\mu = \$100$ in the amount charged, at the $\alpha = 0.01$ significance level. Perhaps a sample of only $n = 50$ customers would accomplish this. Find the approximate power of the test with $n = 50$ against the alternative $\mu = \$100$ as follows.

(a) What is the critical value t^* for the one-sided test with $\alpha = 0.01$ and $n = 50$?

(b) Write the rule for rejecting H_0: $\mu = 0$ in terms of the t statistic. Then take $s = 108$ (an estimate based on the data in Exercise 7.12) and state the rejection rule in terms of $\bar{x}$.

(c) Assume that $\mu = 100$ (the given alternative) and that $\sigma = 108$ (an estimate from the data in Exercise 7.12). The approximate power is the probability of the event you found in (b), calculated under these assumptions. Find the power. Would you recommend that the bank do a test on 50 customers, or should more customers be included?

7.27 Optional: The power of a t test. Exercise 7.22 reports a small study comparing ARSMA and BI, two tests of the acculturation of Mexican Americans. Would this study usually detect a difference in mean

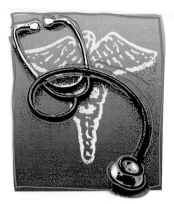

Sounds good—but no comparison

Most women have mammograms to check for breast cancer once they reach middle age. Could a fancier test do a better job of finding cancers early? PET scans are a fancier (and more expensive) test. Doctors used PET scans on 14 women with tumors and got the detailed diagnosis right in 12 cases. That's promising. But there were no controls, and 14 cases are not statistically significant. Medical standards require randomized comparative experiments and statistically significant results. Only then can we be confident that the fancy test really is better.

scores of 0.2? To answer this question, calculate the approximate power of the test (with $n = 22$ subjects and $\alpha = 0.05$) of

$$H_0 : \mu = 0$$

$$H_a : \mu \neq 0$$

against the alternative $\mu = 0.2$. We do this by acting as if σ were known.

(a) From Table C, what is the critical value for $\alpha = 0.05$?

(b) Write the rule for rejecting H_0 at the $\alpha = 0.05$ level. Then take $s = 0.3$, the approximate value observed in Exercise 7.22, and restate the rejection criterion in terms of $\overline{x}$. Note that this is a two-sided test.

(c) Find the probability of this event when $\mu = 0.2$ (the alternative given) and $\sigma = 0.3$ (estimated from the data in Exercise 7.22) by a normal probability calculation. This is the approximate power.

7.2 Comparing Two Means

Comparing two populations or two treatments is one of the most common situations encountered in statistical practice. We call such situations *two-sample problems*.

> **TWO-SAMPLE PROBLEMS**
>
> • The goal of inference is to compare the responses to two treatments or to compare the characteristics of two populations.
>
> • We have a separate sample from each treatment or each population.

Two-sample problems

A two-sample problem can arise from a randomized comparative experiment that randomly divides subjects into two groups and exposes each group to a different treatment. Comparing random samples separately selected from two populations is also a two-sample problem. Unlike the matched pairs designs studied earlier, there is no matching of the units in the two samples and the two samples can be of different sizes. Inference procedures for two-sample data differ from those for matched pairs. Here are some typical two-sample problems.

EXAMPLE 7.6 **Two-sample problems**

(a) A medical researcher is interested in the effect on blood pressure of added calcium in our diet. She conducts a randomized comparative experiment in which one group of subjects receives a calcium supplement and a control group gets a placebo.

(b) A psychologist develops a test that measures social insight. He compares the social insight of male college students with that of female college students by giving the test to a sample of students of each gender.

(c) A bank wants to know which of two incentive plans will most increase the use of its credit cards. It offers each incentive to a random sample of credit card customers and compares the amount charged during the following six months.

APPLY YOUR KNOWLEDGE

7.28   **Which data design?** The following situations require inference about a mean or means. Identify each as (1) single sample, (2) matched pairs, or (3) two samples. The procedures of Section 7.1 apply to cases (1) and (2). We are about to learn procedures for (3).

(a) An education researcher wants to learn whether it is more effective to put questions before or after introducing a new concept in an elementary school mathematics text. He prepares two text segments that teach the concept, one with motivating questions before and the other with review questions after. He uses each text segment to teach a separate group of children. The researcher compares the scores of the groups on a test over the material.

(b) Another researcher approaches the same issue differently. She prepares text segments on two unrelated topics. Each segment comes in two versions, one with questions before and the other with questions after. The subjects are a single group of children. Each child studies both topics, one (chosen at random) with questions before and the other with questions after. The researcher compares test scores for each child on the two topics to see which topic he or she learned better.

7.29   **Which data design?** The following situations require inference about a mean or means. Identify each as (1) single sample, (2) matched pairs, or (3) two samples. The procedures of Section 7.1 apply to cases (1) and (2). We are about to learn procedures for (3).

(a) To check a new analytical method, a chemist obtains a reference specimen of known concentration from the National Institute of Standards and Technology. She then makes 20 measurements of the concentration of this specimen with the new method and checks for bias by comparing the mean result with the known concentration.

(b) Another chemist is checking the same new method. He has no reference specimen, but a familiar analytic method is available. He wants to know if the new and old methods agree. He takes a specimen of unknown concentration and measures the concentration 10 times with the new method and 10 times with the old method.

Comparing two population means

We can examine two-sample data graphically by comparing stemplots (for small samples) or histograms or boxplots (for larger samples). Now we will apply the ideas of formal inference in this setting. When both population distributions are symmetric, and especially when they are at least approximately normal, a comparison of the mean responses in the two populations is the most common goal of inference. Here are the assumptions we will make.

ASSUMPTIONS FOR COMPARING TWO MEANS

- We have **two SRSs**, from two distinct populations. The samples are **independent**. That is, one sample has no influence on the other. Matching violates independence, for example. We measure the same variable for both samples.
- Both populations are **normally distributed**. The means and standard deviations of the populations are unknown.

Call the variable we measure x_1 in the first population and x_2 in the second because the variable may have different distributions in the two populations. Here is the notation we will use to describe the two populations:

Population	Variable	Mean	Standard deviation
1	x_1	μ_1	σ_1
2	x_2	μ_2	σ_2

There are four unknown parameters, the two means and the two standard deviations. The subscripts remind us which population a parameter describes. We want to compare the two population means, either by giving a confidence interval for their difference $\mu_1 - \mu_2$ or by testing the hypothesis of no difference, $H_0: \mu_1 = \mu_2$.

We use the sample means and standard deviations to estimate the unknown parameters. Again, subscripts remind us which sample a statistic comes from. Here is the notation that describes the samples:

Population	Sample size	Sample mean	Sample standard deviation
1	n_1	$\bar{x}_1$	s_1
2	n_2	$\bar{x}_2$	s_2

To do inference about the difference $\mu_1 - \mu_2$ between the means of the two populations, we start from the difference $\bar{x}_1 - \bar{x}_2$ between the means of the two samples.

EXAMPLE 7.7 **Does polyester decay?**

How quickly do synthetic fabrics such as polyester decay in landfills? A researcher buried polyester strips in the soil for different lengths of time, then dug up the strips and measured the force required to break them. Breaking strength is easy to measure and is a good indicator of decay. Lower strength means the fabric has decayed.

Part of the study buried 10 polyester strips in well-drained soil in the summer. Five of the strips, chosen at random, were dug up after 2 weeks; the other 5 were dug up after 16 weeks. Here are the breaking strengths in pounds:[14]

| 2 weeks | 118 | 126 | 126 | 120 | 129 |
| 16 weeks | 124 | 98 | 110 | 140 | 110 |

From the data, calculate the summary statistics:

Group	Treatment	n	$\bar{x}$	s
1	2 weeks	5	123.80	4.60
2	16 weeks	5	116.40	16.09

The fabric that was buried longer has somewhat lower mean strength, along with more variation. The observed difference in mean strengths is

$$\bar{x}_1 - \bar{x}_2 = 123.80 - 116.40 = 7.40 \text{ pounds}$$

Is this good evidence that polyester decays more in 16 weeks than in 2 weeks?

Example 7.7 fits the two-sample setting. We write hypotheses in terms of the mean breaking strengths in the entire population of polyester fabric, μ_1 for fabric buried for 2 weeks and μ_2 for fabric buried for 16 weeks. The hypotheses are

$$H_0 : \mu_1 = \mu_2$$

$$H_a : \mu_1 > \mu_2$$

We want to test these hypotheses and also estimate the amount of the decrease in mean breaking strength, $\mu_1 - \mu_2$.

Are the assumptions satisfied? Because of the randomization, we are willing to regard two groups of fabric strips as two independent SRSs. Although the samples are small, we check for serious nonnormality by examining the data. Here is a back-to-back stemplot of the responses:

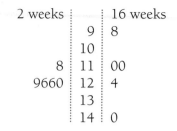

The 16-week group is much more spread out, as its larger standard deviation suggests. As far as we can tell from so few observations, there are no departures from normality that prevent use of t procedures.

Two-sample t procedures

To assess the significance of the observed difference between the means of our two samples, we follow a familiar path. Whether an observed difference between two samples is surprising depends on the spread of the observations as well as on the two means. Widely different means can arise just by chance if the individual observations vary a great deal. To take variation into account, we would like to standardize the observed difference $\bar{x}_1 - \bar{x}_2$ by dividing by its standard deviation. This standard deviation is

$$\sqrt{\frac{\sigma_1^2}{n_1} + \frac{\sigma_2^2}{n_2}}$$

This standard deviation gets larger as either population gets more variable, that is, as σ_1 or σ_2 increases. It gets smaller as the sample sizes n_1 and n_2 increase.

Because we don't know the population standard deviations, we estimate them by the sample standard deviations from our two samples. The result is *standard error* the **standard error**, or estimated standard deviation, of the difference in sample means:

$$SE = \sqrt{\frac{s_1^2}{n_1} + \frac{s_2^2}{n_2}}$$

two-sample t statistic When we standardize the estimate by dividing it by its standard error, the result is the **two-sample t statistic**:

$$t = \frac{\bar{x}_1 - \bar{x}_2}{SE}$$

The statistic t has the same interpretation as any z or t statistic: it says how far $\bar{x}_1 - \bar{x}_2$ is from 0 in standard deviation units.

Unfortunately, the two-sample t statistic does *not* have a t distribution. Nonetheless, the two-sample t statistic is used with t distribution critical values in inference for two-sample problems. There are two ways to do this.

Option 1. Use the statistic t with critical values from a t distribution with degrees of freedom computed from the data. The degrees of freedom are generally not a whole number. This is a very accurate approximation to the distribution of t.

Option 2. Use the statistic t with critical values from the t distribution with degrees of freedom equal to the smaller of $n_1 - 1$ and $n_2 - 1$. These procedures are always conservative for any two normal populations.

Most statistical software systems use Option 1 unless the user requests another method. Using Option 1 without software is a bit complicated. We will therefore present the second, simpler, option first. We recommend that you use Option 2 when doing calculations without software. If you use software, it should automatically do the calculations for Option 1. Some details of Option 1 appear in the optional section beginning on page 402. Here is a description of the Option 2 procedures that includes a statement of just how they are "conservative."

THE TWO-SAMPLE t PROCEDURES

Draw an SRS of size n_1 from a normal population with unknown mean μ_1, and draw an independent SRS of size n_2 from another normal population with unknown mean μ_2. The confidence interval for $\mu_1 - \mu_2$ given by

$$(\bar{x}_1 - \bar{x}_2) \pm t^* \sqrt{\frac{s_1^2}{n_1} + \frac{s_2^2}{n_2}}$$

has confidence level *at least* C no matter what the population standard deviations may be. Here t^* is the upper $(1 - C)/2$ critical value for the $t(k)$ distribution with k the smaller of $n_1 - 1$ and $n_2 - 1$.

To test the hypothesis $H_0 : \mu_1 = \mu_2$, compute the two-sample t statistic

$$t = \frac{\bar{x}_1 - \bar{x}_2}{\sqrt{\frac{s_1^2}{n_1} + \frac{s_2^2}{n_2}}}$$

and use P-values or critical values for the $t(k)$ distribution. The true P-value or fixed significance level will always be *equal to or less than* the value calculated from $t(k)$ no matter what values the unknown population standard deviations have.

Notice that the two-sample t confidence interval again has the form

$$\text{estimate} \pm t^* \text{SE}_{\text{estimate}}$$

These two-sample t procedures always err on the safe side, reporting *higher* P-values and *lower* confidence than are actually true. The gap between what is reported and the truth is quite small unless the sample sizes are both small and unequal. As the sample sizes increase, probability values based on t with degrees of freedom equal to the smaller of $n_1 - 1$ and $n_2 - 1$ become more accurate.[15]

7.30 **Treating scrapie in hamsters.** Scrapie is a degenerative disease of the nervous system. A study of the substance IDX as a treatment for scrapie used as subjects 20 infected hamsters. Ten, chosen at random, were injected with IDX. The other 10 were untreated. The researchers recorded how long each hamster lived. They reported, "Thus, although all infected control hamsters had died by 94 days after infection (mean ± SEM = 88.5 ± 1.9 days), IDX-treated hamsters lived up to 128 days (mean ± SEM = 116 ± 5.6 days)."[16]

(a) Fill in the values in this summary table:

Group	Treatment	n	$\bar{x}$	s
1	IDX	?	?	?
2	Untreated	?	?	?

(b) What degrees of freedom would you use in the conservative two-sample t procedures to compare the two treatments?

7.31 **Going bankrupt.** A business school study compared a sample of Greek firms that went bankrupt with a sample of healthy Greek businesses. One measure of a firm's financial health is the ratio of current assets to current liabilities, called *CA/CL*. For the year before bankruptcy, the study found the mean *CA/CL* to be 1.72565 in the healthy group and 0.78640 in the group that failed. The paper reporting the study says that $t = 7.36$.[17]

(a) You can draw a conclusion from this t without using a table and even without knowing the sizes of the samples (as long as the samples are not tiny). What is your conclusion? Why don't you need the sample sizes and a table?

(b) In fact, the study looked at 33 firms that failed and 68 healthy firms. What degrees of freedom would you use for the t test if you follow the conservative approach recommended for use without software?

Examples of the two-sample *t* procedures

EXAMPLE 7.8   Does polyester decay?

We will use the two-sample t procedures to compare the fabric strengths in Example 7.7. The test statistic for the null hypothesis $H_0 : \mu_1 = \mu_2$ is

$$t = \frac{\bar{x}_1 - \bar{x}_2}{\sqrt{\dfrac{s_1^2}{n_1} + \dfrac{s_2^2}{n_2}}}$$

$$= \frac{123.8 - 116.4}{\sqrt{\dfrac{4.60^2}{5} + \dfrac{16.09^2}{5}}}$$

$$= \frac{7.4}{7.484} = 0.989$$

There are 4 degrees of freedom, the smaller of $n_1 - 1 = 4$ and $n_2 - 1 = 4$. Because H_a is one-sided on the high side, the P-value is the area to the right of $t = 0.989$ under the $t(4)$ curve. Figure 7.7 illustrates this P-value. Table C shows that it lies between 0.15 and 0.20. Software tells us that the actual value is $P = 0.189$. The experiment did not find convincing evidence that polyester decays more in 16 weeks than in 2 weeks.

For a 90% confidence interval, Table C shows that the $t(4)$ critical value is $t^* = 2.132$. We are 90% confident that the mean strength change between 2 and 16 weeks, $\mu_1 - \mu_2$, lies in the interval

$$(\bar{x}_1 - \bar{x}_2) \pm t^* \sqrt{\frac{s_1^2}{n_1} + \frac{s_2^2}{n_2}} = (123.8 - 116.4) \pm 2.132 \sqrt{\frac{4.60^2}{5} + \frac{16.09^2}{5}}$$

$$= 7.40 \pm 15.96$$

$$= -8.56 \text{ to } 23.36$$

That the 90% confidence interval covers 0 tells us that we cannot reject $H_0: \mu_1 = \mu_2$ against the two-sided alternative at the $\alpha = 0.10$ level of significance.

df = 4		
p	.20	.15
t^*	0.941	1.190

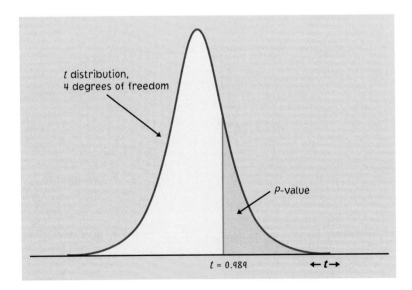

Figure 7.7 *The P-value in Example 7.8. This example uses the conservative method, which leads to the t distribution with 4 degrees of freedom.*

Catching cheaters

A certification test for surgeons asks 277 multiple-choice questions. Smith and Jones have 193 common right answers and 53 identical wrong choices. The computer flags their 246 identical answers as evidence of possible cheating. They sue. The court wants to know how unlikely it is that exams this similar would occur just by chance. That is, the courts want a P-value. Statisticians offer several P-values based on different models for the exam-taking process. They all say that results this similar would almost never happen just by chance. Smith and Jones fail the exam.

Sample size strongly influences the P-value of a test. An effect that fails to be significant at a level α in a small sample will be significant in a larger sample. In the light of the small samples in Example 7.7, we suspect that more data might show that longer burial time does significantly reduce strength. Even if significant, the reduction may be quite small. Our data suggest that buried polyester decays slowly, as the mean breaking strength dropped only from 123.8 pounds to 116.4 pounds between 2 and 16 weeks.

EXAMPLE 7.9   Social insight among men and women

The Chapin Social Insight Test is a psychological test designed to measure how accurately a person appraises other people. The possible scores on the test range from 0 to 41. During the development of the Chapin test, it was given to several different groups of people. Here are the results for male and female college students majoring in the liberal arts:[18]

Group	Sex	n	$\bar{x}$	s
1	Male	133	25.34	5.05
2	Female	162	24.94	5.44

Do these data support the contention that female and male students differ in average social insight?

Step 1: Hypotheses. Because we had no specific direction for the male/female difference in mind before looking at the data, we choose the two-sided alternative. The hypotheses are

$$H_0 : \mu_1 = \mu_2$$
$$H_a : \mu_1 \neq \mu_2$$

Step 2: Test statistic. The two-sample t statistic is

$$t = \frac{\bar{x}_1 - \bar{x}_2}{\sqrt{\frac{s_1^2}{n_1} + \frac{s_2^2}{n_2}}}$$

$$= \frac{25.34 - 24.94}{\sqrt{\frac{5.05^2}{133} + \frac{5.44^2}{162}}}$$

$$= 0.654$$

Step 3: P-value. There are 132 degrees of freedom, the smaller of

$$n_1 - 1 = 133 - 1 = 132 \quad \text{and} \quad n_2 - 1 = 162 - 1 = 161$$

Figure 7.8 illustrates the P-value. Find it by comparing 0.654 to critical values for the $t(132)$ distribution and then doubling p because the alternative is two-sided. Degrees of freedom 132 do not appear in Table C, so we use the next smaller table value, degrees of freedom 100.

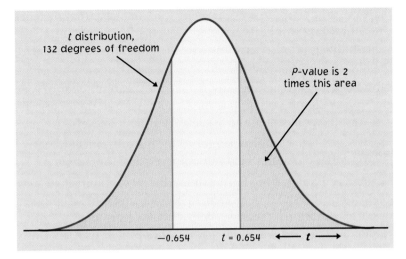

Figure 7.8 *The P-value in Example 7.9. To find P, find the area above* $t = 0.654$ *and double it because the alternative is two-sided.*

Table C shows that 0.654 does not reach the 0.25 critical value, which is the largest upper tail probability in Table C. The P-value is therefore greater than 0.50. The data give no evidence of a male/female difference in mean social insight score ($t = 0.654$, df $= 132$, $P > 0.5$).

df $= 100$		
p	.25	.20
t^*	0.677	0.845

The researcher in Example 7.9 did not do an experiment but compared samples from two populations. The large samples imply that the assumption that the populations have normal distributions is of little importance. The sample means will be nearly normal in any case. The major question concerns the population to which the conclusions apply. The student subjects are certainly not an SRS of all liberal arts majors in the country. If they are volunteers from a single college, the sample results may not extend to a wider population.

APPLY YOUR KNOWLEDGE

7.32 **The effect of logging.** How badly does logging damage tropical rainforests? One study compared forest plots in Borneo that had never been logged with similar plots nearby that had been logged 8 years earlier. The study found that the effects of logging were somewhat less severe than expected. Here are the data on the number of tree species in 12 unlogged plots and 9 logged plots:[19]

Unlogged	22	18	22	20	15	21	13	13	19	13	19	15
Logged	17	4	18	14	18	15	15	10	12			

(DAVID AUSTEN/STOCK, BOSTON.)

(a) The study report says, "Loggers were unaware that the effects of logging would be assessed." Why is this important? The study report also explains why the plots can be considered to be randomly assigned.

(b) Does logging significantly reduce the mean number of species in a plot after 8 years? State the hypotheses and do a *t* test. Is the result significant at the 5% level? At the 1% level?

(c) Give a 90% confidence interval for the difference in mean number of species between unlogged and logged plots.

7.33 **Each day I am getting better in math.** A "subliminal" message is below our threshold of awareness but may nonetheless influence us. Can subliminal messages help students learn math? A group of students who had failed the mathematics part of the City University of New York Skills Assessment Test agreed to participate in a study to find out.

All received a daily subliminal message, flashed on a screen too rapidly to be consciously read. The treatment group of 10 students (chosen at random) was exposed to "Each day I am getting better in math." The control group of 8 students was exposed to a neutral message, "People are walking on the street." All students participated in a summer program designed to raise their math skills, and all took the assessment test again at the end of the program. Table 7.6 gives data on the subjects' scores before and after the program.[20]

(a) Is there good evidence that the treatment brought about a greater improvement in math scores than the neutral message? State hypotheses, carry out a test, and state your conclusion. Is your result significant at the 5% level? At the 10% level?

Table 7.6 Mathematics skills scores before and after a subliminal message

Treatment Group		Control Group	
Pre-test	Post-test	Pre-test	Post-test
18	24	18	29
18	25	24	29
21	33	20	24
18	29	18	26
18	33	24	38
20	36	22	27
23	34	15	22
23	36	19	31
21	34		
17	27		

(b) Give a 90% confidence interval for the mean difference in gains between treatment and control.

7.34 **Beetles in oats.** In a study of cereal leaf beetle damage on oats, researchers measured the number of beetle larvae per stem in small plots of oats after randomly applying one of two treatments: no pesticide or malathion at the rate of 0.25 pound per acre. The data appear roughly normal. Here are the summary statistics:[21]

Group	Treatment	n	$\overline{x}$	s
1	Control	13	3.47	1.21
2	Malathion	14	1.36	0.52

Is there significant evidence at the 1% level that malathion reduces the mean number of larvae per stem? Be sure to state H_0 and H_a.

Robustness again

The two-sample t procedures are more robust than the one-sample t methods, particularly when the distributions are not symmetric. When the sizes of the two samples are equal and the two populations being compared have distributions with similar shapes, probability values from the t table are quite accurate for a broad range of distributions when the sample sizes are as small as $n_1 = n_2 = 5$.[22] When the two population distributions have different shapes, larger samples are needed.

As a guide to practice, adapt the guidelines given on page 380 for the use of one-sample t procedures to two-sample procedures by replacing "sample size" with the "sum of the sample sizes," $n_1 + n_2$. These guidelines err on the side of safety, especially when the two samples are of equal size. In planning a two-sample study, you should usually choose equal sample sizes. The two-sample t procedures are most robust against nonnormality in this case, and the conservative probability values are most accurate.

APPLY YOUR KNOWLEDGE

7.35 **More nutritious corn.** Ordinary corn doesn't have as much of the amino acid lysine as animals need in their feed. Plant scientists have developed varieties of corn that have increased amounts of lysine. In a test of the quality of high-lysine corn as animal feed, an experimental group of 20 one-day-old male chicks ate a ration containing the new corn. A control group of another 20 chicks received a ration that was identical except that it contained normal corn. Here are the weight gains (in grams) after 21 days:[23]

Control				Experimental			
380	321	366	356	361	447	401	375
283	349	402	462	434	403	393	426
356	410	329	399	406	318	467	407
350	384	316	272	427	420	477	392
345	455	360	431	430	339	410	326

(a) Present the data graphically. Are there outliers or strong skewness that might prevent the use of t procedures?

(b) Is there good evidence that chicks fed high-lysine corn gain weight faster? Carry out a test and report your conclusions.

(c) Give a 95% confidence interval for the mean extra weight gain in chicks fed high-lysine corn.

7.36 **Students' attitudes.** The Survey of Study Habits and Attitudes (SSHA) is a psychological test that measures the motivation, attitude toward school, and study habits of students. Scores range from 0 to 200. A selective private college gives the SSHA to an SRS of both male and female first-year students. The data for the women are as follows:

154	109	137	115	152	140	154	178	101
103	126	126	137	165	165	129	200	148

Here are the scores of the men:

108	140	114	91	180	115	126	92	169	146
109	132	75	88	113	151	70	115	187	104

(a) Examine each sample graphically, with special attention to outliers and skewness. Is use of a t procedure acceptable for these data?

(b) Most studies have found that the mean SSHA score for men is lower than the mean score in a comparable group of women. Is this true for first-year students at this college? Carry out a test and give your conclusions.

(c) Give a 90% confidence interval for the mean difference between the SSHA scores of male and female first-year students at this college.

More accurate levels in the t procedures*

The two-sample t statistic does not have a t distribution. Moreover, the exact distribution changes as the unknown population standard deviations σ_1 and σ_2 change. However, an excellent approximation is available.

---

*This section can be omitted unless you are using statistical software and wish to understand what the software does.

APPROXIMATE DISTRIBUTION OF THE TWO-SAMPLE t STATISTIC

The distribution of the two-sample t statistic is close to the t distribution with degrees of freedom df given by

$$df = \frac{\left(\dfrac{s_1^2}{n_1} + \dfrac{s_2^2}{n_2}\right)^2}{\dfrac{1}{n_1 - 1}\left(\dfrac{s_1^2}{n_1}\right)^2 + \dfrac{1}{n_2 - 1}\left(\dfrac{s_2^2}{n_2}\right)^2}$$

This approximation is quite accurate when both sample sizes n_1 and n_2 are 5 or larger.

The t procedures remain exactly as before except that we use the t distribution with df degrees of freedom to give critical values and P-values.

EXAMPLE 7.10 **Does polyester decay?**

In the experiment of Examples 7.7 and 7.8, the data on buried polyester fabric gave

Group	Treatment	n	$\bar{x}$	s
1	2 weeks	5	123.80	4.60
2	16 weeks	5	116.40	16.09

For improved accuracy, we can use critical points from the t distribution with degrees of freedom df given by

$$df = \frac{\left(\dfrac{4.60^2}{5} + \dfrac{16.09^2}{5}\right)^2}{\dfrac{1}{4}\left(\dfrac{4.60^2}{5}\right)^2 + \dfrac{1}{4}\left(\dfrac{16.09^2}{5}\right)^2}$$

$$= \frac{3137.08}{674.71} = 4.65$$

Notice that the degrees of freedom df is not a whole number. In this example, there is little difference between the conservative degrees of freedom 4 and the more elaborate result 4.65.

The two-sample t procedures are exactly as before, except that we use a t distribution with more degrees of freedom. The number df is always at least as large as the smaller of $n_1 - 1$ and $n_2 - 1$. On the other hand, df is never larger than the sum $n_1 + n_2 - 2$ of the two individual degrees of freedom. The degrees of freedom df is generally not a whole number. There is a t distribution for any positive degrees of freedom, even though Table C contains entries only for whole-number degrees of freedom. Some statistical software packages find df and then use the t distribution with the next smaller whole-number degrees of freedom. Others use $t(df)$ even when df is not a whole number. We do not recommend regular use of this method unless a computer is doing the arithmetic. With a computer, the more accurate procedures are painless. Here is the output from the SAS statistical software for the fabric strength data:[24]

```
                      TTEST  PROCEDURE


Variable:  STRENGTH

WEEKS     N                Mean          Std Dev          Std Error
-----------------------------------------------------------------------
    2     5        123.80000000        4.60434577        2.05912603
   16     5        116.40000000       16.08726204        7.19444230


Variances    T     DF    Prob>|T|
-------------------------------------------
Unequal    0.9889   4.7    0.3718
Equal      0.9889   8.0    0.3517
```

SAS reports the results of two t procedures: the general two-sample procedure ("Unequal" variances) and a special procedure that assumes that the two populations have the same standard deviation. We are interested in the first of these procedures. The two-sample t statistic has the value $t = 0.9889$, agreeing with our result in Example 7.8. The degrees of freedom are df $= 4.7$, agreeing up to roundoff error with our result in Example 7.10. The two-sided P-value from the $t(4.7)$ distribution is 0.3718. Because SAS always gives two-sided P-values, we must divide by 2 to find that the one-sided result is $P = 0.1859$.

The difference between the t procedures using the conservative and the approximately correct distributions is rarely of practical importance. That is why we recommend the simpler, conservative procedure for inference without a computer.

APPLY YOUR KNOWLEDGE

7.37 **DDT poisoning.** In a randomized comparative experiment, researchers compared 6 white rats poisoned with DDT with a control group of 6 unpoisoned rats. Electrical measurements of nerve activity are the main clue to the nature of DDT poisoning. When a nerve is stimulated, its electrical response shows a sharp spike followed by a much smaller second spike. The experiment found that the second spike is larger in rats fed DDT than in normal rats.[25]

The researchers measured the height of the second spike as a percent of the first spike when a nerve in the rat's leg was stimulated. For the poisoned rats the results were

12.207   16.869   25.050   22.429   8.456   20.589

The control group data were

11.074   9.686   12.064   9.351   8.182   6.642

Both populations are reasonably normal, as far as can be judged from 6 observations. Here is the output from the SAS statistical software system for these data:

```
                   TTEST PROCEDURE

Variable: SPIKE

GROUP      N            Mean          Std Dev        Std Error
-------------------------------------------------------------
DDT        6      17.60000000     6.34014839      2.58835474
CONTROL    6       9.49983333     1.95005932      0.79610839


Variances        T      DF   Prob>|T|
-------------------------------------
Unequal      2.9912     5.9    0.0247
Equal        2.9912    10.0    0.0135
```

(a) The researchers wondered if poisoned rats differ from unpoisoned rats. State H_0 and H_a.

(b) What is the value of the two-sample t statistic and its P-value? (Note that SAS provides two-sided P-values. If you need a one-sided P-value, divide the two-sided value by 2.) What do you conclude?

(c) Give a 90% confidence interval for the mean difference between poisoned and unpoisoned rats. (When software gives a degrees of freedom that is not a whole number, use the next smaller degrees of freedom in Table C.)

7.38   Exercise 7.37 reports the analysis of data on the effects of DDT poisoning. The software uses the two-sample t test with degrees of freedom df given in the box on page 403. Starting from the computer's results for $\bar{x}_i$ and s_i, verify the computer's values for the test statistic $t = 2.99$ and the degrees of freedom df $= 5.9$.

7.39 **Students' self-concept.** Here is SAS output for a study of the self-concept of seventh-grade students. The variable SC is the score on the Piers-Harris Self Concept Scale. The analysis was done to see if male and female students differ in mean self-concept score.[26]

```
                            TTEST PROCEDURE

Variable: SC

SEX     N              Mean           Std Dev          Std Error
---------------------------------------------------------------
F       31       55.51612903       12.69611743       2.28029001
M       47       57.91489362       12.26488410       1.78901722


Variances       T      DF    Prob>|T|
-------------------------------------
Unequal     -0.8276   62.8    0.4110
Equal       -0.8336   76.0    0.4071
```

Write a sentence or two summarizing the comparison of females and males, as if you were preparing a report for publication.

The pooled two-sample t procedures*

In Example 7.10, the software offered a choice between two t tests. One is labeled for "unequal" variances, the other for "equal" variances. The "unequal" variance procedure is our two-sample t. This test is valid whether or not the population variances are equal. The other choice is a special version of the two-sample t statistic that assumes that the two populations have the same variance. This procedure averages (the statistical term is "pools") the two sample variances to estimate the common population variance. The resulting statistic is called the *pooled two-sample t statistic*. It is equal to our t statistic if the two sample sizes are the same, but not otherwise. We could choose to use the pooled t for both tests and confidence intervals.

The pooled t statistic has the advantage that it has exactly the t distribution with $n_1 + n_2 - 2$ degrees of freedom *if* the two population variances really are equal. Of course, the population variances are often not equal. Moreover, the assumption of equal variances is hard to check from the data. The pooled t was in common use before software made it easy to use the accurate approximation to the distribution of our two-sample t statistic. Now it is useful only in special situations. We cannot use the pooled t in Example 7.10, for example, because it is clear that the variance is much larger among fabric strips buried for 16 weeks.

---

*This is a special topic that is optional.

SECTION 7.2 S u m m a r y

The data in a **two-sample problem** are two independent SRSs, each drawn from a separate normally distributed population.

Tests and confidence intervals for the difference between the means μ_1 and μ_2 of the two populations start from the difference $\bar{x}_1 - \bar{x}_2$ of the two sample means. Because of the central limit theorem, the resulting procedures are approximately correct for other population distributions when the sample sizes are large.

Draw independent SRSs of sizes n_1 and n_2 from two normal populations with parameters μ_1, σ_1 and μ_2, σ_2. The **two-sample t statistic** is

$$t = \frac{(\bar{x}_1 - \bar{x}_2) - (\mu_1 - \mu_2)}{\sqrt{\dfrac{s_1^2}{n_1} + \dfrac{s_2^2}{n_2}}}$$

The statistic t does *not* have exactly a t distribution.

For conservative inference procedures to compare μ_1 and μ_2, use the two-sample t statistic with the $t(k)$ distribution. The degrees of freedom k is the smaller of $n_1 - 1$ and $n_2 - 1$. For more accurate probability values, use the $t(\text{df})$ distribution with degrees of freedom df estimated from the data. This is the usual procedure in statistical software.

The **confidence interval** for $\mu_1 - \mu_2$ given by

$$(\bar{x}_1 - \bar{x}_2) \pm t^* \sqrt{\frac{s_1^2}{n_1} + \frac{s_2^2}{n_2}}$$

has confidence level at least C if t^* is the upper $(1 - C)/2$ critical value for $t(k)$ with k the smaller of $n_1 - 1$ and $n_2 - 1$.

Significance tests for $H_0 : \mu_1 = \mu_2$ based on

$$t = \frac{\bar{x}_1 - \bar{x}_2}{\sqrt{\dfrac{s_1^2}{n_1} + \dfrac{s_2^2}{n_2}}}$$

have a true P-value no higher than that calculated from $t(k)$.

The guidelines for practical use of two-sample t procedures are similar to those for one-sample t procedures. Equal sample sizes are recommended.

SECTION 7.2 E x e r c i s e s

In exercises that call for two-sample t procedures, you may use as the degrees of freedom either the smaller of $n_1 - 1$ and $n_2 - 1$ or the more exact value df given in the box on page 403. We recommend the first choice unless you are using a computer. Many of these exercises ask you to think about issues of statistical practice as well as to carry out t procedures.

7.40 **Treating scrapie.** Exercise 7.30 (page 396) contains the results of a study to determine whether IDX is an effective treatment of scrapie.

(a) Is there good evidence that hamsters treated with IDX live longer on the average?

(b) Give a 95% confidence interval for the mean amount by which IDX prolongs life.

7.41 **Talented 13-year-olds.** The Johns Hopkins Regional Talent Searches give the SAT college entrance exams to 13-year-olds. In all, 19,883 males and 19,937 females took the tests between 1980 and 1982. The mean scores of males and females on the verbal test are nearly equal, but there is a clear difference between the sexes on the mathematics test. The reason for this difference is not understood. Here are the data:[27]

Group	$\bar{x}$	s
Males	416	87
Females	386	74

Give a 99% confidence interval for the difference between the mean score for males and the mean score for females in the population that Johns Hopkins searches. Must SAT scores have a normal distribution in order for your confidence interval to be valid? Why?

7.42 **Extraterrestrial handedness?** Molecules often have "left-handed" and "right-handed" versions. Some classes of molecules found in life on earth are almost entirely left-handed. Did this left-handedness precede the origin of life? To find out, scientists analyze meteorites from space. To correct for bias in the sensitive analysis, they also analyzed standard compounds known to be even-handed. Here are the results for the percents of left-handed forms of one molecule in two analyses:[28]

	Meteorite			Standard		
Analysis	n	$\bar{x}$	s	n	$\bar{x}$	s
1	5	52.6	0.5	14	48.8	1.9
2	10	51.7	0.4	13	49.0	1.3

The researchers used the t test to see if the meteorite had a significantly higher percent than the standard. Carry out the tests and report the results. The researchers concluded, "The observations suggest that organic matter of extraterrestrial origin could have played an essential role in the origin of terrestrial life."

7.43 **Active versus passive learning.** A study of computer-assisted learning examined the learning of "Blissymbols" by children. Blissymbols are pictographs (think of Egyptian hieroglyphs) that are sometimes used to help learning-impaired children communicate. The researcher designed two computer lessons that taught the same content using the same examples. One lesson required the children to interact with the material, while in the other the children controlled only the pace of the lesson. Call these two styles "Active" and "Passive." After the lesson, the computer presented a quiz that asked the children to identify 56 Blissymbols. Here are the numbers of correct identifications by the 24 children in the Active group:[29]

29	28	24	31	15	24	27	23	20	22	23	21
24	35	21	24	44	28	17	21	21	20	28	16

The 24 children in the Passive group had these counts of correct identifications:

16	14	17	15	26	17	12	25	21	20	18	21
20	16	18	15	26	15	13	17	21	19	15	12

(a) Is there good evidence that active learning is superior to passive learning? State hypotheses, give a test and its *P*-value, and state your conclusion.

(b) Give a 90% confidence interval for the mean number of Blissymbols identified correctly in a large population of children after the Active computer lesson.

(c) What assumptions do your procedures in (a) and (b) require? Which of these assumptions can you use the data to check? Examine the data to check the assumptions and report your results.

7.44 **IQ scores for boys and girls.** Here are the IQ test scores of 31 seventh-grade girls in a Midwest school district:[30]

114	100	104	89	102	91	114	114	103	105	
108	130	120	132	111	128	118	119	86	72	
111	103	74	112	107	103	98	96	112	112	93

The IQ test scores of 47 seventh-grade boys in the same district are

111	107	100	107	115	111	97	112	104	106	113
109	113	128	128	118	113	124	127	136	106	123
124	126	116	127	119	97	102	110	120	103	115
93	123	79	119	110	110	107	105	105	110	77
90	114	106								

(a) Find the mean scores for girls and for boys. It is common for boys to have somewhat higher scores on standardized tests. Is that true here?

(b) Make stemplots or histograms of both sets of data. Because the distributions are reasonably symmetric with no extreme outliers, the t procedures will work well.

(c) Treat these data as SRSs from all seventh-grade students in the district. Is there good evidence that girls and boys differ in their mean IQ scores?

(d) Give a 90% confidence interval for the difference between the mean IQ scores of all boys and girls in the district.

(e) What other information would you ask for before accepting the results as describing all seventh-graders in the school district?

7.45 Fitness and personality. Physical fitness is related to personality characteristics. In one study of this relationship, middle-aged college faculty who had volunteered for a fitness program were divided into low-fitness and high-fitness groups based on a physical examination. The subjects then took the Cattell Sixteen Personality Factor Questionnaire. Here are the data for the "ego strength" personality factor:[31]

Group	Fitness	n	$\bar{x}$	s
1	Low	14	4.64	0.69
2	High	14	6.43	0.43

(a) Is the difference in mean ego strength significant at the 5% level? At the 1% level? Be sure to state H_0 and H_a.

(b) You should be hesitant to generalize these results to the population of all middle-aged men. Explain why.

7.46 Market research. A market research firm supplies manufacturers with estimates of the retail sales of their products from samples of retail stores. Marketing managers are prone to look at the estimate and ignore sampling error. An SRS of 75 stores this month shows mean sales of 52 units of a small appliance, with standard deviation 13 units. During the same month last year, an SRS of 53 stores gave mean sales of 49 units, with standard deviation 11 units. An increase from 49 to 52 is a rise of 6%. The marketing manager is happy, because sales are up 6%.

(a) Use the two-sample t procedure to give a 95% confidence interval for the difference between this year and last year in the mean number of units sold at all retail stores.

(b) Explain in language that the manager can understand why he cannot be confident that sales rose by 6%, and that in fact sales may even have dropped.

7.47 **Will they charge more?** A bank compares two proposals to increase the amount that its credit card customers charge on their cards. (The bank earns a percentage of the amount charged, paid by the stores that accept the card.) Proposal A offers to eliminate the annual fee for customers who charge $2400 or more during the year. Proposal B offers a small percent of the total amount charged as a cash rebate at the end of the year. The bank offers each proposal to an SRS of 150 of its credit card customers. At the end of the year, the total amount charged by each customer is recorded. Here are the summary statistics:

Group	n	$\bar{x}$	s
A	150	$1987	$392
B	150	$2056	$413

(a) Do the data show a significant difference between the mean amounts charged by customers offered the two plans? Give the null and alternative hypotheses, and calculate the two-sample t statistic. Obtain the P-value. State your practical conclusions.

(b) The distributions of amounts charged are skewed to the right, but outliers are prevented by the limits that the bank imposes on credit balances. Do you think that skewness threatens the validity of the test that you used in (a)? Explain your answer.

(c) Is the bank's study an experiment? Why? How does this affect the conclusions the bank can draw from the study?

7.48 **Learning to solve a maze (EESEE).** Table 7.1 (page 375) contains the times required to complete a maze for 21 subjects wearing scented and unscented masks. Example 7.3 used the matched pairs t test to show that the scent makes no significant difference in the time. Now we ask whether there is a learning effect, so that subjects complete the maze faster on their second trial. All of the odd-numbered subjects in Table 7.1 first worked the maze wearing the unscented mask. Even-numbered subjects wore the scented mask first. The numbers were assigned at random.

(a) We will compare the unscented times for "unscented first" subjects with the unscented times for the "scented first" subjects. Explain why this comparison requires two-sample procedures.

(b) We suspect that on the average subjects are slower when the unscented time is their first trial. Make a back-to-back stemplot of unscented times for "scented first" and "unscented first" subjects. Find the mean unscented times for these two groups. Do the data appear to support our suspicion? Do the data have features that prevent use of the t procedures?

(c) Do the data give statistically significant support to our suspicion? State hypotheses, carry out a test, and report your conclusion.

7.49 **Studying speech.** Researchers studying the learning of speech often compare measurements made on the recorded speech of adults and children. One variable of interest is called the voice onset time (VOT). Here are the results for 6-year-old children and adults asked to pronounce the word "bees." The VOT is measured in milliseconds and can be either positive or negative.[32]

Group	n	$\bar{x}$	s
Children	10	-3.67	33.89
Adults	20	-23.17	50.74

(a) The researchers were investigating whether VOT distinguishes adults from children. State H_0 and H_a and carry out a two-sample t test. Give a P-value and report your conclusions.

(b) Give a 95% confidence interval for the difference in mean VOTs when pronouncing the word "bees." Explain why you knew from your result in (a) that this interval would contain 0 (no difference).

7.50 The researchers in the study discussed in Exercise 7.49 looked at VOTs for adults and children pronouncing many different words. Explain why they should not do a separate two-sample t test for each word and conclude that those words with a significant difference (say $P < 0.05$) distinguish children from adults. (The researchers did not make this mistake.)

7.51 **Optional: The power of a two-sample t test.** A bank asks you to compare two ways to increase the use of their credit cards. Plan A would offer customers a cash-back rebate based on their total amount charged. Plan B would reduce the interest rate charged on card balances. The bank thinks that Plan B will be more effective. The response variable is the total amount a customer charges during the test period. You decide to offer each of Plan A and Plan B to a separate SRS of the bank's credit card customers. In the past, the mean amount charged in a six month period has been about $1100, with a standard deviation of $400. Will a two-sample t test based on SRSs of 350 customers in each group detect a difference of $100 in the mean amounts charged under the two plans?

We will compute the approximate power of the two-sample t test of

$$H_0 : \mu_B = \mu_A$$

$$H_a : \mu_B > \mu_A$$

against the specific alternative $\mu_B - \mu_A = 100$. We will use the past value \$400 as a rough estimate of both the population σ's and future sample s's.

(a) What is the approximate value of the $\alpha = 0.05$ critical value t^* for the two-sample t statistic when $n_1 = n_2 = 350$?

(b) **Step 1.** *Write the rule for rejecting H_0 in terms of $\bar{x}_B - \bar{x}_A$.* The test rejects H_0 when

$$\frac{\bar{x}_B - \bar{x}_A}{\sqrt{\dfrac{s_B^2}{n_B} + \dfrac{s_A^2}{n_A}}} \geq t^*$$

Take both s_B and s_A to be 400, and n_B and n_A to be 350. Find the number c such that the test rejects H_0 when $\bar{x}_B - \bar{x}_A \geq c$.

(c) **Step 2.** *The power is the probability of rejecting H_0 when the alternative is true.* Suppose that $\mu_B - \mu_A = 100$ and that both σ_B and σ_A are 400. The power we seek is the probability that $\bar{x}_B - \bar{x}_A \geq c$ under these assumptions. Calculate the power.

7.3 Inference for Population Spread*

The two most basic descriptive features of a distribution are its center and spread. In a normal population, we measure center by the mean and spread by the standard deviation. We use the t procedures for inference about population means for normal populations, and we know that t procedures are often useful for nonnormal populations as well. It is natural to turn next to inference about the standard deviations of normal populations. Our advice here is short and clear: Don't do it without expert advice.

Avoid inference about standard deviations

There are methods for inference about the standard deviations of normal populations. We will describe the most common such method, the F test for comparing the spread of two normal populations. Unlike the t procedures for means, the F test and other procedures for standard deviations are extremely sensitive to nonnormal distributions. This lack of robustness does not improve in large samples. It is difficult in practice to tell whether a significant F-value is evidence of unequal population spreads or simply a sign that the populations are not normal.

---

*This section is not required for an understanding of later material, except for Chapter 10.

The deeper difficulty underlying the very poor robustness of normal population procedures for inference about spread already appeared in our work on describing data. The standard deviation is a natural measure of spread for normal distributions but not for distributions in general. In fact, because skewed distributions have unequally spread tails, no single numerical measure does a good job of describing the spread of a skewed distribution. In summary, the standard deviation is not always a useful parameter, and even when it is (for symmetric distributions), the results of inference are not trustworthy. Consequently, we do not recommend trying to do inference about population standard deviations in basic statistical practice.[33]

It was once common to test equality of standard deviations as a preliminary to performing the pooled two-sample t test for equality of two population means. It is better practice to check the distributions graphically, with special attention to skewness and outliers, and to use the version of the two-sample t featured in Section 7.2. This test does not require equal standard deviations.

The *F* test for comparing two standard deviations

Because of the limited usefulness of procedures for inference about the standard deviations of normal distributions, we will present only one such procedure. Suppose that we have independent SRSs from two normal populations, a sample of size n_1 from $N(\mu_1, \sigma_1)$ and a sample of size n_2 from $N(\mu_2, \sigma_2)$. The population means and standard deviations are all unknown. The two-sample t test examines whether the means are equal in this setting. To test the hypothesis of equal spread,

$$H_0 : \sigma_1 = \sigma_2$$

$$H_a : \sigma_1 \neq \sigma_2$$

we use the ratio of sample variances. This is the F *statistic*.

THE *F* STATISTIC AND *F* DISTRIBUTIONS

When s_1^2 and s_2^2 are sample variances from independent SRSs of sizes n_1 and n_2 drawn from normal populations, the **F statistic**

$$F = \frac{s_1^2}{s_2^2}$$

has the **F distribution** with $n_1 - 1$ and $n_2 - 1$ degrees of freedom when $H_0 : \sigma_1 = \sigma_2$ is true.

The F distributions are a family of distributions with two parameters. The parameters are the degrees of freedom of the sample variances in the numerator and denominator of the F statistic. The numerator degrees of freedom are

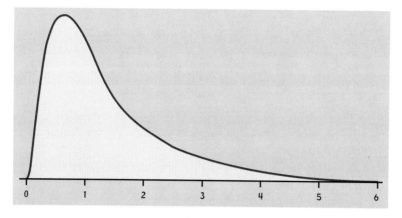

Figure 7.9 *The density curve for the F(9, 10) distribution. The F distributions are skewed to the right.*

always mentioned first. Interchanging the degrees of freedom changes the distribution, so the order is important. Our brief notation will be $F(j, k)$ for the F distribution with j degrees of freedom in the numerator and k in the denominator. The F distributions are right-skewed. The density curve in Figure 7.9 illustrates the shape. Because sample variances cannot be negative, the F statistic takes only positive values, and the F distribution has no probability below 0. The peak of the F density curve is near 1. When the two populations have the same standard deviation, we expect the two sample variances to be close in size, so that F takes a value near 1. Values of F far from 1 in either direction provide evidence against the hypothesis of equal standard deviations.

Tables of F critical points are awkward, because we need a separate table for every pair of degrees of freedom j and k. Table D in the back of the book gives upper p critical points of the F distributions for $p = 0.10, 0.05, 0.025, 0.01$, and 0.001. For example, these critical points for the $F(9, 10)$ distribution shown in Figure 7.9 are

p	.10	.05	.025	.01	.001
F^*	2.35	3.02	3.78	4.94	8.96

The skewness of the F distributions causes additional complications. In the symmetric normal and t distributions, the point with probability 0.05 below it is just the negative of the point with probability 0.05 above it. This is not true for F distributions. We therefore need either tables of both the upper and lower tails or some way to eliminate the need for lower tail critical values. Statistical software that does away with the need for tables is very convenient. If you do not use statistical software, arrange the two-sided F test as follows.

CARRYING OUT THE *F* TEST

Step 1. Take the test statistic to be

$$F = \frac{\text{larger } s^2}{\text{smaller } s^2}$$

This amounts to naming the populations so that Population 1 has the larger of the observed sample variances. The resulting F is always 1 or greater.

Step 2. Compare the value of F with critical values from Table D. Then *double* the significance levels from the table to obtain the significance level for the two-sided F test.

The idea is that we calculate the probability in the upper tail and double it to obtain the probability of all ratios on either side of 1 that are at least as improbable as that observed. Remember that the order of the degrees of freedom is important in using Table D.

EXAMPLE 7.11 **Comparing variability**

Example 7.7 describes an experiment to compare the mean breaking strengths of polyester fabric after being buried for 2 weeks and for 16 weeks. The data summaries are

Group	Treatment	n	$\bar{x}$	s
2	2 weeks	5	123.80	4.60
1	16 weeks	5	116.40	16.09

We might also compare the standard deviations to see whether strength loss is more or less variable after 16 weeks. We want to test

$$H_0 : \sigma_1 = \sigma_2$$

$$H_a : \sigma_1 \neq \sigma_2$$

Note that we relabeled the groups so that Group 1 (16 weeks) has the larger standard deviation. The F test statistic is

$$F = \frac{\text{larger } s^2}{\text{smaller } s^2} = \frac{16.09^2}{4.60^2} = 12.23$$

Compare the calculated value $F = 12.23$ with critical points for the $F(4, 4)$ distribution. Table D shows that 12.23 lies between the 0.025 and 0.01 critical values of the $F(4, 4)$ distribution. So the two-sided P-value lies between 0.05 and 0.02. The data show significantly unequal spreads at the 5% level. The P-value depends heavily on the assumption that both samples come from normally distributed populations.

SECTION 7.3 S u m m a r y

Inference procedures for comparing the standard deviations of two normal populations are based on the **F statistic**, which is the ratio of sample variances

$$F = \frac{s_1^2}{s_2^2}$$

If an SRS of size n_1 is drawn from Population 1 and an independent SRS of size n_2 is drawn from Population 2, the F statistic has the **F distribution** $F(n_1 - 1, n_2 - 1)$ if the two population standard deviations σ_1 and σ_2 are in fact equal.

The **F distributions** are skewed to the right and take only values greater than 0. A specific F distribution $F(j, k)$ is fixed by the two **degrees of freedom** j and k.

The two-sided **F test** of $H_0 : \sigma_1 = \sigma_2$ uses the statistic

$$F = \frac{\text{larger } s^2}{\text{smaller } s^2}$$

and doubles the upper tail probability to obtain the P-value.

The F tests and other procedures for inference on the spread of one or more normal distributions are so strongly affected by lack of normality that we do not recommend them for regular use.

SECTION 7.3 E x e r c i s e s

In all exercises calling for use of the F test, assume that both population distributions are very close to normal. The actual data are not always sufficiently normal to justify use of the F test.

7.52   The F statistic $F = s_1^2/s_2^2$ is calculated from samples of sizes $n_1 = 10$ and $n_2 = 8$. (Remember that n_1 is the numerator sample size.)

(a) What is the upper 5% critical value for this F?

(b) In a test of equality of standard deviations against the two-sided alternative, this statistic has the value $F = 3.45$. Is this value significant at the 10% level? Is it significant at the 5% level?

7.53   The F statistic for equality of standard deviations based on samples of sizes $n_1 = 21$ and $n_2 = 16$ takes the value $F = 2.78$.

(a) Is this significant evidence of unequal population standard deviations at the 5% level? At the 1% level?

(b) Between which two values obtained from Table D does the P-value of the test fall?

7.54 **DDT poisoning.** The sample variance for the treatment group in the DDT experiment of Exercise 7.37 (page 404) is more than 10 times as large as the sample variance for the control group. Calculate the F statistic. Can you reject the hypothesis of equal population standard deviations at the 5% significance level? At the 1% level?

7.55 **Treating scrapie.** The report in Exercise 7.30 (page 396) suggests that hamsters in the longer-lived group also had more variation in their length of life. It is common to see s increase along with $\bar{x}$ when we compare groups. Do the data give significant evidence of unequal standard deviations?

7.56 **Social insight among men and women.** The data in Example 7.9 (page 398) show that scores of men and women on the Chapin Social Insight Test have similar sample standard deviations. The samples are large, however, so we might nonetheless find evidence of a significant difference between the population standard deviations. Is this the case? (Use the $F(120, 100)$ critical values from Table D.)

7.57 **Studying speech.** The data for VOTs of children and adults in Exercise 7.49 (page 412) show quite different sample standard deviations. How statistically significant is the observed inequality?

7.58 **Students' attitudes.** Return to the SSHA data in Exercise 7.36 (page 402). We want to know if the spread of SSHA scores is different among women and among men at this college. Use the F test to obtain a conclusion.

STATISTICS IN SUMMARY

This chapter presents t tests and confidence intervals for inference about the mean of a single population and for comparing the means of two populations. The one-sample t procedures do inference about one mean and the two-sample t procedures compare two means. Matched pairs studies use one-sample procedures because you first create a single sample by taking the differences in the responses within each pair. These t procedures are among the most common methods of statistical inference. The Statistics in Summary figure on the next page helps you decide when to use them. Before you use any inference method, think about the design of the study and examine the data for outliers and other problems.

The t procedures require that the data be random samples and that the distribution of the population or populations be normal. One reason for the wide use of t procedures is that they are not very strongly affected by lack of normality. If you can't regard your data as a random sample, however, the results of inference may be of little value.

The chapter exercises are important in this and later chapters. You must now recognize problem settings and decide which of the methods presented in the chapter fits. In this chapter, you must recognize one-sample studies, matched

STATISTICS IN SUMMARY
The *t* Procedures for Means

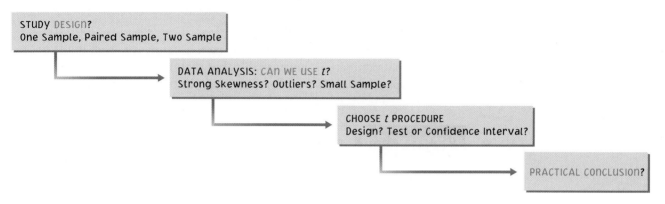

pairs studies, and two-sample studies. Here are the most important skills you should have after reading this chapter.

A. RECOGNITION

1. Recognize when a problem requires inference about a mean or comparing two means.
2. Recognize from the design of a study whether one-sample, matched pairs, or two-sample procedures are needed.

B. ONE-SAMPLE *t* PROCEDURES

1. Recognize when the *t* procedures are appropriate in practice, in particular that they are quite robust against lack of normality but are influenced by outliers.
2. Also recognize when the design of the study, outliers, or a small sample from a skewed distribution make the *t* procedures risky.
3. Use *t* to obtain a confidence interval at a stated level of confidence for the mean μ of a population.
4. Carry out a *t* test for the hypothesis that a population mean μ has a specified value against either a one-sided or a two-sided alternative. Use Table C of *t* distribution critical values to approximate the *P*-value or carry out a fixed α test.
5. Recognize matched pairs data and use the *t* procedures to obtain confidence intervals and to perform tests of significance for such data.

C. TWO-SAMPLE *t* PROCEDURES

1. Recognize when the two-sample *t* procedures are appropriate in practice.
2. Give a confidence interval for the difference between two means. Use the two-sample *t* statistic with conservative degrees of freedom if you do not have statistical software. Use software if you have it.

3. Test the hypothesis that two populations have equal means against either a one-sided or a two-sided alternative. Use the two-sample t test with conservative degrees of freedom if you do not have statistical software. Use software if you have it.

4. Know that procedures for comparing the standard deviations of two normal populations are available, but that these procedures are risky because they are not at all robust against nonnormal distributions.

CHAPTER 7 Review Exercises

7.59 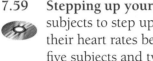 **Stepping up your heart rate (EESEE).** A student project asked subjects to step up and down for three minutes and measured their heart rates before and after the exercise. Here are data for five subjects and two treatments: stepping at a low rate (14 steps per minute) and at a medium rate (21 steps per minute). For each subject, we give the resting heart rate (beats per minute) and the heart rate at the end of the exercise.[34]

	Low Rate		Medium Rate	
Subject	Resting	Final	Resting	Final
1	60	75	63	84
2	90	99	69	93
3	87	93	81	96
4	78	87	75	90
5	84	84	90	108

(a) Does exercise at the low rate raise heart rate significantly? By how much? (Use a 90% confidence interval to answer the second question.)

(b) Does exercise at the medium rate raise heart rate significantly? By how much?

(c) Does medium exercise raise heart rate more than low exercise?

7.60 **Weight-loss programs.** In a study of the effectiveness of a weight-loss program, 47 subjects who were at least 20% overweight took part in the program for 10 weeks. Private weighings determined each subject's weight at the beginning of the program and 6 months after the program's end. The matched pairs t test was used to assess the significance of the average weight loss. The paper reporting the study said, "The subjects lost a significant amount of weight over time, $t(46) = 4.68$, $p < 0.01$." It is common to report the results of statistical tests in this abbreviated style.[35]

(a) Why was the matched pairs t test appropriate?

(b) Explain to someone who knows no statistics but is interested in weight-loss programs what the practical conclusion is.

(c) The paper follows the tradition of reporting significance only at fixed levels such as $\alpha = 0.01$. In fact, the results are more significant than "$p < 0.01$" suggests. Use Table C to say more about the P-value of the t test.

7.61 **Expensive ads.** Consumers who think a product's advertising is expensive often also think the product must be of high quality. Can other information undermine this effect? To find out, marketing researchers did an experiment. The subjects were 90 women from the clerical and administrative staff of a large organization. All subjects read an ad that described a fictional line of food products called "Five Chefs." The ad also described the major TV commercials that would soon be shown, an unusual expense for this type of product. The 45 women in the control group read nothing else. The 45 in the "undermine group" also read a news story headlined "No Link between Advertising Spending and New Product Quality."

   All the subjects then rated the quality of Five Chefs products on a seven-point scale. The study report said, "The mean quality ratings were significantly lower in the undermine treatment ($\overline{X}_A = 4.56$) than in the control treatment ($\overline{X}_C = 5.05$; $t = 2.64$, $p < 0.01$)."[36]

(a) Is the matched pairs t test or the two-sample t test the right test in this setting? Why?

(b) What degrees of freedom would you use for the t statistic you chose in (a)?

(c) The distribution of individual responses is not normal, because there is only a seven-point scale. Why is it nonetheless proper to use a t test?

7.62 **New welfare programs.** A major study of alternative welfare programs randomly assigned women on welfare to one of two programs, called "WIN" and "Options." WIN was the existing program. The new Options program gave more incentives to work. An important question was how much more (on the average) women in Options earned than those in WIN. Here is Minitab output for earnings in dollars over a three-year period:[37]

```
TWOSAMPLE T FOR 'OPT' VS 'WIN'

           N     MEAN      STDEV     SE MEAN
OPT     1362     7638       289      7.8309
WIN     1395     6595       247      6.6132

95 PCT CI FOR MU OPT - MU WIN: (1022.90, 1063.10)
```

(a) Give a 99% confidence interval for the amount by which the mean earnings of Options participants exceeded the mean earnings of WIN subjects. (Minitab will give a 99% confidence interval if you instruct it to do so. Here we have only the basic output, which includes the 95% confidence interval.)

(b) The distribution of incomes is strongly skewed to the right but includes no extreme outliers because all the subjects were on welfare. What fact about these data allows us to use t procedures despite the strong skewness?

7.63 **Fertilizing pasture.** An agricultural researcher thinks that applying potassium fertilizer to grasslands at several times during the growing season may give higher yields than applying only in the spring. He therefore compares two treatments: 100 pounds per acre of potassium in the spring (Treatment 1) and 50, 25, and 25 pounds per acre applied in the spring, early summer, and late summer (Treatment 2). He continues the experiment over several years because growing conditions vary from year to year.

The table below gives the yields, in pounds of dry matter per acre. It is known from long experience that yields vary roughly normally.[38]

Treatment	Year 1	Year 2	Year 3	Year 4	Year 5
1	3902	4281	5135	5350	5746
2	3970	4271	5440	5490	6028

(a) Do the data give good evidence that Treatment 2 leads to higher average yields? (State hypotheses, carry out a test, give a P-value as exact as the tables in the text allow, and state your conclusions in words.)

(b) Give a 98% confidence interval for the mean increase in yield due to spreading potassium applications over the growing season.

7.64 **Nitrites and bacteria.** Nitrites are often added to meat products as preservatives. In a study of the effect of nitrites on bacteria, researchers measured the rate of uptake of an amino acid for 60 cultures of bacteria: 30 growing in a medium to which nitrites had been added and another 30 growing in a standard medium as a control group. Table 7.7 gives the data from this study. Examine each of the two samples and briefly describe their distribution. Carry out a test of the research hypothesis that nitrites decrease amino acid uptake, and report your results.

7.65 **Ancient air.** The composition of the earth's atmosphere may have changed over time. To try to discover the nature of the atmosphere long ago, we can examine the gas in bubbles inside ancient amber. Amber is tree resin that has hardened and been trapped in rocks. The

Table 7.7 Amino acid uptake by bacteria under two conditions

Control			Nitrite		
6,450	8,709	9,361	8,303	8,252	6,594
9,011	9,036	8,195	8,534	10,227	6,642
7,821	9,996	8,202	7,688	6,811	8,766
6,579	10,333	7,859	8,568	7,708	9,893
8,066	7,408	7,885	8,100	6,281	7,689
6,679	8,621	7,688	8,040	9,489	7,360
9,032	7,128	5,593	5,589	9,460	8,874
7,061	8,128	7,150	6,529	6,201	7,605
8,368	8,516	8,100	8,106	4,972	7,259
7,238	8,830	9,145	7,901	8,226	8,552

gas in bubbles within amber should be a sample of the atmosphere at the time the amber was formed. Measurements on specimens of amber from the late Cretaceous era (75 to 95 million years ago) give these percents of nitrogen:

63.4   65.0   64.4   63.3   54.8   64.5   60.8   49.1   51.0

These values are quite different from the present 78.1% of nitrogen in the atmosphere. Assume (this is not yet agreed on by experts) that these observations are an SRS from the late Cretaceous atmosphere.[39]

(a) Graph the data, and comment on skewness and outliers.

(b) The t procedures will be only approximate in this case. Give a 95% t confidence interval for the mean percent of nitrogen in ancient air.

7.66 **Testing pharmaceuticals.** A pharmaceutical manufacturer does a chemical analysis to check the potency of products. The standard release potency for cephalothin crystals is 910. An assay of 16 lots gives the following potency data:

897   914   913   906   916   918   905   921
918   906   895   893   908   906   907   901

(a) Check the data for outliers or strong skewness that might threaten the validity of the t procedures.

(b) Give a 95% confidence interval for the mean potency.

(c) Is there significant evidence at the 5% level that the mean potency is not equal to the standard release potency?

7.67 **Double-decker diets.** A British study compared the food and drink intake of 98 drivers and 83 conductors of London double-decker buses. The conductors' jobs require more physical activity. The article reporting the study gives the data as "Mean daily consumption (± s. e.)." Some of the study results appear below.[40]

	Drivers	Conductors
Total calories	2821 ± 44	2844 ± 48
Alcohol (grams)	0.24 ± 0.06	0.39 ± 0.11

(a) What does "s. e." stand for? Give $\bar{x}$ and s for each of the four sets of measurements.

(b) Is there significant evidence at the 5% level that conductors and drivers consume different numbers of calories per day?

(c) How significant is the observed difference in mean alcohol consumption?

(d) Give a 90% confidence interval for the mean daily alcohol consumption of London double-decker bus conductors.

(e) Give an 80% confidence interval for the difference in mean daily alcohol consumption between drivers and conductors.

7.68 **California counties.** You look up a census report that gives the populations of all 58 counties in the state of California. Is it proper to apply the one-sample t method to these data to give a 95% confidence interval for the mean population of a California county? Explain your answer.

7.69 **Lead in soil.** The amount of lead in a type of soil, measured by a standard method, averages 86 parts per million (ppm). A new method is tried on 40 specimens of the soil, yielding a mean of 83 ppm lead and a standard deviation of 10 ppm.

(a) Is there significant evidence at the 1% level that the new method frees less lead from the soil?

(b) A critic argues that because of variations in the soil, the effectiveness of the new method is confounded with characteristics of the particular soil specimens used. Briefly describe a better data production design that avoids this criticism.

7.70 **Cholesterol in dogs.** High levels of cholesterol in the blood are not healthy in either humans or dogs. Because a diet rich in saturated fats raises the cholesterol level, it is plausible that dogs owned as pets have higher cholesterol levels than dogs owned by a veterinary research clinic. "Normal" levels of cholesterol based on the clinic's dogs would then be misleading. A clinic compared healthy dogs it owned with healthy pets brought to the clinic to be neutered. The summary statistics for blood cholesterol levels (milligrams per deciliter of blood) appear below.[41]

Group	n	$\bar{x}$	s
Pets	26	193	68
Clinic	23	174	44

(a) Is there strong evidence that pets have higher mean cholesterol level than clinic dogs? State the H_0 and H_a and carry out an appropriate test. Give the P-value and state your conclusion.

(b) Give a 95% confidence interval for the difference in mean cholesterol levels between pets and clinic dogs.

(c) Give a 95% confidence interval for the mean cholesterol level in pets.

(d) What assumptions must be satisfied to justify the procedures you used in (a), (b), and (c)? Assuming that the cholesterol measurements have no outliers and are not strongly skewed, what is the chief threat to the validity of the results of this study?

7.71 **The density of the earth.** Exercise 1.41 (page 43) gives 29 measurements of the density of the earth, made in 1798 by Henry Cavendish. Display the data graphically to check for skewness and outliers. Then give an estimate for the density of the earth from Cavendish's data and a margin of error for your estimate.

7.72 **Mouse genes.** A study of genetic influences on diabetes compared normal mice with similar mice genetically altered to remove the gene called $aP2$. Mice of both types were allowed to become obese by eating a high-fat diet. The researchers then measured the levels of insulin and glucose in their blood plasma. Here are some excerpts from their findings.[42] The normal mice are called "wild-type" and the altered mice are called "$aP2^{-/-}$."

Each value is the mean $\pm$ SEM of measurements on at least 10 mice. Mean values of each plasma component are compared between $aP2^{-/-}$ mice and wild-type controls by Student's t test (*P < 0.05 and **P < 0.005).

Parameter	Wild-type	$aP2^{-/-}$
Insulin (ng/ml)	5.9 ± 0.9	0.75 ± 0.2**
Glucose (mg/dl)	230 ± 25	150 ± 17*

Despite much greater circulating amounts of insulin, the wild-type mice had higher blood glucose than the $aP2^{-/-}$ animals. These results indicate that the absence of $aP2$ interferes with the development of dietary obesity-induced insulin resistance.

Other biologists are supposed to understand the statistics reported so tersely.

(a) What does "SEM" mean? What is the expression for SEM based on $n, \bar{x}$, and s from a sample?

(b) Which of the t tests we have studied did the researchers apply?

(c) Explain to a biologist who knows no statistics what $P < 0.05$ and $P < 0.005$ mean. Which is stronger evidence of a difference between the two types of mice?

(d) The report says only that the sample sizes were "at least 10." Suppose that the results are based on exactly 10 mice of each type. Use the values in the table to find $\bar{x}$ and s for the two insulin concentrations and carry out a test to assess the significance of the difference in means. What P-value do you obtain?

(e) Do the same thing for the glucose concentrations.

7.73 **(Optional)** Do the data in Example 7.9 (page 398) provide evidence of different standard deviations for Chapin test scores in the populations of female and male college liberal arts majors?

(a) State the hypotheses and carry out the test. Software can assess significance exactly, but inspection of the proper table is enough to draw a conclusion.

(b) Do the large sample sizes allow us to ignore the assumption that the population distributions are normal?

The remaining exercises concern a study of air pollution. One component of air pollution is airborne particulate matter such as dust and smoke. To measure particulate pollution, a vacuum motor draws air through a filter for 24 hours. Weigh the filter at the beginning and end of the period. The weight gained is a measure of the concentration of particles in the air. A study of air pollution made measurements every 6 days with identical instruments in the center of a small city and at a rural location 10 miles southwest of the city. Because the prevailing winds blow from the west, we suspect that the rural readings will generally be lower than the city readings, but that the city readings can be predicted from the rural readings. Table 7.8 gives readings taken every 6 days over a 7-month period. The entry NA means that the reading for that date is not available, usually because of equipment failure.[43]

Missing data are common, especially in field studies like this one. We think that equipment failures are not related to pollution levels. If that is true, the missing data do not introduce bias. We can work with the data that are not missing as if they are a random sample of days. We can analyze these data in different ways to answer different questions. For each of the three exercises below, do a careful descriptive analysis with graphs and summary statistics and whatever formal inference is called for. Then present and interpret your findings.

7.74 **City pollution.** We want to assess the level of particulate pollution in the city center. Describe the distribution of city pollution levels,

Table 7.8 Particulate levels (grams) in two nearby locations

Day	Rural	City	Day	Rural	City
1	NA	39	19	43	42
2	67	68	20	39	38
3	42	42	21	NA	NA
4	33	34	22	52	57
5	46	48	23	48	50
6	NA	82	24	56	58
7	43	45	25	44	45
8	54	NA	26	51	69
9	NA	NA	27	21	23
10	NA	60	28	74	72
11	NA	57	29	48	49
12	NA	NA	30	84	86
13	38	39	31	51	51
14	88	NA	32	43	42
15	108	123	33	45	46
16	57	59	34	41	NA
17	70	71	35	47	44
18	42	41	36	35	42

and estimate the mean particulate level in the city center. (All estimates should include a statistically justified margin of error.)

7.75 **City versus country.** We want to compare the level of particulates in the city with the rural level on the same day. We suspect that pollution is higher in the city, and we hope that a statistical test will show that there is significant evidence to confirm this suspicion. Make a graph to check for conditions that might prevent the use of the test you plan to employ. Your graph should reflect the type of procedure that you will use. Then carry out a significance test and report your conclusion. Also estimate the mean amount by which the city particulate level exceeds the rural level on the same day.

7.76 **Predicting city from country.** We hope to use the rural particulate level to predict the city level on the same day. Make a graph to examine the relationship. Does the graph suggest that using the least-squares regression line for prediction will give approximately correct results over the range of values appearing in the data? Calculate the least-squares line for predicting city pollution from rural pollution. What percent of the observed variation in city pollution levels does this straight-line relationship account for? On the fourteenth date in the series, the rural reading was 88 and the city reading was not available. What do you estimate the city reading to be for that date? (In Chapter 11, we will learn how to give a margin of error for the predictions we make from the regression line.)

Janet Norwood

(LANA HARRIS/AP/WIDE WORLD PHOTOS.)

The Government's Statistician

Modern governments run on statistics. They need data on economic and social trends that are accurate, timely, and free of political influence. Unlike most nations, the United States does not have a single statistical agency such as Statistics Canada. The Bureau of Labor Statistics is one of the government's major statistical offices, and its head, the commissioner of labor statistics, is one of the nation's most influential statisticians.

The data collected by the Bureau of Labor Statistics are often politically sensitive, as when a report released just before an election shows rising unemployment.

The data collected by the Bureau of Labor Statistics are often politically sensitive, as when a report released just before an election shows rising unemployment. To safeguard the bureau's independence, the commissioner is appointed by the President and confirmed by the Senate for a fixed term of four years. The commissioner must have statistical skill, administrative ability, and a talent for working with both Congress and the President.

Janet Norwood served three terms as commissioner, from 1979 to 1991, under three presidents. When she retired, the *New York Times* said (December 31, 1991) that she left with "a near-legendary reputation for nonpartisanship and plaudits that include one senator's designation of her as a 'national treasure.'" Norwood says, "There have been times in the past when commissioners have been in open disagreement with the Secretary of Labor or, in some cases, with the President. We have guarded our professionalism with great care."

Some of the most important statistics produced by the Bureau of Labor Statistics are proportions. The monthly unemployment rate, for example, is the proportion of the labor force that is unemployed this month. Methods for inference about proportions are the topic of this chapter.

Inference for Proportions

The government's data on employment and unemployment are produced
by a statistically designed sample survey.

Introduction

Our discussion of statistical inference to this point has concerned making inferences about population *means*. Now we turn to questions about the *proportion* of some outcome in a population. Here are some examples that call for inference about population proportions.

| EXAMPLE 8.1 | **Risky behavior in the age of AIDS** |

How common is behavior that puts people at risk of AIDS? The National AIDS Behavioral Surveys interviewed a random sample of 2673 adult heterosexuals. Of these, 170 had more than one sexual partner in the past year. That's 6.36% of the sample.[1] Based on these data, what can we say about the percent of all adult heterosexuals who have multiple partners? We want to *estimate a single population proportion*.

| EXAMPLE 8.2 | **Does preschool make a difference?** |

Do preschool programs for poor children make a difference in later life? A study looked at 62 children who were enrolled in a Michigan preschool in the late 1960s and at a control group of 61 similar children who were not enrolled. At 27 years of age, 61% of the preschool group and 80% of the control group had required the help of a social service agency (mainly welfare) in the previous ten years.[2] Is this significant evidence that preschool for poor children reduces later use of social services? We want to *compare two population proportions*.

To do inference about a population mean μ, we use the mean $\bar{x}$ of a random sample from the population. The reasoning of inference starts with the sampling distribution of $\bar{x}$. Now we follow the same pattern, replacing means by proportions.

We are interested in the unknown proportion p of a population that has some outcome. For convenience, call the outcome we are looking for a "success." In Example 8.1, the population is adult heterosexuals, and the parameter p is the proportion who have had more than one sexual partner in the past year. To estimate p, the National AIDS Behavioral Surveys used random dialing of telephone numbers to contact a sample of 2673 people. Of these, 170 said they had multiple sexual partners. The statistic that estimates the parameter p is the *sample proportion* **sample proportion**

$$\hat{p} = \frac{\text{count of successes in the sample}}{\text{count of observations in the sample}}$$

$$= \frac{170}{2673} = 0.0636$$

Read the sample proportion $\hat{p}$ as "p-hat."

You just don't understand

A sample survey of journalists and scientists found quite a communications gap. Journalists think that scientists are arrogant, while scientists think that journalists are ignorant. We won't take sides, but here is one interesting $\hat{p}$ from the survey: 82% of the scientists agree that the "media do not understand statistics well enough to explain new findings" in medicine and other fields.

In each of the following settings:

(a) Describe the population and explain in words what the parameter p is.

(b) Give the numerical value of the statistic $\hat{p}$ that estimates p.

8.1   Tonya wants to estimate what proportion of the students in her dormitory like the dorm food. She interviews an SRS of 50 of the 175 students living in the dormitory. She finds that 14 think the dorm food is good.

8.2   Glenn wonders what proportion of the students at his school think that tuition is too high. He interviews an SRS of 50 of the 2400 students at his college. Thirty-eight of those interviewed think tuition is too high.

8.3   A college president says, "99% of the alumni support my firing of Coach Boggs." You contact an SRS of 200 of the college's 15,000 living alumni and find that 76 of them support firing the coach.

8.1 Inference for a Population Proportion

How good is the statistic $\hat{p}$ as an estimate of the parameter p? To find out, we ask, "What would happen if we took many samples?" The sampling distribution of $\hat{p}$ answers this question. Here are the facts.

The sampling distribution of $\hat{p}$

SAMPLING DISTRIBUTION OF A SAMPLE PROPORTION

Choose an SRS of size n from a large population that contains population proportion p of "successes." Let $\hat{p}$ be the **sample proportion** of successes,

$$\hat{p} = \frac{\text{count of successes in the sample}}{n}$$

Then:

- As the sample size increases, the sampling distribution of $\hat{p}$ becomes **approximately normal**.
- The **mean** of the sampling distribution is p.
- The **standard deviation** of the sampling distribution is

$$\sqrt{\frac{p(1-p)}{n}}$$

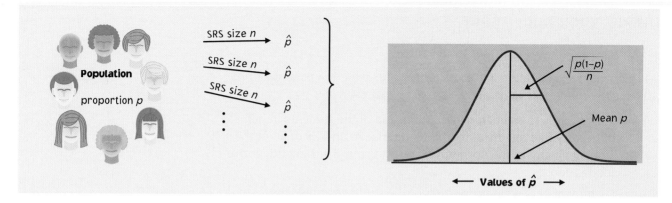

Figure 8.1 *Select a large SRS from a population of which the proportion p are successes. The sampling distribution of the proportion $\hat{p}$ of successes in the sample is approximately normal. The mean is p and the standard deviation is $\sqrt{p(1-p)/n}$.*

Figure 8.1 summarizes these facts in a form that helps you recall the big idea of a sampling distribution. The behavior of sample proportions $\hat{p}$ is similar to the behavior of sample means $\overline{x}$. When the sample size n is large, the sampling distribution is approximately normal. The larger the sample, the more nearly normal the distribution is. The mean of the sampling distribution is the true value of the population proportion p. That is, $\hat{p}$ is an unbiased estimator of p. The standard deviation of $\hat{p}$ gets smaller as the sample size n gets larger, so that estimation is likely to be more accurate when the sample is larger. But the standard deviation gets smaller only at the rate $\sqrt{n}$. We need four times as many observations to cut the standard deviation in half.

You should not use the normal approximation to the distribution of $\hat{p}$ when the sample size n is very small. What is more, the formula given for the standard deviation of $\hat{p}$ is not accurate unless the population is much larger than the sample—say, at least ten times larger.[3] We will give guidelines to help you decide when methods for inference based on this sampling distribution are trustworthy.

EXAMPLE 8.3 **Asking about risky behavior**

Suppose that in fact 6% of all adult heterosexuals had more than one sexual partner in the past year (and would admit it when asked). The National AIDS Behavioral Surveys interviewed a random sample of 2673 people from this population. What is the probability that at least 5% of such a sample admit to having more than one partner?

If the sample size is $n = 2673$ and the population proportion is $p = 0.06$, the sample proportion $\hat{p}$ has mean 0.06 and standard deviation

$$\sqrt{\frac{p(1-p)}{n}} = \sqrt{\frac{(0.06)(0.94)}{2673}}$$

$$= \sqrt{0.0000211} = 0.00459$$

We want the probability that $\hat{p}$ is 0.05 or greater.

Standardize $\hat{p}$ by subtracting the mean 0.06 and dividing by the standard deviation 0.00459. This produces a new statistic that has the standard normal distribution.

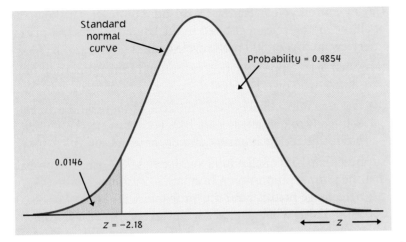

Figure 8.2 *Probabilities in Example 8.3 as areas under the standard normal curve.*

As usual, we call this statistic z:

$$z = \frac{\hat{p} - 0.06}{0.00459}$$

Figure 8.2 shows the probability we want as an area under the standard normal curve.

$$P(\hat{p} \geq 0.05) = P\left(\frac{\hat{p} - 0.06}{0.00459} \geq \frac{0.05 - 0.06}{0.00459}\right)$$

$$= P(Z \geq -2.18)$$

$$= 1 - 0.0146 = 0.9854$$

If we repeat the National AIDS Behavioral Surveys many times, more than 98% of all the samples will contain at least 5% of respondents who admit to more than one sexual partner.

APPLY YOUR KNOWLEDGE

8.4 **Do you jog?** The Gallup Poll once asked a random sample of 1540 adults, "Do you happen to jog?" Suppose that in fact 15% of all adults jog.

(a) Find the mean and standard deviation of the proportion $\hat{p}$ of the sample who jog. (Assume the sample is an SRS.)

(b) What sample size would be required to reduce the standard deviation of the sample proportion to one-half the value you found in (a)?

(c) Use the normal approximation to find the probability that between 13% and 17% of the sample jog.

(PETER SAUNDERS/INDEX STOCK IMAGERY)

8.5 **Harley motorcycles.** Harley-Davidson motorcycles make up 14% of all the motorcycles registered in the United States. You plan to interview an SRS of 500 motorcycle owners.

(a) What is the approximate distribution of the proportion of your sample who own Harleys?

(b) Is your sample likely to contain 20% or more who own Harleys? Is it likely to contain at least 15% Harley owners? Do normal probability calculations to answer these questions.

8.6 Suppose that 15% of all adults jog. Exercise 8.4 asks the probability that the sample proportion $\hat{p}$ from an SRS estimates $p = 0.15$ within ± 2 percentage points. Find this probability for SRSs of sizes 200, 800, and 3200. What general conclusion can you draw from your calculations?

Assumptions for inference

To do inference about a population proportion p, we use the z statistic that results from standardizing the sample proportion $\hat{p}$:

$$z = \frac{\hat{p} - p}{\sqrt{\dfrac{p(1 - p)}{n}}}$$

The statistic z has approximately the standard normal distribution $N(0, 1)$ if the sample is not too small and the sample is not a large part of the entire population. Here are rough rules of thumb that tell us when we can safely use z:

Rule of thumb 1. The population is at least 10 times as large as the sample.

Rule of thumb 2. The sample size n is large enough that both np and $n(1 - p)$ are at least 10.

The first rule of thumb tells us, for example, that we cannot use inference based on z if we interview 50 of the 75 students in a course. The second rule of thumb reflects the fact that for a fixed sample size n, the normal approximation is most accurate when p is close to 1/2 and least accurate when p is near 0 or 1. You can see that the approximation is no good at all when $p = 1$ or $p = 0$. If $p = 1$, for example, the population consists entirely of successes and $\hat{p}$ is always 1. Inference is still possible when the rules of thumb are not satisfied, but more elaborate methods are needed. See a statistician for help.[4]

In practice, of course, we don't know the value of p, so we cannot calculate z or apply rule of thumb 2. Here's what we do:

• To test the null hypothesis H_0: $p = p_0$ that the unknown p has a specific value p_0, just replace p by p_0 in the z statistic and in rule of thumb 2.

• In a confidence interval for p, we have no specific value to substitute. In large samples, $\hat{p}$ will be close to p. So we replace p by $\hat{p}$ in rule of thumb 2. We also replace the standard deviation by the **standard error of $\hat{p}$**

standard error of $\hat{p}$

$$SE = \sqrt{\dfrac{\hat{p}(1 - \hat{p})}{n}}$$

to get a confidence interval of the form

$$\text{estimate} \pm z^* SE_{\text{estimate}}$$

ASSUMPTIONS FOR INFERENCE ABOUT A PROPORTION

- The data are an SRS from the population of interest.
- The population is at least 10 times as large as the sample.
- For a test of $H_0: p = p_0$, the sample size n is so large that both np_0 and $n(1 - p_0)$ are 10 or more. For a confidence interval, n is so large that both the count of successes $n\hat{p}$ and the count of failures $n(1 - \hat{p})$ are 10 or more.

EXAMPLE 8.4   **Are the assumptions met?**

We want to use the National AIDS Behavioral Surveys data to give a confidence interval for the proportion of adult heterosexuals who have had multiple sexual partners. Does the sample meet the requirements for inference?

- The sampling design was a complex stratified sample, and the survey used inference procedures for that design. The overall effect is close to an SRS, however.
- The number of adult heterosexuals (the population) is much larger than 10 times the sample size, $n = 2673$.
- The counts of "Yes" and "No" responses are much greater than 10:

$$n\hat{p} = (2673)(0.0636) = 170$$

$$n(1 - \hat{p}) = (2673)(0.9364) = 2503$$

The second and third requirements are easily met. The first requirement, that the sample be an SRS, is only approximately met.

As usual, the practical problems of a large sample survey pose a greater threat to the AIDS survey's conclusions. Only people in households with telephones could be reached. This is acceptable for surveys of the general population, because about 94% of American households have telephones. However, some groups at high risk for AIDS, like intravenous drug users, often don't live in settled households and are under-represented in the sample. About 30% of the people reached refused to cooperate. A nonresponse rate of 30% is not unusual in large sample surveys, but it may cause some bias if those who refuse differ systematically from those who cooperate. The survey used statistical methods that adjust for unequal response rates in different groups. Finally, some respondents may not have told the truth when asked about their sexual behavior. The survey team tried hard to make respondents feel comfortable. For example, Hispanic women were interviewed only by Hispanic women, and Spanish speakers were interviewed by Spanish speakers with the same regional accent

(Cuban, Mexican, or Puerto Rican). Nonetheless, the survey report says that some bias is probably present:

> It is more likely that the present figures are underestimates; some respondents may underreport their numbers of sexual partners and intravenous drug use because of embarrassment and fear of reprisal, or they may forget or not know details of their own or of their partner's HIV risk and their antibody testing history.[5]

Reading the report of a large study like the National AIDS Behavioral Surveys reminds us that statistics in practice involves much more than recipes for inference.

APPLY YOUR KNOWLEDGE

8.7   In which of the following situations can you safely use the methods of this section to get a confidence interval for the population proportion p? Explain your answers.

    (a) Tonya wants to estimate what proportion of the students in her dormitory like the dorm food. She interviews an SRS of 50 of the 175 students living in the dormitory. She finds that 14 think the dorm food is good.

    (b) Glenn wonders what proportion of the students at his school think that tuition is too high. He interviews an SRS of 50 of the 2400 students at his college. Thirty-eight of those interviewed think tuition is too high.

    (c) In the National AIDS Behavioral Surveys sample of 2673 adult heterosexuals, 0.2% (that's 0.002 as a decimal fraction) had both received a blood transfusion and had a sexual partner from a group at high risk of AIDS. We want to estimate the proportion p in the population who share these two risk factors.

8.8   In which of the following situations can you safely use the methods of this section for a significance test? Explain your answers.

    (a) You toss a coin 10 times in order to test the hypothesis $H_0: p = 0.5$ that the coin is balanced.

    (b) A college president says, "99% of the alumni support my firing of Coach Boggs." You contact an SRS of 200 of the college's 15,000 living alumni to test the hypothesis $H_0: p = 0.99$.

    (c) Do a majority of the 250 students in a statistics course agree that knowing statistics will help them in their future careers? You interview an SRS of 20 students to test $H_0: p = 0.5$.

The z procedures

Here are the z procedures for inference about p when our assumptions are satisfied.

INFERENCE FOR A POPULATION PROPORTION

Draw an SRS of size n from a population with unknown proportion p of successes. An approximate level C confidence interval for p is

$$\hat{p} \pm z^* \sqrt{\frac{\hat{p}(1 - \hat{p})}{n}}$$

where z^* is the upper $(1 - C)/2$ standard normal critical value.
To test the hypothesis $H_0: p = p_0$, compute the z statistic

$$z = \frac{\hat{p} - p_0}{\sqrt{\dfrac{p_0(1 - p_0)}{n}}}$$

In terms of a variable Z having the standard normal distribution, the approximate P-value for a test of H_0 against

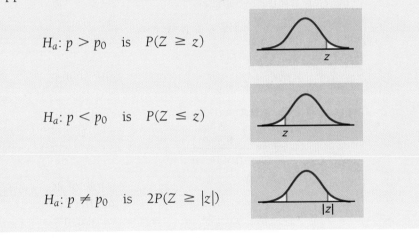

$H_a: p > p_0$ is $P(Z \geq z)$

$H_a: p < p_0$ is $P(Z \leq z)$

$H_a: p \neq p_0$ is $2P(Z \geq |z|)$

EXAMPLE 8.5	**Estimating risky behavior**

The National AIDS Behavioral Surveys found that 170 of a sample of 2673 adult heterosexuals had multiple partners. That is, $\hat{p} = 0.0636$. We will act as if the sample were an SRS.

A 99% confidence interval for the proportion p of all adult heterosexuals with multiple partners uses the standard normal critical value $z^* = 2.576$. (Look in the bottom row of Table C for standard normal critical values.) The confidence interval is

$$\hat{p} \pm z^* \sqrt{\frac{\hat{p}(1 - \hat{p})}{n}} = 0.0636 \pm 2.576 \sqrt{\frac{(0.0636)(0.9364)}{2673}}$$

$$= 0.0636 \pm 0.0122$$

$$= 0.0514 \text{ to } 0.0758$$

We are 99% confident that the percent of adult heterosexuals who had more than one sexual partner in the past year lies between about 5% and 7.6%.

EXAMPLE 8.6 **Is this coin fair?**

A coin that is balanced should come up heads half the time in the long run. The population for coin tossing contains the results of tossing the coin forever. The parameter p is the probability of a head, which is the proportion of all tosses that give a head. The tosses we actually make are an SRS from this population.

The French naturalist Count Buffon (1707–1788) tossed a coin 4040 times. He got 2048 heads. The sample proportion of heads is

$$\hat{p} = \frac{2048}{4040} = 0.5069$$

That's a bit more than one-half. Is this evidence that Buffon's coin was not balanced?

Step 1: Hypotheses. The null hypothesis says that the coin is balanced ($p = 0.5$). The alternative hypothesis is two-sided, because we did not suspect before seeing the data that the coin favored either heads or tails. We therefore test the hypotheses

$$H_0: p = 0.5$$

$$H_a: p \neq 0.5$$

The null hypothesis gives p the value $p_0 = 0.5$.

Step 2: Test statistic. The z test statistic is

$$z = \frac{\hat{p} - p_0}{\sqrt{\dfrac{p_0(1 - p_0)}{n}}}$$

$$= \frac{0.5069 - 0.5}{\sqrt{\dfrac{(0.5)(0.5)}{4040}}} = 0.88$$

Step 3: P-value. Because the test is two-sided, the P-value is the area under the standard normal curve more than 0.88 away from 0 in either direction. Figure 8.3 shows this area. In Table A we read that the area below -0.88 is 0.1894. The P-value is twice this area:

$$P = 2(0.1894) = 0.3788$$

Conclusion. A proportion of heads as far from one-half as Buffon's would happen 38% of the time when a balanced coin is tossed 4040 times. Buffon's result doesn't show that his coin is unbalanced.

In Example 8.6, we failed to find good evidence against $H_0: p = 0.5$. We *cannot* conclude that H_0 is true, that is, that the coin is perfectly balanced. No doubt p is not exactly 0.5. The test of significance only shows that the results of Buffon's 4040 tosses do not distinguish this coin from one that is perfectly

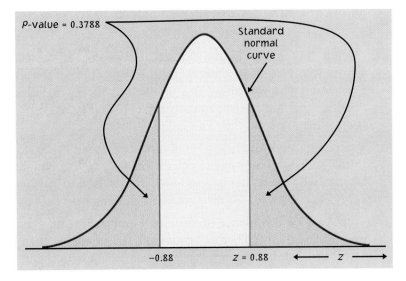

Figure 8.3 *The P-value for the two-sided test of Example 8.6.*

balanced. To see what values of p are consistent with the sample results, use a confidence interval.

EXAMPLE 8.7   Estimating the chance of a head

The 95% confidence interval for the probability p that Buffon's coin gives a head is

$$\hat{p} \pm z^* \sqrt{\frac{\hat{p}(1-\hat{p})}{n}} = 0.5069 \pm 1.960 \sqrt{\frac{(0.5069)(0.4931)}{4040}}$$

$$= 0.5069 \pm 0.0154$$

$$= 0.4915 \text{ to } 0.5223$$

We are 95% confident that the probability of a head is between 0.4915 and 0.5223.

The confidence interval is more informative than the test in Example 8.6. We would not be surprised if the true probability of a head for Buffon's coin were something like 0.51.

APPLY YOUR KNOWLEDGE

8.9   **Equality for women?** Have efforts to promote equality for women gone far enough in the United States? A poll on this issue by the cable network MSNBC contacted 1019 adults. A newspaper article about the poll said, "Results have a margin of sampling error of plus or minus 3 percentage points."[6]

(a) Overall, 54% of the sample (550 of 1019 people) answered "Yes." Find a 95% confidence interval for the proportion in the adult population who would say "Yes" if asked. Is the report's claim about the margin of error roughly right? (Assume that the sample is an SRS.)

(b) The news article said that 65% of men, but only 43% of women, think that efforts to promote equality have gone far enough. Explain why we do not have enough information to give confidence intervals for men and women separately.

(c) Would a 95% confidence interval for women alone have a margin of error less than 0.03, about equal to 0.03, or greater than 0.03? Why? You see that the news article's statement about the margin of error for poll results is a bit misleading.

8.10 **Teens and their TV sets.** The *New York Times* and CBS News conducted a nationwide poll of 1048 randomly selected 13- to 17-year-olds. Of these teenagers, 692 had a television in their room and 189 named Fox as their favorite television network.[7] We will act as if the sample were an SRS.

(a) Give 95% confidence intervals for the proportion of all people in this age group who have a TV in their room and the proportion who would choose Fox as their favorite network. Check that we can use our methods.

(b) The news article says, "In theory, in 19 cases out of 20, the poll results will differ by no more than three percentage points in either direction from what would have been obtained by seeking out all American teenagers." Explain how your results agree with this statement.

(c) Is there good evidence that more than half of all teenagers have a TV in their room? State hypotheses, give a test statistic, and state your conclusion about the strength of the evidence.

8.11 **We want to be rich.** In a recent year, 73% of first-year college students responding to a national survey identified "being very well-off financially" as an important personal goal. A state university finds that 132 of an SRS of 200 of its first-year students say that this goal is important.

(a) Give a 95% confidence interval for the proportion of all first-year students at the university who would identify being well-off as an important personal goal.

(b) Is there good evidence that the proportion of first-year students at this university who think being very well-off is important differs from the national value, 73%? (Be sure to state hypotheses, give the *P*-value, and state your conclusion.)

(c) Check that you can safely use the methods of this section in both (a) and (b).

Choosing the sample size

In planning a study, we may want to choose a sample size that will allow us to estimate the parameter within a given margin of error. We saw earlier (page 311) how to do this for a population mean. The method is similar for estimating a population proportion.

The margin of error in the confidence interval for *p* is

$$m = z^* \sqrt{\frac{\hat{p}(1 - \hat{p})}{n}}$$

Here z^* is the standard normal critical value for the level of confidence we want. Because the margin of error involves the sample proportion of successes $\hat{p}$, we need to guess this value when choosing *n*. Call our guess p^*. Here are two ways to get p^*:

1. Use a guess p^* based on a pilot study or on past experience with similar studies. You should do several calculations that cover the range of $\hat{p}$-values you might get.

2. Use $p^* = 0.5$ as the guess. The margin of error *m* is largest when $\hat{p} = 0.5$, so this guess is conservative in the sense that if we get any other $\hat{p}$ when we do our study, we will get a margin of error smaller than planned.

Once you have a guess p^*, the recipe for the margin of error can be solved to give the sample size *n* needed. Here is the result.

SAMPLE SIZE FOR DESIRED MARGIN OF ERROR

The level *C* confidence interval for a population proportion *p* will have margin of error approximately equal to a specified value *m* when the sample size is

$$n = \left(\frac{z^*}{m}\right)^2 p^*(1 - p^*)$$

where p^* is a guessed value for the sample proportion. The margin of error will be less than or equal to *m* if you take the guess p^* to be 0.5.

Which method for finding the guess p^* should you use? The *n* you get doesn't change much when you change p^* as long as p^* is not too far from 0.5.

Meta-analysis

Small samples have large margins of error. Large samples are expensive. Often we can find several studies of the same issue—if we could combine their results we would have a large sample with a small margin of error. That is the idea of "meta-analysis." Of course, we can't just lump the studies together, because of the differences in design and quality. Statisticians have more sophisticated ways of combining the results. Meta-analysis has been applied to issues from the effect of secondhand smoke to whether coaching improves SAT scores.

So use the conservative guess $p^* = 0.5$ if you expect the true $\hat{p}$ to be roughly between 0.3 and 0.7. If the true $\hat{p}$ is close to 0 or 1, using $p^* = 0.5$ as your guess will give a sample much larger than you need. So try to use a better guess from a pilot study when you suspect that $\hat{p}$ will be less than 0.3 or greater than 0.7.

EXAMPLE 8.8 **Planning a poll**

Gloria Chavez and Ronald Flynn are the candidates for mayor in a large city. You are planning a sample survey to determine what percent of the voters plan to vote for Chavez. This is a population proportion p. You will contact an SRS of registered voters in the city. You want to estimate p with 95% confidence and a margin of error no greater than 3%, or 0.03. How large a sample do you need?

The winner's share in all but the most lopsided elections is between 30% and 70% of the vote. So use the guess $p^* = 0.5$. The sample size you need is

$$n = \left(\frac{1.96}{0.03}\right)^2 (0.5)(1 - 0.5) = 1067.1$$

You should round the result up to $n = 1068$. (Rounding down would give a margin of error slightly greater than 0.03.) If you want a 2.5% margin of error, we have (after rounding up)

$$n = \left(\frac{1.96}{0.025}\right)^2 (0.5)(1 - 0.5) = 1537$$

For a 2% margin of error the sample size you need is

$$n = \left(\frac{1.96}{0.02}\right)^2 (0.5)(1 - 0.5) = 2401$$

As usual, smaller margins of error call for larger samples.

APPLY YOUR KNOWLEDGE

8.12 **School vouchers.** A national opinion poll found that 44% of all American adults agree that parents should be given vouchers good for education at any public or private school of their choice. The result was based on a small sample. How large an SRS is required to obtain a margin of error of 0.03 (that is, ±3%) in a 95% confidence interval?

(a) Answer this question using the previous poll's result as the guessed value p^*.

(b) Do the problem again using the conservative guess $p^* = 0.5$. By how much do the two sample sizes differ?

8.13 **Can you taste PTC?** PTC is a substance that has a strong bitter taste for some people and is tasteless for others. The ability to taste PTC is inherited. About 75% of Italians can taste PTC, for example. You want to estimate the proportion of Americans with at least one Italian grandparent who can taste PTC. Starting with the 75% estimate for Italians, how large a sample must you test in order to estimate the proportion of PTC tasters within ±0.04 with 95% confidence?

SECTION 8.1 Summary

Tests and confidence intervals for a population proportion p when the data are an SRS of size n are based on the **sample proportion** $\hat{p}$.

When n is large, $\hat{p}$ has approximately the normal distribution with mean p and standard deviation $\sqrt{p(1-p)/n}$.

The level C **confidence interval** for p is

$$\hat{p} \pm z^* \sqrt{\frac{\hat{p}(1-\hat{p})}{n}}$$

where z^* is the upper $(1-C)/2$ standard normal critical value.

Tests of $H_0: p = p_0$ are based on the **z statistic**

$$z = \frac{\hat{p} - p_0}{\sqrt{\frac{p_0(1-p_0)}{n}}}$$

with P-values calculated from the standard normal distribution.

These inference procedures are approximately correct when the population is at least 10 times as large as the sample and the sample is large enough to satisfy $n\hat{p} \geq 10$ and $n(1-\hat{p}) \geq 10$ for a confidence interval or $np_0 \geq 10$ and $n(1-p_0) \geq 10$ for a test of $H_0: p = p_0$.

The **sample size** needed to obtain a confidence interval with approximate margin of error m for a population proportion is

$$n = \left(\frac{z^*}{m}\right)^2 p^*(1-p^*)$$

where p^* is a guessed value for the sample proportion $\hat{p}$, and z^* is the standard normal critical point for the level of confidence you want. If you use $p^* = 0.5$ in this formula, the margin of error of the interval will be less than or equal to m no matter what the value of $\hat{p}$ is.

SECTION 8.1 Exercises

8.14 The IRS plans an SRS. The Internal Revenue Service plans to examine an SRS of individual federal income tax returns from each state. One variable of interest is the proportion of returns claiming itemized deductions. The total number of tax returns in a state varies from more than 13 million in California to fewer than 220,000 in Wyoming.

(a) Will the sampling variability of the sample proportion change from state to state if an SRS of 2000 tax returns is selected in each state? Explain your answer.

(b) Will the sampling variability of the sample proportion change from state to state if an SRS of 1% of all tax returns is selected in each state? Explain your answer.

8.15 Stolen Harleys. Harley-Davidson motorcycles make up 14% of all motorcycles registered in the United States. In 1995, 9224 motorcycles were reported stolen; 2490 of these were Harleys. We can think of motorcycles stolen in 1995 as an SRS of motorcycles stolen in recent years.

(a) If Harleys made up 14% of motorcycles stolen, what would be the sampling distribution of the proportion of Harleys in a sample of 9224 stolen motorcycles?

(b) Is the proportion of Harleys among stolen bikes significantly higher than their share of all motorcycles?

8.16 Side effects. An experiment on the side effects of pain relievers assigned arthritis patients to one of several over-the-counter pain medications. Of the 440 patients who took one brand of pain reliever, 23 suffered some "adverse symptom."

(a) If 10% of all patients suffer adverse symptoms, what would be the sampling distribution of the proportion with adverse symptoms in a sample of 440 patients?

(b) Does the experiment provide strong evidence that fewer than 10% of patients who take this medication have adverse symptoms?

8.17 Condom usage. The National AIDS Behavioral Surveys (Example 8.1) also interviewed a sample of adults in the cities where AIDS is most common. This sample included 803 heterosexuals who reported having more than one sexual partner in the past year. We can consider this an SRS of size 803 from the population of all heterosexuals in high-risk cities who have multiple partners. These people risk infection with the AIDS virus. Yet 304 of the respondents said they never use condoms. Is this strong evidence that more than one-third of this population never use condoms?

8.18 Unhappy HMO patients. How likely are patients who file complaints with a health maintenance organization (HMO) to leave the HMO? In one recent year, 639 of the more than 400,000 members of a large New England HMO filed complaints. Fifty-four of the complainers left the HMO voluntarily. (That is, they were not forced to leave by a move or a job change.)[8] Consider this year's complainers as an SRS of all patients who will complain in the future. Give a 90% confidence interval for the proportion of complainers who voluntarily leave the HMO.

8.19 **Do you go to church?** The Gallup Poll asked a sample of 1785 adults, "Did you, yourself, happen to attend church or synagogue in the last 7 days?" Of the respondents, 750 said "Yes." Treat Gallup's sample as an SRS of all American adults.

(a) Give a 99% confidence interval for the proportion of all adults who attended church or synagogue during the week preceding the poll.

(b) Do the results provide good evidence that fewer than half of the population attended church or synagogue?

(c) How large a sample would be required to obtain a margin of error of 0.01 in a 99% confidence interval for the proportion who attend church or synagogue? (Use the conservative guess $p^* = 0.5$, and explain why this method is reasonable in this situation.)

8.20 **Small-business failures.** A study of the survival of small businesses chose an SRS from the telephone directory's Yellow Pages listings of food-and-drink businesses in 12 counties in central Indiana. For various reasons, the study got no response from 45% of the businesses chosen. Interviews were completed with 148 businesses. Three years later, 22 of these businesses had failed.[9]

(a) Give a 95% confidence interval for the percent of all small businesses in this class that fail within three years.

(b) Based on the results of this study, how large a sample would you need to reduce the margin of error to 0.04?

(c) The authors hope that their findings describe the population of all small businesses. What about the study makes this unlikely? What population do you think the study findings describe?

8.21 **Matched pairs.** One-sample procedures for proportions, like those for means, are used to analyze data from matched pairs designs. Here is an example.

Each of 50 subjects tastes two unmarked cups of coffee and says which he or she prefers. One cup in each pair contains instant coffee; the other, fresh-brewed coffee. Thirty-one of the subjects prefer the fresh-brewed coffee. Take p to be the proportion of the population who would prefer fresh-brewed coffee in a blind tasting.

(a) Test the claim that a majority of people prefer the taste of fresh-brewed coffee. State hypotheses and report the z statistic and its P-value. Is your result significant at the 5% level? What is your practical conclusion?

(b) Find a 90% confidence interval for p.

(c) When you do an experiment like this, in what order should you present the two cups of coffee to the subjects?

8.22 Customer satisfaction. An automobile manufacturer would like to know what proportion of its customers are not satisfied with the service provided by the local dealer. The customer relations department will survey a random sample of customers and compute a 99% confidence interval for the proportion who are not satisfied.

(a) Past studies suggest that this proportion will be about 0.2. Find the sample size needed if the margin of error of the confidence interval is to be about 0.015.

(b) When the sample is actually contacted, 10% of the sample say they are not satisfied. What is the margin of error of the 99% confidence interval?

8.23 Surveying students. You are planning a survey of students at a large university to determine what proportion favor an increase in student fees to support an expansion of the student newspaper. Using records provided by the registrar, you can select a random sample of students. You will ask each student in the sample whether he or she is in favor of the proposed increase. Your budget will allow a sample of 100 students.

(a) For a sample of size 100, construct a table of the margins of error for 95% confidence intervals when $\hat{p}$ takes the values 0.1, 0.2, 0.3, 0.4, 0.5, 0.6, 0.7, 0.8, and 0.9.

(b) A former editor of the student newspaper offers to provide funds for a sample of size 500. Repeat the margin of error calculations in (a) for the larger sample size. Then write a short thank-you note to the former editor describing how the larger sample size will improve the results of the survey.

8.2 Comparing Two Proportions

two-sample problem

In a **two-sample problem,** we want to compare two populations or the responses to two treatments based on two independent samples. When the comparison involves the mean of a quantitative variable, we use the two-sample t methods of Section 7.2. To compare the standard deviations of a variable in two groups, we use (under restrictive conditions) the F statistic described in the optional Section 7.3. Now we turn to methods to compare the proportions of successes in two groups.

We will use notation similar to that used in our study of two-sample t statistics. The groups we want to compare are Population 1 and Population 2. We have a separate SRS from each population or responses from two treatments in a randomized comparative experiment. A subscript shows which group a parameter or statistic describes. Here is our notation:

Population	Population proportion	Sample size	Sample proportion
1	p_1	n_1	$\hat{p}_1$
2	p_2	n_2	$\hat{p}_2$

We compare the populations by doing inference about the difference $p_1 - p_2$ between the population proportions. The statistic that estimates this difference is the difference between the two sample proportions, $\hat{p}_1 - \hat{p}_2$.

EXAMPLE 8.9   **Does preschool help?**

To study the long-term effects of preschool programs for poor children, the High/Scope Educational Research Foundation has followed two groups of Michigan children since early childhood. One group of 62 attended preschool as 3- and 4-year-olds. This is a sample from Population 2, poor children who attend preschool. A control group of 61 children from the same area and similar backgrounds represents Population 1, poor children with no preschool. Thus the sample sizes are $n_1 = 61$ and $n_2 = 62$.

One response variable of interest is the need for social services as adults. In the past ten years, 38 of the preschool sample and 49 of the control sample have needed social services (mainly welfare). The sample proportions are

$$\hat{p}_1 = \frac{49}{61} = 0.803$$

$$\hat{p}_2 = \frac{38}{62} = 0.613$$

That is, about 80% of the control group uses social services, as opposed to about 61% of the preschool group.

To see if the study provides significant evidence that preschool reduces the later need for social services, we test the hypotheses

$$H_0: p_1 - p_2 = 0 \quad \text{or} \quad H_0: p_1 = p_2$$

$$H_a: p_1 - p_2 > 0 \quad \text{or} \quad H_a: p_1 > p_2$$

To estimate how large the reduction is, we give a confidence interval for the difference, $p_1 - p_2$. Both the test and the confidence interval start from the difference of sample proportions:

$$\hat{p}_1 - \hat{p}_2 = 0.803 - 0.613 = 0.190$$

The sampling distribution of $\hat{p}_1 - \hat{p}_2$

To use $\hat{p}_1 - \hat{p}_2$ for inference, we must know its sampling distribution. Here are the facts we need:

- When the samples are large, the distribution of $\hat{p}_1 - \hat{p}_2$ is approximately normal.
- The mean of $\hat{p}_1 - \hat{p}_2$ is $p_1 - p_2$. That is, the difference of sample proportions is an unbiased estimator of the difference of population proportions.
- The standard deviation of the difference is

$$\sqrt{\frac{p_1(1 - p_1)}{n_1} + \frac{p_2(1 - p_2)}{n_2}}$$

Figure 8.4 displays the distribution of $\hat{p}_1 - \hat{p}_2$. The standard deviation of $\hat{p}_1 - \hat{p}_2$ involves the unknown parameters p_1 and p_2. Just as in the previous section, we must replace these by estimates in order to do inference. And just as in the previous section, we do this a bit differently for confidence intervals and for tests.

Confidence intervals for $p_1 - p_2$

To obtain a confidence interval, replace the population proportions p_1 and p_2 in the standard deviation by the sample proportions. The result is the **standard** *standard error* **error** of the statistic $\hat{p}_1 - \hat{p}_2$:

$$\text{SE} = \sqrt{\frac{\hat{p}_1(1 - \hat{p}_1)}{n_1} + \frac{\hat{p}_2(1 - \hat{p}_2)}{n_2}}$$

The confidence interval again has the form

$$\text{estimate} \pm z^* \text{SE}_{\text{estimate}}$$

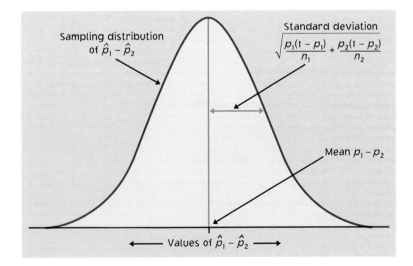

Figure 8.4 *Select independent SRSs from two populations having proportions of successes p_1 and p_2. The proportions of successes in the two samples are $\hat{p}_1$ and $\hat{p}_2$. When the samples are large, the sampling distribution of the difference $\hat{p}_1 - \hat{p}_2$ is approximately normal.*

CONFIDENCE INTERVALS FOR COMPARING TWO PROPORTIONS

Draw an SRS of size n_1 from a population having proportion p_1 of successes and draw an independent SRS of size n_2 from another population having proportion p_2 of successes. When n_1 and n_2 are large, an approximate level C confidence interval for $p_1 - p_2$ is

$$(\hat{p}_1 - \hat{p}_2) \pm z^*\text{SE}$$

In this formula the standard error SE of $\hat{p}_1 - \hat{p}_2$ is

$$\text{SE} = \sqrt{\frac{\hat{p}_1(1 - \hat{p}_1)}{n_1} + \frac{\hat{p}_2(1 - \hat{p}_2)}{n_2}}$$

and z^* is the upper $(1-C)/2$ standard normal critical value. In practice, use this confidence interval when the populations are at least 10 times as large as the samples and the counts of successes and failures are 5 or more in both samples.

EXAMPLE 8.10   **How much does preschool help?**

Example 8.9 describes a study of the effect of preschool on later use of social services. The facts are:

Population	Population description	Sample size	Count of successes	Sample proportion
1	control	$n_1 = 61$	49	$\hat{p}_1 = 0.803$
2	preschool	$n_2 = 62$	38	$\hat{p}_2 = 0.613$

To check that our approximate confidence interval is safe, look at the counts of successes and failures in the two samples. The smallest of these four quantities is the count of failures in the control group, $61 - 49 = 12$. This is larger than 5, so the interval will be accurate.

The difference $p_1 - p_2$ measures the effect of preschool in reducing the proportion of people who later need social services. To compute a 95% confidence interval for $p_1 - p_2$, first find the standard error:

$$\begin{aligned}
\text{SE} &= \sqrt{\frac{\hat{p}_1(1 - \hat{p}_1)}{n_1} + \frac{\hat{p}_2(1 - \hat{p}_2)}{n_2}} \\
&= \sqrt{\frac{(0.803)(0.197)}{61} + \frac{(0.613)(0.387)}{62}} \\
&= \sqrt{0.00642} = 0.0801
\end{aligned}$$

Statisticians honest and dishonest

Developed nations rely on government statisticians to produce honest data. We trust the monthly unemployment rate and Consumer Price Index to guide both public and private decisions. Honesty can't be taken for granted everywhere, however. In June 1998, the Russian government arrested the top statisticians in the State Committee for Statistics. They were accused of taking bribes to fudge data to help companies avoid taxes. "It means that we know nothing about the performance of Russian companies," said one newspaper editor.

Figure 8.5 *Output for Example 8.10. (a) TI-83 graphing calculator. (b) Minitab.*

```
2-PropZInt
 (.03337, .34738)
 p̂1=.80328
 p̂2=.61290
 n1=61.00000
 n2=62.00000
```

(a)

```
Test and Confidence Interval for Two Proportions

Sample      X      N     Sample p
1          49     61     0.803279
2          38     62     0.612903


Estimate for p(1) - p(2):  0.190375
95% CI for p(1) - p(2):   (0.0333680, 0.347383)
Test for p(1) - p(2)=0(vs not=0): Z=2.32 P-Value=0.020
```

(b)

The 95% confidence interval is

$$(\hat{p}_1 - \hat{p}_2) \pm z^*\text{SE} = (0.803 - 0.613) \pm (1.960)(0.0801)$$

$$= 0.190 \pm 0.157$$

$$= 0.033 \text{ to } 0.347$$

We are 95% confident that the percent needing social services is somewhere between 3.3% and 34.7% lower among people who attended preschool. The confidence interval is wide because the samples are quite small. Figure 8.5 displays software output for this example. As usual, you can understand the output even without knowledge of the program that produced it.

The researchers in the study of Example 8.9 selected two separate samples from the two populations they wanted to compare. Many comparative studies start with just one sample, then divide it into two groups based on data gathered from the subjects. Exercises 8.24 and 8.26 are examples of this approach. The two-sample z procedures for comparing proportions are valid in such situations. This is an important fact about these methods.

APPLY YOUR KNOWLEDGE

8.24 **In-line skaters.** A study of injuries to in-line skaters used data from the National Electronic Injury Surveillance System, which collects data from a random sample of hospital emergency rooms. In the six-month study period, 206 people came to the sample hospitals with injuries from in-line skating. We can think of these people as an SRS of all people injured while skating. Researchers were able to interview 161 of these people. Wrist injuries (mostly fractures) were the most common.[10]

(a) The interviews found that 53 people were wearing wrist guards and 6 of these had wrist injuries. Of the 108 who did not wear wrist guards, 45 had wrist injuries. What are the two sample proportions of wrist injuries?

(b) Give a 95% confidence interval for the difference between the two population proportions of wrist injuries. State carefully what populations your inference compares. We would like to draw conclusions about all in-line skaters, but we have data only for injured skaters.

(c) What was the percent of nonresponse among the original sample of 206 injured skaters? Explain why nonresponse may bias your conclusions.

8.25 **Lyme disease.** Lyme disease is spread in the northeastern United States by infected ticks. The ticks are infected mainly by feeding on mice, so more mice result in more infected ticks. The mouse population in turn rises and falls with the abundance of acorns, their favored food. Experimenters studied two similar forest areas in a year when the acorn crop failed. They added hundreds of thousands of acorns to one area to imitate an abundant acorn crop, while leaving the other area untouched. The next spring, 54 of the 72 mice trapped in the first area were in breeding condition, versus 10 of the 17 mice trapped in the second area.[11] Give a 90% confidence interval for the difference between the proportion of mice ready to breed in good acorn years and bad acorn years. Explain why we can apply the z methods.

8.26 **Free speech?** The 1958 Detroit Area Study was an important investigation of the influence of religion on everyday life. The sample "was basically a simple random sample of the population of the metropolitan area" of Detroit, Michigan. Of the 656 respondents, 267 were white Protestants and 230 were white Catholics.

The study took place at the height of the cold war. One question asked if the right of free speech included the right to make speeches in favor of communism. Of the 267 white Protestants, 104 said "Yes," while 75 of the 230 white Catholics said "Yes."[12]

(a) Give a 95% confidence interval for the difference between the proportion of Protestants who agreed that communist speeches are protected and the proportion of Catholics who held this opinion.

(b) Check that it is safe to use the z confidence interval.

Significance tests for $p_1 - p_2$

An observed difference between two sample proportions can reflect a difference in the populations, or it may just be due to chance variation in random sampling. Significance tests help us decide if the effect we see in the samples is really there in the populations. The null hypothesis says that there is no difference between the two populations:

$$H_0: p_1 = p_2$$

The alternative hypothesis says what kind of difference we expect.

EXAMPLE 8.11   **Cholesterol and heart attacks**

High levels of cholesterol in the blood are associated with higher risk of heart attacks. Will using a drug to lower blood cholesterol reduce heart attacks? The Helsinki Heart Study looked at this question. Middle-aged men were assigned at random to one of two treatments: 2051 men took the drug gemfibrozil to reduce their cholesterol levels, and a control group of 2030 men took a placebo. During the next five years, 56 men in the gemfibrozil group and 84 men in the placebo group had heart attacks.

The sample proportions who had heart attacks are

$$\hat{p}_1 = \frac{56}{2051} = 0.0273 \qquad \text{(gemfibrozil group)}$$

$$\hat{p}_2 = \frac{84}{2030} = 0.0414 \qquad \text{(placebo group)}$$

That is, about 4.1% of the men in the placebo group had heart attacks, against only about 2.7% of the men who took the drug. Is the apparent benefit of gemfibrozil statistically significant? We hope to show that gemfibrozil reduces heart attacks, so we have a one-sided alternative:

$$H_0: p_1 = p_2$$

$$H_a: p_1 < p_2$$

To do a test, standardize $\hat{p}_1 - \hat{p}_2$ to get a z statistic. If H_0 is true, all the observations in both samples really come from a single population of men of whom a single unknown proportion p will have a heart attack in a five-year period. So instead of estimating p_1 and p_2 separately, we pool the two samples and use the overall sample proportion to estimate the single population parameter p. *pooled sample proportion*   Call this the **pooled sample proportion.** It is

$$\hat{p} = \frac{\text{count of successes in both samples combined}}{\text{count of observations in both samples combined}}$$

Use $\hat{p}$ in place of both $\hat{p}_1$ and $\hat{p}_2$ in the expression for the standard error SE of $\hat{p}_1 - \hat{p}_2$ to get a z statistic that has the standard normal distribution when H_0 is true. Here is the test.

SIGNIFICANCE TEST FOR COMPARING TWO PROPORTIONS

To test the hypothesis

$$H_0: p_1 = p_2$$

first find the pooled proportion $\hat{p}$ of successes in both samples combined. Then compute the z statistic

$$z = \frac{\hat{p}_1 - \hat{p}_2}{\sqrt{\hat{p}(1 - \hat{p})\left(\dfrac{1}{n_1} + \dfrac{1}{n_2}\right)}}$$

In terms of a variable Z having the standard normal distribution, the P-value for a test of H_0 against

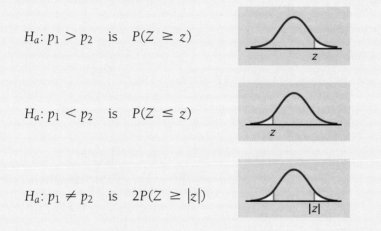

$$H_a: p_1 > p_2 \quad \text{is} \quad P(Z \geq z)$$

$$H_a: p_1 < p_2 \quad \text{is} \quad P(Z \leq z)$$

$$H_a: p_1 \neq p_2 \quad \text{is} \quad 2P(Z \geq |z|)$$

In practice, use this test when the populations are at least 10 times as large as the samples and the counts of successes and failures are 5 or more in both samples.

EXAMPLE 8.12 *Cholesterol and heart attacks, continued*

The pooled proportion of heart attacks for the two groups in the Helsinki Heart Study is

$$\hat{p} = \frac{\text{count of heart attacks in both samples combined}}{\text{count of subjects in both samples combined}}$$

$$= \frac{56 + 84}{2051 + 2030}$$

$$= \frac{140}{4081} = 0.0343$$

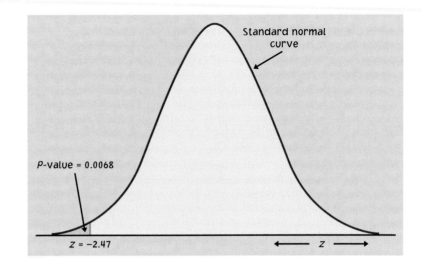

Figure 8.6 *The P-value for the one-sided test of Example 8.12.*

The z test statistic is

$$z = \frac{\hat{p}_1 - \hat{p}_2}{\sqrt{\hat{p}(1 - \hat{p})\left(\dfrac{1}{n_1} + \dfrac{1}{n_2}\right)}}$$

$$= \frac{0.0273 - 0.0414}{\sqrt{(0.0343)(0.9657)\left(\dfrac{1}{2051} + \dfrac{1}{2030}\right)}}$$

$$= \frac{-0.0141}{0.005698} = -2.47$$

The one-sided P-value is the area under the standard normal curve to the left of -2.47. Figure 8.6 shows this area. Table A gives $P = 0.0068$. Because $P < 0.01$, the results are statistically significant at the $\alpha = 0.01$ level. There is strong evidence that gemfibrozil reduced the rate of heart attacks. The large samples in the Helsinki Heart Study helped the study get highly significant results.

APPLY YOUR KNOWLEDGE

8.27 **The Gold Coast.** A historian examining British colonial records for the Gold Coast in Africa suspects that the death rate was higher among African miners than among European miners. In the year 1936, there were 223 deaths among 33,809 African miners and 7 deaths among 1541 European miners on the Gold Coast.[13]

Consider this year as a sample from the pre-war era in Africa. Is there good evidence that the proportion of African miners who died was higher than the proportion of European miners who died? (State hypotheses, calculate a test statistic, give a P-value, and state your conclusion.)

8.28   **Preventing strokes.** Aspirin prevents blood from clotting and so helps prevent strokes. The Second European Stroke Prevention Study asked whether adding another anticlotting drug named dipyridamole would be more effective for patients who had already had a stroke. Here are the data on strokes and deaths during the two years of the study:[14]

	Number of patients	Number of strokes	Number of deaths
Aspirin alone	1649	206	182
Aspirin + dipyridamole	1650	157	185

(a) The study was a randomized comparative experiment. Outline the design of the study.

(b) Is there a significant difference in the proportion of strokes in the two groups? State hypotheses, find the P-value, and state your conclusion.

(c) Is there a significant difference in death rates for the two groups?

8.29   **Access to computers.** A sample survey by Nielsen Media Research looked at computer access and use of the Internet. Whites were significantly more likely than blacks to own a home computer, but the black-white difference in computer access at work was not significant. The study team then looked separately at the households with at least $40,000 income. The sample contained 1916 white and 131 black households in this class. Here are the sample counts for these households:[15]

	Blacks	Whites
Own home computer	86	1173
PC access at work	100	1132

Do higher-income blacks and whites differ significantly at the 5% level in the proportion who own home computers? Do they differ significantly in the proportion who have PC access at work?

Section 8.2 Summary

We want to compare the proportions p_1 and p_2 of successes in two populations. The comparison is based on the difference $\hat{p}_1 - \hat{p}_2$ between the sample proportions of successes. When the sample sizes n_1 and n_2 are large enough, we can use z procedures because the sampling distribution of $\hat{p}_1 - \hat{p}_2$ is close to normal.

An approximate level C **confidence interval** for $p_1 - p_2$ is

$$(\hat{p}_1 - \hat{p}_2) \pm z^*\text{SE}$$

where the **standard error** of $\hat{p}_1 - \hat{p}_2$ is

$$\text{SE} = \sqrt{\frac{\hat{p}_1(1 - \hat{p}_1)}{n_1} + \frac{\hat{p}_2(1 - \hat{p}_2)}{n_2}}$$

and z^* is a standard normal critical value.

Significance tests of H_0: $p_1 = p_2$ use the **pooled sample proportion**

$$\hat{p} = \frac{\text{count of successes in both samples combined}}{\text{count of observations in both samples combined}}$$

and the z **statistic**

$$z = \frac{\hat{p}_1 - \hat{p}_2}{\sqrt{\hat{p}(1 - \hat{p})(\dfrac{1}{n_1} + \dfrac{1}{n_2})}}$$

P-values come from the standard normal table.

SECTION 8.2 Exercises

8.30 **Reducing nonresponse (EESEE)?** Telephone surveys often have high rates of nonresponse. When the call is handled by an answering machine, perhaps leaving a message on the machine will encourage people to respond when they are called again. Here are data from a study in which (at random) a message was or was not left when an answering machine picked up the first call from a survey:[16]

	Total households	Eventual contact	Completed survey
No message	100	58	33
Message	291	200	134

(a) Is there good evidence that leaving a message increases the proportion of households that are eventually contacted?

(b) Is there good evidence that leaving a message increases the proportion who complete the survey?

(c) If you find significant effects, look at their size. Do you think these effects are large enough to be important to survey takers?

8.31 **Unhappy HMO patients.** Exercise 8.18 describes a study of whether patients who file complaints leave a health maintenance organization (HMO). We want to know whether complainers are more likely to leave than patients who do not file complaints. In the year of the study, 639 patients filed complaints, and 54 of these patients left the HMO voluntarily. For comparison, the HMO chose an SRS of 743 patients who had not filed complaints. Twenty-two of these patients left voluntarily.

(a) How much higher is the proportion of complainers who leave? Give a 90% confidence interval.

(b) The HMO has more than 400,000 members. Check that you can safely use the methods of this section.

8.32 **Treating AIDS.** The drug AZT was the first drug that seemed effective in delaying the onset of AIDS. Evidence for AZT's effectiveness came from a large randomized comparative experiment. The subjects were 1300 volunteers who were infected with HIV, the virus that causes AIDS, but did not yet have AIDS. The study assigned 435 of the subjects at random to take 500 milligrams of AZT each day, and another 435 to take a placebo. (The others were assigned to a third treatment, a higher dose of AZT. We will compare only two groups.) At the end of the study, 38 of the placebo subjects and 17 of the AZT subjects had developed AIDS. We want to test the claim that taking AZT lowers the proportion of infected people who will develop AIDS in a given period of time.

(a) State hypotheses, and check that you can safely use the z procedures.

(b) How significant is the evidence that AZT is effective?

(c) The experiment was double-blind. Explain what this means.

Comment: Medical experiments on treatments for AIDS and other fatal diseases raise hard ethical questions. Some people argue that because AIDS is always fatal, infected people should get any drug that has any hope of helping them. The counterargument is that we will then never find out which drugs really work. The placebo patients in this study were given AZT as soon as the results were in.

8.33 **Are urban students more successful?** North Carolina State University looked at the factors that affect the success of students in a required chemical engineering course. Students must get a C or better in the course in order to continue as chemical engineering majors. There were 65 students from urban or suburban backgrounds, and 52 of these students succeeded. Another 55 students were from rural or small-town backgrounds; 30 of these students succeeded in the course.[17]

(a) Is there good evidence that the proportion of students who succeed is different for urban/suburban versus rural/small-town backgrounds? (State hypotheses, give the *P*-value of a test, and state your conclusion.)

(b) Give a 90% confidence interval for the size of the difference.

8.34 **Small-business failures.** The study of small-business failures described in Exercise 8.20 looked at 148 food-and-drink businesses in central Indiana. Of these, 106 were headed by men and 42 were headed by women. During a three-year period, 15 of the men's businesses and 7 of the women's businesses failed. Is there a significant difference between the rate at which businesses headed by men and those headed by women fail?

8.35 **Female and male students.** The North Carolina State University study (Exercise 8.33) also looked at possible differences in the proportions of female and male students who succeeded in the course. They found that 23 of the 34 women and 60 of the 89 men succeeded. Is there evidence of a difference between the proportions of women and men who succeed?

8.36 **Who gets stock options?** Different kinds of companies compensate their key employees in different ways. Established companies may pay higher salaries, while new companies may offer stock options that will be valuable if the company succeeds. Do high-tech companies tend to offer stock options more often than other companies? One study looked at a random sample of 200 companies. Of these, 91 were listed in the *Directory of Public High Technology Corporations* and 109 were not listed. Treat these two groups as SRSs of high-tech and non-high-tech companies. Seventy-three of the high-tech companies and 75 of the non-high-tech companies offered incentive stock options to key employees.[18]

(a) Is there evidence that a higher proportion of high-tech companies offer stock options?

(b) Give a 95% confidence interval for the difference in the proportions of the two types of companies that offer stock options.

8.37 **Nobody is home in July.** Nonresponse to sample surveys may differ with the season of the year. In Italy, for example, many people leave town during the summer. The Italian National Statistical Institute called random samples of telephone numbers between 7 p.m. and 10 p.m. at several seasons of the year. Here are the results for two seasons:[19]

Dates	Number of calls	No answer	Total nonresponse
Jan. 1 to Apr. 13	1558	333	491
July 1 to Aug. 31	2075	861	1174

(a) How much higher is the proportion of "no answers" in July and August compared with the early part of the year? Give a 99% confidence interval.

(b) The difference between the proportions of "no answers" is so large that it is clearly statistically significant. How can you tell from your work in (a) that the difference is significant at the $\alpha = 0.01$ level?

(c) Use the information given to find the counts of calls that had nonresponse for some reason other than "no answer." Do the rates of nonresponse due to other causes also differ significantly for the two seasons?

8.38 **Aspirin and heart attacks.** The Physicians' Health Study examined the effects of taking an aspirin every other day. Earlier studies suggested that aspirin might reduce the risk of heart attacks. The subjects were 22,071 healthy male physicians at least 40 years old. The study assigned 11,037 of the subjects at random to take aspirin. The others took a placebo pill. The study was double-blind. Here are the counts for some of the outcomes of interest to the researchers:

	Aspirin group	Placebo group
Fatal heart attacks	10	26
Nonfatal heart attacks	129	213
Strokes	119	98

For which outcomes is the difference between the aspirin and placebo groups significant? (Use two-sided alternatives. Check that you can apply the z test. Write a brief summary of your conclusions.)

8.39 **Men versus women.** The National Assessment of Educational Progress (NAEP) Young Adult Literacy Assessment Survey interviewed a random sample of 1917 people 21 to 25 years old. The sample contained 840 men, of whom 775 were fully employed. There were 1077 women, and 680 of them were fully employed.[20]

(a) Use a 99% confidence interval to describe the difference between the proportions of young men and young women who are fully employed. Is the difference statistically significant at the 1% significance level?

(b) The mean and standard deviation of scores on the NAEP's test of quantitative skills were $\bar{x}_1 = 272.40$ and $s_1 = 59.2$ for the men in the sample. For the women, the results were $\bar{x}_2 = 274.73$ and $s_2 = 57.5$. Is the difference between the mean scores for men and women significant at the 1% level?

8.40 **Child-care workers.** The Current Population Survey (CPS) is the monthly government sample survey of 50,000 households that provides data on employment in the United States. A study of child-care workers drew a sample from the CPS data tapes. We can consider this sample to be an SRS from the population of child-care workers.[21]

(a) Out of 2455 child-care workers in private households, 7% were black. Of 1191 nonhousehold child-care workers, 14% were black. Give a 99% confidence interval for the difference in the percents of these groups of workers who are black. Is the difference statistically significant at the $\alpha = 0.01$ level?

(b) The study also examined how many years of school child-care workers had. For household workers, the mean and standard deviation were $\bar{x}_1 = 11.6$ years and $s_1 = 2.2$ years. For nonhousehold workers, $\bar{x}_2 = 12.2$ years and $s_2 = 2.1$ years. Give a 99% confidence interval for the difference in mean years of education for the two groups. Is the difference significant at the $\alpha = 0.01$ level?

8.41 **Significant does not mean important.** Never forget that even small effects can be statistically significant if the samples are large. To illustrate this fact, return to the study of 148 small businesses in Exercise 8.34.

(a) Find the proportions of failures for businesses headed by women and businesses headed by men. These sample proportions are quite close to each other. Give the P-value for the z test of the hypothesis that the same proportion of women's and men's businesses fail. (Use the two-sided alternative.) The test is very far from being significant.

(b) Now suppose that the same sample proportions came from a sample 30 times as large. That is, 210 out of 1260 businesses headed by women and 450 out of 3180 businesses headed by men fail. Verify that the proportions of failures are exactly the same as in (a). Repeat the z test for the new data, and show that it is now significant at the $\alpha = 0.05$ level.

(c) It is wise to use a confidence interval to estimate the size of an effect, rather than just giving a P-value. Give 95% confidence intervals for the difference between the proportions of women's and men's businesses that fail for the settings of both (a) and (b). What is the effect of larger samples on the confidence interval?

STATISTICS IN SUMMARY

Inference about population proportions is based on sample proportions. We rely on the fact that a sample proportion has a distribution that is close to normal unless the sample is small. Review Figure 8.1 (page 432) for the most important facts. All the z procedures in this chapter work well when the samples are large enough. You must check this before using them. Here are the things you should now be able to do.

A. RECOGNITION

1. Recognize from the design of a study whether one-sample, matched pairs, or two-sample procedures are needed.
2. Recognize what parameter or parameters an inference problem concerns. In particular, distinguish among settings that require inference about a mean, comparing two means, inference about a proportion, or comparing two proportions.
3. Calculate from sample counts the sample proportion or proportions that estimate the parameters of interest.

B. INFERENCE ABOUT ONE PROPORTION

1. Use the z procedure to give a confidence interval for a population proportion p.
2. Use the z statistic to carry out a test of significance for the hypothesis $H_0: p = p_0$ about a population proportion p against either a one-sided or a two-sided alternative.
3. Check that you can safely use these z procedures in a particular setting.

C. COMPARING TWO PROPORTIONS

1. Use the two-sample z procedure to give a confidence interval for the difference $p_1 - p_2$ between proportions in two populations based on independent samples from the populations.
2. Use a z statistic to test the hypothesis $H_0: p_1 = p_2$ that proportions in two distinct populations are equal.
3. Check that you can safely use these z procedures in a particular setting.

Statistical inference always draws conclusions about one or more parameters of a population. When you think about doing inference, ask first what the population is and what parameter you are interested in. The t procedures of Chapter 7 allow us to give confidence intervals and carry out tests about population means. We use the z procedures of this chapter for inference about population proportions. The Statistics in Summary figure on the next page outlines the decisions you must make in choosing among the

STATISTICS IN SUMMARY
Inference Procedures from Chapters 7 and 8

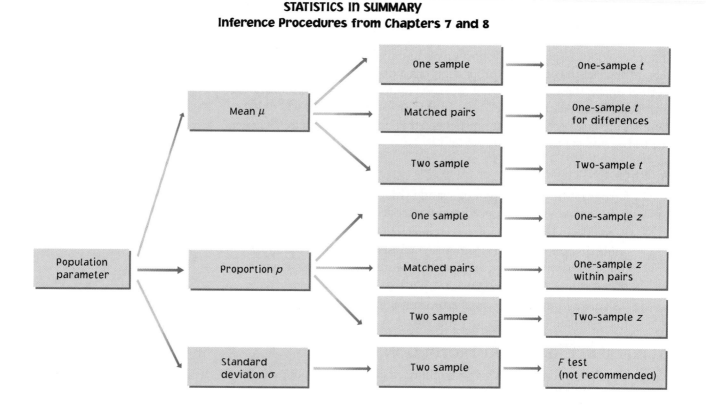

procedures we have met. First ask *what type of population parameter* does the inference concern. Then ask *what type of design* produced the data.

CHAPTER 8 Review Exercises

(WILLIE L. HILL/STOCK, BOSTON.)

8.42 Police radar and speeding (EESEE). Do drivers reduce excessive speed when they encounter police radar? Researchers studied the behavior of drivers on a rural interstate highway in Maryland where the speed limit was 55 miles per hour. They measured speed with an electronic device hidden in the pavement and, to eliminate large trucks, considered only vehicles less than 20 feet long. During some time periods, police radar was set up at the measurement location. Here are some of the data:[22]

	Number of vehicles	Number over 65 mph
No radar	12,931	5,690
Radar	3,285	1,051

(a) Give a 95% confidence interval for the proportion of vehicles going faster than 65 miles per hour when no radar is present.

(b) Give a 95% confidence interval for the effect of radar, as measured by the difference in proportions of vehicles going faster than 65 miles per hour with and without radar.

(c) The researchers chose a rural highway so that cars would be separated rather than in clusters where some cars might slow because they see other cars slowing. Explain why such clusters might make inference invalid.

8.43 **Steroids in high school.** A study by the National Athletic Trainers Association surveyed 1679 high school freshmen and 1366 high school seniors in Illinois. Results showed that 34 of the freshmen and 24 of the seniors had used anabolic steroids. Steroids, which are dangerous, are sometimes used to improve athletic performance.[23]

(a) In order to draw conclusions about all Illinois freshmen and seniors, how should the study samples be chosen?

(b) Give a 95% confidence interval for the proportion of all high school freshmen in Illinois who have used steroids.

(c) Is there a significant difference between the proportions of freshman and seniors who have used steroids?

8.44 **Do our athletes graduate?** The National Collegiate Athletic Association (NCAA) requires colleges to report the graduation rates of their athletes. Here are data from a Big Ten university's report.[24]

(a) Ninety-five of the 147 athletes admitted in 1989–1991 graduated within six years. Does the proportion of athletes who graduate differ significantly from the all-university proportion, which was 70%?

(b) The graduation rates were 37 of 45 female athletes and 58 of 102 male athletes. Is there evidence that a smaller proportion of male athletes than of female athletes graduate within six years?

(c) We are willing to regard athletes admitted in this period as an SRS from the large population of athletes the university will admit under its present standards. Explain why you can safely use the z procedures in parts (a) and (b). Then explain why you cannot use these procedures to test whether male basketball players (4 out of 8 admitted graduated) differ from other athletes.

8.45 **A television poll.** A television news program conducts a call-in poll about a proposed city ban on handgun ownership. Of the 2372 calls, 1921 oppose the ban. The station, following recommended practice, makes a confidence statement: "81% of the Channel 13 Pulse Poll sample opposed the ban. We can be 95% confident that the true proportion of citizens opposing a handgun ban is within 1.6% of the sample result." In this conclusion justified?

8.46 **Alternative medicine.** A nationwide random survey of 1500 adults asked about attitudes toward "alternative medicine" such as acupuncture, massage therapy, and herbal therapy. Among the respondents, 660 said they would use alternative medicine if traditional medicine was not producing the results they wanted.[25]

(a) Give a 95% confidence interval for the proportion of all adults who would use alternative medicine.

(b) Write a short paragraph for a news report based on the survey results.

8.47 **Do chemists have more girls?** Some people think that chemists are more likely than other parents to have female children. (Perhaps chemists are exposed to something in their laboratories that affects the sex of their children.) The Washington State Department of Health lists the parents' occupations on birth certificates. Between 1980 and 1990, 555 children were born to fathers who were chemists. Of these births, 273 were girls. During this period, 48.8% of all births in Washington State were girls.[26] Is there evidence that the proportion of girls born to chemists is higher than the state proportion?

8.48 **Sickle-cell and malaria.** Sickle-cell trait is a hereditary condition that is common among blacks and can cause medical problems. Some biologists suggest that sickle-cell trait protects against malaria. That would explain why it is found in people who originally came from Africa, where malaria is common. A study in Africa tested 543 children for the sickle-cell trait and also for malaria. In all, 136 of the children had the sickle-cell trait, and 36 of these had heavy malaria infections. The other 407 children lacked the sickle-cell trait, and 152 of them had heavy malaria infections.[27]

(a) Give a 95% confidence interval for the proportion of all children in the population studied who have the sickle-cell trait.

(b) Is there good evidence that the proportion of heavy malaria infections is lower among children with the sickle-cell trait?

8.49 **Side effects of medication.** A study of "adverse symptoms" in users of over-the-counter pain relief medications assigned subjects at random to one of two common pain relievers: acetaminophen and ibuprofen. (Both of these pain relievers are sold under various brand names, sometimes combined with other ingredients.) In all, 650 subjects took acetaminophen, and 44 experienced some adverse symptom. Of the 347 subjects who took ibuprofen, 49 had an adverse symptom. How strong is the evidence that the two pain relievers differ in the proportion of people who experience an adverse symptom?

(a) State hypotheses and check that you can use the z test.

(b) Find the P-value of the test and give your conclusion.

8.50 **Australia versus the U.S.** A study comparing American and Australian corporations examined a sample of 133 American and 63 Australian corporations. There are the usual practical difficulties involving nonresponse and the question of what population the samples represent. Ignore these issues and treat the samples as SRSs from the United States and Australia. The average percent of revenues from "highly regulated businesses" was 27% among the Australian companies and 41% among the American companies.[28]

(a) The data are given as percents. Explain carefully why comparing the percent of revenues from highly regulated businesses for U.S. and Australian corporations is *not* a comparison of two population proportions.

(b) What test would you use to make the comparison? (Don't try to carry out a test.)

Topics in Inference

S tatistical inference offers more methods than anyone can know well, as a glance at the offerings of any large statistical software package demonstrates. In an introduction, we must be selective. In Parts I and II we laid a foundation for understanding statistics:

- The nature and purpose of data analysis.
- The central ideas of designs for data production.
- The reasoning behind confidence intervals and significance tests.
- Experience applying these ideas in simple settings.

Each of the three chapters of Part III offers a short introduction to a more advanced topic in statistical inference. You may choose to read any or all of them, in any order.

What makes a statistical method "more advanced"? More complex data, for one thing. Our introduction to inference in Part II looked only at methods for inference about a single population parameter and for comparing two parameters. The next step is to compare more than two parameters. Chapter 9 shows how to compare more than two proportions, and Chapter 10 discusses comparing more than two means. In these chapters we have data from three or more samples or experimental treatments, not just one or two. Another more complex setting concerns the association between two variables. We describe association by a two-way table if the variables are categorical, or by correlation and regression if they are quantitative. Chapter 9 presents inference for two-way tables. Inference for regression is the topic of Chapter 11. Chapters 9, 10, and 11 together bring our knowledge of inference to the same point that our study of data analysis reached in Chapters 1 and 2.

With greater complexity comes greater reliance on software. In these final three chapters you will more often be interpreting the output of

(REPRODUCED BY PERMISSION OF BOB SCHOCHET.)

"It was a numbers explosion."

statistical software or using software yourself. You can do the calculations in Chapter 9 without software or a specialized calculator. In Chapters 10 and 11, the pain is too great and the contribution to learning too small. Fortunately, you can grasp the ideas without step-by-step arithmetic.

Another aspect of "more advanced" methods is new concepts and ideas. This is where we draw the line in deciding what statistical topics we can master in a first course. Chapters 9, 10, and 11 build more elaborate methods on the foundation we have laid without introducing fundamentally new concepts. You can see that statistical practice does need additional big ideas by reading the sections on "the problem of multiple comparisons" in Chapters 9 and 10. But the ideas you already know place you among the world's statistical sophisticates.

Karl Pearson

(UNIVERSITY COLLEGE, LONDON.)

The First Inference Procedure

K arl Pearson (1857–1936), a professor at University College in London, had already published nine books before he turned his abundant energy to statistics in 1893.

After Pearson, statistics was a field of study.

Of course, Pearson didn't really take up statistics, which was not yet a separate field of study. He took up problems of heredity and evolution, which led him into statistics.

Pearson developed a family of curves—we would call them density curves—for describing biological data that don't follow a normal distribution. He then asked how he could test whether one of these curves actually fit a set of data well. In 1900 he invented a method, the chi-square test. Pearson's chi-square test has the honor of being the oldest inference procedure still in use. It is now most often used for problems somewhat different from the one that motivated Pearson, as we will see in this chapter.

After Pearson, statistics was a field of study. Fisher and Neyman in the 1920s and 1930s would provide much of its present form, but here is what the leading historian of statistics says about the origins: "Before 1900 we see many scientists of different fields developing and using techniques we now recognize as belonging to modern statistics. After 1900 we begin to see identifiable statisticians developing such techniques into a unified logic of empirical science that goes far beyond its component parts. There was no sharp moment of birth; but with Pearson and Yule and the growing numbers of students in Pearson's laboratory, the infant discipline may be said to have arrived."[1]

Inference for Two-Way Tables

(JERRY COOKE/PHOTO RESEARCHERS.)

Heredity works statistically, as breeders of animals from show dogs to beef cattle know.

Introduction

The two-sample z procedures of Chapter 8 allow us to compare the proportions of successes in two groups, either two populations or two treatment groups in an experiment. What if we want to compare more than two groups? We need a new statistical test. The new test starts by presenting the data in a new way, as a two-way table. Two-way tables have more general uses than comparing the proportions of successes in several groups. As we saw in Section 5 of Chapter 2, they describe relationships between any two categorical variables. The same test that compares several proportions also tests whether the row and column variables are related in any two-way table. We will start with the problem of comparing several proportions.

EXAMPLE 9.1 **Treating cocaine addiction**

Cocaine addicts need the drug to feel pleasure. Perhaps giving them a medication that fights depression will help them stay off cocaine. A three-year study compared an antidepressant called desipramine with lithium (a standard treatment for cocaine addiction) and a placebo. The subjects were 72 chronic users of cocaine who wanted to break their drug habit. Twenty-four of the subjects were randomly assigned to each treatment. Here are the counts and proportions of the subjects who avoided relapse into cocaine use during the study:[2]

Group	Treatment	Subjects	No relapse	Proportion
1	Desipramine	24	14	0.583
2	Lithium	24	6	0.250
3	Placebo	24	4	0.167

The sample proportions of subjects who stayed off cocaine are quite different. The bar graph in Figure 9.1 compares the results visually. Are these data good evidence that the proportions of successes for the three treatments differ in the population of all cocaine addicts?

The problem of multiple comparisons

Call the population proportions of successes in the three groups p_1, p_2, and p_3. We again use a subscript to remind us which group a parameter or statistic describes. To compare these three population proportions, we might use the two-sample z procedures several times:

- Test H_0: $p_1 = p_2$ to see if the success rate of desipramine differs from that of lithium.
- Test H_0: $p_1 = p_3$ to see if desipramine differs from a placebo.
- Test H_0: $p_2 = p_3$ to see if lithium differs from a placebo.

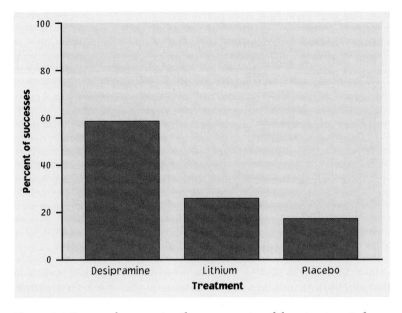

Figure 9.1 *Bar graph comparing the success rates of three treatments for cocaine addiction, for Example 9.1.*

The weakness of doing three tests is that we get three *P*-values, one for each test alone. That doesn't tell us how likely it is that *three* sample proportions are spread apart as far as these are. It may be that $\hat{p}_1 = 0.583$ and $\hat{p}_3 = 0.167$ are significantly different if we look at just two groups, but not significantly different if we know that they are the largest and smallest proportions among three groups. As we look at more groups, we expect the gap between the smallest and largest sample proportions to get larger. (Think of comparing the tallest and shortest person in larger and larger groups of people.) We can't safely compare many parameters by doing tests or confidence intervals for two parameters at a time.

The problem of how to do many comparisons at once with some overall measure of confidence in all our conclusions is common in statistics. This is the problem of **multiple comparisons**. Statistical methods for dealing with multiple comparisons usually have two steps:

multiple comparisons

1. An *overall test* to see if there is good evidence of *any* differences among the parameters that we want to compare.
2. A detailed *follow-up analysis* to decide which of the parameters differ and to estimate how large the differences are.

The overall test, though more complex than the tests we met earlier, is often reasonably straightforward. The follow-up analysis can be quite elaborate. In our basic introduction to statistical practice, we will look only at some overall tests. In this chapter we present a test for comparing several population proportions. The next chapter shows how to compare several population means.

two-way table

9.1 Two-Way Tables

The first step in the overall test for comparing several proportions is to arrange the data in a **two-way table** that gives counts for both successes and failures. Here is the two-way table of the cocaine addiction data:

	Relapse	
	No	Yes
Desipramine	14	10
Lithium	6	18
Placebo	4	20

r × c table

cell

We call this a 3 × 2 table because it has 3 rows and 2 columns. A table with *r* rows and *c* columns is an ***r × c* table.** The table shows the relationship between two categorical variables. The explanatory variable is the treatment (one of three drugs). The response variable is success (no relapse) or failure (relapse). The two-way table gives the counts for all 6 combinations of values of these variables. Each of the 6 counts occupies a **cell** of the table.

Expected counts

We want to test the null hypothesis that there are *no differences* among the proportions of successes for addicts given the three treatments:

$$H_0: p_1 = p_2 = p_3$$

The alternative hypothesis is that there *is* some difference, that not all three proportions are equal:

$$H_a: \text{not all of } p_1, p_2, \text{ and } p_3 \text{ are equal}$$

The alternative hypothesis is no longer one-sided or two-sided. It is "many-sided," because it allows any relationship other than "all three equal." For example, H_a includes the situation in which $p_2 = p_3$ but p_1 has a different value.

To test H_0, we compare the observed counts in a two-way table with the *expected counts*, the counts we would expect—except for random variation—if H_0 were true. If the observed counts are far from the expected counts, that is evidence against H_0. It is easy to find the expected counts.

> **EXPECTED COUNTS**
>
> The **expected count** in any cell of a two-way table when H_0 is true is
>
> $$\text{expected count} = \frac{\text{row total} \times \text{column total}}{\text{table total}}$$

To understand why this recipe works, think first about just one proportion.

EXAMPLE 9.2   Making free throws

Corinne is a basketball player who makes 70% of her free throws. If she shoots 10 free throws in a game, we expect her to make 70% of them, or 7 of the 10. Of course, she won't make exactly 7 every time she shoots 10 free throws in a game. There is chance variation from game to game. But in the long run, 7 of 10 is what we expect. It is, in fact, the *mean* number of shots Corinne makes when she shoots 10 times.

In more formal language, if we have n independent tries and the probability of a success on each try is p, we expect np successes. If we draw an SRS of n individuals from a population in which the proportion of successes is p, we expect np successes in the sample. That's the fact behind the formula for expected counts in a two-way table.

Let's apply this fact to the cocaine study. The two-way table with row and column totals is

| | Relapse | | |
	No	Yes	Total
Desipramine	14	10	24
Lithium	6	18	24
Placebo	4	20	24
Total	24	48	72

He started it!

A study of deaths in bar fights showed that in 90% of the cases, the person who died started the fight. You shouldn't believe this. If you killed someone in a fight, what would you say when the police ask you who started the fight? After all, dead men tell no tales.

We will find the expected count for the cell in row 1 (desipramine) and column 1 (no relapse). The proportion of all 72 subjects who succeed in avoiding a relapse is

$$\frac{\text{count of successes}}{\text{table total}} = \frac{\text{column 1 total}}{\text{table total}} = \frac{24}{72} = \frac{1}{3}$$

Think of this as p, the overall proportion of successes. If H_0 is true, we expect (except for random variation) this same proportion of successes in all three groups. So the expected count of successes among the 24 subjects who took desipramine is

$$np = (24)\left(\frac{1}{3}\right) = 8.00$$

This expected count has the form announced in the box:

$$\frac{\text{row 1 total} \times \text{column 1 total}}{\text{table total}} = \frac{(24)(24)}{72}$$

EXAMPLE 9.3 **Observed versus expected counts**

Here are the observed and expected counts side by side:

	Observed		Expected	
	No	Yes	No	Yes
Desipramine	14	10	8	16
Lithium	6	18	8	16
Placebo	4	20	8	16

Because 1/3 of all subjects succeeded, we expect 1/3 of the 24 subjects in each group to avoid a relapse if there are no differences among the treatments. In fact, desipramine has more successes (14) and fewer relapses (10) than expected. The placebo has fewer successes (4) and more relapses (20). That's another way of saying what the sample proportions in Example 9.1 say more directly: desipramine does much better than the placebo, with lithium in between.

APPLY YOUR KNOWLEDGE

9.1 **Extracurricular activities and grades.** North Carolina State University studied student performance in a course required by its chemical engineering major. One question of interest is the relationship between time spent in extracurricular activities and whether a student earned a C or better in the course. Here are the data for the 119 students who answered a question about extracurricular activities:[3]

	Extracurricular activities (hours per week)		
	< 2	2 to 12	> 12
C or better	11	68	3
D or F	9	23	5

(a) This is an $r \times c$ table. What are the numbers r and c?

(b) Find the proportion of successful students (C or better) in each of the three extracurricular activity groups. What kind of relationship between extracurricular activities and succeeding in the course do these proportions seem to show?

(c) Make a bar graph to compare the three proportions of successes.

(d) The null hypothesis says that the proportions of successes are the same in all three groups if we look at the population of all students. Find the expected counts if this hypothesis is true, and display them in a two-way table.

(e) Compare the observed counts with the expected counts. Are there large deviations between them? These deviations are another way of describing the relationship you described in (b).

9.2 **Smoking by students and their parents.** How are the smoking habits of students related to their parents' smoking? Here are data from a survey of students in eight Arizona high schools:[4]

	Student smokes	Student does not smoke
Both parents smoke	400	1380
One parent smokes	416	1823
Neither parent smokes	188	1168

(a) This is an $r \times c$ table. What are the numbers r and c?

(b) Calculate the proportion of students who smoke in each of the three parent groups. Then describe in words the association between parent smoking and student smoking.

(c) Make a graph to display the association.

(d) Explain in words what the null hypothesis $H_0: p_1 = p_2 = p_3$ says about student smoking.

(e) Find the expected counts if H_0 is true, and display them in a two-way table similar to the table of observed counts.

(f) Compare the tables of observed and expected counts. Explain how the comparison expresses the same association you see in (b) and (c).

9.2 The Chi-Square Test

Comparing the sample proportions of successes describes the differences among the three treatments for cocaine addiction. But the statistical test that tells us whether those differences are statistically significant doesn't use the sample proportions. It compares the observed and expected counts. The test statistic that makes the comparison is the *chi-square statistic*.

CHI-SQUARE STATISTIC

The **chi-square statistic** is a measure of how far the observed counts in a two-way table are from the expected counts. The formula for the statistic is

$$X^2 = \sum \frac{(\text{observed count} - \text{expected count})^2}{\text{expected count}}$$

The sum is over all $r \times c$ cells in the table.

The chi-square statistic is a sum of terms, one for each cell in the table. In the cocaine example, 14 of the desipramine group succeeded in avoiding a relapse. The expected count for this cell is 8. So the component of the chi-square statistic from this cell is

$$\frac{(\text{observed count} - \text{expected count})^2}{\text{expected count}} = \frac{(14 - 8)^2}{8}$$

$$= \frac{36}{8} = 4.5$$

Think of the chi-square statistic X^2 as a measure of the distance of the observed counts from the expected counts. Like any distance, it is always zero or positive, and it is zero only when the observed counts are exactly equal to the expected counts. Large values of X^2 are evidence against H_0 because they say that the observed counts are far from what we would expect if H_0 were true. Although the alternative hypothesis H_a is many-sided, the chi-square test is one-sided because any violation of H_0 tends to produce a large value of X^2. Small values of X^2 are not evidence against H_0.

Calculating the expected counts and then the chi-square statistic by hand is a bit time-consuming. As usual, statistical software saves time and always gets the arithmetic right. Some calculators will also find the chi-square statistic from keyed-in data.

EXAMPLE 9.4   Chi-square from software

Enter the two-way table (the 6 counts) for the cocaine study into the Minitab statistical software and request the chi-square test. The output appears in Figure 9.2. Most statistical software packages produce chi-square output similar to this.

Minitab repeats the two-way table of observed counts, puts the expected count for each cell below the observed count, and inserts the row and column totals. Then the software calculates the chi-square statistic X^2. For these data, $X^2 = 10.500$. The statistic is a sum of 6 terms, one for each cell in the table. The "Chi-Sq" display in the output shows the individual terms, as well as their sum. The first term is 4.500, just as we calculated.

The P-value is the probability that X^2 would take a value as large as 10.500 if H_0 were really true. We see from Figure 9.2 that $P = 0.005$. The small P-value gives

Chi-Square Test

Expected counts are printed below observed counts

	Success	Relapse	Total
D	14	10	24
	8.00	16.00	
L	6	18	24
	8.00	16.00	
P	4	20	24
	8.00	16.00	
Total	24	48	72

```
Chi-Sq  =   4.500 + 2.250   +
            0.500 + 0.250   +
            2.000 + 1.000   =   10.500
DF   =   2, P-Value = 0.005
```

Figure 9.2 *Minitab output for the two-way table in the cocaine study. The output gives the observed counts, the expected counts, the value 10.500 of chi-square statistic, and the P-value P = 0.005.*

us good reason to conclude that there *are* differences among the effects of the three treatments.

The chi-square test is the overall test for comparing any number of population proportions. If the test allows us to reject the null hypothesis that all the proportions are equal, we then want to do a follow-up analysis that examines the differences in detail. We won't describe how to do a formal follow-up analysis, but you should look at the data to see what specific effects they suggest.

EXAMPLE 9.5   The cocaine study: conclusions

The cocaine study found significant differences among the proportions of successes for three treatments for cocaine addiction. We can see the specific differences in three ways.

Look first at the *sample proportions*:

$$\hat{p}_1 = 0.583 \quad \hat{p}_2 = 0.250 \quad \hat{p}_3 = 0.167$$

These suggest that the major difference between the proportions is that desipramine has a much higher success rate than either lithium or a placebo. That is the effect that the study hoped to find.

Next, *compare the observed and expected counts* in Figure 9.2. Treatment D (desipramine) has more successes and fewer failures than we would expect if all three treatments had the same success rate in the population. The other two treatments had fewer successes and more failures than expected.

Finally, Minitab prints under the table the 6 individual "distances" between the observed and expected counts that are added to get X^2. The arrangement of these **components of X^2** is the same as the 3 × 2 arrangement of the table. The largest components show which cells contribute the most to the overall distance X^2. The largest component by far is for the top left cell in the table: desipramine has more successes than would be expected.

All three ways of examining the data point to the same conclusion: desipramine works better than the other treatments. This is an informal conclusion. More advanced methods provide tests and confidence intervals that make this follow-up analysis formal.

components of chi-square

APPLY YOUR KNOWLEDGE · · · · · · · · · ·

9.3 **Extracurricular activities and grades.** In Exercise 9.1, you began to analyze data on the relationship between time spent on extracurricular activities and success in a tough course. Figure 9.3 gives Minitab output for the two-way table in Exercise 9.1.

(a) Starting from the table of expected counts, find the 6 components of the chi-square statistic and then the statistic X^2 itself. Check your work against the computer output.

```
Chi-Square Test

Expected counts are printed below observed counts

              < 2      2 to 12       >12      Total
A, B, C,       11          68          3         82
             13.78       62.71       5.51

D or F          9          23          5         37
              6.22       28.29       2.49

Total          20          91          8        119

Chi-Sq = 0.561  +    0.447  +   1.145  +
          1.244  +    0.991  +   2.538  = 6.926
DF = 2,  P-Value = 0.031
1 cells with expected counts less than 5.0
```

Figure 9.3 *Minitab output for the study of extracurricular activity and success in a tough course, for Exercise 9.3.*

(b) What is the *P*-value for the test? Explain in simple language what it means to reject H_0 in this setting.

(c) Which term contributes the most to X^2? What specific relation between extracurricular activities and academic success does this term point to?

(d) Does the North Carolina State study convince you that spending more or less time on extracurricular activities *causes* changes in academic success? Explain your answer.

9.4 **Smoking by students and their parents.** In Exercise 9.2, you began to analyze data on the relationship between smoking by parents and smoking by high school students. Figure 9.4 gives the Minitab output for the two-way table in Exercise 9.2.

(a) Starting from the table of expected counts, find the 6 components of the chi-square statistic and then the statistic X^2 itself. Check your work against the computer output.

(b) What is the *P*-value for the test? Explain in simple language what it means to reject H_0 in this setting.

(c) Which two terms contribute the most to X^2? What specific relation between parent smoking and student smoking do these terms point to?

```
Chi-Square Test

Expected counts are printed below observed counts

           Smokes        NoSmoke        Total
Both          400           1380         1780
           332.49        1447.51

One           416           1823         2239
           418.22        1820.78

None          188           1168         1356
           253.29        1102.71

Total        1004           4371         5375

Chi-Sq = 13.709    +    3.149 +
          0.012    +    0.003 +
         16.829    +    3.866 = 37.566
DF = 2,   P-Value = 0.000
```

Figure 9.4 *Minitab output for the study of parent smoking and student smoking, for Exercise 9.4.*

(d) Does this study convince you that parent smoking *causes* student smoking? Explain your answer.

The chi-square distributions

Software usually finds *P*-values for us. The *P*-value for a chi-square test comes from comparing the value of the chi-square statistic with critical values for a *chi-square distribution*.

> ### THE CHI-SQUARE DISTRIBUTIONS
>
> The **chi-square distributions** are a family of distributions that take only positive values and are skewed to the right. A specific chi-square distribution is specified by giving its **degrees of freedom**.
>
> The chi-square test for a two-way table with r rows and c columns uses critical values from the chi-square distribution with $(r-1)(c-1)$ degrees of freedom. The *P*-value is the area to the right of X^2 under the chi-square density curve.

Figure 9.5 shows the density curves for three members of the chi-square family of distributions. As the degrees of freedom increase, the density curves

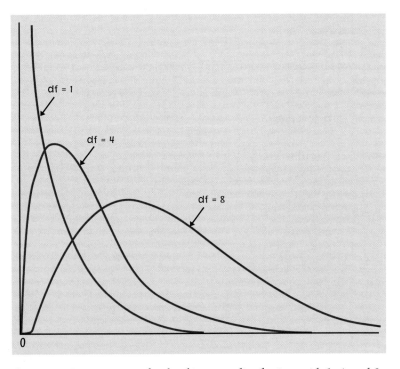

Figure 9.5 *Density curves for the chi-square distributions with 1, 4, and 8 degrees of freedom. Chi-square distributions take only positive values and are right-skewed.*

become less skewed and larger values become more probable. Table E in the back of the book gives critical values for chi-square distributions. You can use Table E if software does not give you P-values for a chi-square test.

| **EXAMPLE 9.6** | **Using the chi-square table** |

The two-way table of 3 treatments by 2 outcomes for the cocaine study has 3 rows and 2 columns. That is, $r = 3$ and $c = 2$. The chi-square statistic therefore has degrees of freedom

$$(r - 1)(c - 1) = (3 - 1)(2 - 1) = (2)(1) = 2$$

The computer output in Figure 9.2 gives 2 as the degrees of freedom.

The observed value of the chi-square statistic is $X^2 = 10.500$. Look in the df $= 2$ row of Table E. The value $X^2 = 10.500$ falls between the 0.01 and 0.005 critical values of the chi-square distribution with 2 degrees of freedom. Remember that the chi-square test is always one-sided. So the P-value of $X^2 = 10.500$ is between 0.01 and 0.005. The software output in Figure 9.2 shows that the P-value is in fact equal to 0.005 when rounded to three decimal places.

df $= 2$

p	.01	.005
x^*	9.21	10.60

We know that all z and t statistics measure the size of an effect in the standard scale centered at zero. We can roughly assess the size of any z or t statistic by the 68–95–99.7 rule, though this is exact only for z. The chi-square statistic does not have any such natural interpretation. But here is a helpful fact: *the mean of any chi-square distribution is equal to its degrees of freedom*. In Example 9.6, X^2 would have mean 2 if the null hypothesis were true. The observed value $X^2 = 10.500$ is so much larger than 2 that we suspect it is significant even before we look at the chi-square table.

APPLY YOUR KNOWLEDGE

9.5   **Extracurricular activities and grades.** The computer output in Figure 9.3 gives the degrees of freedom for the table in Exercise 9.1 as 2.

(a) Verify that this is correct.

(b) The computer gives the value of the chi-square statistic as $X^2 = 6.926$. Between what two entries in Table E does this value lie? What does the table tell you about the P-value?

(c) What would be the mean value of the statistic X^2 if the null hypothesis were true? How does the observed value of X^2 compare with the mean?

9.6   **Smoking by students and their parents.** The computer output in Figure 9.4 gives the degrees of freedom for the table in Exercise 9.2 as 2.

(a) Verify that this is correct.

(b) The computer gives the value of the chi-square statistic as $X^2 = 37.566$. Where in Table E does this value lie? What does the table tell you about the P-value?

(c) What would be the mean value of the statistic X^2 if the null hypothesis were true? How does the observed value of X^2 compare with the mean?

More uses of the chi-square test

Two-way tables can arise in several ways. The cocaine study is an experiment that assigned 24 addicts to each of three groups. Each group is a sample from a separate population corresponding to a separate treatment. The study design fixes the size of each sample in advance, and the data record which of two outcomes occurred for each subject. The null hypothesis of "no difference" among the treatments takes the form of "equal proportions of successes" in the three populations. The next example illustrates a different setting for a two-way table.

EXAMPLE 9.7   Marital status and job level

A study of the relationship between men's marital status and the level of their jobs used data on all 8235 male managers and professionals employed by a large manufacturing firm. Each man's job has a grade set by the company that reflects the value of that particular job to the company. The authors of the study grouped the many job grades into quarters. Grade 1 contains jobs in the lowest quarter of job grades, and grade 4 contains those in the highest quarter. Here are the data:[5]

	Marital Status			
Job Grade	Single	Married	Divorced	Widowed
1	58	874	15	8
2	222	3927	70	20
3	50	2396	34	10
4	7	533	7	4

Do these data show a statistically significant relationship between marital status and job grade?

In Example 9.7 we do not have four separate samples from the four marital statuses. We have a single group of 8235 men, each classified in two ways, by marital status and job grade. The number of men in each marital status is not fixed in advance but is known only after we have the data. Both marital status and job grade have four levels, so a careful statement of the null hypothesis

H_0: there is no relationship between marital status and job grade

in terms of population parameters is complicated.

In fact, we might regard these 8235 men as an entire population rather than as a sample. The data include all the managers and professionals employed by this company. These employees are not necessarily a random sample from any larger population. Nevertheless, we would still like to decide if the relationship between marital status and job grade is statistically significant in the sense that it is too strong to happen just by chance if job grades were handed out at random to men of all marital statuses. "Not likely to happen just by chance if H_0 is true" is the usual meaning of statistical significance. In this example, that meaning makes sense even though we have data on an entire population.

The setting of Example 9.7 is very different from a comparison of several proportions. Nevertheless, we can apply the chi-square test. One of the most useful properties of chi-square is that it tests the hypothesis "the row and column variables are not related to each other" whenever this hypothesis makes sense for a two-way table.

USES OF THE CHI-SQUARE TEST

Use the chi-square test to test the null hypothesis

H_0: there is no relationship between two categorical variables

when you have a two-way table from one of these situations:

- Independent SRSs from each of several populations, with each individual classified according to one categorical variable. (The other variable says which sample the individual comes from.)

- A single SRS, with each individual classified according to both of two categorical variables.

EXAMPLE 9.8 **Marital status and job level**

To analyze the job grade data in Example 9.7, first do the overall chi-square test. The Minitab chi-square output appears in Figure 9.6. The observed chi-square is very large, $X^2 = 67.397$. The P-value, rounded to three decimal places, is 0. We have overwhelming evidence that job grade is related to marital status. Because the sample is so large, we are not surprised to find very significant results.

Table E gives a similar conclusion. For a 4×4 table, the degrees of freedom are $(r - 1)(c - 1) = 9$. Look in the df = 9 row of Table E. The largest critical value is 29.67, corresponding to the P-value 0.0005. The observed $X^2 = 67.397$ is beyond that value, so $P < 0.0005$.

Next, do a follow-up analysis to describe the relationship. As in Section 5 of Chapter 2, we describe a relationship between two categorical variables by comparing percents. Here is a table of the percent of men in each marital status whose jobs have each grade. Each column of this table gives the **conditional distribution** of job grade among men with a specific marital status. Each column adds to 100% (except for roundoff error) because it accounts for all the men in one marital status.

conditional distribution

Figure 9.6 *Minitab output for the job grade data of Example 9.7.*

```
Chi-Square Test

Expected counts are printed below observed counts

          Single    Married   Divorced    Widowed      Total
   1         58        874         15          8         955
           39.08     896.44      14.61       4.87

   2        222       3927         70         20        4239
          173.47    3979.05      64.86      21.62

   3         50       2396         34         10        2490
          101.90    2337.30      38.10      12.70

   4          7        533          7          4         551
           22.55     517.21       8.43       2.81

Total       337       7730        126         42        8235

Chi-Sq = 9.158   +   0.562   +   0.010   +   2.011
        13.575   +   0.681   +   0.407   +   0.121
        26.432   +   1.474   +   0.441   +   0.574
        10.722   +   0.482   +   0.243   +   0.504  = 67.397
DF = 9,  P-Value = 0.000
2 cells with expected counts less than 5.0
```

| | Marital Status | | | |
Job Grade	Single	Married	Divorced	Widowed
1	17.2%	11.3%	11.9%	19.1%
2	65.9%	50.8%	55.5%	47.7%
3	14.9%	31.0%	26.9%	23.8%
4	2.0%	6.9%	5.6%	9.6%
Total	100.0%	100.0%	99.9%	100.2%

The bar graphs in Figure 9.7 help us compare these four conditional distributions. We see at once that smaller percents of single men have jobs in the higher grades 3 and 4. Not only married men but men who were once married and are now divorced or widowed are more likely to hold higher-grade jobs. Look at the 16 components of the chi-square sum in the computer output for confirmation. The four cells for single men have the four largest components of X^2. Minitab's table of counts shows that the observed counts for single men are higher than expected in grades 1 and 2 and lower than expected in grades 3 and 4.

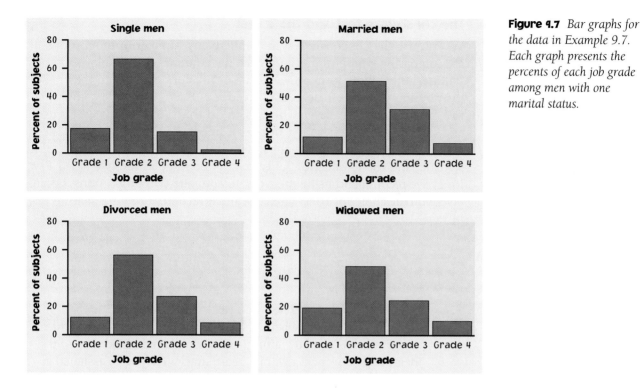

Figure 9.7 *Bar graphs for the data in Example 9.7. Each graph presents the percents of each job grade among men with one marital status.*

Of course, this association between marital status and job grade does not show that being single *causes* lower-grade jobs. The explanation might be as simple as the fact that single men tend to be younger and so have not yet advanced to higher grades.

Cell counts required for the chi-square test

The computer output in Figure 9.6 has one more feature. It warns us that the expected counts in two of the 16 cells are less than 5. The chi-square test, like the z procedures for comparing two proportions, is an approximate method that becomes more accurate as the counts in the cells of the table get larger. Fortunately, the approximation is accurate for quite modest counts. Here is a practical guideline.[6]

> CELL COUNTS REQUIRED FOR THE CHI-SQUARE TEST
>
> You can safely use the chi-square test with critical values from the chi-square distribution when no more than 20% of the expected counts are less than 5 and all individual expected counts are 1 or greater. In particular, all four expected counts in a 2 × 2 table should be 5 or greater.

Example 9.8 easily passes this test. All the expected counts are greater than 1, and only 2 out of 16 (12.5%) are less than 5.

More chi-square tests

There are other chi-square tests for hypotheses more specific than "no relationship." A sociologist places people in classes by social status, waits ten years, then classifies the same people again. The row and column variables are the classes at the two times. She might test the hypothesis that there has been no change in the overall distribution of social status in the group. Or she might ask if moves up in status are balanced by matching moves down. These and other null hypotheses can be tested by variations of the chi-square test.

The chi-square test and the *z* test

We can use the chi-square test to compare any number of proportions. If we are comparing r proportions and make the columns of the table "success" and "failure," the counts form an $r \times 2$ table. P-values come from the chi-square distribution with $r - 1$ degrees of freedom. If $r = 2$, we are comparing just two proportions. We have two ways to do this: the z test from Section 8.2 and the chi-square test with 1 degree of freedom for a 2×2 table. *These two tests always agree.* In fact, the chi-square statistic X^2 is just the square of the z statistic, and the P-value for X^2 is exactly the same as the two-sided P-value for z. We recommend using the z test to compare two proportions, because it gives you the choice of a one-sided test and is related to a confidence interval for $p_1 - p_2$.

APPLY YOUR KNOWLEDGE

9.7 **Stress and heart attacks.** You read a newspaper article that describes a study of whether stress management can help reduce heart attacks. The 107 subjects all had reduced blood flow to the heart and so were at risk of a heart attack. They were assigned at random to three groups. The article goes on to say:

> One group took a four-month stress management program, another underwent a four-month exercise program, and the third received usual heart care from their personal physicians.
>
> In the next three years, only three of the 33 people in the stress management group suffered "cardiac events," defined as a fatal or non-fatal heart attack or a surgical procedure such as a bypass or angioplasty. In the same period, seven of the 34 people in the exercise group and 12 out of the 40 patients in usual care suffered such events.[7]

(a) Use the information in the news article to make a two-way table that describes the study results.

(b) What are the success rates of the three treatments in avoiding cardiac events?

(c) Find the expected cell counts under the null hypothesis that there is no difference among the treatments. Verify that the expected counts meet our guideline for use of the chi-square test.

(d) Is there a significant difference among the success rates for the three treatments?

9.8 **Unhappy HMO patients.** Exercise 8.18 (page 444) compared HMO members who filed complaints with an SRS of members who did not complain. The study actually broke the complainers into two subgroups: those who filed complaints about medical treatment and those who filed nonmedical complaints. Here are the data on the

(AP PHOTO/CHARLES KRUPA.)

total number in each group and the number who voluntarily left the HMO:

	No complaint	Medical complaint	Nonmedical complaint
Total	743	199	440
Left	22	26	28

(a) Find the percent of each group who left.

(b) Make a two-way table of complaint status by left or not.

(c) Find the expected counts and check that you can safely use the chi-square test.

(d) The chi-square statistic for this table is $X^2 = 31.765$. What null and alternative hypotheses does this statistic test? What are its degrees of freedom? Why is it clear without looking at a table that this value of X^2 is highly significant?

(e) Use Table E to approximate the P-value. What do you conclude from these data?

9.9 **Treating ulcers.** Gastric freezing was once a recommended treatment for ulcers in the upper intestine. Use of gastric freezing stopped after experiments showed it had no effect. One randomized comparative experiment found that 28 of the 82 gastric freezing patients improved, while 30 of the 78 patients in the placebo group improved.[8] We can test the hypothesis of "no difference" between the two groups in two ways: using the two-sample z statistic or using the chi-square statistic.

(a) State the null hypothesis with a two-sided alternative and carry out the z test. What is the P-value from Table A?

(b) Present the data in a 2×2 table. Use the chi-square test to test the hypothesis from (a). Verify that the X^2 statistic is the square of the z statistic. Use Table E to verify that the chi-square P-value agrees with the z result up to the accuracy of the table.

(c) What do you conclude about the effectiveness of gastric freezing as a treatment for ulcers?

Summary

The **chi-square test** for a two-way table tests the null hypothesis that there is no relationship between the row variable and the column variable.

One common use of the chi-square test is to **compare several population proportions**. The null hypothesis states that all of the population proportions are equal. The alternative hypothesis states that they are not all equal but allows any other relationship among the population proportions.

The **expected count** in any cell of a two-way table when H_0 is true is

$$\text{expected count} = \frac{\text{row total} \times \text{column total}}{\text{table total}}$$

The **chi-square statistic** is

$$X^2 = \sum \frac{(\text{observed count} - \text{expected count})^2}{\text{expected count}}$$

The chi-square test compares the value of the statistic X^2 with critical values from the **chi-square distribution** with $(r-1)(c-1)$ **degrees of freedom**. Large values of X^2 are evidence against H_0, so the P-value is the area under the chi-square density curve to the right of X^2.

The chi-square distribution is an approximation to the distribution of the statistic X^2. You can safely use this approximation when all expected cell counts are at least 1 and no more than 20% are less than 5.

If the chi-square test finds a statistically significant relationship between the row and column variables in a two-way table, do a follow-up analysis to describe the nature of the relationship. An informal follow-up analysis compares well-chosen percents, compares the observed counts with the expected counts, and looks for the largest **components of chi-square**.

STATISTICS IN SUMMARY

Advanced statistical inference often concerns relationships among several parameters. This chapter begins with the chi-square test for one such relationship: equality of the proportions of successes in any number of populations. The alternative to this hypothesis is "many-sided," because it allows any relationship other than "all equal." The chi-square test is an overall test that tells us whether the data give good reason to reject the hypothesis that all the population proportions are equal. You should always accompany the chi-square test by data analysis to see what kind of inequality is present.

To do a chi-square test, arrange the counts of successes and failures in a two-way table. Two-way tables can display counts for any two categorical variables, not just successes and failures in r populations. The chi-square test is also more general than just a test for equal population proportions of successes. It tests the null hypothesis that there is "no relationship" between the row variable and the column variable in a two-way table. The chi-square test is an approximate test that becomes more accurate as the cell counts in the two-way table increase. Fortunately, chi-square P-values are quite accurate even for small counts.

After studying this chapter, you should be able to do the following.

A. TWO-WAY TABLES

1. Arrange data on successes and failures in several groups into a two-way table of counts of successes and failures in all groups.

2. Use percents to describe the relationship between any two categorical variables starting from the counts in a two-way table.

B. INTERPRETING CHI-SQUARE TESTS

1. Locate expected cell counts, the chi-square statistic, and its *P*-value in output from your software or calculator.

2. Explain what null hypothesis the chi-square statistic tests in a specific two-way table.

3. If the test is significant, use percents, comparison of expected and observed counts, and the components of the chi-square statistic to see what deviations from the null hypothesis are most important.

C. DOING CHI-SQUARE TESTS

1. Calculate the expected count for any cell from the observed counts in a two-way table.

2. Calculate the component of the chi-square statistic for any cell, as well as the overall statistic.

3. Give the degrees of freedom of a chi-square statistic. Make a quick assessment of the significance of the statistic by comparing the observed value to the degrees of freedom.

4. Use the chi-square critical values in Table E to approximate the *P*-value of a chi-square test.

CHAPTER 9 Review Exercises

If you have access to statistical software or a statistical calculator, use it to speed your analysis of the data in these exercises.

9.10 Python eggs. How is the hatching of water python eggs influenced by the temperature of the snake's nest? Researchers assigned newly laid eggs to one of three temperatures: hot, neutral, or cold. Hot duplicates the extra warmth provided by the mother python, and cold duplicates the absence of the mother. Here are the data on the number of eggs and the number that hatched:[9]

	Eggs	Hatched
Cold	27	16
Neutral	56	38
Hot	104	75

(a) Make a two-way table of temperature by outcome (hatched or not).

(HOLT STUDIOS INTERNATIONAL/PHOTO RESEARCHERS)

(b) Calculate the percent of eggs in each group that hatched. The researchers anticipated that eggs would not hatch in cold water. Do the data support that anticipation?

(c) Are there significant differences among the proportions of eggs that hatched in the three groups?

9.11 **The Mediterranean diet.** Cancer of the colon and rectum is less common in the Mediterranean region than in other Western countries. The Mediterranean diet contains little animal fat and lots of olive oil. Italian researchers compared 1953 patients with colon or rectal cancer with a control group of 4154 patients admitted to the same hospitals for unrelated reasons. They estimated consumption of various foods from a detailed interview, then divided the patients into three groups according to their consumption of olive oil. Here are some of the data:[10]

	Olive Oil			
	Low	Medium	High	Total
Colon cancer	398	397	430	1225
Rectal cancer	250	241	237	728
Controls	1368	1377	1409	4154

(a) Is this study an experiment? Explain your answer.

(b) Is high olive oil consumption more common among patients without cancer than in patients with colon cancer or rectal cancer?

(c) Find the chi-square statistic X^2. What would be the mean of X^2 if the null hypothesis (no relationship) were true? What does comparing the observed value of X^2 with this mean suggest? What is the P-value? What do you conclude?

(d) The investigators report that "less than 4% of cases or controls refused to participate." Why does this fact strengthen our confidence in the results?

9.12 **Preventing strokes.** Exercise 8.28 compared aspirin plus another drug with aspirin alone as treatments for patients who had suffered a stroke. The study actually assigned stroke patients at random to four treatments. Here are the data:[11]

Treatment	Number of patients	Number of strokes	Number of deaths
Placebo	1649	250	202
Aspirin	1649	206	182
Dipyridamole	1654	211	188
Both	1650	157	185

(a) Make a two-way table of treatment by whether or not a patient had a stroke during the two-year study period. Compare the rates of strokes for the four treatments. Which treatment appears most effective in preventing strokes? Is there a significant difference among the four rates of strokes? Which components of chi-square account for most of the total?

(b) The data report two response variables: whether the patient had a stroke and whether the patient died. Repeat your analysis for patient deaths.

(c) Write a careful summary of your overall findings.

9.13 **Majors for men and women in business.** A study of the career plans of young women and men sent questionnaires to all 722 members of the senior class in the College of Business Administration at the University of Illinois. One question asked which major within the business program the student had chosen. Here are the data from the students who responded:[12]

	Female	Male
Accounting	68	56
Administration	91	40
Economics	5	6
Finance	61	59

(a) Test the null hypothesis that there is no relation between the gender of students and their choice of major. Give a P-value and state your conclusion.

(b) Describe the differences between the distributions of majors for women and men with percents, with a graph, and in words.

(c) Which two cells have the largest components of the chi-square statistic? How do the observed and expected counts differ in these cells? (This should strengthen your conclusions in (b).)

(d) Two of the observed cell counts are small. Do these data satisfy our guidelines for safe use of the chi-square test?

(e) What percent of the students did not respond to the questionnaire? The nonresponse weakens conclusions drawn from these data.

9.14 **Survey response rates.** To study the export activity of manufacturing firms on Taiwan, researchers mailed questionnaires to an SRS of firms in each of five industries that export many of their products. The response rate was only 12.5%, because private companies don't like to fill out long questionnaires from academic researchers. Here are data on the planned sample sizes and the actual number of responses received from each industry:[13]

	Sample size	Responses
Metal products	185	17
Machinery	301	35
Electrical equipment	552	75
Transportation equipment	100	15
Precision instruments	90	12

If the response rates differ greatly, comparisons among the industries may be difficult. Is there good evidence of unequal response rates among the five industries? (Start by creating a two-way table of response or nonresponse by industry.)

9.15 **Secondhand stores.** Shopping at secondhand stores is becoming more popular and has even attracted the attention of business schools. A study of customers' attitudes toward secondhand stores interviewed samples of shoppers at two secondhand stores of the same chain in two cities. The breakdown of the respondents by gender is as follows:[14]

	City 1	City 2
Men	38	68
Women	203	150
Total	241	218

Is there a significant difference between the proportions of women customers in the two cities?

(a) State the null hypothesis, find the sample proportions of women in both cities, do a two-sided z test, and give a P-value using Table A.

(b) Calculate the chi-square statistic X^2 and show that it is the square of the z statistic. Show that the P-value from Table E agrees (up to the accuracy of the table) with your result from (a).

(c) Give a 95% confidence interval for the difference between the proportions of women customers in the two cities.

9.16 **More secondhand stores.** The study of shoppers in secondhand stores cited in the previous exercise also compared the income distributions of shoppers in the two stores. Here is the two-way table of counts:

Income	City 1	City 2
Under $10,000	70	62
$10,000 to $19,999	52	63
$20,000 to $24,999	69	50
$25,000 to $34,999	22	19
$35,000 or more	28	24

A statistical calculator gives the chi-square statistic for this table as $X^2 = 3.955$. Is there good evidence that customers at the two stores have different income distributions? (Give the degrees of freedom, the P-value, and your conclusion.)

9.17 **Child-care workers.** A large study of child care used samples from the data tapes of the Current Population Survey over a period of several years. The result is close to an SRS of child-care workers. The Current Population Survey has three classes of child-care workers: private household, nonhousehold, and preschool teacher. Here are data on the number of blacks among women workers in these three classes:[15]

	Total	Black
Household	2455	172
Nonhousehold	1191	167
Teachers	659	86

(a) What percent of each class of child-care workers is black?

(b) Make a two-way table of class of worker by race (black or other).

(c) Can we safely use the chi-square test? What null and alternative hypotheses does X^2 test?

(d) The chi-square statistic for this table is $X^2 = 53.194$. What are its degrees of freedom? Use Table E to approximate the P-value.

(e) What do you conclude from these data?

9.18 **Unhappy rats and tumors.** It seems that the attitude of cancer patients can influence the progress of their disease. We can't experiment with humans, but here is a rat experiment on this theme. Inject 60 rats with tumor cells and then divide them at random into two groups of 30. All the rats receive electric shocks, but rats in Group 1 can end the shock by pressing a lever. (Rats learn this sort of thing quickly.) The rats in Group 2 cannot control the shocks, which presumably makes them feel helpless and unhappy. We suspect that the rats in Group 1 will develop fewer tumors. The results: 11 of the Group 1 rats and 22 of the Group 2 rats developed tumors.[16]

(a) State the null and alternative hypotheses for this investigation. Explain why the z test rather than the chi-square test for a 2×2 table is the proper test.

(b) Carry out the test and report your conclusion.

9.19 **Standards for child care.** Do unregulated providers of child care in their homes follow different health and safety practices in different cities? A study looked at people who regularly provided care for someone else's children in poor areas of three cities. The numbers who required medical releases from parents to allow medical care in an emergency were 42 of 73 providers in Newark, N.J., 29 of 101 in Camden, N.J., and 48 of 107 in South Chicago, Ill.[17]

(a) Use the chi-square test to see if there are significant differences among the proportions of child-care providers who require medical releases in the three cities.

(b) How should the data be produced in order for your test to be valid? (In fact, the samples came in part from asking parents who were subjects in another study who provided their child care. The author of the study wisely did not use a statistical test. He wrote: "Application of conventional statistical procedures appropriate for random samples may produce biased and misleading results.")

9.20 **Do you use cocaine?** Sample surveys on sensitive issues can give different results depending on how the question is asked. A University of Wisconsin study divided 2400 respondents into 3 groups at random. All were asked if they had ever used cocaine. One group of 800 was interviewed by phone; 21% said they had used cocaine. Another 800 people were asked the question in a one-on-one personal interview; 25% said "Yes." The remaining 800 were allowed to make an anonymous written response; 28% said "Yes."[18] Are there statistically significant differences among these proportions? (State the hypotheses, convert the information given into a two-way table, give the test statistic and its P-value, and state your conclusions.)

9.21 **Nobody is home in July.** The success of a sample survey can depend on the season of the year. The Italian National Statistical Institute kept records of nonresponse to one of its national telephone surveys. All calls were made between 7 p.m. and 10 p.m. Here is a table of the percents of responses and of three types of nonresponse at different seasons. The percents in each row add to 100%.[19]

Season	Calls made	Successful interviews	Nonresponse		
			No answer	Busy signal	Refusal
Jan. 1 to Apr. 13	1558	68.5%	21.4%	5.8%	4.3%
Apr. 21 to June 20	1589	52.4%	35.8%	6.4%	5.4%
July 1 to Aug. 31	2075	43.4%	41.5%	8.6%	6.5%
Sept. 1 to Dec. 15	2638	60.0%	30.0%	5.3%	4.7%

(a) What are the degrees of freedom for the chi-square test of the hypothesis that the distribution of responses varies with the season? (Don't do the test. The sample sizes are so large that the results are sure to be highly significant.)

(b) Consider just the proportion of successful interviews. Describe how this proportion varies with the seasons, and assess the statistical significance of the changes. What do you think explains the changes? (Look at the full table for ideas.)

(c) **(Optional)** It is incorrect to apply the chi-square test to percents rather than to counts. If you enter the 4×4 table of percents above into statistical software and ask for a chi-square test, well-written software should give an error message. (Counts must be whole numbers, so the software should check that.) Try this using your software or calculator, and report the result.

9.22 Continue the analysis of the data in the previous exercise by considering just the proportion of people called who refused to participate. We might think that the refusal rate changes less with the season than, for example, the rate of "no answer." State the hypothesis that the refusal rate does not change with the season. Check that you can safely use the chi-square test. Carry out the test. What do you conclude?

9.23 **Titanic!** In 1912 the luxury liner *Titanic*, on its first voyage across the Atlantic, struck an iceberg and sank. Some passengers got off the ship in lifeboats, but many died. Think of the *Titanic* disaster as an experiment in how the people of that time behaved when faced with death in a situation where only some can escape. The passengers are a sample from the population of their peers. Here is information about who lived and who died, by gender and economic status. (The data leave out a few passengers whose economic status is unknown.)[20]

(STOCK MONTAGE, INC.)

Men			Women		
Status	Died	Survived	Status	Died	Survived
Highest	111	61	Highest	6	126
Middle	150	22	Middle	13	90
Lowest	419	85	Lowest	107	101
Total	680	168	Total	126	317

(a) Compare the percents of men and of women who died. Is there strong evidence that a higher proportion of men die in such situations? Why do you think this happened?

(b) Look only at the women. Describe how the three economic classes differ in the percent of women who died. Are these differences statistically significant?

(c) Now look only at the men and answer the same questions.

9.24 **Simpson's paradox.** Example 2.23 (page 147) presents artificial data that illustrate Simpson's paradox. The data concern the survival rates of surgery patients at two hospitals.

(a) Apply the chi-square test to the data for all patients combined and summarize the results.

(b) Do separate chi-square tests for the patients in good condition and for those in poor condition. Summarize these results.

(c) Are the effects that illustrate Simpson's paradox in this example statistically significant?

9.25 **Alcoholism in twins (EESEE).** A study of possible genetic influences on alcoholism studied pairs of adult female twins. The subjects were identified from the Virginia Twin Registry, which lists all twins born in Virginia. Each pair of twins was classified as identical or fraternal. Only identical twins share exactly the same genes. Based on an interview, each woman was classified as a problem drinker or not. Here are the data for the 1030 pairs of twins for which information was available:[21]

Problem drinker	Identical	Fraternal
Neither	443	301
One	102	113
Both	45	26
Total	590	440

(a) Is there a significant relationship between type of twin and the presence of problem drinking in the twin pair? Which cells contribute heavily to the chi-square value?

(b) Your result in (a) suggests a clearer analysis. Make a 2 × 2 table of "same or different" problem-drinking behavior within a twin pair by type of twin. To do this, combine the "Neither" and "Both" categories to form the "Same behavior" category. If heredity influences behavior, we would expect a higher proportion of identical twins to show the same behavior. Is there a significant effect of this kind?

W. Edwards Deming

(KIP BRUNDAGE/WOODFIN CAMP & ASSOCIATES.)

Statistics in the Service of Quality

From one point of view, statistics is about understanding variation. Reducing variation in products and processes is the central theme of statistical quality control. So it is not surprising that a statistician should become the leading guru of quality management. In the final decades of his long life, W. Edwards Deming (1900–1993) was one of the world's most influential consultants to management.

It is not surprising that a statistician should become the leading guru of quality management.

Deming grew up in Wyoming and earned a doctorate in physics at Yale. Working for the U.S. Department of Agriculture in the 1930s, he became acquainted with Neyman's work on sampling design and with the new methods of statistical quality control invented by Walter Shewhart of AT&T. In 1939 he moved to the Census Bureau as an expert on sampling. In 1943, he coined the term "P-value" for reporting the result of a significance test.

The work that made Deming famous began after he left the government in 1946. He visited Japan to advise on a census but returned to lecture on quality control. He earned a large following in Japan, which named its premier prize for industrial quality after him. As Japan's reputation for manufacturing excellence rose, Deming's fame rose with it. Blunt-spoken and even abrasive, he told corporate leaders that most quality problems are system problems for which management is responsible. He urged breaking down barriers to worker involvement and constant search for causes of variation. Finding the sources of variation is the theme of analysis of variance, the statistical method we introduce in this chapter.

One-Way Analysis of Variance: Comparing Several Means

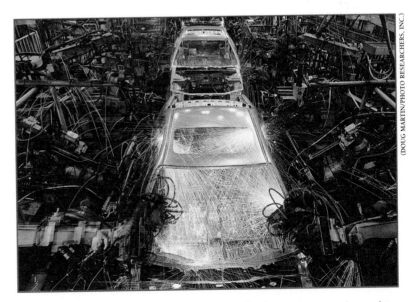

(DOUG MARTIN/PHOTO RESEARCHERS, INC.)

Manufacturers use statistics as well as new technology to improve the quality of their products.

Introduction

The two-sample t procedures of Chapter 7 compare the means of two populations or the mean responses to two treatments in an experiment. Of course, studies don't always compare just two groups. We need a method for comparing any number of means.

> **EXAMPLE 10.1** **Cars, pickup trucks, and SUVs**
>
> Pickup trucks and four-wheel-drive sport utility vehicles are replacing cars in American driveways. Do trucks and SUVs have lower gas mileage than cars? Table 10.1

Table 10.1 Highway gas mileage for 1998 model vehicles

Midsize Cars		Pickup Trucks		Sport Utility Vehicles	
Model	MPG	Model	MPG	Model	MPG
Acura 3.5RL	25	Chevrolet C1500	20	Acura SLX	19
Audi A6 Quattro	26	Dodge Dakota	25	Chevrolet Blazer	20
BMW 740i	24	Dodge Ram	20	Chevrolet Tahoe	19
Buick Century	29	Ford F150	21	Chrysler Town & Country	23
Cadillac Catera	24	Ford Ranger	27	Dodge Durango	17
Cadillac Seville	26	Mazda B2000	25	Ford Expedition	18
Chevrolet Lumina	29	Nissan Frontier	24	Ford Explorer	19
Chevrolet Malibu	32	Toyota T100	23	Geo Tracker	26
Chrysler Cirrus	30			GMC Jimmy	21
Ford Taurus	28			Infiniti QX4	19
Honda Accord	29			Isuzu Rodeo	20
Hyundai Sonata	27			Isuzu Trooper	19
Infiniti I30	28			Jeep Grand Cherokee	21
Infiniti Q45	23			Jeep Wrangler	19
Jaguar XJ8L	24			Kia Sportage	23
Lexus GS300	25			Land Rover Discovery	17
Lexus LS400	25			Lincoln Navigator	16
Lincoln Mark VIII	26			Mazda MPV	19
Mazda 626	29			Mercedes ML320	21
Mercedes-Benz E320	29			Mitsubishi Montero	20
Mitsubishi Diamante	24			Nissan Pathfinder	19
Nissan Maxima	28			Range Rover	17
Oldsmobile Aurora	26			Suburu Forester	27
Oldsmobile Intrigue	30			Suzuki Sidekick	24
Plymouth Breeze	33			Toyota RAV4	26
Saab 900S	25			Toyota 4Runner	22
Toyota Camry	30				
Volvo S70	25				

Source: Environmental Protection Agency, Model Year 1998 Fuel Economy Guide.

contains data on the highway gas mileage (in miles per gallon) for 28 midsize cars, 8 standard size pickup trucks, and 26 SUVs. The gas mileages and vehicle classifications are reported by the Environmental Protection Agency.[1]

Figure 10.1 shows side-by-side stemplots of the gas mileages for the three types of vehicles. We used the same stems in all three for easier comparison. It does appear that gas mileage decreases as we move from cars to pickups to SUVs.

Here are the means, standard deviations, and five-number summaries for the three car types, from statistical software:

	N	MEAN	MEDIAN	STDEV	MIN	MAX	Q1	Q3
Midsize	28	27.107	26.500	2.629	23.000	33.000	25.000	29.000
Pickup	8	23.125	23.500	2.588	20.000	27.000	20.250	25.000
SUV	26	20.423	19.500	2.914	16.000	27.000	19.000	22.250

We will use the mean to describe the center of the gas mileage distributions. As we expect, mean gas mileage goes down as we move from midsize cars to pickup trucks to sport utility vehicles. The differences among the means are not large. Are they statistically significant?

The problem of multiple comparisons

Call the mean highway gas mileages for the three populations of vehicles μ_1 for midsize cars, μ_2 for pickups, and μ_3 for SUVs. The subscript reminds us which

Midsize

```
16 |
17 |
18 |
19 |
20 |
21 |
22 |
23 | 0
24 | 0000
25 | 00000
26 | 0000
27 | 0
28 | 000
29 | 00000
30 | 000
31 |
32 | 0
33 | 0
```

Pickup

```
16 |
17 |
18 |
19 |
20 | 00
21 | 0
22 |
23 | 0
24 | 0
25 | 00
26 |
27 | 0
28 |
29 |
30 |
31 |
32 |
33 |
```

SUV

```
16 | 0
17 | 000
18 | 0
19 | 00000000
20 | 000
21 | 000
22 | 0
23 | 00
24 | 0
25 |
26 | 00
27 | 0
28 |
29 |
30 |
31 |
32 |
33 |
```

Figure 10.1 *Side-by-side stemplots comparing the highway gas mileages of midsize cars, standard pickup trucks, and sport utility vehicles from Table 10.1.*

group a parameter or statistic describes. To compare these three population means, we might use the two-sample t test several times:

- Test H_0: $\mu_1 = \mu_2$ to see if the mean miles per gallon for midsize cars differs from the mean for pickup trucks.
- Test H_0: $\mu_1 = \mu_3$ to see if midsize cars differ from SUVs.
- Test H_0: $\mu_2 = \mu_3$ to see if pickups differ from SUVs.

The weakness of doing three tests is that we get three P-values, one for each test alone. That doesn't tell us how likely it is that *three* sample means are spread apart as far as these are. It may be that $\bar{x}_1 = 27.107$ and $\bar{x}_3 = 20.423$ are significantly different if we look at just two groups but not significantly different if we know that they are the smallest and largest means in three groups. As we look at more groups, we expect the gap between the smallest and largest sample mean to get larger. (Think of comparing the tallest and shortest person in larger and larger groups of people.) We can't safely compare many parameters by doing tests or confidence intervals for two parameters at a time.

multiple comparisons

The problem of how to do many comparisons at once with some overall measure of confidence in all our conclusions is common in statistics. This is the problem of **multiple comparisons**. Statistical methods for dealing with multiple comparisons usually have two steps:

1. An *overall test* to see if there is good evidence of *any* differences among the parameters that we want to compare.
2. A detailed *follow-up analysis* to decide which of the parameters differ and to estimate how large the differences are.

The overall test, though more complex than the tests we met earlier, is often reasonably straightforward. The follow-up analysis can be quite elaborate. In our basic introduction to statistical practice, we will look only at some overall tests. Chapter 9 describes an overall test to compare several population proportions. In this chapter we present a test for comparing several population means.

10.1 The Analysis of Variance F Test

We want to test the null hypothesis that there are *no differences* among the mean highway gas mileages for the three vehicle types:

$$H_0: \mu_1 = \mu_2 = \mu_3$$

The alternative hypothesis is that there is some difference, that not all three population means are equal:

$$H_a: \text{ not all of } \mu_1, \mu_2, \text{ and } \mu_3 \text{ are equal}$$

The alternative hypothesis is no longer one-sided or two-sided. It is "many-sided," because it allows any relationship other than "all three equal." For example, H_a includes the case in which $\mu_2 = \mu_3$ but μ_1 has a different value.

analysis of variance F test

The test of H_0 against H_a is called the **analysis of variance F test**. Analysis of variance is usually abbreviated as ANOVA.

```
Analysis of Variance for Mileage
Source      DF        SS        MS        F        P
Vehicle      2     606.37    303.19    40.12    0.000
Error       59     445.90      7.56
Total       61    1052.27

                                  Individual 95% CIs For Mean
                                  Based on Pooled StDev

Level        N      Mean      StDev  --+---------+---------+---------+----
Midsize     28    27.107      2.629                              (---*-----)
Pickup       8    23.125      2.588              (-------*------)
SUV         26    20.423      2.914    (----*---)

                                       --+---------+---------+---------+----
Pooled StDev  =    2.749             20.0      22.5      25.0      27.5
```

Figure 10.2 *Minitab output for analysis of variance of the gas mileage data, for Example 10.2.*

EXAMPLE 10.2 Interpreting ANOVA from software

Enter the gas mileage data from Table 10.1 into the Minitab software package and request analysis of variance. The output appears in Figure 10.2. Most statistical software packages produce an ANOVA output similar to this one.

First, check that the sample sizes, means, and standard deviations agree with those in Example 10.1. Then find the F test statistic, $F = 40.12$, and its P-value. The P-value is given as 0.000. This means that P is zero to three decimal places, or $P < 0.0005$. There is very strong evidence that the three types of vehicles do not all have the same mean gas mileage.

The F test does not say *which* of the three means are significantly different. It appears from our preliminary analysis of the data that all three may differ. The computer output includes confidence intervals for all three means that suggest the same conclusion. The SUV and pickup intervals overlap only slightly, and the midsize interval lies well above them on the mileage scale. These are 95% confidence intervals for each mean separately. We are not 95% confident that *all three* intervals cover the three means. There are follow-up procedures that provide 95% confidence that we have caught all three means at once, but we won't study them.

Our conclusion: There is strong evidence ($P < 0.0005$) that the means are not all equal. The most important difference among the means is that midsize cars have better gas mileage than pickups and SUVs.

Example 10.2 illustrates our approach to comparing means. The ANOVA F test (often done by software) assesses the evidence for *some* difference among the population means. In most cases, we expect the F test to be significant. We would not undertake a study if we did not expect to find some effect. The formal test is nonetheless important to guard against being misled by chance variation. We will not do the formal follow-up analysis that is often the most useful part of an ANOVA study. Follow-up analysis would allow us to say which means ___ differ and by how much, with (say) 95% confidence that all our conclusions are correct. We rely instead on examination of the data to show what differences are present and whether they are large enough to be interesting. The gap of almost 7 miles per gallon between midsize cars and SUVs is large enough to be of practical interest.

(JOHN ELK III, STOCK, BOSTON.)

APPLY YOUR KNOWLEDGE

10.1 **Dogs, friends, and stress. (EESEE)** If you are a dog lover, perhaps having your dog along reduces the effect of stress. To examine the effect of pets in stressful situations, researchers recruited 45 women who said they were dog lovers. Fifteen of the subjects were randomly assigned to each of three groups to do a stressful task alone, with a good friend present, or with their dog present. (The stressful task was to count backward by 13s or 17s.) The subject's mean heart rate during the task is one measure of the effect of stress. Table 10.2 contains the data.[2]

(a) Make stemplots of the heart rates for the three groups (round to the nearest whole number of beats). Do any of the groups show outliers or extreme skewness?

Table 10.2 Mean heart rates during stress with a pet (P), with a friend (F), and for the control group (C)

Group	Rate	Group	Rate	Group	Rate
P	69.169	P	68.862	C	84.738
F	99.692	C	87.231	C	84.877
P	70.169	P	64.169	P	58.692
C	80.369	C	91.754	P	79.662
C	87.446	C	87.785	P	69.231
P	75.985	F	91.354	C	73.277
F	83.400	F	100.877	C	84.523
F	102.154	C	77.800	C	70.877
P	86.446	P	97.538	F	89.815
F	80.277	P	85.000	F	98.200
C	90.015	F	101.062	F	76.908
C	99.046	F	97.046	P	69.538
C	75.477	C	62.646	P	70.077
F	88.015	F	81.600	F	86.985
F	92.492	P	72.262	P	65.446

```
Analysis of Variance for Beats
Source     DF        SS         MS        F        P
Group       2     2387.7     1193.8    14.08    0.000
Error      42     3561.3       84.8
Total      44     5949.0

                                   Individual 95% CIs For Mean
                                   Based on Pooled StDev

Level       N      Mean      StDev ----+---------+---------+---------+--
Control    15    82.524      9.242                (-----*-----)
Friend     15    91.325      8.341                      (-----*------·)
Pet        15    73.483      9.970      (-----*-----)
                                   ---+---------+---------+---------+--
Pooled StDev =    9.208              72.0      80.0      88.0      96.0
```

Figure 10.3 *Minitab output for the data in Table 10.2 on heart rates (beats per minute) during stress, for Exercise 10.1. The control group worked alone, the "Friend" group had a friend present, and the "Pet" group had a pet dog present.*

(b) Figure 10.3 gives the Minitab ANOVA output for these data. Do the mean heart rates for the groups appear to show that the presence of a pet or a friend reduces heart rate during a stressful task?

(c) What are the values of the ANOVA F statistic and its P-value? What hypotheses does F test? Briefly describe the conclusions you draw from these data. Did you find anything surprising?

10.2 **How much corn should I plant?** How much corn per acre should a farmer plant to obtain the highest yield? Too few plants will give a low yield. On the other hand, if there are too many plants, they will compete with each other for moisture and nutrients, and yields will fall. To find out, plant at different rates on several plots of ground and measure the harvest. (Be sure to treat all the plots the same except for the planting rate.) Here are data from such an experiment:[3]

Plants per acre	Yield (bushels per acre)			
12,000	150.1	113.0	118.4	142.6
16,000	166.9	120.7	135.2	149.8
20,000	165.3	130.1	139.6	149.9
24,000	134.7	138.4	156.1	
28,000	119.0	150.5		

(a) Make side-by-side stemplots of yield for each number of plants per acre. What do the data appear to show about the influence of plants per acre on yield?

```
Analysis of Variance for Yield
Source      DF         SS         MS         F          P
Group        4        600        150       0.50      0.736
Error       12       3597        300
Total       16       4197


                                   Individual 95% CIs For Mean
                                   Based on Pooled StDev

Level       N       Mean       StDev  ----+---------+---------+---------+---
12,000       4     131.03      18.09      (-----------*------------ )
16,000       4     143.15      19.79           (-----------*------------ )
20,000       4     146.22      15.07            (-----------*-----------)
24,000       3     143.07      11.44             (------------*-------------)
28,000       2     134.75      22.27        (------------------*------------------·)

                                       ---+---------+---------+---------+---
Pooled StDev =      17.31              112       128       144       160
```

Figure 10.4 *Minitab output for yields of corn at five planting rates, for Exercise 10.2.*

(b) ANOVA will assess the statistical significance of the observed differences in yield. What are H_0 and H_a for the ANOVA F test in this situation?

(c) The Minitab ANOVA output for these data appears in Figure 10.4. What is the sample mean yield for each planting rate? What does the ANOVA F test say about the significance of the effects you observe?

(d) The observed differences among the mean yields in the sample are quite large. Why are they not statistically significant?

The idea of analysis of variance

Here is the main idea for comparing means: what matters is not how far apart the sample means are but how far apart they are *relative to the variability of individual observations*. Look at the two sets of boxplots in Figure 10.5. For simplicity, these distributions are all symmetric, so that the mean and median are the same. The centerline in each boxplot is therefore the sample mean. Both figures compare three samples with the same three means. Like the three vehicle types in Example 10.1, the means are different but not very different. Could differences this large easily arise just due to chance, or are they statistically significant?

• The boxplots in Figure 10.5(a) have tall boxes, which show lots of variation among the individuals in each group. With this much variation among individuals, we would not be surprised if another set of samples gave quite different sample means. The observed differences among the sample means could easily happen just by chance.

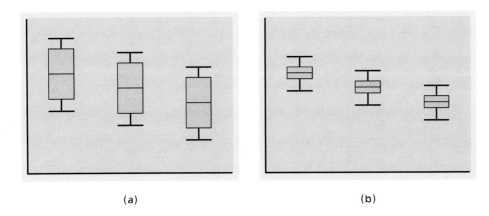

(a)   (b)

Figure 10.5 *Boxplots for two sets of three samples each. The sample means are the same in (a) and (b). Analysis of variance will find a more significant difference among the means in (b) because there is less variation among the individuals within those samples.*

- The boxplots in Figure 10.5(b) have the same centers as those in Figure 10.5(a), but the boxes are much shorter. That is, there is much less variation among the individuals in each group. It is unlikely that any sample from the first group would have a mean as small as the mean of the second group. Because means as far apart as those observed would rarely arise just by chance in repeated sampling, they are good evidence of real differences among the means of the three populations we are sampling from.

This comparison of the two parts of Figure 10.5 is too simple in one way. It ignores the effect of the sample sizes, an effect that boxplots do not show. Small differences among sample means can be significant if the samples are very large. Large differences among sample means may fail to be significant if the samples are very small. All we can be sure of is that for the same sample size, Figure 10.5(b) will give a much smaller *P*-value than Figure 10.5(a). Despite this qualification, the big idea remains: if sample means are far apart relative to the variation among individuals in the same group, that's evidence that something other than chance is at work.

THE ANALYSIS OF VARIANCE IDEA

Analysis of variance compares the variation due to specific sources with the variation among individuals who should be similar. In particular, ANOVA tests whether several populations have the same mean by comparing how far apart the sample means are with how much variation there is within the samples.

It is one of the oddities of statistical language that methods for comparing means are named after the variance. The reason is that the test works by comparing two kinds of variation. Analysis of variance is a general method for studying sources of variation in responses. Comparing several means is the

one-way ANOVA simplest form of ANOVA, called **one-way ANOVA**. One-way ANOVA is the only form of ANOVA that we will study.

> ## THE ANOVA F STATISTIC
>
> The **analysis of variance F statistic** for testing the equality of several means has this form:
>
> $$F = \frac{\text{variation among the sample means}}{\text{variation among individuals in the same sample}}$$

We give more detail later. Because ANOVA is in practice done by software, the idea is more important than the detail. The F statistic can only take values that are zero or positive. It is zero only when all the sample means are identical and gets larger as they move farther apart. Large values of F are evidence against the null hypothesis H_0 that all population means are the same. Although the alternative hypothesis H_a is many-sided, the ANOVA F test is one-sided because any violation of H_0 tends to produce a large value of F.

How large must F be to provide significant evidence against H_0? To answer questions of statistical significance, compare the F statistic with critical values from an **F distribution**. The F distributions are described on page 414 in Chapter 7. A specific F distribution is specified by two parameters: a numerator degrees of freedom and a denominator degrees of freedom. Table D in the back of the book contains critical values for F distributions with various degrees of freedom.

F distribution

> ### EXAMPLE 10.3   Using the F table
>
> Look again at the computer output for the gas mileage data in Figure 10.2. The degrees of freedom for the F test appear in the first two rows of the "DF" column. There are 2 degrees of freedom in the numerator and 59 in the denominator.
>
> In Table D, find the numerator degrees of freedom 2 at the top of the table. Then look for the denominator degrees of freedom 59 at the left of the table. There is no entry for 59, so we use the next smaller entry, 50 degrees of freedom. The upper critical values for 2 and 50 degrees of freedom are
>
> | p | Critical value |
> | --- | --- |
> | 0.100 | 2.41 |
> | 0.050 | 3.18 |
> | 0.025 | 3.97 |
> | 0.010 | 5.06 |
> | 0.001 | 7.96 |

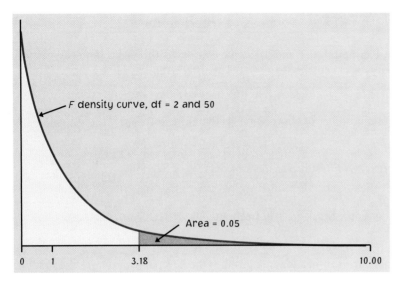

Figure 10.6 *The density curve of the F distribution with 2 degrees of freedom in the numerator and 50 degrees of freedom in the denominator, for Example 10.3. The value $F^* = 3.18$ is the 0.05 critical value.*

Figure 10.6 shows the F density curve with 2 and 50 degrees of freedom and the upper 5% critical value 3.18. The observed $F = 40.12$ lies far to the right on this curve. We see from the table that $F = 40.12$ is larger than the 0.001 critical value, so $P < 0.001$.

The degrees of freedom of the F statistic depend on the number of means we are comparing and the number of observations in each sample. That is, the F test does take into account the number of observations. Here are the details.

DEGREES OF FREEDOM FOR THE F TEST

We want to compare the means of I populations. We have an SRS of size n_i from the ith population, so that the total number of observations in all samples combined is

$$N = n_1 + n_2 + \cdots + n_I$$

If the null hypothesis that all population means are equal is true, the ANOVA F statistic has the F distribution with $I - 1$ degrees of freedom in the numerator and $N - I$ degrees of freedom in the denominator.

EXAMPLE 10.4 **Degrees of freedom for *F***

In Examples 10.1 and 10.2, we compared the mean highway gas mileage for three types of vehicles, so $I = 3$. The three sample sizes are

$$n_1 = 28 \quad n_2 = 8 \quad n_3 = 26$$

The total number of observations is therefore

$$N = 28 + 8 + 26 = 62$$

The ANOVA *F* test has numerator degrees of freedom

$$I - 1 = 3 - 1 = 2$$

and denominator degrees of freedom

$$N - I = 62 - 3 = 59$$

These degrees of freedom are given in the computer output as the first two entries in the "DF" column in Figure 10.2.

APPLY YOUR KNOWLEDGE

10.3 **Dogs, friends, and stress.** Exercise 10.1 compares the mean heart rates for women performing a stressful task under three conditions.

(a) What are I, the n_i, and N for these data? State in words what each of these numbers means.

(b) Find both degrees of freedom for the ANOVA *F* statistic. Check your results against the output in Figure 10.3.

(c) The output shows that $F = 14.08$. What does Table D tell you about the significance of this result?

10.4 **How much corn should I plant?** Exercise 10.2 compares the yields for several planting rates for corn.

(a) What are I, the n_i, and N for these data? Identify these quantities in words and give their numerical values.

(b) Find the degrees of freedom for the ANOVA *F* statistic. Check your work against the computer output in Figure 10.4.

(c) For these data, $F = 0.50$. What does Table D tell you about the *P*-value of this statistic?

10.5 In each of the following situations, we want to compare the mean response in several populations. For each setting, identify the populations and the response variable. Then give I, the n_i, and N. Finally, give the degrees of freedom of the ANOVA *F* statistic.

(a) Do four tomato varieties differ in mean yield? Grow ten plants of each variety and record the yield of each plant in pounds of tomatoes.

(b) A maker of detergents wants to compare the attractiveness to consumers of six package designs. Each package is shown to 120 different consumers who rate the attractiveness of the design on a 1 to 10 scale.

(c) An experiment to compare the effectiveness of three weight-loss programs has 32 subjects who want to lose weight. Ten subjects are assigned at random to each of two programs, and the remaining 12 subjects follow the third program. After six months, each subject's change in weight is recorded.

Assumptions for ANOVA

Like all inference procedures, ANOVA is valid only in some circumstances. Here are the requirements for using ANOVA to compare population means.

> ### ANOVA ASSUMPTIONS
>
> - We have *I* **independent SRSs**, one from each of *I* populations.
> - The *i*th population has a **normal distribution** with unknown mean μ_i. The means may be different in the different populations. The ANOVA *F* statistic tests the null hypothesis that all of the populations have the same mean:
>
> $$H_0: \quad \mu_1 = \mu_2 = \cdots = \mu_I$$
> $$H_a: \quad \text{not all of the } \mu_i \text{ are equal}$$
>
> - All of the populations have the **same standard deviation** σ, whose value is unknown.

Dropping out

An experiment found that weight loss is significantly more effective than exercise for reducing high cholesterol and high blood pressure. The 170 subjects were randomly assigned to a weight-loss program, an exercise program, or a control group. Only 111 of the 170 subjects completed their assigned treatment, and the ANOVA used data from these 111. Always ask about details of the data before trusting inference. Because the percents who dropped out were similar in all three groups, the study authors believe that the dropouts did not cause a strong bias.

The first two requirements are familiar from our study of the two-sample *t* procedures for comparing two means. As usual, the design of the data production is the most important foundation for inference. Biased sampling or confounding can make any inference meaningless. If we do not actually draw separate SRSs from each population or carry out a randomized comparative experiment, it is often unclear to what population the conclusions of inference apply. This is the case in Example 10.1, for example. ANOVA, like other inference procedures, is often used when random samples are not available. You must judge each use on its merits, a judgment that usually requires some knowledge of the subject of the study in addition to some knowledge of statistics. We might consider the 1998 vehicles in Example 10.1 to be samples from vehicles of their types produced in recent years.

Because no real population has exactly a normal distribution, the usefulness of inference procedures that assume normality depends on how sensitive they are to departures from normality. Fortunately, procedures for comparing means are not very sensitive to lack of normality. The ANOVA *F* test, like the *t* procedures, *robustness* is **robust**. What matters is normality of the sample means, so ANOVA becomes safer as the sample sizes get larger, because of the central limit theorem effect. Remember to check for outliers that change the value of sample means and for extreme skewness. When there are no outliers and the distributions are roughly symmetric, you can safely use ANOVA for sample sizes as small as 4 or 5. (Don't confuse the ANOVA F, which compares several means, with the F statistic of Section 7.3, which compares two standard deviations and is not robust against nonnormality.)

The third assumption is annoying: ANOVA assumes that the variability of observations, measured by the standard deviation, is the same in all populations. You may recall from Chapter 7 (page 406) that there is a special version of the two-sample t test that assumes equal standard deviations in both populations. The ANOVA F for comparing two means is exactly the square of this special t statistic. We prefer the t test that does not assume equal standard deviations, but for comparing more than two means there is no general alternative to the ANOVA F. It is not easy to check the assumption that the populations have equal standard deviations. Statistical tests for equality of standard deviations are very sensitive to lack of normality, so much so that they are of little practical value. You must either seek expert advice or rely on the robustness of ANOVA.

How serious are unequal standard deviations? ANOVA is not too sensitive to violations of the assumption, especially when all samples have the same or similar sizes and no sample is very small. When designing a study, try to take samples of the same size from all the groups you want to compare. The sample standard deviations estimate the population standard deviations, so check before doing ANOVA that the sample standard deviations are similar to each other. We expect some variation among them due to chance. Here is a rule of thumb that is safe in almost all situations.

CHECKING STANDARD DEVIATIONS IN ANOVA

The results of the ANOVA F test are approximately correct when the largest sample standard deviation is no more than twice as large as the smallest sample standard deviation.

EXAMPLE 10.5 **Do the standard deviations allow ANOVA?**

In the gas mileage study, the sample standard deviations for midsize cars, pickups, and SUVs are

$$s_1 = 2.629 \quad s_2 = 2.588 \quad s_3 = 2.914$$

These standard deviations easily satisfy our rule of thumb. We can safely use ANOVA to compare the mean gas mileage for the three vehicle types.

The report from which Table 10.1 was taken also contained data on 25 subcompact cars. Can we use ANOVA to compare the means for all four types of vehicles?

The standard deviation for subcompact cars is $s_4 = 5.240$ miles per gallon. This is slightly more than twice the smallest standard deviation:

$$\frac{\text{largest } s}{\text{smallest } s} = \frac{5.240}{2.588} = 2.02$$

A large standard deviation is often due to skewness or an outlier. Here is a stemplot of the subcompact car gas mileages:

```
2 ┊ 44
2 ┊ 5778899999
3 ┊ 00011334
3 ┊ 6667
4 ┊
4 ┊ 9
```

The Geo Metro gets 49 miles per gallon. The standard deviation drops from 5.24 to 3.73 if we omit this one car. The Metro is a subcompact car based on its interior volume but has a much smaller engine than any other car in the class. If we decide to drop the Metro, ANOVA is safe.

EXAMPLE 10.6   Which color attracts beetles best?

To detect the presence of harmful insects in farm fields, we can put up boards covered with a sticky material and examine the insects trapped on the boards. Which colors attract insects best? Experimenters placed six boards of each of four colors at random locations in a field of oats and measured the number of cereal leaf beetles trapped. Here are the data:[4]

Board color	Insects trapped					
Blue	16	11	20	21	14	7
Green	37	32	20	29	37	32
White	21	12	14	17	13	20
Yellow	45	59	48	46	38	47

We would like to use ANOVA to compare the mean numbers of beetles that would be trapped by all boards of each color. Because the samples are small, we plot the data in side-by-side stemplots in Figure 10.7. Computer output for ANOVA appears in Figure 10.8. The yellow boards attract by far the most insects ($\bar{x}_4 = 47.167$), with green next ($\bar{x}_2 = 31.167$) and blue and white far behind.

Check that we can safely use ANOVA to test equality of the four means. The largest of the four sample standard deviations is 6.795 and the smallest is 3.764. The ratio

$$\frac{\text{largest } s}{\text{smallest } s} = \frac{6.795}{3.764} = 1.8$$

Blue		Green		White		Yellow	
0	7	0		0		0	
1	146	1		1	2347	1	
2	01	2	09	2	01	2	
3		3	2277	3		3	8
4		4		4		4	5678
5		5		5		5	9

Figure 10.7 *Side-by-side stemplots comparing the counts of insects attracted by six boards for each of four board colors, for Example 10.6.*

is less than 2, so these data satisfy our rule of thumb. The shapes of the four distributions are irregular, as we expect with only 6 observations in each group, but there are no outliers. The ANOVA results will be approximately correct.

There are $I = 4$ groups and $N = 24$ observations overall, so the degrees of freedom for F are

$$\text{numerator:} \quad I - 1 = 4 - 1 = 3$$
$$\text{denominator:} \quad N - I = 24 - 4 = 20$$

This agrees with the computer results. The F statistic is $F = 42.84$, a large F with P-value $P < 0.001$. Despite the small samples, the experiment gives very strong evidence of differences among the colors. Yellow boards appear best at attracting leaf beetles.

```
Analysis of Variance for Beetles
Source       DF        SS         MS        F         P
Color         3     4134.0     1378.0     42.84     0.000
Error        20      643.3       32.2
Total        23     4777.3


                                   Individual 95% CIs For Mean
                                   Based on Pooled StDev
Level        N      Mean      StDev ----+---------+---------+---------+----
Blue         6    14.833      5.345  (---*---)
Green        6    31.167      6.306                 (---*---)
White        6    16.167      3.764    (---*---)
Yellow       6    47.167      6.795                              (---*---)
                                      ----+---------+---------+---------+----
Pooled StDev =    5.672                 12        24        36        48
```

Figure 10.8 *Minitab ANOVA output for comparing the four board colors in Example 10.6.*

10.6 Verify that the sample standard deviations for these sets of data do allow use of ANOVA to compare the population means.

(a) The heart rates of Exercise 10.1 and Figure 10.3.

(b) The corn yields of Exercise 10.2 and Figure 10.4.

10.7 **Marital status and salary.** Married men tend to earn more than single men. An investigation of the relationship between marital status and income collected data on all 8235 men employed as managers or professionals by a large manufacturing firm in 1976. Suppose (this is risky) we regard these men as a random sample from the population of all men employed in managerial or professional positions in large companies. Here are descriptive statistics for the salaries of these men:[5]

	Single	Married	Divorced	Widowed
n_i	337	7,730	126	42
$\bar{x}_i$	$21,384	$26,873	$25,594	$26,936
s_i	$5,731	$7,159	$6,347	$8,119

(a) Briefly describe the relationship between marital status and salary.

(b) Do the sample standard deviations allow use of the ANOVA F test? (The distributions are skewed to the right. We expect right skewness in income distributions. The investigators actually applied ANOVA to the logarithms of the salaries, which are more symmetric.)

(c) What are the degrees of freedom of the ANOVA F test?

(d) The F test is a formality for these data, because we are sure that the P-value will be very small. Why are we sure?

(e) Single men earn less on the average than men who are or have been married. Do the highly significant differences in mean salary show that getting married raises men's mean income? Explain your answer.

10.8 **Who succeeds in college?** What factors influence the success of college students? Look at all 256 students who entered a university planning to study computer science (CS) in a specific year. We are willing to regard these students as a random sample of the students the university CS program will attract in subsequent years. After three semesters of study, some of these students were CS majors, some were majors in another field of science or engineering, and some had left science and engineering or left the university. The table below gives the sample means and standard deviations and the ANOVA F statistics for three variables that describe the students' high school performance. These are three separate ANOVA F tests.[6]

The first variable is a student's rank in the high school class, given as a percentile (so rank 50 is the middle of the class and rank 100 is the top). The next variable is the number of semester courses in mathematics the student took in high school. The third variable is the student's average grade in high school mathematics. The mean and standard deviation appear in a form common in published reports, with the standard deviation in parentheses following the mean.

| | | Mean (standard deviation) | | |
| | | High school | Semesters of | Average grade |
Group	n	class rank	HS math	in HS math
CS majors	103	88.0 (10.5)	8.74 (1.28)	3.61 (0.46)
Sci./eng. majors	31	89.2 (10.8)	8.65 (1.31)	3.62 (0.40)
Other	122	85.8 (10.8)	8.25 (1.17)	3.35 (0.55)
F statistic		1.95	4.56	9.38

(a) What null and alternative hypotheses does F test for rank in the high school class? Express the hypotheses both in symbols and in words. The hypotheses are similar for the other two variables.

(b) What are the degrees of freedom for each F?

(c) Check that the standard deviations allow use of all three F tests. The shapes of the distributions also allow use of F. How significant is F for each of these variables?

(d) Write a brief summary of the differences among the three groups of students, taking into account both the significance of the F tests and the values of the means.

10.2 Some Details of Anova*

Now we will give the actual recipe for the ANOVA F statistic. We have SRSs from each of I populations. Subscripts from 1 to I tell us which sample a statistic refers to:

---

*This more advanced section is optional if you are using computer software to find the F statistic.

Population	Sample size	Sample mean	Sample std. dev.
1	n_1	$\bar{x}_1$	s_1
2	n_2	$\bar{x}_2$	s_2
$\vdots$	$\vdots$	$\vdots$	$\vdots$
I	n_I	$\bar{x}_I$	s_I

You can find the F statistic from just the sample sizes n_i, the sample means $\bar{x}_i$, and the sample standard deviations s_i. You don't need to go back to the individual observations.

The ANOVA F statistic has the form

$$F = \frac{\text{variation among the sample means}}{\text{variation among individuals within samples}}$$

The measures of variation in the numerator and denominator of F are called **mean squares**. A mean square is a more general form of a sample variance. An ordinary sample variance s^2 is an average (or mean) of the squared deviations of observations from their mean, so it qualifies as a "mean square."

mean squares

The numerator of F is a mean square that measures variation among the I sample means $\bar{x}_1, \bar{x}_2, \ldots, \bar{x}_I$. Call the overall mean response (the mean of all N observations together) $\bar{x}$. You can find $\bar{x}$ from the I sample means by

$$\bar{x} = \frac{n_1\bar{x}_1 + n_2\bar{x}_2 + \cdots + n_I\bar{x}_I}{N}$$

The sum of each mean multiplied by the number of observations it represents is the sum of all the individual observations. Dividing this sum by N, the total number of observations, gives the overall mean $\bar{x}$. The numerator mean square in F is an average of the I squared deviations of the means of the samples from $\bar{x}$. It is called the **mean square for groups**, abbreviated as MSG.

MSG

$$\text{MSG} = \frac{n_1(\bar{x}_1 - \bar{x})^2 + n_2(\bar{x}_2 - \bar{x})^2 + \cdots + n_I(\bar{x}_I - \bar{x})^2}{I - 1}$$

Each squared deviation is weighted by n_i, the number of observations it represents.

The mean square in the denominator of F measures variation among individual observations in the same sample. For any one sample, the sample variance s_i^2 does this job. For all I samples together, we use an average of the individual sample variances. It is again a weighted average in which each s_i^2 is weighted by one fewer than the number of observations it represents, $n_i - 1$. Another way to put this is that each s_i^2 is weighted by its degrees of freedom $n_i - 1$. The resulting mean square is called the **mean square for error**, MSE.

MSE

$$\text{MSE} = \frac{(n_1 - 1)s_1^2 + (n_2 - 1)s_2^2 + \cdots + (n_I - 1)s_I^2}{N - I}$$

Here is a summary of the ANOVA test.

THE ANOVA F TEST

Draw an independent SRS from each of I populations. The ith population has the $N(\mu_i, \sigma)$ distribution, where σ is the common standard deviation in all the populations. The ith sample has size n_i, sample mean $\bar{x}_i$, and sample standard deviation s_i.

The **ANOVA F statistic** tests the null hypothesis that all I populations have the same mean:

$$H_0: \mu_1 = \mu_2 = \cdots = \mu_I$$
$$H_a: \text{not all of the } \mu_i \text{ are equal}$$

The statistic is

$$F = \frac{\text{MSG}}{\text{MSE}}$$

The numerator of F is the **mean square for groups**

$$\text{MSG} = \frac{n_1(\bar{x}_1 - \bar{x})^2 + n_2(\bar{x}_2 - \bar{x})^2 + \cdots + n_I(\bar{x}_I - \bar{x})^2}{I - 1}$$

The denominator of F is the **mean square for error**

$$\text{MSE} = \frac{(n_1 - 1)s_1^2 + (n_2 - 1)s_2^2 + \cdots + (n_I - 1)s_I^2}{N - I}$$

When H_0 is true, F has the **F distribution** with $I - 1$ and $N - I$ degrees of freedom.

The denominators in the recipes for MSG and MSE are the two degrees of freedom $I - 1$ and $N - I$ of the F test. The numerators are called *sums of squares*, from their algebraic form. It is usual to present the results of ANOVA

ANOVA table   in an **ANOVA table** like that in the Minitab output. The table has columns for degrees of freedom (DF), sums of squares (SS), and mean squares (MS). Check that each MS entry in Figure 10.8, for example, is the sum of squares SS divided by the degrees of freedom DF in the same row. The F statistic in the "F" column is MSG/MSE. The rows are labeled by sources of variation. In Figure 10.8, we called the group variable "Color." Variation among observations in the same group is called "Error" by most software. This doesn't mean a mistake has been made. It's a traditional term for variation among individual subjects.

Because MSE is an average of the individual sample variances, it is also called the *pooled sample variance*, written as s_p^2. When all I populations have the same population variance σ^2 (ANOVA assumes that they do), s_p^2 estimates the com-

pooled standard deviation   mon variance σ^2. The square root of MSE is the **pooled standard deviation** s_p. It estimates the common standard deviation σ of observations in each group.

Minitab, like most ANOVA programs, gives the value of s_p as well as MSE. It is the "Pooled StDev" value in Figure 10.8.

The pooled standard deviation s_p is a better estimator of the common σ than any individual sample standard deviation s_i because it combines (pools) the information in all I samples. We can get a confidence interval for any of the means μ_i from the usual form

$$\text{estimate} \pm t^*\text{SE}_{\text{estimate}}$$

using s_p to estimate σ. The confidence interval for μ_i is

$$\overline{x}_i \pm t^* \frac{s_p}{\sqrt{n_i}}$$

Use the critical value t^* from the t distribution with $N - I$ degrees of freedom, because s_p has $N - I$ degrees of freedom. These are the confidence intervals that appear in the Minitab ANOVA output.

Where are the details?

Papers reporting scientific research must be short. Brevity allows researchers to hide details about their data. Did they choose their subjects in a biased way? Did they report data on only some of their subjects? Did they try several statistical analyses and report the one that looked best? The statistician John Bailar screened more than 4000 papers for the *The New England Journal of Medicine*. He says, "When it came to the statistical review, it was often clear that critical information was lacking, and the gaps nearly always had the practical effect of making the authors' conclusions look stronger than they should have."

EXAMPLE 10.7 **Anova calculations**

We can do the ANOVA test comparing board colors in Example 10.6 using only the sample sizes, sample means, and sample standard deviations. These appear in Figure 10.8, but it is easy to find them with a calculator.

The overall mean of the 24 counts is

$$\overline{x} = \frac{n_1\overline{x}_1 + n_2\overline{x}_2 + \cdots + n_I\overline{x}_I}{N}$$

$$= \frac{(6)(14.833) + (6)(31.167) + (6)(16.167) + (6)(47.167)}{24}$$

$$= \frac{656}{24} = 27.333$$

The mean square for groups is

$$\text{MSG} = \frac{n_1(\overline{x}_1 - \overline{x})^2 + n_2(\overline{x}_2 - \overline{x})^2 + \cdots + n_I(\overline{x}_I - \overline{x})^2}{I - 1}$$

$$= \frac{1}{4 - 1}[(6)(14.833 - 27.333)^2 + (6)(31.167 - 27.333)^2$$

$$+ (6)(16.167 - 27.333)^2 + (6)(47.167 - 27.333)^2]$$

$$= \frac{4134.100}{3} = 1378.033$$

The mean square for error is

$$\text{MSE} = \frac{(n_1 - 1)s_1^2 + (n_2 - 1)s_2^2 + \cdots + (n_I - 1)s_I^2}{N - I}$$

$$= \frac{(5)(5.345^2) + (5)(6.306^2) + (5)(3.764^2) + (5)(6.795^2)}{24 - 4}$$

$$= \frac{643.372}{20} = 32.169$$

Finally, the ANOVA test statistic is

$$F = \frac{\text{MSG}}{\text{MSE}} = \frac{1378.033}{32.169} = 42.84$$

Our work agrees with the computer output in Figure 10.8. We don't recommend doing these calculations, because tedium and roundoff errors cause frequent mistakes.

The pooled estimate of the standard deviation σ in any group is

$$s_p = \sqrt{\text{MSE}} = \sqrt{32.169} = 5.672$$

A 95% confidence interval for the mean count of insects trapped by yellow boards, using s_p and 20 degrees of freedom, is

$$\bar{x}_4 \pm t^* \frac{s_p}{\sqrt{n_4}} = 47.167 \pm 2.086\frac{5.672}{\sqrt{6}}$$

$$= 47.167 \pm 4.830$$

$$= 42.34 \text{ to } 52.00$$

This confidence interval appears in the graph in the Minitab ANOVA output in Figure 10.8.

APPLY YOUR KNOWLEDGE

10.9 Weights of newly hatched pythons. A study of the effect of nest temperature on the development of water pythons separated python eggs at random into nests at three temperatures: cold, neutral, and hot. Exercise 9.10 (page 489) shows that the proportions of eggs that hatched at each temperature did not differ significantly. Now we will examine the little pythons. In all, 16 eggs hatched at the cold temperature, 38 at the neutral temperature, and 75 at the hot temperature. The report of the study summarizes the data in the common form "mean ± standard error" as follows:[7]

Temperature	n	Weight (grams) at hatching	Propensity to strike
Cold	16	28.89 ± 8.08	6.40 ± 5.67
Neutral	38	32.93 ± 5.61	5.82 ± 4.24
Hot	75	32.27 ± 4.10	4.30 ± 2.70

(a) We will compare the mean weights at hatching. Recall that the standard error of the mean is $s/\sqrt{n}$. Find the standard deviations of the weights in the three groups and verify that they satisfy our rule of thumb for using ANOVA.

(b) Starting from the sample sizes n_i, the means $\bar{x}_i$, and the standard deviations s_i, carry out an ANOVA. That is, find MSG, MSE, and the F statistic, and use Table D to approximate the P-value. Is there evidence that nest temperature affects the mean weight of newly hatched pythons?

10.10 Python strikes. The data in the previous exercise also describe the "propensity to strike" of the hatched pythons at 30 days of age. This is the number of taps on the head with a small brush until the python launches a strike. (Don't try this with adult pythons.) The data are again summarized in the form "sample mean ± standard error of the mean." Follow the outline in (a) and (b) of the previous exercise for propensity to strike. Does nest temperature appear to influence propensity to strike?

10.11 How much corn should I plant? Return to the data in Exercise 10.2 on corn yields for different planting rates.

(a) Starting from the sample means and standard deviations for the five groups (Figure 10.4), calculate MSE, the overall mean yield $\bar{x}$, and MSG. Use the computer output in Figure 10.4 to check your work.

(b) Give a 90% confidence interval for the mean yield of corn planted at 20,000 plants per acre. Use the pooled standard deviation s_p to estimate σ in the standard error.

Summary

One-way analysis of variance (ANOVA) compares the means of several populations. The **ANOVA F test** tests the overall H_0 that all the populations have the same mean. If the F test shows significant differences, examine the data to see where the differences lie and whether they are large enough to be important.

ANOVA assumes that we have an **independent SRS** from each population; that each population has a **normal distribution**; and that all populations have the **same standard deviation**.

In practice, ANOVA is relatively **robust** when the populations are nonnormal, especially when the samples are large. Before doing the F test, check the observations in each sample for outliers or strong skewness. Also verify that the largest sample standard deviation is no more than twice as large as the smallest standard deviation.

When the null hypothesis is true, the **ANOVA F statistic** for comparing I means from a total of N observations in all samples combined has the **F distribution** with $I - 1$ and $N - I$ degrees of freedom.

ANOVA calculations are reported in an **ANOVA table** that gives sums of squares, mean squares, and degrees of freedom for variation among groups and for variation within groups. In practice, we use software to do the calculations.

STATISTICS in SUMMARY

Advanced statistical inference often concerns relationships among several parameters. This chapter introduces the ANOVA F test for one such relationship: equality of the means of any number of populations. The alternative to this hypothesis is "many-sided," because it allows any relationship other than "all equal." The ANOVA F test is an overall test that tells us whether the data give good reason to reject the hypothesis that all the population means are equal. You should always accompany the test by data analysis to see what kind of inequality is present. Plotting the data in all groups side by side is particularly helpful. After studying this chapter, you should be able to do the following.

A. RECOGNITION

1. Recognize when testing the equality of several means is helpful in understanding data.
2. Recognize that the statistical significance of differences among sample means depends on the sizes of the samples and on how much variation there is within the samples.
3. Recognize when you can safely use ANOVA to compare means. Check the data production, the presence of outliers, and the sample standard deviations for the groups you want to compare.

B. INTERPRETING ANOVA

1. Explain what null hypothesis F tests in a specific setting.
2. Locate the F statistic and its P-value on the output of a computer analysis of variance program.
3. Find the degrees of freedom for the F statistic from the number and sizes of the samples. Use Table D of the F distributions to approximate the P-value when software does not give it.
4. If the test is significant, use graphs and descriptive statistics to see what differences among the means are most important.

CHAPTER 10 **R e v i e w   E x e r c i s e s**

10.12 In each of the following situations, we want to compare the mean response in several populations. For each setting, identify the populations and the response variable. Then give I, the n_i, and N. Finally, give the degrees of freedom of the ANOVA F test.

(a) A study of the effects of smoking classifies subjects as nonsmokers, moderate smokers, or heavy smokers. The investigators interview a sample of 200 people in each group. Among the questions is "How many hours do you sleep on a typical night?"

(b) The strength of concrete depends on the mixture of sand, gravel, and cement used to prepare it. A study compares five different mixtures. Workers prepare six batches of each mixture and measure the strength of the concrete made from each batch.

(c) Which of four methods of teaching American Sign Language is most effective? Assign 10 of the 42 students in a class at random to each of three methods. Teach the remaining 12 students by the fourth method. Record the students' scores on a standard test of sign language after a semester's study.

10.13 Does polyester decay? How quickly do synthetic fabrics such as polyester decay in landfills? A researcher buried polyester strips in the soil for different lengths of time, then dug up the strips and measured the force required to break them. Breaking strength is easy to measure and is a good indicator of decay; lower strength means the fabric has decayed.

Part of the study buried 20 polyester strips in well-drained soil in the summer. Five of the strips, chosen at random, were dug up after 2 weeks; another 5 were dug up after 4 weeks, 8 weeks, and 16 weeks. Here are the breaking strengths in pounds:[8]

2 weeks	118	126	126	120	129
4 weeks	130	120	114	126	128
8 weeks	122	136	128	146	140
16 weeks	124	98	110	140	110

(a) Find the mean strength for each group and plot the means against time. Does it appear that polyester loses strength consistently over time after it is buried?

(b) Find the standard deviations for each group. Do they meet our criterion for applying ANOVA?

(c) In Examples 7.7 and 7.8 (pages 393 and 396), we used the two-sample t test to compare the mean breaking strengths for strips buried for 2 weeks and for 16 weeks. The ANOVA F test extends the two-sample t to more than two groups. Explain carefully why use of the two-sample t for two of the groups was acceptable but using the F test on all four groups is not acceptable.

10.14 Can you hear these words? To test whether a hearing aid is right for a patient, audiologists play a tape on which words are pronounced at low volume. The patient tries to repeat the words. There are several different lists of words that are supposed to be equally difficult. Are the lists equally difficult when there is background noise? To find out, an experimenter had subjects with normal hearing listen to four lists with a noisy background. The response variable was the percent of the 50 words in a list that the subject repeated correctly. The data set contains 96 responses.[9]

(a) Here are two study designs that could produce these data:

Design A. The experimenter assigns 96 subjects to 4 groups at random. Each group of 24 subjects listens to one of the lists. All individuals listen and respond separately.

Design B. The experimenter has 24 subjects. Each subject listens to all four lists in random order. All individuals listen and respond separately.

Does Design A allow use of one-way ANOVA to compare the lists? Does Design B allow use of one-way ANOVA to compare the lists? Briefly explain your answers.

(b) Figure 10.9 displays Minitab output for one-way ANOVA. The response variable is "Percent," and "List" identifies the four lists of words. Based on this analysis, is there good reason to think that the four lists are not all equally difficult? Write a brief summary of the study findings.

10.15 Logging in the rainforest. "Conservationists have despaired over destruction of tropical rainforest by logging, clearing, and burning." These words begin a report on a statistical study of the effects of

```
Analysis of Variance for Percent
Source      DF        SS        MS        F         P
List         3     920.5     306.8      4.92     0.003
Error       92    5738.2      62.4
Total       95    6658.6


                                    Individual 95% CIs For Mean
                                    Based on Pooled StDev
Level        N      Mean     StDev ----+---------+---------+---------+----
1           24    32.750     7.409                         (-------*------)
2           24    29.667     8.058                  (-------*------)
3           24    25.250     8.316     (-------*------)
4           24    25.583     7.779       (-------*------)

                                     ---+---------+---------+---------+----
Pooled StDev =     7.898              24.0      28.0      32.0      36.0
```

Figure 10.9 *Minitab ANOVA output for comparing the percents heard correctly in four lists of words, for Exercise 10.14.*

Table 10.3 The effects of logging in a tropical rainforest

Never logged		Logged 1 year ago		Logged 8 years ago	
Trees per plot	Species per plot	Trees per plot	Species per plot	Trees per plot	Species per plot
27	22	12	11	18	17
22	18	12	11	4	4
29	22	15	14	22	18
21	20	9	7	15	14
19	15	20	18	18	18
33	21	18	15	19	15
16	13	17	15	22	15
20	13	14	12	12	10
24	19	14	13	12	12
27	13	2	2		
28	19	17	15		
19	15	19	8		

Source: Charles Cannon, Duke University.

logging in Borneo.[10] The study compared forest plots that had never been logged with similar plots nearby that had been logged 1 year earlier and 8 years earlier. Although the study was not an experiment, the authors explain why we can consider the plots to be randomly selected. Table 10.3 contains the data.

(a) Use side-by-side stemplots to compare the distributions of number of trees per plot for the three groups of plots. Are there features that might prevent use of ANOVA?

(b) Figure 10.10 displays Minitab output for one-way ANOVA on the trees per plot data. Do the standard deviations satisfy our rule of thumb for safe use of ANOVA? What do the means suggest about the effect of logging on the number of trees per plot? Report the F statistic and its P-value and state your overall conclusion.

10.16 Logging in the rainforest, continued. Table 10.3 gives data on the number of tree species per forest plot as well as on the number of individual trees. In the previous exercise, you examined the effect of logging on the number of trees. Use software to analyze the effect of logging on the number of species.

(a) Make a table of the group means and standard deviations. Do the standard deviations satisfy our rule of thumb for safe use of ANOVA? What do the means suggest about the effect of logging on the number of species?

(b) Carry out the ANOVA. Report the F statistic and its P-value and state your conclusion.

(DAVID AUSTEN/STOCK, BOSTON.)

```
Analysis of Variance for Trees
Source      DF        SS        MS        F         P
Logging      2      625.2     312.6     11.43     0.000
Error       30      820.7      27.4
Total       32     1445.9

                                    Individual 95% CIs For Mean
                                    Based on Pooled StDev

Level        N      Mean      StDev --------+----------+----------+--------
Unlogged    12     23.750     5.065                        (-----*-----)
1 year      12     14.083     4.981  (-----*-----)
3 years      9     15.778     5.761
                                       (-----*-----)

                                    --------+----------+----------+--------
Pooled StDev =     5.230                 15.0       20.0       25.0
```

Figure 10.10 *Minitab ANOVA output for comparing the number of trees per plot in unlogged forest plots, plots logged 1 year ago, and plots logged 8 years ago, for Exercise 10.15.*

10.17 **Nematodes and tomato plants.** How do nematodes (microscopic worms) affect plant growth? A botanist prepares 16 identical planting pots and then introduces different numbers of nematodes into the pots. He transplants a tomato seedling into each plot. Here are data on the increase in height of the seedlings (in centimeters) 16 days after planting:[11]

Nematodes	Seedling growth			
0	10.8	9.1	13.5	9.2
1,000	11.1	11.1	8.2	11.3
5,000	5.4	4.6	7.4	5.0
10,000	5.8	5.3	3.2	7.5

(a) Make a table of means and standard deviations for the four treatments. Make side-by-side stemplots to compare the treatments. What do the data appear to show about the effect of nematodes on growth?

(b) State H_0 and H_a for the ANOVA test for these data, and explain in words what ANOVA tests in this setting.

(c) Use software to carry out the ANOVA. Report your overall conclusions about the effect of nematodes on plant growth.

10.18 **Optional: F versus t.** We have two methods to compare the means of two groups: the two-sample t test of Section 7.2 and the ANOVA F test with $I = 2$. We prefer the t test because it allows one-sided alternatives and does not assume that both populations have the same standard deviation. Let us apply both tests to the same data.

There are two types of life insurance companies. "Stock" companies have shareholders, and "mutual" companies are owned by their policyholders. Take an SRS of each type of company from those listed in a directory of the industry. Then ask the annual cost per $1000 of insurance for a $50,000 policy insuring the life of a 35-year-old man who does not smoke. Here are the data summaries:[12]

	Stock companies	Mutual companies
n_i	13	17
$\bar{x}_i$	$2.31	$2.37
s_i	$0.38	$0.58

(a) Calculate the two-sample t statistic for testing H_0: $\mu_1 = \mu_2$ against the two-sided alternative. Use the conservative method to find the P-value.

(b) Calculate MSG, MSE, and the ANOVA F statistic for the same hypotheses. What is the P-value of F?

(c) How close are the two P-values? (The square root of the F statistic is a t statistic with $N - I = n_1 + n_2 - 2$ degrees of freedom. This is the "pooled two-sample t" mentioned on page 406. So F for $I = 2$ is exactly equivalent to a t statistic, but it is a slightly different t from the one we use.)

10.19 Optional: Do it the hard way. Carry out the ANOVA calculations (MSG, MSE, and F) required for part (b) of Exercise 10.16. Find the degrees of freedom for F and report its P-value as closely as Table D allows.

Sir Francis Galton

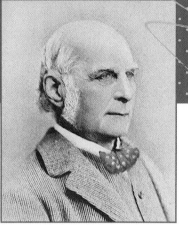

(THE GRANGER COLLECTION, NEW YORK.)

Correlation, Regression, and Heredity

Galton was full of ideas but was no mathematician. He didn't even use least squares, preferring to avoid unpleasant computations.

The least-squares method will happily fit a straight line to any two-variable data. It is an old method, going back to the French mathematician Legendre in about 1805. Legendre invented least squares for use on data from astronomy and surveying. It was Sir Francis Galton (1822–1911), however, who turned "regression" into a general method for understanding relationships. He even invented the word. While he was at it, he also invented "correlation," both the word and the definition of r.

Galton was one of the last gentleman scientists, an upper-class Englishman who studied medicine at Cambridge and explored Africa before turning to the study of heredity. He was well connected here also: Charles Darwin, who published *The Origin of Species* in 1859, was his cousin.

Galton was full of ideas but was no mathematician. He didn't even use least squares, preferring to avoid unpleasant computations. But Galton's ideas led eventually to the machinery for inference about regression that we will meet in this chapter. He asked: If people's heights are distributed normally in every generation, and height is inherited, what is the relationship between generations? He discovered a straight-line relationship between the heights of parent and child and found that tall parents tended to have children who were taller than average but less tall than their parents. He called this "regression toward mediocrity." Galton went further: he described inheritance by a straight-line relationship with responses y that have a normal distribution about the line for every fixed input x. This is the model for regression we use in this chapter.

Inference for Regression

(PAUL BARTON/THE STOCK MARKET.)

Galton used regression ideas to study how heights change from one generation of a family to the next.

Introduction

When a scatterplot shows a linear relationship between a quantitative explanatory variable x and a quantitative response variable y, we can use the least-squares line fitted to the data to predict y for a given value of x. Now we want to do tests and confidence intervals in this setting.

EXAMPLE 11.1   **Crying and IQ**

Infants who cry easily may be more easily stimulated than others and this may be a sign of higher IQ. Child development researchers explored the relationship between the crying of infants four to ten days old and their later IQ test scores. A snap of a rubber band on the sole of the foot caused the infants to cry. The researchers recorded the crying and measured its intensity by the number of peaks in the most active 20 seconds. They later measured the children's IQ at age three years using the Stanford-Binet IQ test. Table 11-1 contains data on 38 infants.[1]

scatterplot

Plot and interpret. As always, we first examine the data. Figure 11.1 is a **scatterplot** of the crying data. Plot the explanatory variable (count of crying peaks) horizontally and the response variable (IQ) vertically. Look for the form, direction, and strength of the relationship as well as for outliers or other deviations. There is a moderate positive linear relationship, with no extreme outliers or potentially influential observations.

correlation

Numerical summary. Because the scatterplot shows a roughly linear (straight-line) pattern, the **correlation** describes the direction and strength of the relationship. The correlation between crying and IQ is $r = 0.455$.

least-squares line

Mathematical model. We are interested in predicting the response from information about the explanatory variable. So we find the **least-squares**

Table 11.1 Infants' crying and IQ scores

Crying	IQ	Crying	IQ	Crying	IQ	Crying	IQ
10	87	20	90	17	94	12	94
12	97	16	100	19	103	12	103
9	103	23	103	13	104	14	106
16	106	27	108	18	109	10	109
18	109	15	112	18	112	23	113
15	114	21	114	16	118	9	119
12	119	12	120	19	120	16	124
20	132	15	133	22	135	31	135
16	136	17	141	30	155	22	157
33	159	13	162				

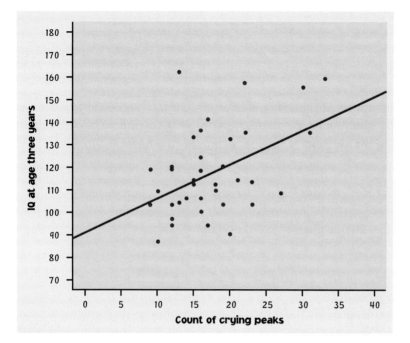

Figure 11.1 *Scatterplot of the IQ score of infants at age three years against the intensity of their crying soon after birth, for Example 11.1.*

regression line for predicting IQ from crying. This line lies as close as possible to the points (in the sense of least squares) in the vertical (y) direction. The equation of the least-squares regression line is

$$\hat{y} = a + bx$$
$$= 91.27 + 1.493x$$

We use the notation $\hat{y}$ to remind ourselves that the regression line gives *predictions* of IQ. The predictions usually won't agree exactly with the actual values of the IQ measured several years later. Drawing the least-squares line on the scatterplot helps us see the overall pattern. Because $r^2 = 0.207$, only about 21% of the variation in IQ scores is explained by crying intensity. Prediction of IQ will not be very accurate. It is nonetheless impressive that behavior soon after birth can even partly predict IQ several years later.

The regression model

The slope b and intercept a of the least-squares line are *statistics*. That is, we calculated them from the sample data. These statistics would take somewhat different values if we repeated the study with different infants. To do formal inference, we think of a and b as estimates of unknown *parameters*. The parameters appear in a mathematical model of the process that produces our data. Here are the assumptions that describe the regression model.

May the longer name win!

A writer in the early 1960s noted a simple method for predicting presidential elections: just choose the candidate with the longer name. In the 22 elections from 1876 to 1960, this method failed only once. Lets hope that the writer didn't bet the family silver on this idea. The nine elections from 1964 to 1996 presented eight tests of the "long name wins" method (the 1980 candidates, Reagan and Carter, had names of the same length). The longer name lost five of the eight.

ASSUMPTIONS FOR REGRESSION INFERENCE

We have n observations on an explanatory variable x and a response variable y. Our goal is to study or predict the behavior of y for given values of x.

- For any fixed value of x, the response y varies according to a normal distribution. Repeated responses y are independent of each other.
- The mean response μ_y has a straight-line relationship with x:

$$\mu_y = \alpha + \beta x$$

The slope β and intercept α are unknown parameters.

- The standard deviation of y (call it σ) is the same for all values of x. The value of σ is unknown.

true regression line

The heart of this model is that there is an "on the average" straight-line relationship between y and x. The **true regression line** $\mu_y = \alpha + \beta x$ says that the *mean* response μ_y moves along a straight line as the explanatory variable x changes. We can't observe the true regression line. The values of y that we do observe vary about their means according to a normal distribution. If we hold x fixed and take many observations on y, the normal pattern will eventually appear in a stemplot or histogram. In practice, we observe y for many different values of x, so that we see an overall linear pattern formed by points scattered about the true line. The standard deviation σ determines whether the points fall close to the true regression line (small σ) or are widely scattered (large σ).

Figure 11.2 shows the regression model in picture form. The line in the figure is the true regression line. The mean of the response y moves along this line as the explanatory variable x takes different values. The normal curves show how y will vary when x is held fixed at different values. All of the curves have the same σ, so the variability of y is the same for all values of x. You should check the assumptions for inference when you do inference about regression. We will see later how to do that.

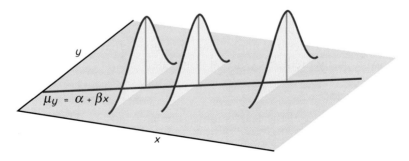

Figure 11.2 *The regression model. The line is the true regression line, which shows how the mean response μ_y changes as the explanatory variable x changes. For any fixed value of x, the observed response y varies according to a normal distribution having mean μ_y.*

11.1 Inference about the Model

The first step in inference is to estimate the unknown parameters α, β, and σ. When the regression model describes our data and we calculate the least-squares line $\hat{y} = a + bx$, **the slope b of the least-squares line is an unbiased estimator of the true slope β, and the intercept a of the least-squares line is an unbiased estimator of the true intercept α.**

EXAMPLE 11.2   **Slope and intercept**

The data in Figure 11.1 fit the regression model of scatter about an invisible true regression line reasonably well. The least-squares line is $\hat{y} = 91.27 + 1.493x$. The slope is particularly important. *A slope is a rate of change.* The true slope β says how much higher average IQ is for children with one more peak in their crying measurement. Because $b = 1.493$ estimates the unknown β, we estimate that on the average IQ is about 1.5 points higher for each added crying peak.

We need the intercept $a = 91.27$ to draw the line, but it has no statistical meaning in this example. No child had fewer than 9 crying peaks, so we have no data near $x = 0$. We suspect that all normal children would cry when snapped with a rubber band, so that we will never observe $x = 0$.

The remaining parameter of the model is the standard deviation σ, which describes the variability of the response y about the true regression line. The least-squares line estimates the true regression line. So the **residuals** estimate how much y varies about the true line. Recall that the residuals are the vertical deviations of the data points from the least-squares line:

residuals

$$\text{residual} = \text{observed } y - \text{predicted } y$$
$$= y - \hat{y}$$

There are n residuals, one for each data point. Because σ is the standard deviation of responses about the true regression line, we estimate it by a sample standard deviation of the residuals. We call this sample standard deviation a *standard error* to emphasize that it is estimated from data. The residuals from a least-squares line always have mean zero. That simplifies their standard error.

STANDARD ERROR ABOUT THE LEAST-SQUARES LINE

The **standard error about the line** is

$$s = \sqrt{\frac{1}{n-2}\sum \text{residual}^2}$$

$$= \sqrt{\frac{1}{n-2}\sum (y - \hat{y})^2}$$

Use s to estimate the unknown σ in the regression model.

degrees of freedom

Because we use the standard error about the line so often in regression inference, we just call it s. Notice that s^2 is an average of the squared deviations of the data points from the line, so it qualifies as a variance. We average the squared deviations by dividing by $n - 2$, the number of data points less 2. It turns out that if we know $n - 2$ of the n residuals, the other two are determined. That is, $n - 2$ is the **degrees of freedom** of s. We first met the idea of degrees of freedom in the case of the ordinary sample standard deviation of n observations, which has $n - 1$ degrees of freedom. Now we observe two variables rather than one, and the proper degrees of freedom is $n - 2$ rather than $n - 1$.

Calculating s is unpleasant. You must find the predicted response for each x in your data set, then the residuals, and then s. In practice you will use software that does this arithmetic instantly. Nonetheless, here is an example to help you understand the standard error s.

EXAMPLE 11.3 **Residuals and standard error**

Table 11.1 shows that the first infant studied had 10 crying peaks and a later IQ of 87. The predicted IQ for $x = 10$ is

$$\hat{y} = 91.27 + 1.493x$$
$$= 91.27 + 1.493(10) = 106.2$$

The residual for this observation is

$$\text{residual} = y - \hat{y}$$
$$= 87 - 106.2 = -19.2$$

That is, the observed IQ for this infant lies 19.2 points below the least-squares line on the scatterplot.

Repeat this calculation 37 more times, once for each subject. The 38 residuals are

−19.20	−31.13	−22.65	−15.18	−12.18	−15.15	−16.63	−6.18
−1.70	−22.60	−6.68	−6.17	−9.15	−23.58	−9.14	2.80
−9.14	−1.66	−6.14	−12.60	0.34	−8.62	2.85	14.30
9.82	10.82	0.37	8.85	10.87	19.34	10.89	−2.55
20.85	24.35	18.94	32.89	18.47	51.32		

Check the calculations by verifying that the sum of the residuals is zero. It is 0.04, not quite zero, because of roundoff error. Another reason to use software in regression is that roundoff errors in hand calculation can accumulate to make the results inaccurate.

The variance about the line is

$$s^2 = \frac{1}{n-2} \sum \text{residual}^2$$

$$= \frac{1}{38-2}[(-19.20)^2 + (-31.13)^2 + \cdots + 51.32^2]$$

$$= \frac{1}{36}(11{,}023.3) = 306.20$$

Finally, the standard error about the line is

$$s = \sqrt{306.20} = 17.50$$

Software gives 17.4987 to four decimal places, so the error resulting from rounding in this hand calculation is small.

We will study several kinds of inference in the regression setting. The standard error s about the line is the key measure of the variability of the responses in regression. It is part of the standard error of all the statistics we will use for inference.

APPLY YOUR KNOWLEDGE

11.1 **An extinct beast.** *Archaeopteryx* is an extinct beast having feathers like a bird but teeth and a long bony tail like a reptile. Here are the lengths in centimeters of the femur (a leg bone) and the humerus (a bone in the upper arm) for the five fossil specimens that preserve both bones:[2]

Femur	38	56	59	64	74
Humerus	41	63	70	72	84

The strong linear relationship between the lengths of the two bones helped persuade scientists that all five specimens belong to the same species.

(a) Examine the data. Make a scatterplot with femur length as the explanatory variable. Use your calculator to obtain the correlation r and the equation of the least-squares regression line. Do you think that femur length will allow good prediction of humerus length?

(b) Explain in words what the slope β of the true regression line says about *Archaeopteryx*. What is the estimate of β from the data? What is your estimate of the intercept α of the true regression line?

(c) Calculate the residuals for the five data points. Check that their sum is 0 (up to roundoff error). Use the residuals to estimate the standard deviation σ in the regression model. You have now estimated all three parameters in the model.

11.2 **Natural gas consumption.** Table 11.2 contains data on the natural gas consumption of the Sanchez household for 16 months.[3] Gas

Table 11.2 Average degree-days and natural gas consumption for the Sanchez household

Month	Degree-days	Gas (100 cu. ft.)	Month	Degree-days	Gas (100 cu.ft.)
Nov.	24	6.3	July	0	1.2
Dec.	51	10.9	Aug.	1	1.2
Jan.	43	8.9	Sept.	6	2.1
Feb.	33	7.5	Oct.	12	3.1
Mar.	26	5.3	Nov.	30	6.4
Apr.	13	4.0	Dec.	32	7.2
May	4	1.7	Jan.	52	11.0
June	0	1.2	Feb.	30	6.9

consumption is higher in cold weather. The table gives the average amount of natural gas consumed each day during the month, in hundreds of cubic feet, and the average number of heating degree-days each day during the month. (Degree-days are the number of degrees the average daily temperature falls below 65° F.)

(a) We want to predict gas used from degree-days. Make a scatterplot of the data with this goal in mind. Use your calculator to find the correlation r and the equation of the least-squares regression line. Describe the form and strength of the relationship.

(b) Find the residuals for all 16 data points. Check that their sum is 0 (up to roundoff error).

(c) The model for regression inference has three parameters, which we call α, β, and σ. Estimate these parameters from the data.

Confidence intervals for the regression slope

The slope β of the true regression line is usually the most important parameter in a regression problem. The slope is the rate of change of the mean response as the explanatory variable increases. We often want to estimate β. The slope b of the least-squares line is an unbiased estimator of β. A confidence interval is more useful because it shows how accurate the estimate b is likely to be. The confidence interval for β has the familiar form

$$\text{estimate} \pm t^* SE_{\text{estimate}}$$

Because b is our estimate, the confidence interval becomes

$$b \pm t^* SE_b$$

Here are the details.

CONFIDENCE INTERVAL FOR REGRESSION SLOPE

A level C confidence interval for the slope β of the true regression line is

$$b \pm t^* SE_b$$

In this recipe, the standard error of the least-squares slope b is

$$SE_b = \frac{s}{\sqrt{\sum(x - \bar{x})^2}}$$

and t^* is the upper $(1 - C)/2$ critical value from the t distribution with $n - 2$ degrees of freedom.

As advertised, the standard error of b is a multiple of s. Although we give the recipe for this standard error, you should rarely have to calculate it by hand. Regression software gives the standard error SE_b along with b itself.

EXAMPLE 11.4 **Regression output: crying and IQ**

Figure 11.3 shows the basic output for the crying study from the regression command in the Minitab software package. Most statistical software provides similar output. (Minitab, like other software, produces more than this basic output. When you use software, just ignore the parts you don't need.)

The first line gives the equation of the least-squares regression line. The slope and intercept are rounded off there, so look in the "Coef" column of the table that follows for more accurate values. The intercept $a = 91.268$ appears in the "Constant" row. The slope $b = 1.4929$ appears in the "Crycount" row because we named the x variable "Crycount" when we entered the data.

The next column of output, headed "StDev," gives standard errors. In particular, $SE_b = 0.4870$. The standard error about the line, $s = 17.50$, appears below the table.

Regression Analysis
The regression equation is
IQ = 91.3 + 1.49 Crycount

Predictor	Coef	StDev	T	P
Constant	91.268	8.934	10.22	0.000
Crycount	1.4929	0.4870	3.07	0.004

S = 17.50   R-Sq = 20.7%

Figure 11.3 *Minitab regression output for the crying and IQ data, Example 11.4.*

There are 38 data points, so the degrees of freedom are $n - 2 = 36$. For a 95% confidence interval for the true slope β, we will use the critical value $t^* = 2.042$ from the df = 30 row of Table C. This is the table degrees of freedom next smaller than 36. The interval is

$$b \pm t^* SE_b = 1.4929 \pm (2.042)(0.4870)$$
$$= 1.4929 \pm 0.9944$$
$$= 0.4985 \text{ to } 2.4873$$

We are 95% confident that mean IQ increases by between about 0.5 and 2.5 points for each additional peak in crying.

You can find a confidence interval for the intercept α of the true regression line in the same way, using a and SE_a from the "Constant" line of the printout. We rarely need to estimate α.

APPLY YOUR KNOWLEDGE

11.3 **Time at the table.** Does how long young children remain at the lunch table help predict how much they eat? Here are data on 20 toddlers observed over several months at a nursery school.[4] "Time" is the average number of minutes a child spent at the table when lunch was served. "Calories" is the average number of calories the child consumed during lunch, calculated from careful observation of what the child ate each day.

Time	21.4	30.8	37.7	33.5	32.8	39.5	22.8	34.1	33.9	43.8
Calories	472	498	465	456	423	437	508	431	479	454

Time	42.4	43.1	29.2	31.3	28.6	32.9	30.6	35.1	33.0	43.7
Calories	450	410	504	437	489	436	480	439	444	408

Make a scatterplot of the data and find the equation of the least-squares line for predicting calories consumed from time at the table. Describe briefly what the data show about the behavior of children. Then give a 95% confidence interval for the slope of the true regression line.

11.4 **Natural gas consumption.** Figure 11.4 gives computer output for the regression of natural gas consumed by the Sanchez household on degree-days. Use the information in this output to give a 90% confidence interval for the slope β of the true regression line. Explain clearly what your result tells us about how gas usage responds to falling temperatures.

```
The regression equation is
Gas = 1.09 + 0.189 D-days


Predictor        Coef        StDev           T          P
Constant       1.0892       0.1389        7.84      0.000
D-days       0.188999     0.004934       38.31      0.000


S = 0.3389        R-Sq = 99.1%
```

Figure 11.4 *Minitab output for the natural gas consumption data, Exercise 11.4.*

11.5 **The professor swims.** Here are data on the time (in minutes) Professor Moore takes to swim 2000 yards and his pulse rate (beats per minute) after swimming:

Time	34.12	35.72	34.72	34.05	34.13	35.72	36.17	35.57
Pulse	152	124	140	152	146	128	136	144

Time	35.37	35.57	35.43	36.05	34.85	34.70	34.75	33.93
Pulse	148	144	136	124	148	144	140	156

Time	34.60	34.00	34.35	35.62	35.68	35.28	35.97
Pulse	136	148	148	132	124	132	139

A scatterplot shows a negative linear relationship: a faster time (fewer minutes) is associated with a higher heart rate. Here is part of the output from the regression function in the Excel spreadsheet:

	Coefficients	Standard Error	t Stat	P-value
Intercept	479.9341457	66.22779275	7.246718119	3.87075E-07
X Variable	-9.694903394	1.888664503	-5.1332057	4.37908E-05

Give a 90% confidence interval for the slope of the true regression line. Explain what your result tells us about the relationship between the professor's swimming time and heart rate.

Testing the hypothesis of no linear relationship

We can also test hypotheses about the slope β. The most common hypothesis is

$$H_0 : \beta = 0$$

A regression line with slope 0 is horizontal. That is, the mean of y does not change at all when x changes. So this H_0 says that there is *no true linear relationship* between x and y. Put another way, H_0 says that *straight-line dependence on x is of no value for predicting y*. Put yet another way, H_0 says that there is *no correlation* between x and y in the population from which we drew our data.

You can use the test for zero slope to test the hypothesis of zero correlation between any two quantitative variables. That's a useful trick. Do notice that testing correlation makes sense only if the observations are a random sample. That is often not the case in regression settings, where researchers may fix the values of x they want to study.

The test statistic is just the standardized version of the least-squares slope b. It is another t statistic. Here are the details.

SIGNIFICANCE TESTS FOR REGRESSION SLOPE

To test the hypothesis $H_0 : \beta = 0$, compute the t statistic

$$t = \frac{b}{SE_b}$$

In terms of a random variable T having the $t(n-2)$ distribution, the P-value for a test of H_0 against

$H_a : \beta > 0$ is $P(T \geq t)$

$H_a : \beta < 0$ is $P(T \leq t)$

$H_a : \beta \neq 0$ is $2P(T \geq |t|)$

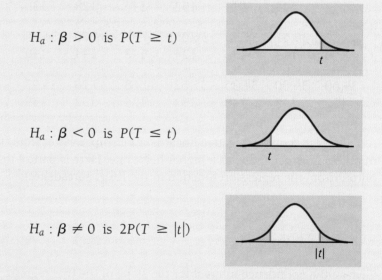

This test is also a test of the hypothesis that the correlation is 0 in the population.

Regression output from statistical software usually gives t and its *two-sided* P-value. For a one-sided test, divide the P-value in the output by 2.

EXAMPLE 11.5 Testing regression slope

The hypothesis $H_0 : \beta = 0$ says that crying has no straight-line relationship with IQ. Figure 11.1 shows that there is a relationship, so it is not surprising that the computer output in Figure 11.3 gives $t = 3.07$ with two-sided P-value 0.004. There is very strong evidence that IQ is correlated with crying.

EXAMPLE 11.6   Beer and blood alcohol (EESEE)

(DOUG MARTIN/PHOTO RESEARCHERS)

How well does the number of beers a student drinks predict his or her blood alcohol content? Sixteen student volunteers at Ohio State University drank a randomly assigned number of cans of beer. Thirty minutes later, a police officer measured their blood alcohol content (BAC). Here are the data:[5]

Student	1	2	3	4	5	6	7	8
Beers	5	2	9	8	3	7	3	5
BAC	0.10	0.03	0.19	0.12	0.04	0.095	0.07	0.06

Student	9	10	11	12	13	14	15	16
Beers	3	5	4	6	5	7	1	4
BAC	0.02	0.05	0.07	0.10	0.085	0.09	0.01	0.05

The students were equally divided between men and women and differed in weight and usual drinking habits. Because of this variation, many students don't believe that number of drinks predicts blood alcohol well. What do the data say?

The scatterplot in Figure 11.5 shows a clear linear relationship. Figure 11.6 gives part of the Minitab regression output. The solid line on the scatterplot is the least-squares line

$$\hat{y} = -0.0127 + 0.0180x$$

Because $r^2 = 0.800$, number of drinks accounts for 80% of the observed variation in BAC. That is, the data say that student opinion is wrong: the number of beers you drink predicts blood alcohol level quite well. Five beers produce an average BAC of

$$\hat{y} = -0.0127 + 0.0180(5) = 0.077$$

perilously close to the legal driving limit of 0.08 in many states.

We can test the hypothesis that the number of beers has *no* effect on blood alcohol versus the one-sided alternative that more beers increases BAC. The hypotheses are

$$H_0 : \beta = 0$$
$$H_a : \beta > 0$$

It is no surprise that the t statistic is $t = 7.48$ with two-sided P-value $P = 0.000$ to three decimal places. The one-sided P-value is half this value, so it is also close to 0. Check that t is the slope $b = 0.01796$ divided by its standard error, $SE_b = 0.0024$.

The scatterplot shows one unusual point: student number 3, who drank 9 beers. You can see from Figure 11.5 that this observation lies farthest from the fitted line in the y direction. That is, this point has the largest residual. Student number 3 may also be influential, though the point is not extreme in the x direction. To verify that our results are not too dependent on this one observation, do the regression again omitting student 3. The new regression line is the dashed line in Figure 11.5. Omitting student 3 decreases r^2 from 80% to 77%, and it changes the predicted BAC after 5 beers from 0.077 to 0.073. These small changes show that this observation is not very influential.

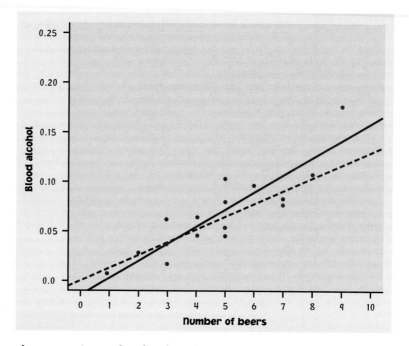

Figure 11.5 *Scatterplot of students' blood alcohol content against the numbers of cans of beer consumed, for Example 11.6.*

Figure 11.6 *Minitab output for the blood alcohol content data, Example 11.6.*

```
The regression equation is
BAC = - 0.0127 + 0.0180 Beers


Predictor        Coef        StDev             T             P
Constant     -0.01270      0.01264         -1.00         0.332
Beers         0.017964     0.002402          7.48         0.000


S = 0.02044      R-Sq = 80.0%
```

APPLY YOUR KNOWLEDGE

11.6 **An extinct beast.** Exercise 11.1 presents data on the lengths of two bones in five fossil specimens of the extinct beast *Archaeopteryx*. Here is part of the output from the S-PLUS statistical software when we regress the length y of the humerus on the length x of the femur.

```
Coefficients:
                 Value   Std. Error   t value   Pr(>|t|)
(Intercept)    -3.6596       4.4590   -0.8207     0.4719
      Femur     1.1969       0.0751
```

(a) What is the equation of the least-squares regression line?

(b) We left out the t statistic for testing $H_0 : \beta = 0$ and its P-value. Use the output to find t.

(c) How many degrees of freedom does t have? Use Table C to approximate the P-value of t against the one-sided alternative $H_a : \beta > 0$.

11.7 **Is wine good for your heart?** There is some evidence that drinking moderate amounts of wine helps prevent heart attacks. Table 2.3 (page 94) gives data on yearly wine consumption (liters of alcohol from drinking wine, per person) and yearly deaths from heart disease (deaths per 100,000 people) in 19 developed nations. Is there statistically significant evidence that the correlation between wine consumption and heart disease deaths is negative?

11.8 **Does fast driving waste fuel?** Exercise 2.6 (page 87) gives data on the fuel consumption of a small car at various speeds from 10 to 150 kilometers per hour. Is there evidence of straight-line dependence between speed and fuel use? Make a scatterplot and use it to explain the result of your test.

11.2 Inference about Prediction

One of the most common reasons to fit a line to data is to predict the response to a particular value of the explanatory variable. The method is simple: just substitute the value of x into the equation of the line. We saw in Example 11.6 that drinking 5 beers produces an average BAC of

$$\hat{y} = -0.0127 + 0.0180(5) = 0.077$$

We would like to give a confidence interval that describes how accurate this prediction is. To do that, you must answer these questions: Do you want to predict the *mean* blood alcohol level for *all students* who drink 5 beers? Or do you want to predict the BAC of *one individual student* who drinks 5 beers? Both of these predictions may be interesting, but they are two different problems. The actual prediction is the same, $\hat{y} = 0.077$. But the margin of error is different for the two kinds of prediction. Individual students who drink 5 beers don't all have the same BAC. So we need a larger margin of error to pin down one student's result than to estimate the mean BAC for all students who have 5 beers.

Write the given value of the explanatory variable x as x^*. In the example, $x^* = 5$. The distinction between predicting a single outcome and predicting the mean of all outcomes when $x = x^*$ determines what margin of error is correct. To emphasize the distinction, we use different terms for the two intervals.

- To estimate the *mean* response, we use a *confidence interval*. It is an ordinary confidence interval for the parameter

$$\mu_y = \alpha + \beta x^*$$

Regression for lawyers

Jury Verdict Research makes money from regression. The company collects data on more than 20,000 jury verdicts in personal injury lawsuits each year, recording more than 30 variables describing each case. Multiple regression using this mass of data allows Jury Verdict Research to predict how much a jury will award in a new case. Lawyers pay for these predictions and use them to negotiate settlements with insurance companies.

The regression model says that μ_y is the mean of responses y when x has the value x^*. It is a fixed number whose value we don't know.

prediction interval
- To estimate an *individual* response y, we use a **prediction interval**. A prediction interval estimates a single random response y rather than a parameter like μ_y. The response y is not a fixed number. If we took more observations with $x = x^*$, we would get different responses.

Fortunately, the meaning of a prediction interval is very much like the meaning of a confidence interval. A 95% prediction interval, like a 95% confidence interval, is right 95% of the time in repeated use. "Repeated use" now means that we take an observation on y for each of the n values of x in the original data, and then take one more observation y with $x = x^*$. Form the prediction interval from the n observations, then see if it covers the one more y. It will in 95% of all repetitions.

The interpretation of prediction intervals is a minor point. The main point is that it is harder to predict one response than to predict a mean response. Both intervals have the usual form

$$\hat{y} \pm t^* \text{SE}$$

but the prediction interval is wider than the confidence interval. Here are the details.

CONFIDENCE AND PREDICTION INTERVALS FOR REGRESSION RESPONSE

A level C **confidence interval for the mean response** μ_y when x takes the value x^* is

$$\hat{y} \pm t^* \text{SE}_{\hat{\mu}}$$

The standard error $\text{SE}_{\hat{\mu}}$ is

$$\text{SE}_{\hat{\mu}} = s\sqrt{\frac{1}{n} + \frac{(x^* - \bar{x})^2}{\sum(x - \bar{x})^2}}$$

The sum runs over all the observations on the explanatory variable x.

A level C **prediction interval for a single observation** on y when x takes the value x^* is

$$\hat{y} \pm t^* \text{SE}_{\hat{y}}$$

The standard error for prediction $\text{SE}_{\hat{y}}$ is[6]

$$\text{SE}_{\hat{y}} = s\sqrt{1 + \frac{1}{n} + \frac{(x^* - \bar{x})^2}{\sum(x - \bar{x})^2}}$$

In both recipes, t^* is the upper $(1 - C)/2$ critical value of the t distribution with $n - 2$ degrees of freedom.

There are two standard errors: $SE_{\hat{\mu}}$ for estimating the mean response μ_y and $SE_{\hat{y}}$ for predicting an individual response y. The only difference between the two standard errors is the extra 1 under the square root sign in the standard error for prediction. The extra 1 makes the prediction interval wider. Both standard errors are multiples of s. The degrees of freedom are again $n - 2$, the degrees of freedom of s. Calculating these standard errors by hand is a nuisance, which software spares us.

EXAMPLE 11.7 **Predicting blood alcohol**

Steve thinks he can drive legally 30 minutes after he finishes drinking 5 beers. We want to predict Steve's blood alcohol content, using no information except that he drinks 5 beers. Here is the output from the prediction option in the Minitab regression command for $x^* = 5$ when we ask for 95% intervals:

```
Predicted Values

     Fit   StDev Fit          95.0% CI              95.0% PI
  0.07712     0.00513   ( 0.06612, 0.08812)   ( 0.03192, 0.12232)
```

The "Fit" entry gives the predicted BAC, 0.07712. This agrees with our result in Example 11.6. Minitab gives both 95% intervals. You must choose which one you want. We are predicting a single response, so the prediction interval "95.0% PI" is the right choice. We are 95% confident that Steve's blood alcohol content will fall between about 0.032 and 0.122. The upper part of that range will get him arrested if he drives. The 95% confidence interval for the mean BAC of all students after 5 beers, given as "95.0% CI," is much narrower.

APPLY YOUR KNOWLEDGE

11.9 **The professor swims.** Exercise 11.5 gives data on a swimmer's time and heart rate. One day the swimmer completes his laps in 34.3 minutes but forgets to take his pulse. Minitab gives this prediction for heart rate when $x^* = 34.3$:

```
   Fit   StDev Fit        90.0% CI              90.0% PI
 147.40       1.97   ( 144.02, 150.78)   ( 135.79, 159.01)
```

(a) Verify that "Fit" is the predicted heart rate from the least-squares line found in Exercise 11.5. Then choose one of the intervals from the output to estimate the swimmer's heart rate that day and explain why you chose this interval.

(b) Minitab gives only one of the two standard errors used in prediction. It is $SE_{\hat{\mu}}$, the standard error for estimating the mean response. Use this fact and a critical value from Table C to verify

Minitab's 90% confidence interval for the mean heart rate on days when the swimming time is 34.3 minutes.

11.10 **Natural gas consumption.** Figure 11.4 (page 539) gives computer output for the regression of natural gas consumed by the Sanchez household on degree-days. After these data were collected, the Sanchez family installed solar panels.

(a) In the month of January after solar panels were installed, there were 40 degree-days per day. How much gas do you predict the Sanchez household would have used per day without the solar panels? They actually used 7.5 hundred cubic feet per day. How much gas per day did the solar panels save?

(b) Here is the prediction output from Minitab for 40 degree-days per day. Give a 95% interval for the amount of gas that would have been used this January without the solar panels.

```
    Fit  StDev Fit           95.0% CI            95.0% PI
 8.6492      0.1216  ( 8.3883,   8.9100)  ( 7.8768,  9.4215)
```

(c) Give a 95% interval for the mean gas consumption per day in all months with 40 degree-days per day.

(d) Minitab gives only one of the two standard errors used in prediction. It is $SE_{\hat{\mu}}$, the standard error for estimating the mean response. Use Table C and this standard error to give a 90% confidence interval for mean gas consumption per day in all months with 40 degree-days per day.

11.3 Checking the Regression Assumptions

You can fit a least-squares line to any set of explanatory-response data when both variables are quantitative. If the scatterplot doesn't show a roughly linear pattern, the fitted line may be almost useless. But it is still the line that fits the data best in the least-squares sense. To use regression inference, however, the data must satisfy the regression model assumptions. Before we do inference, we must check these assumptions one by one.

The observations are independent. In particular, repeated observations on the same individual are not allowed. So we can't use ordinary regression to make inferences about the growth of a single child over time, for example.

The true relationship is linear. We can't observe the true regression line, so we will almost never see a perfect straight-line relationship in our data. Look at the scatterplot to check that the overall pattern is roughly linear. A plot of the residuals against x magnifies any unusual pattern. Draw a horizontal line at zero on the residual plot to orient your eye. Because the sum of the residuals is always zero, zero is also the mean of the residuals.

The standard deviation of the response about the true line is the same everywhere. Look at the scatterplot again. The scatter of the data points about the line should be roughly the same over the entire range of the data. A plot of the residuals against x, with a horizontal line at zero, makes this easier to check. It is quite common to find that as the response y gets larger, so does the scatter of the points about the fitted line. Rather than remaining fixed, the standard deviation σ about the line is changing with x as the mean response changes with x. You cannot safely use our inference recipes when this happens. There is no fixed σ for s to estimate.

The response varies normally about the true regression line. We can't observe the true regression line. We can observe the least-squares line and the residuals, which show the variation of the response about the fitted line. The residuals estimate the deviations of the response from the true regression line, so they should follow a normal distribution. Make a histogram or stemplot of the residuals and check for clear skewness or other major departures from normality. Like other t procedures, inference for regression is (with one exception) not very sensitive to minor lack of normality, especially when we have many observations. Do beware of influential observations, which move the regression line and can greatly affect the results of inference.

The exception is the prediction interval for a single response y. This interval relies on normality of individual observations, not just on the approximate normality of statistics like the slope a and intercept b of the least-squares line. The statistics a and b become more normal as we take more observations. This contributes to the robustness of regression inference, but it isn't enough for the prediction interval. We will not study methods that carefully check normality of the residuals, so you should regard prediction intervals as rough approximations.

The assumptions for regression inference are a bit elaborate. Fortunately, it is not hard to check for gross violations. There are ways to deal with violations of any of the regression model assumptions. If your data don't fit the regression model, get expert advice. Checking assumptions uses the residuals. Most regression software will calculate and save the residuals for you.

EXAMPLE 11.8 **Blood alcohol residuals**

Example 11.6 shows the regression of the blood alcohol content of 16 students on the number of beers they drink. The statistical software that did the regression calculations also calculates the 16 residuals. Here they are:

0.0229	0.0068	0.0410	−0.0110	−0.0012	−0.0180	0.0288	−0.0171
−0.0212	−0.0271	0.0108	0.0049	0.0079	−0.0230	0.0047	−0.0092

A residual plot appears in Figure 11.7. The values of x are on the horizontal axis. The residuals are on the vertical axis, with a horizontal line at zero.

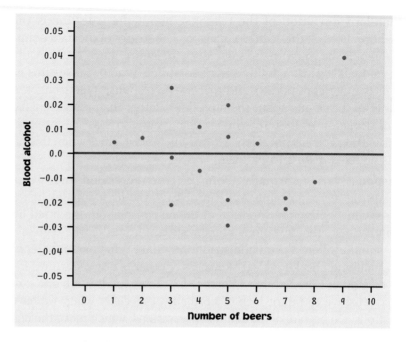

Figure 11.7 *Plot of the regression residuals for the blood alcohol data against the explanatory variable, number of beers consumed. The mean of the residuals is always 0.*

Examine the residual plot to check that the relationship is roughly linear and that the scatter about the line is about the same from end to end. Overall, there is no clear deviation from the even scatter about the line that should occur (except for chance variation) when the regression assumptions hold.

Now examine the distribution of the residuals for signs of strong nonnormality. Here is a stemplot of the residuals after rounding to three decimal places:

```
-2 | 731
-1 | 871
-0 | 91
 0 | 5578
 1 | 1
 2 | 39
 3 |
 4 | 1
```

Student number 3 is a mild outlier. We saw in Example 11.6 that omitting this observation has little effect on r^2 or the fitted line. It also has little effect on inference. For example, $t = 7.58$ for the slope becomes $t = 6.57$, a change of no practical importance.

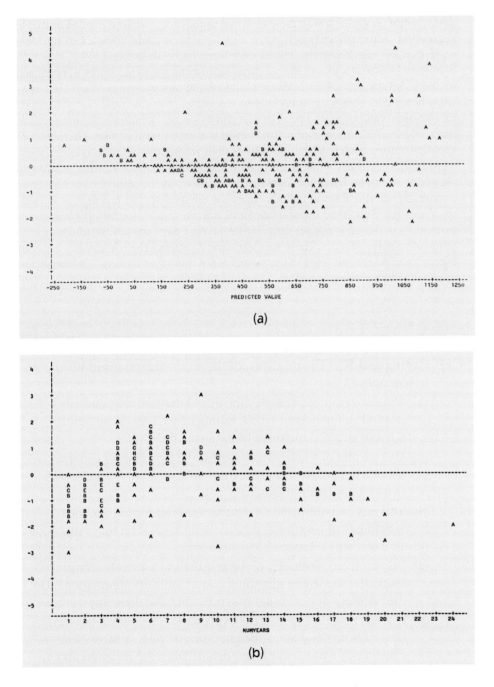

Figure 11.8 *Two residual plots that illustrate violations of the regression assumptions. (a) The variation of the residuals is not constant. (b) There is a curved relationship between the response variable and the explanatory variable.*

EXAMPLE 11.9 **Using residual plots**

The residual plots in Figure 11.8 illustrate violations of the regression assumptions that require corrective action before using regression. Both plots come from a study of the salaries of major-league baseball players.[7] Salary is the response variable. There are several explanatory variables that measure the players' past

multiple regression

Is regression garbage?

No—but garbage can be the setting for regression. The Census Bureau once asked if weighing a neighborhood's garbage would help count its people. So 63 households had their garbage sorted and weighed. It turned out that pounds of plastic in the trash gave the best garbage prediction of the number of people in a neighborhood. The margin of error for a 95% prediction interval in a neighborhood of about 100 households, based on five weeks' worth of garbage, was about ±2.5 people. Alas, that is not accurate enough to help the Census Bureau.

performance. Regression with more than one explanatory variable is called **multiple regression**. Although interpreting the fitted model is more complex in multiple regression, we check assumptions by examining residuals as usual.

Figure 11.8(a) is a plot of the residuals against the predicted salary $\hat{y}$, produced by the SAS statistical software. When points on the plot overlap, SAS uses letters to show how many observations each point represents. A is one observation, B stands for two observations, and so on. The plot shows a clear violation of the assumption that the spread of responses about the model is everywhere the same. There is more variation among players with high salaries than among players with lower salaries.

Although we don't show a histogram, the distribution of salaries is strongly skewed to the right. Using the *logarithm* of the salary as the response variable gives a more normal distribution and also fixes the unequal-spread problem. It is common to work with some transformation of data in order to satisfy the regression assumptions. But all is not yet well. Figure 11.8(b) plots the new residuals against years in the major leagues. There is a clear curved pattern. The relationship between logarithm of salary and years in the majors is not linear but curved. The statistician must take more corrective action.

APPLY YOUR KNOWLEDGE

11.11 Crying and IQ. The residuals for the study of crying and IQ appear in Example 11.3.
   (a) Make a stemplot to display the distribution of the residuals. (Round to the nearest whole number first.) Are there outliers or signs of strong departures from normality?
   (b) Make a plot of the residuals against the explanatory variable. Draw a horizontal line at height 0 on your plot. Does the plot shows a nonrandom pattern?

11.12 Natural gas consumption. Find the residuals for the Sanchez household gas consumption data in Table 11.2. (You may have done this in Exercise 11.2, page 535.)
   (a) Display the distribution of the residuals in a plot. It is hard to assess the shape of a distribution from only 16 observations. Do the residuals appear roughly symmetric? Are there any outliers?
   (b) Plot the residuals against the explanatory variable, degree-days. Draw a horizontal line at height 0 on the plot. Is there clear evidence of a nonlinear relationship? Does the variation about the line appear roughly the same as the number of degree-days changes?

CHAPTER 11 Summary

Least-squares regression fits a straight line to data in order to predict a response variable y from an explanatory variable x. Inference about regression requires more assumptions.

The **regression model** says that there is a **true regression line** $\mu_y = \alpha + \beta x$ that describes how the mean response varies as x changes. The observed response y for any x has a normal distribution with mean given by the true regression line and with the same standard deviation σ for any value of x. The parameters of the regression model are the intercept α, the slope β, and the standard deviation σ.

The slope a and intercept b of the least-squares line estimate the slope α and intercept β of the true regression line. To estimate σ, use the **standard error about the line s**.

The standard error s has $n - 2$ **degrees of freedom**. All t procedures in regression inference have $n - 2$ degrees of freedom.

Confidence intervals for the slope of the true regression line have the form $b \pm t^* SE_b$. In practice, use software to find the slope b of the least-squares line and its standard error SE_b.

To test the hypothesis that the true slope is zero, use the **t statistic** $t = b/SE_b$, also given by software. This null hypothesis says that straight-line dependence on x has no value for predicting y. It also says that the population correlation between x and y is zero.

Confidence intervals for the mean response when x has value x^* have the form $\hat{y} \pm t^* SE_{\hat{\mu}}$. **Prediction intervals** for an individual future response y have a similar form with a larger standard error, $\hat{y} \pm t^* SE_{\hat{y}}$. Software often gives these intervals.

STATISTICS IN SUMMARY

The methods of data analysis apply to any set of data. We can make a scatterplot and calculate the correlation and the least-squares regression line whenever we have data on two quantitative variables. Statistical inference makes sense only in more restrictive circumstances. The regression model describes the circumstances in which we can do inference about regression. The regression model includes a new parameter, the standard deviation σ that describes how much variation there is in responses y when x is held fixed. Estimating σ is the key to inference about regression. We use the standard error s (roughly, the sample standard deviation of the residuals) to estimate σ. Here are the skills you should develop from studying this chapter.

A. PRELIMINARIES

1. Make a scatterplot to show the relationship between an explanatory and a response variable.

2. Use a calculator or software to find the correlation and the equation of the least-squares regression line.

B. RECOGNITION

1. Recognize the regression setting: a straight-line relationship between an explanatory variable x and a response variable y.

2. Recognize which type of inference you need in a particular regression setting.

3. Inspect the data to recognize situations in which inference isn't safe: a nonlinear relationship, influential observations, strongly skewed residuals in a small sample, or nonconstant variation of the data points about the regression line.

C. DOING INFERENCE USING SOFTWARE OUTPUT

1. Explain in any specific regression setting the meaning of the slope β of the true regression line.

2. Understand software output for regression. Find in the output the slope and intercept of the least-squares line, their standard errors, and the standard error about the line.

3. Use that information to carry out tests and calculate confidence intervals for β.

4. Explain the distinction between a confidence interval for the mean response and a prediction interval for an individual response.

5. If software gives output for prediction, use that output to give either confidence or prediction intervals.

CHAPTER 11 Review Exercises

(DOUGLAS FAULKNER/PHOTO RESEARCHERS)

11.13 The endangered manatee. Manatees are large, gentle sea creatures that live along the Florida coast. Many manatees are killed or injured by powerboats. Here are data on powerboat registrations (in thousands) and the number of manatees killed by boats in Florida in the years 1977 to 1990.

Year	Powerboat registrations	Manatees killed	Year	Powerboat registrations	Manatees killed
1977	447	13	1984	559	34
1978	460	21	1985	585	33
1979	481	24	1986	614	33
1980	498	16	1987	645	39
1981	513	24	1988	675	43
1982	512	20	1989	711	50
1983	526	15	1990	719	47

(a) Make a scatterplot showing the relationship between powerboats registered and manatees killed. Is the overall pattern roughly linear? Are there clear outliers or strongly influential data points?

(b) Here is part of the output from the Minitab regression command:

```
Predictor       Coef        Stdev           T          P
Constant     -41.430        7.412       -5.59      0.000
Boats        0.12486      0.01290        9.68      0.000

S = 4.276           R-Sq = 88.6%
```

What does $r^2 = 0.886$ tell you about the relationship between boats and manatees killed?

(c) Explain what the slope β of the true regression line means in this setting. Then give a 90% confidence interval for β.

11.14 Manatee predictions. The previous exercise gives data on the Florida manatee for the years 1977 to 1990.

(a) Based on these data, you want to predict the number of manatees killed in a year when 716,000 powerboats are registered. Use the regression line from the software output in the previous problem to give a prediction.

(b) Here is the result of asking Minitab to do prediction for $x^* = 716$:

```
  Fit   Stdev.Fit           95% C.I.              95% P.I.
47.97        2.23    ( 43.11,   52.83)    ( 37.46,   58.48)
```

Check that the prediction 47.97 agrees with your result in (a). Then give a 95% interval for the number of manatees that would be killed in a future year if boat registrations remained at 716,000.

(c) Here are four more years of data:

| 1991 | 716 | 53 | 1993 | 716 | 35 |
| 1992 | 716 | 38 | 1994 | 735 | 49 |

It happens that powerboat registrations remained at 716,000 for the next three years. Did the numbers of manatees killed in those years fall within your prediction interval?

11.15 More manatee predictions. Exercise 11.14 gives 95% intervals for predicting manatee deaths when 716,000 powerboats are registered. The output gives only one of the two standard errors used in prediction. It is $SE_{\hat{\mu}}$, the standard error for estimating the mean response. Use Table C and this standard error to give a 90%

confidence interval for the mean number of manatees killed in all years when 716,000 boats are registered.

11.16 **Beavers and beetles.** Ecologists sometimes find rather strange relationships in our environment. One study seems to show that beavers benefit beetles. The researchers laid out 23 circular plots, each four meters in diameter, in an area where beavers were cutting down cottonwood trees. In each plot, they measured the number of stumps from trees cut by beavers and the number of clusters of beetle larvae. Here are the data:[8]

Stumps	2	2	1	3	3	4	3	1	2	5	1	3
Beetle larvae	10	30	12	24	36	40	43	11	27	56	18	40

Stumps	2	1	2	2	1	1	4	1	2	1	4
Beetle larvae	25	8	21	14	16	6	54	9	13	14	50

(CARLYN GALATI/VISUALS UNLIMITED.)

(a) Make a scatterplot that shows how the number of beaver-caused stumps influences the number of beetle larvae clusters. What does your plot show?

(b) Here is part of the Minitab regression output for these data:

```
Predictor        Coef        StDev           T           P
Constant       -1.286        2.853       -0.45       0.657
Stumps         11.894        1.136       10.47       0.000

S = 6.419        R-Sq = 83.9%
```

Find the least-squares regression line and draw it on your plot. What percent of the observed variation in beetle larvae counts can be explained by straight-line dependence on beaver stump counts?

(c) Is there strong evidence that beaver stumps help explain beetle larvae counts? State hypotheses, give a test statistic and its *P*-value, and state your conclusion.

11.17 **Beaver and beetle residuals.** Software often calculates *standardized residuals* as well as the actual residuals from regression. Because the standardized residuals have the standard *z*-score scale, it is easier to judge whether any are extreme. Here are the standardized residuals from the previous exercise, rounded to 2 decimal places:

standardized residuals

−1.99	1.20	0.23	−1.67	0.26	−1.06	1.38	0.06	0.72	−0.40	1.21	0.90
0.40	−0.43	−0.24	−1.36	0.88	−0.75	1.30	−0.26	−1.51	0.55	0.62	

(a) Find the mean and standard deviation of the standardized residuals. Why do you expect values close to those you obtain?

(b) Make a stemplot of the standardized residuals. Are there any striking deviations from normality? The most extreme residual is $z = -1.99$. Would this be surprisingly large if the 23 observations had a normal distribution? Explain your answer.

(c) Plot the standardized residuals against the explanatory variable. Are there any suspicious patterns?

11.18 Investing at home and overseas. Investors ask about the relationship between returns on investments in the United States and investments overseas. Table 2.7 (page 127) gives the percent returns on U.S. and overseas common stocks over a 27-year period.

(a) Make a scatterplot suitable for predicting overseas returns from U.S. returns.

(b) Here is part of the output from the Minitab regression command:

```
Predictor         Coef          StDev           T          P
Constant         5.683         5.144         1.10      0.280
USreturn        0.6181        0.2369

S = 19.90         R-Sq = 21.4%
```

We have omitted the t statistic for β and its P-value. What is the value of t? What are its degrees of freedom? From Table C, how strong is the evidence for a linear relationship between U.S. and overseas returns?

(c) Here is the output for prediction of overseas returns when U.S. stocks return 15%:

```
    Fit   StDev Fit          90.0% CI                  90.0% PI
  14.95        3.83     (   8.41,    21.50)    (  -19.65,    49.56)
```

Verify the "Fit" by using the least-squares line from the output in (b). You think U.S. stocks will return 15% next year. Give a 90% interval for the return on foreign stocks next year if you are right about U.S. stocks.

(d) Is the regression prediction useful in practice? Use the r^2-value for this regression to help explain your finding.

11.19 Stock return residuals. Exercise 11.18 presents a regression of overseas stock returns on U.S. stock returns based on 27 years' data. The residuals for this regression (in order by years across the rows) are

14.89	18.93	−11.44	−12.57	6.72	−17.77	16.99	22.96	−12.13
−3.05	−4.89	−20.87	4.17	−2.05	30.98	52.22	15.76	12.36
−14.78	−27.17	−12.04	−22.18	20.97	−0.29	−17.72	−13.80	−24.23

(a) Plot the residuals against x, the U.S. return. The plot suggests a mild violation of one of the regression assumptions. Which one?

(b) Display the distribution of the residuals in a graph. In what way is the shape somewhat nonnormal? There is one possible outlier. Circle that point on the residual plot in (a). What year is this? This point is not very influential: redoing the regression without it does not greatly change the results. With 27 observations, we are willing to do regression inference for these data.

11.20 Optional: Confidence interval for the intercept. Figure 11.4 (page 539) gives Minitab output for the regression of household natural gas consumption on degree-days. In Exercise 11.4 you used the output to find a 90% confidence interval for the slope β of the true regression line. The intercept α of the true regression line is the average amount of gas the household consumes when there are zero degree-days. (There are zero degree-days when the average temperature is 65° F or above.) Confidence intervals for α have the form

$$a \pm t^* SE_a$$

Use the intercept a of the least-squares line and its standard error, given in the computer output, to find a 95% confidence interval for α. (The degrees of freedom are $n - 2$ once again.)

11.21 Do heavy people burn more energy? Metabolic rate, the rate at which the body consumes energy, is important in studies of weight gain, dieting, and exercise. Lean body mass is an important influence on metabolic rate. Table 2.2 (page 90) gives data for 19 people. Because men and women showed a similar pattern, we will now ignore gender. Here are the data on mass (in kilograms) and metabolic rate (in calories):

Mass	62.0	62.9	36.1	54.6	48.5	42.0	47.4	50.6	42.0	48.7
Rate	1792	1666	995	1425	1396	1418	1362	1502	1256	1614

Mass	40.3	33.1	51.9	42.4	34.5	51.1	41.2	51.9	46.9
Rate	1189	913	1460	1124	1052	1347	1204	1867	1439

Use software to analyze these data. Make a scatterplot and find the least-squares line. Give a 90% confidence interval for the slope β and explain clearly what your interval says about the relationship between lean body mass and metabolic rate. Find the residuals and examine them. Are the assumptions for regression inference met?

11.22 Weeds among the corn. Lamb's-quarter is a common weed that interferes with the growth of corn. An agriculture researcher planted corn at the same rate in 16 small plots of ground, then weeded the plots by hand to allow a fixed number of lamb's-quarter plants to grow in each meter of corn row. No other weeds were allowed to grow. Here are the yields of corn (bushels per acre) in each of the plots:[9]

Weeds per meter	Corn yield	Weeds per meter	Corn yield	Weeds per meter	Corn yield	Weeds per meter	Corn yield
0	166.7	1	166.2	3	158.6	9	162.8
0	172.2	1	157.3	3	176.4	9	142.4
0	165.0	1	166.7	3	153.1	9	162.8
0	176.9	1	161.1	3	156.0	9	162.4

Use software to analyze these data.

(a) Make a scatterplot and find the least-squares line. What percent of the observed variation in corn yield can be explained by a linear relationship between yield and weeds per meter?

(b) Is there good evidence that more weeds reduce corn yield?

(c) Explain from your findings in (a) and (b) why you expect predictions based on this regression to be quite imprecise. Predict the mean corn yield under these experimental conditions when there are 6 weeds per meter of row. If your software allows, give a 95% confidence interval for this mean.

11.23 City and highway mileage. How well does a vehicle's city gas mileage predict its highway gas mileage? The data file *ex11-23.dat* on the CD that accompanies this book gives the city and highway gas mileages for 29 midsize cars and 26 four-wheel-drive sport utility vehicles from the 1998 model year.[10] The first column specifies the size of the car (1 = midsize car, 2 = SUV), the second column gives the city mileage, and the third gives the highway mileage. Use software to analyze these data.

(a) Make a scatterplot of the data. Use a different plot symbol for midsize cars and SUVs. Is the association positive or negative? Is there an overall linear pattern? How do the two types of vehicle differ in your plot?

(b) There is one clear outlier. Circle it in your plot. This is the Volkswagen Passat with diesel engine. Find the correlation between city and highway mileage both with and without this outlier. Why does r decrease when we remove this point? Because this is the only vehicle with a diesel engine, drop it from the data before doing any more work.

(c) Find the equation of the least-squares regression line for predicting highway mileage from city mileage. Draw the line on your plot from (a).

(d) Find the residuals. The two most extreme residuals are -6.43 and 4.78. Locate and circle on your plot in (a) the data points that produce these residuals. In what way are these vehicles unusual? Make a graph of the distribution of the residuals and a plot of the residuals against city mileage. The distribution is reasonably symmetric (with the two possible outliers we have noted) and the residual plot is acceptable.

(e) Explain the meaning of the slope β of the true regression line in this setting. Give a 95% confidence interval for β.

11.24 Fish sizes. Table 11.3 contains data on the size of perch caught in a lake in Finland.[11] Statistical software will help you analyze these data.

(a) We want to know how well we can predict the width of a perch from its length. Make a scatterplot of width against length. There is a strong linear pattern, as expected. Perch number 143 had six newly eaten fish in its stomach. Find this fish on your scatterplot and circle the point. Is this fish an outlier in your plot of width against length?

(b) Find the least-squares regression line to predict width from length.

(c) The length of a typical perch is about $x^* = 27$ centimeters. Predict the mean width of such fish and give a 95% confidence interval.

(d) Examine the residuals. Is there any reason to mistrust inference? Does fish number 143 have an unusually large residual?

11.25 Fish weights. We can also use the data in Table 11.3 to study the prediction of the weight of a perch from its length.

(a) Make a scatterplot of weight versus length, with length as the explanatory variable. Describe the pattern of the data and any clear outliers.

(b) It is more reasonable to expect the one-third power of the weight to have a straight-line relationship with length than to expect weight itself to have a straight-line relationship with length. Explain why this is true. (Hint: What happens to weight if length, width, and height all double?)

(c) Use your software to create a new variable that is the one-third power of weight. Make a scatterplot of this new response variable against length. Describe the pattern and any clear outliers.

Table 11.3 Measurements on 56 perch

Obs. number	Weight (grams)	Length (cm)	Width (cm)	Obs. number	Weight (grams)	Length (cm)	Width (cm)
104	5.9	8.8	1.4	132	197.0	27.0	4.2
105	32.0	14.7	2.0	133	218.0	28.0	4.1
106	40.0	16.0	2.4	134	300.0	28.7	5.1
107	51.5	17.2	2.6	135	260.0	28.9	4.3
108	70.0	18.5	2.9	136	265.0	28.9	4.3
109	100.0	19.2	3.3	137	250.0	28.9	4.6
110	78.0	19.4	3.1	138	250.0	29.4	4.2
111	80.0	20.2	3.1	139	300.0	30.1	4.6
112	85.0	20.8	3.0	140	320.0	31.6	4.8
113	85.0	21.0	2.8	141	514.0	34.0	6.0
114	110.0	22.5	3.6	142	556.0	36.5	6.4
115	115.0	22.5	3.3	143	840.0	37.3	7.8
116	125.0	22.5	3.7	144	685.0	39.0	6.9
117	130.0	22.8	3.5	145	700.0	38.3	6.7
118	120.0	23.5	3.4	146	700.0	39.4	6.3
119	120.0	23.5	3.5	147	690.0	39.3	6.4
120	130.0	23.5	3.5	148	900.0	41.4	7.5
121	135.0	23.5	3.5	149	650.0	41.4	6.0
122	110.0	23.5	4.0	150	820.0	41.3	7.4
123	130.0	24.0	3.6	151	850.0	42.3	7.1
124	150.0	24.0	3.6	152	900.0	42.5	7.2
125	145.0	24.2	3.6	153	1015.0	42.4	7.5
126	150.0	24.5	3.6	154	820.0	42.5	6.6
127	170.0	25.0	3.7	155	1100.0	44.6	6.9
128	225.0	25.5	3.7	156	1000.0	45.2	7.3
129	145.0	25.5	3.8	157	1100.0	45.5	7.4
130	188.0	26.2	4.2	158	1000.0	46.0	8.1
131	180.0	26.5	3.7	159	1000.0	46.6	7.6

(d) Is the straight-line pattern in (c) stronger or weaker than that in (a)? Compare the plots and also the values of r^2.

(e) Find the least-squares regression line to predict the new weight variable from length. Predict the mean of the new variable for perch 27 centimeters long, and give a 95% confidence interval.

(f) Examine the residuals from your regressions. Does it appear that any of the regression assumptions are not met?

The Thinking Person's Guide to Basic Statistics

We began our study of statistics with a look at "Statistical Thinking." We end with a review in outline form of the most important content of basic statistics, combining statistical thinking with your new knowledge of statistical practice.

STATISTICS IN SUMMARY
Overview of basic inference methods

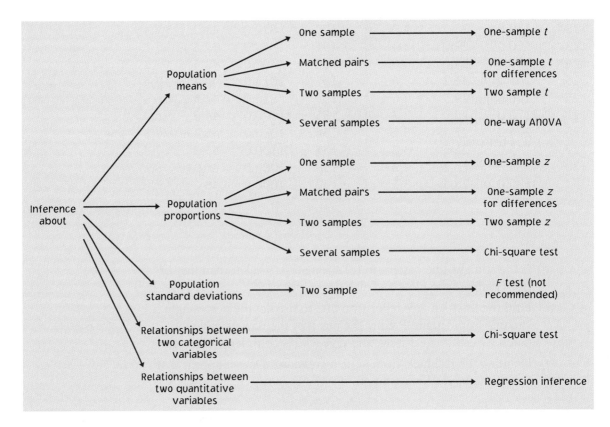

Data Production
- Data basics:

    Individuals (subjects, cases).

    Variables: categorical versus quantitative, units of measurement, explanatory versus response.

    Purpose of study.

- Basic designs:

    Observation versus experiment.

    Simple random samples.

    Completely randomized experiments.

- Inference acts as if your data come from properly randomized data production.

- Really bad data production (voluntary response, confounding) can make interpretation impossible.

- Weaknesses in data production (e.g., sampling students at only one campus) can make generalizing conclusions difficult.

Data Analysis
- Always examine your data before doing inference. Inference often requires a regular pattern and no outliers.

- Start with graphs, look for overall pattern and striking deviations.

- Add numerical descriptions based on what you see.

- One quantitative variable:

    Graphs: stemplot, histogram, boxplot.

    Pattern: distribution shape, center, spread. Outliers?

    Numerical descriptions: five-number summary or $\bar{x}$ and s.

- Relationships between two quantitative variables:

    Graph: scatterplot.

    Pattern: relationship form, direction, strength. Outliers? Influential observations?

    Numerical description for linear relationships: correlation, regression line.

The Reasoning of Inference
- Inference uses data to infer conclusions about a wider population.

- Inference depends on good data production, reasonably regular data pattern (such as roughly normal with no strong outliers).

- Key idea: "What would happen if we did this many times?"

- Confidence intervals: estimate a population parameter.

    95% confidence: I used a method that captures the true parameter 95% of the time in repeated use.

- Tests: assess evidence against H_0 in favor of H_a.

    P-value: If H_0 were true, how often would I get an outcome favoring the alternative this strongly?

    Statistical significance at, e.g., the 5% level: $P < 0.05$, an outcome this extreme would occur less than 5% of the time if H_0 were true.

    Power: If a specific alternative were true, how often would my test produce a significant result?

Inference Procedures

- Population means (quantitative response):

    One sample (or matched pairs): one-sample t procedures.

    Two independent samples: two-sample t procedures.

    Several independent samples: one-way ANOVA.

- Population standard deviations (quantitative response):

    Two independent samples: F test (not recommended).

- Population proportions (categorical response):

    One sample (or matched pairs): one-sample z procedures.

    Two independent samples: two-sample z procedures.

    Several independent samples: chi-square test.

- Relationships between two variables:

    Two categorical variables: chi-square test.

    Two quantitative variables: regression inference.

Preface Notes

1. D. S. Moore and discussants, "New pedagogy and new content: the case of statistics" *International Statistical Review,* 65 (1997), pp. 123–165. Richard Scheaffer's comment appears on page 156.

2. This summary of the committee's report was unanimously endorsed by the Board of Directors of the American Statistical Association. The full report is George Cobb, "Teaching statistics," in L. A. Steen (ed.), *Heeding the Call for Change: Suggestions for Curricular Action,* Mathematical Association of America, Washington, D.C., 1990, pp. 3–43.

3. David S. Moore and George P. McCabe, *Introduction to the Practice of Statistics,* 3rd ed., W. H. Freeman, New York, 1999.

4. J. D. Emerson and G. A. Colditz, "Use of statistical analysis in the *New England Journal of Medicine,"* in *Medical Uses of Statistics,* 2nd ed., J. C. Bailar and F. Mosteller (eds.), NEJM Press, Boston, 1992, pp. 45–57.

Introduction Notes

1. Reported in the *New York Times,* March 25, 1996.

2. E. W. Campion, "Editorial: power lines, cancer, and fear," *New England Journal of Medicine,* 337, No. 1 (1997). The study report is M. S. Linet et al., "Residential exposure to magnetic fields and acute lymphoblastic leukemia in children" in the same issue. I found these on-line at http://www.nejm.org/. See also G. Taubes, "Magnetic field-cancer link: will it rest in peace?" *Science,* 277 (1997), p. 29.

3. These data, from reports submitted by airlines to the Department of Transportation, appear in A. Barnett, "How numbers can trick you," *Technology Review,* October 1994, pp. 38–45.

4. A. C. Nielsen, Jr., "Statistics in marketing," in *Making Statistics More Effective in Schools of Business,* Graduate School of Business, University of Chicago, 1986.

5. The data in Figure 2 are based on a component of the Consumer Price Index, from the Bureau of Labor Statistics on-line data center at http://stats.bls.gov:80/datahome.htm. To convert the index numbers into dollars, I used the national average price of $1.131 per gallon for January 1998, from the Energy Information Agency Web site, http://www.eia.doe.gov/.

6. H. C. Sox, "Editorial: benefit and harm associated with screening for breast cancer," *New England Journal of Medicine,* 338, No. 16 (1998).

Chapter 1 Notes

1. Data for 1995, collected by the Current Population Survey and reported in the 1996 *Statistical Abstract of the United States.* This annual publication is an essential source for many kinds of data.

2. Our eyes do respond to area, but not quite linearly. It appears that we perceive the ratio of two bars to be about the 0.7 power of the ratio of their actual areas. See W. S. Cleveland, *The Elements of Graphing Data,* Wadsworth, Monterey, Calif., 1985, pp. 278–284.

3. Data from U.S. Department of Energy, *Model Year 1998 Fuel Economy Guide,* Washington, D.C., 1997.

4. The Shakespeare data appear in C. B. Williams, *Style and Vocabulary: Numerical Studies,* Griffin, London, 1970.

5. From John K. Ford, "Diversification: how many stocks will suffice?" *American Association of Individual Investors Journal,* January 1990, pp. 14–16.

6. Data from Albert J. Fredman, "A closer look at money market funds," *American Association of Individual Investors Journal,* February 1997, pp. 22–27.

7. Centers for Disease Control and Prevention, *Births and Deaths: Preliminary Data for 1997,* Monthly Vital Statistics Reports, 47, No. 4, 1998.

8. These data were collected by students as a class project.

9. Hurricane data from H. C. S. Thom, *Some Methods of Climatological Analysis,* World Meteorological Organization, Geneva, Switzerland, 1966.

10. These are some of the data from the EESEE story "Influenza Outbreak of 1918," found on the CD that accompanies this book. The data are taken from A. W. Crosby, *America's Forgotten Pandemic: The Influenza of 1918,* Cambridge University Press, New York, 1989.

11. *Consumer Reports,* June 1986, pp. 366–367. A more recent study of hot dogs appears in *Consumer Reports,* July 1993, pp. 415–419. The newer data cover few brands of poultry hot dogs and take calorie counts mainly from the package labels, resulting in suspiciously round numbers.

12. Cavendish's data and the background information about his work appear in S. M. Stigler, "Do robust estimators work with real data?" *Annals of Statistics,* 5 (1977), pp. 1055–1078.

13. Data from Gary Community School Corporation, courtesy of Celeste Foster, Department of Education, Purdue University.

14. Detailed data appear in P. S. Levy et al., "Total serum cholesterol values for youths 12–17 years," *Vital and Health Statistics Series 11,* No. 150 (1975), U.S. National Center for Health Statistics.

15. Data from Stephen Jay Gould, "Entropic homogeneity isn't why no one hits .400 any more," *Discover,* August 1986, pp. 60–66. Gould does not standardize but gives a speculative discussion instead.

16. Ulric Neisser, "Rising scores on intelligence tests," *American Scientist,* September-October 1997, on-line edition.

17. Based on Antoni Basinski, "Almost never on Sunday: implications of the patterns of admission and discharge for common conditions," Institute for Clinical Evaluative Sciences in Ontario, October 18, 1993.

18. Based on data summaries in G. L. Cromwell et al., "A comparison of the nutritive value of *opaque-2, floury-2* and normal corn for the chick," *Poultry Science,* 57 (1968), pp. 840–847.

19. See Note 3.

20. Data from T. Bjerkedal, "Acquisition of resistance in guinea pigs infected with different doses of virulent tubercle bacilli," *American Journal of Hygiene,* 72 (1960), pp. 130–148.

21. Reported in *MacWorld,* September 1996, p. 145.

22. These are some of the data from the EESEE story "Acorn Size and Oak Tree Range," found on the CD that accompanies this book.

Chapter 2 Notes

1. Data provided by Robert Dale, Purdue University.

2. Based on T. N. Lam, "Estimating fuel consumption from engine size," *Journal of Transportation Engineering,* 111 (1985), pp. 339–357. The data for 10 to 50 km/h are measured; those for 60 and higher are calculated from a model given in the paper and are therefore smoothed.

3. A sophisticated treatment of improvements and additions to scatterplots is W. S. Cleveland and R. McGill, "The many faces of a scatterplot," *Journal of the American Statistical Association,* 79 (1984), pp. 807–822.

4. Data provided by Darlene Gordon, Purdue University.

5. Data from *Consumer Reports,* June 1986, pp. 366–367.

6. Data for 1995, from the 1997 *Statistical Abstract of the United States.*

7. Data from M. H. Criqui, University of California, San Diego, reported in the *New York Times,* December 28, 1994.

8. Data from W. L. Colville and D. P. McGill, "Effect of rate and method of planting on several plant characters and yield of irrigated corn," *Agronomy Journal,* 54 (1962), pp. 235–238.

9. Modified from M. C. Wilson and R. E. Shade, "Relative attractiveness of various luminescent colors to the cereal leaf beetle and the meadow spittlebug," *Journal of Economic Entomology,* 60 (1967), pp. 578–580.

10. A careful study of this phenomenon is W. S. Cleveland, P. Diaconis, and R. McGill, "Variables on scatterplots look more highly correlated when the scales are increased," *Science,* 216 (1982), pp. 1138–1141.

11. Data from M. A. Houck et al. "Allometric scaling in the earliest fossil bird, *Archaeopteryx lithographica,*" *Science,* 247 (1990), pp. 195–198. The authors conclude from a variety of evidence that all specimens represent the same species.

12. From a survey by the Wheat Industry Council reported in *USA Today,* October 20, 1983.

13. These are some of the data from the EESEE story "Brain Size and Intelligence," found on the CD that accompanies this book. The study is described in L. Willerman, R. Schultz, J. N. Rutledge, and E. Bigler, "In vivo brain size and intelligence," *Intelligence,* 15 (1991), pp. 223–228.

14. *T. Rowe Price Report,* winter 1997, p. 4.

15. From W. M. Lewis and M. C. Grant, "Acid precipitation in the western United States," *Science,* 207 (1980), pp. 176–177.

16. Data from E. P. Hubble, "A relation between distance and radial velocity among extra-galactic nebulae," *Proceedings of the National Academy of Sciences,* 15 (1929), pp. 168–173.

17. Based on a plot in G. D. Martinsen, E. M. Driebe, and T. G. Whitham, "Indirect interactions mediated by changing plant chemistry: beaver browsing benefits beetles," *Ecology,* 79 (1998), pp. 192–200.

18. These data were originally collected by L. M. Linde of UCLA but were first published by M. R. Mickey, O. J. Dunn, and V. Clark, "Note on the use of stepwise regression in detecting outliers," *Computers and Biomedical Research,* 1 (1967), pp. 105–111. The data have been used by several authors. I found them in N. R. Draper and J. A. John, "Influential observations and outliers in regression," *Technometrics,* 23 (1981), pp. 21–26.

19. From the *Philadelphia City Paper,* May 23–29, 1997. Because the sodas served vary in size, I have converted soda prices to the price of a 16-ounce soda at each price per ounce.

20. From a presentation by Charles Knauf, Monroe County (New York) Environmental Health Laboratory.

21. The U.S. returns are for the Standard & Poor's 500-stock index. The overseas returns are for the Morgan Stanley Europe, Australasia, Far East (EAFE) index.

22. Frank J. Anscombe, "Graphs in statistical analysis," *The American Statistician,* 27 (1973), pp. 17–21.

23. Gary Smith, "Do statistics test scores regress toward the mean?" *Chance,* 10, No. 4 (1997), pp. 42–45.

24. Data provided by Peter Cook, Purdue University.

25. The quotation is from Dr. Daniel Mark of Duke University, in an article "Age, not bias, may explain differences in treatment," *New York Times,* April 26, 1994.

26. This example is drawn from M. Goldstein, "Preliminary inspection of multivariate data," *The American Statistician,* 36 (1982), pp. 358–362.

27. L. E. Moses and F. Mosteller, "Safety of anesthetics," in J. Tanur et al. (eds.), *Statistics: A Guide to the Unknown,* 3rd ed., Wadsworth, Belmont, Calif., 1989, pp. 15–24.

28. *The Health Consequences of Smoking: 1983,* U.S. Public Health Service, Washington, D.C., 1983.

29. From a Gannett News Service article appearing in the Lafayette, Indiana, *Journal and Courier,* April 23, 1994.

30. See Gary Taubes, "Magnetic field-cancer link: will it rest in peace?" *Science,* 277 (1997), p. 29.

31. From S. V. Zagona (ed.), *Studies and Issues in Smoking Behavior,* University of Arizona Press, Tucson, 1967, pp. 157–180.

32. R. Shine, T. R. L. Madsen, M. J. Elphick, and P. S. Harlow, "The influence of nest temperatures and maternal brooding on hatchling phenotypes in water pythons," *Ecology,* 78 (1997), pp. 1713–1721.

33. From F. D. Blau and M. A. Ferber, "Career plans and expectations of young women and men," *Journal of Human Resources,* 26 (1991), pp. 581–607.

34. These data, from reports submitted by airlines to the Department of Transportation, appear in A. Barnett, "How numbers can trick you," *Technology Review,* October 1994, pp. 38–45.

35. From M. Radelet, "Racial characteristics and imposition of the death penalty," *American Sociological Review,* 46 (1981), pp. 918–927.

36. From *Digest of Education Statistics 1997,* accessed on the National Center for Education Statistics Web site, http://www.ed.gov/NCES.

37. S. W. Hargarten et al., "Characteristics of firearms involved in fatalities," *Journal of the American Medical Association,* 275 (1996), pp. 42–45.

38. From D. M. Barnes, "Breaking the cycle of addiction," *Science,* 241 (1988), pp. 1029–1030.

39. See P. J. Bickel and J. W. O'Connell, "Is there a sex bias in graduate admissions?" *Science,* 187 (1975), pp. 398–404.

40. Modified from data provided by Dana Quade, University of North Carolina.

41. Data provided by Matthew Moore.

42. These are some of the data from the EESEE story "Influenza Outbreak of 1918," found on the CD that accompanies this book. The data are taken from A. W. Crosby, *America's Forgotten Pandemic: The Influenza of 1918,* Cambridge University Press, New York, 1989.

43. National Science Board, *Science and Engineering Indicators, 1991,* U.S. Government Printing Office, Washington, D.C., 1991. The detailed data appear in Appendix Table 3-5, p. 274.

44. Data from many studies compiled in D. F. Greene and E. A. Johnson, "Estimating the mean annual seed production of trees," *Ecology,* 75 (1994), pp. 642–647.

45. P. Velleman, *ActivStats 2.0,* Addison-Wesley Interactive, Reading, Mass., 1997.

46. Reported in the *New York Times,* July 20, 1989, from an article appearing that day in the *New England Journal of Medicine.*

47. Condensed from D. R. Appleton, J. M. French, and M. P. J. Vanderpump, "Ignoring a covariate: an example of Simpson's paradox," *The American Statistician,* 50 (1996), pp. 340–341.

Chapter 3 Notes

1. Reported by D. Horvitz in his contribution to "Pseudo-opinion polls: SLOP or useful data?" *Chance,* 8, No. 2 (1995), pp. 16–25.

2. Based in part on Randall Rothenberger, "The trouble with mall interviewing," *New York Times,* August 16, 1989.

3. The information in this example is taken from *The ASCAP Survey and Your Royalties,* ASCAP, New York, undated.

4. This exercise is based on the EESEE story "What Makes a Pre-Teen Popular?" found on the CD that accompanies this book.

5. For more detail on the material of this section, along with references, see P. E. Converse and M. W. Traugott, "Assessing the accuracy of polls and surveys," *Science,* 234 (1986), pp. 1094–1098.

6. The estimates of the census undercount come from Howard Hogan, "The 1990 post-enumeration survey: operations and results," *Journal of the American Statistical Association,* 88 (1993), pp. 1047–1060. The information about nonresponse appears in Eugene P. Eriksen and Teresa K. DeFonso, "Beyond the net undercount: how to measure census error," *Chance,* 6, No. 4 (1993), pp. 38–43 and 14.

7. For more detail on the limits of memory in surveys, see N. M. Bradburn, L. J. Rips, and S. K. Shevell, "Answering autobiographical questions: the impact of memory and inference on surveys," *Science,* 236 (1987), pp. 157–161.

8. Cynthia Crossen, "Margin of error: studies galore support products and positions, but are they reliable?" *Wall Street Journal,* November 14, 1991.

9. M. R. Kagay, "Poll on doubt of Holocaust is corrected," *New York Times,* July 8, 1994.

10. Giuliana Coccia, "An overview of non-response in Italian telephone surveys," *Proceedings of the 99th Session of the International Statistical Institute, 1993,* Book 3, pp. 271–272.

11. W. Mitofsky, "Mr. Perot, you're no pollster," *New York Times,* March 27, 1993.

12. From the *New York Times,* August 21, 1989.

13. Simplified from Arno J. Rethans, John L. Swasy, and Lawrence J. Marks, "Effects of television commercial repetition, receiver knowledge, and commercial length: a test of the two-factor model," *Journal of Marketing Research,* 23 (February 1986), pp. 50–61.

14. L. L. Miao, "Gastric freezing: an example of the evaluation of medical therapy by randomized clinical trials," in J. P. Bunker, B. A. Barnes, and F. Mosteller (eds.), *Costs, Risks and Benefits of Surgery,* Oxford University Press, New York, 1977, pp. 198–211.

15. Based on Christopher Anderson, "Measuring what works in health care," *Science,* 263 (1994), pp. 1080–1082.

16. Taken from "Advertising: the cola war," *Newsweek,* August 30, 1976, p. 67.

17. This exercise is based on the EESEE story "Blinded Knee Doctors" found on the CD that accompanies this book. The study is reported in W. E. Nelson, R. C. Henderson, L.C. Almekinders, R. A. DeMasi, and T. N. Taft, "An evaluation of pre- and postoperative nonsteroidal antiinflammatory drugs in patients undergoing knee arthroscopy," *Journal of Sports Medicine,* 21 (1994), pp. 510–516.

18. Information from Warren McIsaac and Vivek Goel, "Is access to physician services in Ontario equitable?" Institute for Clinical Evaluative Sciences in Ontario, October 18, 1993.

19. The study is described in G. Kolata, "New study finds vitamins are not cancer preventers," *New York Times,* July 21, 1994. Look in the *Journal of the American Medical Association* of the same date for the details.

20. Steering Committee of the Physicians' Health Study Research Group, "Final report on the aspirin component of the ongoing Physicians' Health Study," *New England Journal of Medicine,* 321 (1989), pp. 129–135.

Chapter 4 Notes

1. U.S. Bureau of the Census, Current Population Reports, P60-200, *Money Income in the United States, 1997.* Government Printing Office, Washington, D.C.,1998.

2. From the EESEE story "Home-Field Advantage," found on the CD that accompanies this book. The study is W. Hurley, "What sort of tournament should the World Series be?" *Chance,* 6, No. 2 (1993), pp. 31–33.

3. Strictly speaking, the recipe $\sigma/\sqrt{n}$ for the standard deviation of $\bar{x}$ assumes that we draw an SRS of size n from an *infinite* population. If the population has finite size N, the standard deviation in the recipe is multiplied by $\sqrt{1 - (n - 1)/(N - 1)}$. This "finite population correction" approaches 1 as N increases. When the population is at least 10 times as large as the sample, the correction factor is between about 0.95 and 1. It is reasonable to use the simpler form $\sigma/\sqrt{n}$ in these settings.

Chapter 5 Notes

1. A. Tversky and D. Kahneman, "Extensional versus intuitive reasoning: the conjunction fallacy in probability judgment," *Psychological Review,* 90 (1983), pp. 293–315.

2. This exercise is based on the EESEE story "Anecdotes of Probability," found on the CD that accompanies this book. The statements about craps appear in J. Gollehon, *Pay the Line!* Putnam, New York, 1988.

3. The survey question is reported in Trish Hall, "Shop? Many say 'Only if I must,' " *New York Times,* November 28, 1990. In fact, 66% (1650 of 2500) in the sample said "Agree."

4. Office of Technology Assessment, *Scientific Validity of Polygraph Testing: A Research Review and Evaluation,* Government Printing Office, Washington, D.C., 1983.

5. These probabilities come from studies by the sociologist Harry Edwards, reported in the *New York Times,* February 25, 1986.

6. These probabilities are estimated from a large national study reported in E. M. Sloand et al., "HIV testing: state of the art," *Journal of the American Medical Association,* 266 (1991), pp. 2861–2866.

Chapter 6 Notes

1. Information from Francisco L. Rivera-Batiz, "Quantitative literacy and the likelihood of employment among young adults," *Journal of Human Resources,* 27 (1992), pp. 313–328.

2. Data provided by Darlene Gordon, Purdue University.

3. P. H. Lewis, "Technology" column, *New York Times,* May 29, 1995.

4. Data provided by John Rousselle and Huei-Ru Shieh, Department of Restaurant, Hotel, and Institutional Management, Purdue University.

5. Data provided by Mugdha Gore and Joseph Thomas, Purdue University School of Pharmacy.

6. Data provided by Drina Iglesia, Department of Biological Sciences, Purdue University. The data are part of a larger study reported in D. D. S. Iglesia, E. J. Cragoe, Jr., and J. W. Vanable, "Electric field strength and epithelization in the newt (*Notophthalmus viridescens*)," *Journal of Experimental Zoology,* 274 (1996), pp. 56–62.

7. Based on G. Salvendy, G. P. McCabe, S. G. Sanders, J. L. Knight, and E. J. McCormick, "Impact of personality and intelligence on job satisfaction of assembly line and bench work—an industrial study," *Applied Ergonomics,* 13 (1982), pp. 293–299.

8. From a study by M. R. Schlatter et al., Division of Financial Aid, Purdue University.

9. See Note 2.

10. William T. Robinson, "Sources of market pioneer advantages: the case of industrial goods industries," *Journal of Marketing Research,* 25 (February 1988), pp. 87–94.

11. Michael B. Maziz et al., "Perceived age and attractiveness of models in cigarette advertisements," *Journal of Marketing,* 56 (January 1992), pp. 22–37.

12. For a discussion of statistical significance in the legal setting, see D. H. Kaye, "Is proof of statistical significance relevant?" *Washington Law Review,* 61 (1986), pp. 1333–1365. Kaye argues: "Presenting the P-value without characterizing the evidence by a significance test is a step in the right direction. Interval estimation, in turn, is an improvement over P-values."

13. These are part of the data from the EESEE story "Radar Detectors and Speeding," found on the CD that accompanies this book. The study is reported in N. Teed, K. L. Adrian, and R. Knoblouch, "The duration of speed reductions attributable to radar detectors," *Accident Analysis and Prevention,* 25 (1991), pp. 131–137.

14. Robert J. Schiller, "The volatility of stock market prices," *Science,* 235 (1987), pp. 33–36.

Chapter 7 Notes

1. This example is based on information in D. L. Shankland et al., "The effect of 5-thio-D-glucose on insect development and its absorption by insects," *Journal of Insect Physiology,* 14 (1968), pp. 63–72.

2. Data from D. L. Shankland, "Involvement of spinal cord and peripheral nerves in DDT-poisoning syndrome in albino rats," *Toxicology and Applied Pharmacology,* 6 (1964), pp. 197–213.

3. This is a simplified version of the EESEE story "Floral Scents and Learning," found on the CD that accompanies this book. The study is reported in A. R. Hirsch and L. H. Johnston, "Odors and learning," *Journal of Neurological and Orthopedic Medicine and Surgery,* 17 (1996), pp. 119–126.

4. Data provided by Timothy Sturm.

5. These recommendations are based on extensive computer work. See, for example, Harry O. Posten, "The robustness of the one-sample *t*-test over the Pearson system," *Journal of Statistical Computation and Simulation,* 9 (1979), pp. 133–149, and E. S. Pearson and N. W. Please, "Relation between the shape of population distribution and the robustness of four simple test statistics," *Biometrika,* 62 (1975), pp. 223–241.

6. These are part of the data from the EESEE story "Is Caffeine Dependence Real?" found on the CD that accompanies this book. The study results appear in E. C. Strain, G. K. Mumford, K. Silverman, and R. R. Griffiths, "Caffeine dependence syndrome: evidence from case histories and experimental evaluation," *Journal of the American Medical Association,* 272 (1994), pp. 1604–1607.

7. Data provided by Chris Olsen, who found the information in scuba diving magazines.

8. Data from the appendix of D. A. Kurtz (ed.), *Trace Residue Analysis,* American Chemical Society Symposium Series, No. 284, 1985.

9. From S. C. Hand and E. Gnaiger, "Anaerobic dormancy quantified in *Artemia* embryos," *Science,* 239 (1988), pp. 1425–1427.

10. Based on I. Cuellar, L. C. Harris, and R. Jasso, "An acculturation scale for Mexican American normal and clinical populations," *Hispanic Journal of Behavioral Sciences,* 2 (1980), pp. 199–217.

11. Data provided by Drina Iglesia, Department of Biological Sciences, Purdue University. The study results are reported in D. D. S. Iglesia, E. J. Cragoe, Jr., and J. W. Vanable, "Electric field strength and epithelization in the newt (*Notophthalmus viridescens*)," *Journal of Experimental Zoology,* 274 (1996), pp. 56–62.

12. Based on Charles W. L. Hill and Phillip Phan, "CEO tenure as a determinant of CEO pay," *Academy of Management Journal,* 34 (1991), pp. 707–717.

13. Data from Lianng Yuh, "A biopharmaceutical example for undergraduate students," unpublished manuscript.

14. From Sapna Aneja, "Biodeterioration of textile fibers in soil," M. S. thesis, Purdue University, 1994.

15. Detailed information about the conservative t procedures can be found in Paul Leaverton and John J. Birch, "Small sample power curves for the two sample location problem," *Technometrics,* 11 (1969), pp. 299–307; in Henry Scheffé, "Practical solutions of the Behrens-Fisher problem," *Journal of the American Statistical Association,* 65 (1970), pp. 1501–1508; and in D. J. Best and J. C. W. Rayner, "Welch's approximate solution for the Behrens-Fisher problem," *Technometrics,* 29 (1987), pp. 205–210.

16. F. Tagliavini et al., "Effectiveness of anthracycline against experimental prion disease in Syrian hamsters," *Science,* 276 (1997), pp. 1119–1121.

17. From Costas Papoulias and Panayiotis Theodossiou, "Analysis and modeling of recent business failures in Greece," *Managerial and Decision Economics,* 13 (1992), pp. 163–169.

18. From H. G. Gough, *The Chapin Social Insight Test,* Consulting Psychologists Press, Palo Alto, Calif., 1968.

19. Data provided by Charles Cannon, Duke University. The study report is C. H. Cannon, D. R. Peart, and M. Leighton, "Tree species diversity in commercially logged Bornean rainforest," *Science,* 281 (1998), pp. 1366–1367.

20. Data provided by Warren Page, New York City Technical College, from a study done by John Hudesman.

21. Based on M. C. Wilson et al., "Impact of cereal leaf beetle larvae on yields of oats," *Journal of Economic Entomology,* 62 (1969), pp. 699–702.

22. See the extensive simulation studies in Harry O. Posten, "The robustness of the two-sample t-test over the Pearson system," *Journal of Statistical Computation and Simulation,* 6 (1978), pp. 295–311, and in Harry O. Posten, H. Yeh, and Donald B. Owen, "Robustness of the two-sample t-test under violations of the homogeneity assumption," *Communications in Statistics,* 11 (1982), pp. 109–126.

23. Based on G. L. Cromwell et al., "A comparison of the nutritive value of *opaque-2, floury-2* and normal corn for the chick," *Poultry Science,* 47 (1968), pp. 840–847.

24. We did not use Minitab or Data Desk in Example 7.10 because these packages shortcut the two-sample t procedure. They calculate the degrees of freedom df using the formula in the box on page 403 but then truncate to the next lower whole-number degrees of freedom to obtain the P-value. The result is slightly less accurate than the P-value from the t(df) distribution.

25. This exercise is loosely based on D. L. Shankland, "Involvement of spinal cord and peripheral nerves in DDT-poisoning syndrome in albino rats," *Toxicology and Applied Pharmacology,* 6 (1964), pp. 197–213.

26. Data provided by Darlene Gordon, School of Education, Purdue University.

27. From a news article in *Science,* 224 (1983), pp. 1029–1031.

28. John R. Cronin and Sandra Pizzarello, "Enantiometric excesses in meteoritic amino acids," *Science,* 275 (1997), pp. 951–955.

29. Data from Orit E. Hetzroni, "The effects of active versus passive computer-assisted instruction on the acquisition, retention, and generalization of Blissymbols while

using elements for teaching compounds," Ph.D. thesis, Purdue University, 1995.

30.   See Note 26.

31.   From A. H. Ismail and R. J. Young,"The effect of chronic exercise on the personality of middle-aged men," *Journal of Human Ergology,* 2 (1973), pp. 47–57.

32.   From M. A. Zlatin and R. A. Koenigsknecht, "Development of the voicing contrast: a comparison of voice onset time in stop perception and production," *Journal of Speech and Hearing Research,* 19 (1976), pp. 93–111.

33.   The problem of comparing spreads is difficult even with advanced methods. Common distribution-free procedures do not offer a satisfactory alternative to the *F* test, because they are sensitive to unequal shapes when comparing two distributions. A good introduction to the available methods is W. J. Conover, M. E. Johnson, and M. M. Johnson, "A comparative study of tests for homogeneity of variances, with applications to outer continental shelf bidding data," *Technometrics,* 23 (1981), pp. 351–361. Modern resampling procedures often work well. See Dennis D. Boos and Colin Brownie, "Bootstrap methods for testing homogeneity of variances," *Technometrics,* 31 (1989), pp. 69–82.

34.   This is a partial and simplified version of the EESEE story "Stepping Up Your Heart Rate," found on the CD that accompanies this book.

35.   Based loosely on D. R. Black et al., "Minimal interventions for weight control: a cost-effective alternative," *Addictive Behaviors,* 9 (1984), pp. 279–285.

36.   Based on Amna Kirmani and Peter Wright, "Money talks: perceived advertising expense and expected product quality," *Journal of Consumer Research,* 16 (1989), pp. 344–353.

37.   Based on D. Friedlander, *Supplemental Report on the Baltimore Options Program,* Manpower Demonstration Research Corporation, New York, 1987.

38.   Data from R. R. Robinson, C. L. Rhykerd, and C. F. Gross, "Potassium uptake by orchardgrass as affected by time, frequency and rate of potassium fertilization," *Agronomy Journal,* 54 (1962). pp. 351–353.

39.   Data from R. A. Berner and G. P. Landis, "Gas bubbles in fossil amber as possible indicators of the major gas composition of ancient air," *Science,* 239 (1988), pp. 1406–1409.

40.   From J. W. Marr and J. A. Heady, "Within- and between-person variation in dietary surveys: number of days needed to classify individuals," *Human Nutrition: Applied Nutrition,* 40A (1986), pp. 347–364.

41.   From V. D. Bass, W. E. Hoffmann, and J. L. Dorner, "Normal canine lipid profiles and effects of experimentally induced pancreatitis and hepatic necrosis on lipids," *American Journal of Veterinary Research,* 37 (1976), pp. 1355–1357.

42.   G. S. Hotamisligil, R. S. Johnson, R. J. Distel, R. Ellis, V. E. Papaioannou, and B. M. Spiegelman, "Uncoupling of obesity from insulin resistance through a targeted mutation in *aP*2, the adipocyte fatty acid binding protein," *Science,* 274 (1996), pp. 1377–1379.

43.   Data provided by Matthew Moore.

Chapter 8 Notes

1. Data from Joseph H. Catania et al., "Prevalence of AIDS-related risk factors and condom use in the United States," *Science,* 258 (1992), pp. 1101–1106.

2. The study is reported in William Celis III, "Study suggests Head Start helps beyond school," *New York Times,* April 20, 1993.

3. Strictly speaking, the recipe $\sqrt{p(1-p)/n}$ for the standard deviation of $\hat{p}$ assumes that an SRS of size n is drawn from an *infinite* population. If the population has finite size N, this standard deviation is multiplied by $\sqrt{1-(n-1)/(N-1)}$. This "finite population correction" approaches 1 as N increases. When the population is at least 10 times as large as the sample, the correction factor is between 0.95 and 1 and can be ignored in practice.

4. For the state of the art in confidence intervals for p, see Alan Agresti and Brent Coull, "Approximate is better than 'exact' for interval estimation of binomial proportions," *The American Statistician,* 52 (1998), pp. 119–126. The authors note that the accuracy of the usual confidence interval for p can be greatly improved by simply "adding 2 successes and 2 failures." That is, replace $\hat{p}$ by

$$\frac{\text{count of successes} + 2}{n + 4}$$

   If this substitution is used, no rules of thumb on $n\hat{p}$ are needed. I have not used this improvement in the text because there is no easily understood explanation for it. Note also that the rule requiring np and $n(1-p)$ to be at least 5, given in many other texts, is *not* adequate for reasonable accuracy.

5. The quotation is from page 1104 of the article cited in Note 1.

6. "Poll: men, women at odds on sexual equality," Associated Press dispatch appearing in the Lafayette, Indiana, *Journal and Courier,* October 20, 1997.

7. Laurie Goodstein and Marjorie Connelly, "Teen-age poll finds support for tradition," *New York Times,* April 30, 1998.

8. Sara J. Solnick and David Hemenway, "Complaints and disenrollment at a health maintenance organization," *The Journal of Consumer Affairs,* 26 (1992), pp. 90–103.

9. Arne L. Kalleberg and Kevin T. Leicht, "Gender and organizational performance: determinants of small business survival and success," *The Academy of Management Journal,* 34 (1991), pp. 136–161.

10. Modified from Richard A. Schieber et al., "Risk factors for injuries from in-line skating and the effectiveness of safety gear," *New England Journal of Medicine,* 335 (1996), Internet summary.

11. Clive G. Jones, Richard S. Ostfeld, Michele P. Richard, Eric M. Schauber, and Jerry O. Wolf, "Chain reactions linking acorns to gypsy moth outbreaks and Lyme disease risk," *Science,* 279 (1998), pp. 1023–1026.

12. The Detroit Area Study is described in Gerhard Lenski, *The Religious Factor,* Doubleday, New York, 1961.

13. Data courtesy of Raymond Dumett, Department of History, Purdue University.

14. Martin Enserink, "Fraud and ethics charges hit stroke drug trial," *Science,* 274 (1996), pp. 2004–2005.

15. Donna L. Hoffman and Thomas P. Novak, "Bridging the racial divide on the Internet," *Science,* 280 (1998), pp. 390–391.

16. These are part of the data from the EESEE story "Leave Survey after the Beep," found on the CD that accompanies this book. The study is reported in M. Xu, B. J. Bates, and J. C. Schweitzer, "The impact of messages on survey participation in answering machine households," *Public Opinion Quarterly,* 57 (1993), pp. 232–237.

17. Data from Richard M. Felder et al., "Who gets it and who doesn't: a study of student performance in an introductory chemical engineering course," *1992 ASEE Annual Conference Proceedings,* American Society for Engineering Education, Washington, D.C., 1992, pp. 1516–1519.

18. Based on Greg Clinch, "Employee compensation and firms' research and development activity," *Journal of Accounting Research,* 29 (1991), pp. 59–78.

19. Giuliana Coccia, "An overview of non-response in Italian telephone surveys," *Proceedings of the 99th Session of the International Statistical Institute,* 1993, Book 3, pp. 271–272.

20. Francisco L. Rivera-Batiz, "Quantitative literacy and the likelihood of employment among young adults," *Journal of Human Resources,* 27 (1992), pp. 313–328.

21. David M. Blau, "The child care labor market," *Journal of Human Resources,* 27 (1992), pp. 9–39.

22. These are part of the data from the EESEE story "Radar Detectors and Speeding," found on the CD that accompanies this book. The study is reported in N. Teed, K. L. Adrian, and R. Knoblouch, "The duration of speed reductions attributable to radar detectors," *Accident Analysis and Prevention,* 25 (1991), pp. 131–137.

23. National Athletic Trainers Association, press release dated September 30, 1994. The study was to be published in the *Journal of Athletic Training.*

24. Office of the Registrar, Purdue University, *Summary of the NCAA Graduation-Rates Disclosure Forms,* West Lafayette, Ind., 1997.

25. Jane E. Brody, "Alternative medicine makes inroads," *New York Times,* April 28, 1998.

26. Eric Ossiander, letter to the editor, *Science,* 257 (1992), p. 1461.

27. A. C. Allison and D. F. Clyde, "Malaria in African children with deficient erythrocyte dehydrogenase," *British Medical Journal,* 1 (1961), pp. 1346–1349.

28. Noel Capon et al., "A comparative analysis of the strategy and structure of United States and Australian corporations," *Journal of International Business Studies,* 18 (1987), pp. 51–74.

Chapter 9 Notes

1. Stephen M. Stigler, *The History of Statistics: The Measurement of Uncertainty before 1900,* Belknap Press, Cambridge, Mass., 1986. The quotation is from p. 361.

2. D. M. Barnes, "Breaking the cycle of addiction," *Science,* 241 (1988), pp. 1029–1030.

3. Richard M. Felder et al., "Who gets it and who doesn't: a study of student performance in an introductory chemical engineering course," *1992 ASEE Annual Conference Proceedings,* American Society for Engineering Education, Washington, D.C., 1992, pp. 1516–1519.

4. S. V. Zagona (ed.), *Studies and Issues in Smoking Behavior,* University of Arizona Press, Tucson, 1967, pp. 157–180.

5. Sanders Korenman and David Neumark, "Does marriage really make men more productive?" *Journal of Human Resources,* 26 (1991), pp. 282–307.

6. There are many computer studies of the accuracy of chi-square critical values for X^2. For a brief discussion and some references, see Section 3.2.5 of David S. Moore, "Tests of chi-squared type," in Ralph B. D'Agostino and Michael A. Stephens (eds.), *Goodness-of-Fit Techniques,* Marcel Dekker, New York, 1986, pp. 63–95. If the expected cell counts are roughly equal, the chi-square approximation is adequate when the average expected counts are as small as 1 or 2. The guideline given in the text protects against unequal expected counts. For a survey of inference for smaller samples, see Alan Agresti, "A survey of exact inference for contingency tables," *Statistical Science,* 7 (1992), pp. 131–177.

7. Brenda C. Coleman, "Study: heart attack risk cut 74% by stress management," Associated Press dispatch appearing in the Lafayette, Indiana, *Journal and Courier,* October 20, 1997.

8. Lillian Lin Miao, "Gastric freezing: an example of the evaluation of medical therapy by randomized clinical trials," in John P. Bunker, Benjamin A. Barnes, and Frederick Mosteller (eds.), *Costs, Risks, and Benefits of Surgery,* Oxford University Press, New York, 1977, pp. 198–211.

9. R. Shine, T. R. L. Madsen, M. J. Elphick, and P. S. Harlow, "The influence of nest temperatures and maternal brooding on hatchling phenotypes in water pythons," *Ecology,* 78 (1997), pp. 1713–1721.

10. Claudia Braga et al., "Olive oil, other seasoning fats, and the risk of colorectal carcinoma," *Cancer,* 82 (1998), pp. 448–453.

11. Martin Enserink, "Fraud and ethics charges hit stroke drug trial," *Science,* 274 (1996), pp. 2004–2005.

12. Francine D. Blau and Marianne A. Ferber, "Career plans and expectations of young women and men," *Journal of Human Resources,* 26 (1991), pp. 581–607.

13. Erdener Kaynak and Wellington Kang-yen Kuan, "Environment, strategy, structure, and performance in the context of export activity: an empirical study of Taiwanese manufacturing firms," *Journal of Business Research,* 27 (1993), pp. 33–49.

14. William D. Darley, "Store-choice behavior for pre-owned merchandise," *Journal of Business Research,* 27 (1993), pp. 17–31.

15. David M. Blau, "The child care labor market," *Journal of Human Resources,* 27 (1992), pp. 9–39.

16. Adapted from M. A. Visintainer, J. R. Volpicelli, and M. E. P. Seligman, "Tumor rejection in rats after inescapable or escapable shock," *Science,* 216 (1982), pp. 437–439.

17. James R. Walker, "New evidence on the supply of child care," *Journal of Human Resources,* 27 (1991), pp. 40–69.

18. Modified from Felicity Barringer, "Measuring sexuality through polls can be shaky," *New York Times,* April 25, 1993.

19. Giuliana Coccia, "An overview of non-response in Italian telephone surveys," *Proceedings of the 99th Session of the International Statistical Institute,* 1993, Book 3, pp. 271–272.

20. Data reported by Robert J. M. Dawson, "The 'unusual episode' data revisited," *Journal of Statistics Education,* 3, No. 3 (1995). Electronic journal available at the American Statistical Association Web site, www.amstat.org.

21. These are part of the data from the EESEE story "Alcoholism in Twins," found on the CD that accompanies this book. The study results appear in K. S. Kendler et al., "A population-based twin study of alcoholism in women," *Journal of the American Medical Association,* 268 (1992), pp. 1877–1882.

Chapter 10 Notes

1. The data in Table 10.1 are from the Environmental Protection Agency's *Model Year 1998 Fuel Economy Guide,* available on-line at www.epa.gov. The table gives data for the basic engine/transmission combination for each model. Models that are essentially identical (such as the Ford Taurus and Mercury Sable) appear only once.

2. These are part of the data from the EESEE story "Stress among Pets and Friends" found on the CD that accompanies this book. The study results appear in K. Allen, J. Blascovich, J. Tomaka, and R. M. Kelsey, "Presence of human friends and pet dogs as moderators of autonomic responses to stress in women," *Journal of Personality and Social Psychology,* 83 (1988), pp. 582–589.

3. Simplified from W. L. Colville and D. P. McGill, "Effect of rate and method of planting on several plant characters and yield of irrigated corn," *Agronomy Journal,* 54 (1962), pp. 235–238.

4. Modified from M. C. Wilson and R. E. Shade, "Relative attractiveness of various luminescent colors to the cereal leaf beetle and the meadow spittlebug," *Journal of Economic Entomology,* 60 (1967), pp. 578–580.

5. Sanders Korenman and David Neumark, "Does marriage really make men more productive?" *Journal of Human Resources,* 26 (1991), pp. 282–307.

6. Patricia F. Campbell and George P. McCabe, "Predicting the success of freshmen in a computer science major," *Communications of the ACM,* 27 (1984), pp. 1108–1113.

7. R. Shine, T. R. L. Madsen, M. J. Elphick, and P. S. Harlow, "The influence of nest temperatures and maternal brooding on hatchling phenotypes in water pythons," *Ecology,* 78 (1997), pp. 1713–1721.

8. From Sapna Aneja, "Biodeterioration of textile fibers in soil," M.S. thesis, Purdue University, 1994.

9. The data and the full story can be found in the Data and Story Library at http://lib.stat.cmu.edu/. The original study is by Faith Loven, "A study of

interlist equivalency of the CID W-22 word list presented in quiet and in noise," M.S. thesis, University of Iowa, 1981.

10. Data provided by Charles Cannon, Duke University. The study report is C. H. Cannon, D. R. Peart, and M. Leighton, "Tree species diversity in commercially logged Bornean rainforest," *Science,* 281 (1998), pp. 1366–1367.

11. Data provided by Matthew Moore.

12. Mark Kroll, Peter Wright, and Pochera Theerathorn, "Whose interests do hired managers pursue? An examination of select mutual and stock life insurers," *Journal of Business Research,* 26 (1993), pp. 133–148.

Chapter 11 Notes

1. Samuel Karelitz et al., "Relation of crying activity in early infancy to speech and intellectual development at age three years," *Child Development,* 35 (1964), pp. 769–777.

2. From M. A. Houck et al., "Allometric scaling in the earliest fossil bird, *Archaeopteryx lithographica,*" *Science,* 247 (1990), pp. 195–198.

3. Data provided by Robert Dale, Purdue University.

4. Based on Marion E. Dunshee, "A study of factors affecting the amount and kind of food eaten by nursery school children," *Child Development,* 2 (1931), pp. 163–183. This article gives the means, standard deviations, and correlation for 37 children but does not give the actual data.

5. These are part of the data from the EESEE story "Blood Alcohol Content," found on the CD that accompanies this book.

6. Strictly speaking, this quantity is the estimated standard deviation of $\hat{y} - y$, where y is the additional observation taken at $x = x^*$.

7. The data are for 1987 salaries and measures of past performance. They were collected and distributed by the Statistical Graphics Section of the American Statistical Association for an annual data analysis contest. The analysis here was done by Crystal Richard of Purdue University.

8. Based on a plot in G. D. Martinsen, E. M. Driebe, and T. G. Whitham, "Indirect interactions mediated by changing plant chemistry: beaver browsing benefits beetles," *Ecology,* 79 (1998), pp. 192–200.

9. Data provided by Samuel Phillips, Purdue University.

10. From the Environmental Protection Agency's *Model Year 1998 Fuel Economy Guide,* available on-line at www.epa.gov.

11. The data in Table 11.3 are part of a larger data set in the *Journal of Statistics Education* archive, accessible via the Internet. The original source is Pekka Brofeldt, "Bidrag till kaennedom on fiskbestondet i vaara sjoear. Laengelmaevesi," in T. H. Jaervi, *Finlands Fiskeriet,* Band 4, *Meddelanden utgivna av fiskerifoereningen i Finland,* Helsinki, 1917. The data were contributed to the archive (with information in English) by Juha Puranen of the University of Helsinki.

APPENDIX

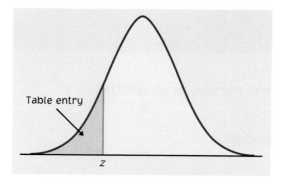

Table entry for z is the area under the standard normal curve left of z.

TABLE A Standard normal probabilities

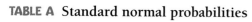

z	.00	.01	.02	.03	.04	.05	.06	.07	.08	.09
−3.4	.0003	.0003	.0003	.0003	.0003	.0003	.0003	.0003	.0003	.0002
−3.3	.0005	.0005	.0005	.0004	.0004	.0004	.0004	.0004	.0004	.0003
−3.2	.0007	.0007	.0006	.0006	.0006	.0006	.0006	.0005	.0005	.0005
−3.1	.0010	.0009	.0009	.0009	.0008	.0008	.0008	.0008	.0007	.0007
−3.0	.0013	.0013	.0013	.0012	.0012	.0011	.0011	.0011	.0010	.0010
−2.9	.0019	.0018	.0018	.0017	.0016	.0016	.0015	.0015	.0014	.0014
−2.8	.0026	.0025	.0024	.0023	.0023	.0022	.0021	.0021	.0020	.0019
−2.7	.0035	.0034	.0033	.0032	.0031	.0030	.0029	.0028	.0027	.0026
−2.6	.0047	.0045	.0044	.0043	.0041	.0040	.0039	.0038	.0037	.0036
−2.5	.0062	.0060	.0059	.0057	.0055	.0054	.0052	.0051	.0049	.0048
−2.4	.0082	.0080	.0078	.0075	.0073	.0071	.0069	.0068	.0066	.0064
−2.3	.0107	.0104	.0102	.0099	.0096	.0094	.0091	.0089	.0087	.0084
−2.2	.0139	.0136	.0132	.0129	.0125	.0122	.0119	.0116	.0113	.0110
−2.1	.0179	.0174	.0170	.0166	.0162	.0158	.0154	.0150	.0146	.0143
−2.0	.0228	.0222	.0217	.0212	.0207	.0202	.0197	.0192	.0188	.0183
−1.9	.0287	.0281	.0274	.0268	.0262	.0256	.0250	.0244	.0239	.0233
−1.8	.0359	.0351	.0344	.0336	.0329	.0322	.0314	.0307	.0301	.0294
−1.7	.0446	.0436	.0427	.0418	.0409	.0401	.0392	.0384	.0375	.0367
−1.6	.0548	.0537	.0526	.0516	.0505	.0495	.0485	.0475	.0465	.0455
−1.5	.0668	.0655	.0643	.0630	.0618	.0606	.0594	.0582	.0571	.0559
−1.4	.0808	.0793	.0778	.0764	.0749	.0735	.0721	.0708	.0694	.0681
−1.3	.0968	.0951	.0934	.0918	.0901	.0885	.0869	.0853	.0838	.0823
−1.2	.1151	.1131	.1112	.1093	.1075	.1056	.1038	.1020	.1003	.0985
−1.1	.1357	.1335	.1314	.1292	.1271	.1251	.1230	.1210	.1190	.1170
−1.0	.1587	.1562	.1539	.1515	.1492	.1469	.1446	.1423	.1401	.1379
−0.9	.1841	.1814	.1788	.1762	.1736	.1711	.1685	.1660	.1635	.1611
−0.8	.2119	.2090	.2061	.2033	.2005	.1977	.1949	.1922	.1894	.1867
−0.7	.2420	.2389	.2358	.2327	.2296	.2266	.2236	.2206	.2177	.2148
−0.6	.2743	.2709	.2676	.2643	.2611	.2578	.2546	.2514	.2483	.2451
−0.5	.3085	.3050	.3015	.2981	.2946	.2912	.2877	.2843	.2810	.2776
−0.4	.3446	.3409	.3372	.3336	.3300	.3264	.3228	.3192	.3156	.3121
−0.3	.3821	.3783	.3745	.3707	.3669	.3632	.3594	.3557	.3520	.3483
−0.2	.4207	.4168	.4129	.4090	.4052	.4013	.3974	.3936	.3897	.3859
−0.1	.4602	.4562	.4522	.4483	.4443	.4404	.4364	.4325	.4286	.4247
−0.0	.5000	.4960	.4920	.4880	.4840	.4801	.4761	.4721	.4681	.4641

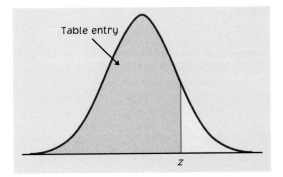

Table entry for z is the area under the standard normal curve to the left of z.

TABLE A Standard normal probabilities (*continued*)

z	.00	.01	.02	.03	.04	.05	.06	.07	.08	.09
0.0	.5000	.5040	.5080	.5120	.5160	.5199	.5239	.5279	.5319	.5359
0.1	.5398	.5438	.5478	.5517	.5557	.5596	.5636	.5675	.5714	.5753
0.2	.5793	.5832	.5871	.5910	.5948	.5987	.6026	.6064	.6103	.6141
0.3	.6179	.6217	.6255	.6293	.6331	.6368	.6406	.6443	.6480	.6517
0.4	.6554	.6591	.6628	.6664	.6700	.6736	.6772	.6808	.6844	.6879
0.5	.6915	.6950	.6985	.7019	.7054	.7088	.7123	.7157	.7190	.7224
0.6	.7257	.7291	.7324	.7357	.7389	.7422	.7454	.7486	.7517	.7549
0.7	.7580	.7611	.7642	.7673	.7704	.7734	.7764	.7794	.7823	.7852
0.8	.7881	.7910	.7939	.7967	.7995	.8023	.8051	.8078	.8106	.8133
0.9	.8159	.8186	.8212	.8238	.8264	.8289	.8315	.8340	.8365	.8389
1.0	.8413	.8438	.8461	.8485	.8508	.8531	.8554	.8577	.8599	.8621
1.1	.8643	.8665	.8686	.8708	.8729	.8749	.8770	.8790	.8810	.8830
1.2	.8849	.8869	.8888	.8907	.8925	.8944	.8962	.8980	.8997	.9015
1.3	.9032	.9049	.9066	.9082	.9099	.9115	.9131	.9147	.9162	.9177
1.4	.9192	.9207	.9222	.9236	.9251	.9265	.9279	.9292	.9306	.9319
1.5	.9332	.9345	.9357	.9370	.9382	.9394	.9406	.9418	.9429	.9441
1.6	.9452	.9463	.9474	.9484	.9495	.9505	.9515	.9525	.9535	.9545
1.7	.9554	.9564	.9573	.9582	.9591	.9599	.9608	.9616	.9625	.9633
1.8	.9641	.9649	.9656	.9664	.9671	.9678	.9686	.9693	.9699	.9706
1.9	.9713	.9719	.9726	.9732	.9738	.9744	.9750	.9756	.9761	.9767
2.0	.9772	.9778	.9783	.9788	.9793	.9798	.9803	.9808	.9812	.9817
2.1	.9821	.9826	.9830	.9834	.9838	.9842	.9846	.9850	.9854	.9857
2.2	.9861	.9864	.9868	.9871	.9875	.9878	.9881	.9884	.9887	.9890
2.3	.9893	.9896	.9898	.9901	.9904	.9906	.9909	.9911	.9913	.9916
2.4	.9918	.9920	.9922	.9925	.9927	.9929	.9931	.9932	.9934	.9936
2.5	.9938	.9940	.9941	.9943	.9945	.9946	.9948	.9949	.9951	.9952
2.6	.9953	.9955	.9956	.9957	.9959	.9960	.9961	.9962	.9963	.9964
2.7	.9965	.9966	.9967	.9968	.9969	.9970	.9971	.9972	.9973	.9974
2.8	.9974	.9975	.9976	.9977	.9977	.9978	.9979	.9979	.9980	.9981
2.9	.9981	.9982	.9982	.9983	.9984	.9984	.9985	.9985	.9986	.9986
3.0	.9987	.9987	.9987	.9988	.9988	.9989	.9989	.9989	.9990	.9990
3.1	.9990	.9991	.9991	.9991	.9992	.9992	.9992	.9992	.9993	.9993
3.2	.9993	.9993	.9994	.9994	.9994	.9994	.9994	.9995	.9995	.9995
3.3	.9995	.9995	.9995	.9996	.9996	.9996	.9996	.9996	.9996	.9997
3.4	.9997	.9997	.9997	.9997	.9997	.9997	.9997	.9997	.9997	.9998

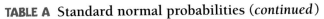

TABLE B Random Digits

Line								
101	19223	95034	05756	28713	96409	12531	42544	82853
102	73676	47150	99400	01927	27754	42648	82425	36290
103	45467	71709	77558	00095	32863	29485	82226	90056
104	52711	38889	93074	60227	40011	85848	48767	52573
105	95592	94007	69971	91481	60779	53791	17297	59335
106	68417	35013	15529	72765	85089	57067	50211	47487
107	82739	57890	20807	47511	81676	55300	94383	14893
108	60940	72024	17868	24943	61790	90656	87964	18883
109	36009	19365	15412	39638	85453	46816	83485	41979
110	38448	48789	18338	24697	39364	42006	76688	08708
111	81486	69487	60513	09297	00412	71238	27649	39950
112	59636	88804	04634	71197	19352	73089	84898	45785
113	62568	70206	40325	03699	71080	22553	11486	11776
114	45149	32992	75730	66280	03819	56202	02938	70915
115	61041	77684	94322	24709	73698	14526	31893	32592
116	14459	26056	31424	80371	65103	62253	50490	61181
117	38167	98532	62183	70632	23417	26185	41448	75532
118	73190	32533	04470	29669	84407	90785	65956	86382
119	95857	07118	87664	92099	58806	66979	98624	84826
120	35476	55972	39421	65850	04266	35435	43742	11937
121	71487	09984	29077	14863	61683	47052	62224	51025
122	13873	81598	95052	90908	73592	75186	87136	95761
123	54580	81507	27102	56027	55892	33063	41842	81868
124	71035	09001	43367	49497	72719	96758	27611	91596
125	96746	12149	37823	71868	18442	35119	62103	39244
126	96927	19931	36809	74192	77567	88741	48409	41903
127	43909	99477	25330	64359	40085	16925	85117	36071
128	15689	14227	06565	14374	13352	49367	81982	87209
129	36759	58984	68288	22913	18638	54303	00795	08727
130	69051	64817	87174	09517	84534	06489	87201	97245
131	05007	16632	81194	14873	04197	85576	45195	96565
132	68732	55259	84292	08796	43165	93739	31685	97150
133	45740	41807	65561	33302	07051	93623	18132	09547
134	27816	78416	18329	21337	35213	37741	04312	68508
135	66925	55658	39100	78458	11206	19876	87151	31260
136	08421	44753	77377	28744	75592	08563	79140	92454
137	53645	66812	61421	47836	12609	15373	98481	14592
138	66831	68908	40772	21558	47781	33586	79177	06928
139	55588	99404	70708	41098	43563	56934	48394	51719
140	12975	13258	13048	45144	72321	81940	00360	02428
141	96767	35964	23822	96012	94591	65194	50842	53372
142	72829	50232	97892	63408	77919	44575	24870	04178
143	88565	42628	17797	49376	61762	16953	88604	12724
144	62964	88145	83083	69453	46109	59505	69680	00900
145	19687	12633	57857	95806	09931	02150	43163	58636
146	37609	59057	66967	83401	60705	02384	90597	93600
147	54973	86278	88737	74351	47500	84552	19909	67181
148	00694	05977	19664	65441	20903	62371	22725	53340
149	71546	05233	53946	68743	72460	27601	45403	88692
150	07511	88915	41267	16853	84569	79367	32337	03316

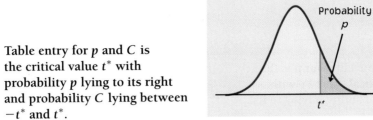

Table entry for p and C is the critical value t^* with probability p lying to its right and probability C lying between $-t^*$ and t^*.

TABLE C *t* distribution critical values

	Upper tail probability p											
df	.25	.20	.15	.10	.05	.025	.02	.01	.005	.0025	.001	.0005
1	1.000	1.376	1.963	3.078	6.314	12.71	15.89	31.82	63.66	127.3	318.3	636.6
2	0.816	1.061	1.386	1.886	2.920	4.303	4.849	6.965	9.925	14.09	22.33	31.60
3	0.765	0.978	1.250	1.638	2.353	3.182	3.482	4.541	5.841	7.453	10.21	12.92
4	0.741	0.941	1.190	1.533	2.132	2.776	2.999	3.747	4.604	5.598	7.173	8.610
5	0.727	0.920	1.156	1.476	2.015	2.571	2.757	3.365	4.032	4.773	5.893	6.869
6	0.718	0.906	1.134	1.440	1.943	2.447	2.612	3.143	3.707	4.317	5.208	5.959
7	0.711	0.896	1.119	1.415	1.895	2.365	2.517	2.998	3.499	4.029	4.785	5.408
8	0.706	0.889	1.108	1.397	1.860	2.306	2.449	2.896	3.355	3.833	4.501	5.041
9	0.703	0.883	1.100	1.383	1.833	2.262	2.398	2.821	3.250	3.690	4.297	4.781
10	0.700	0.879	1.093	1.372	1.812	2.228	2.359	2.764	3.169	3.581	4.144	4.587
11	0.697	0.876	1.088	1.363	1.796	2.201	2.328	2.718	3.106	3.497	4.025	4.437
12	0.695	0.873	1.083	1.356	1.782	2.179	2.303	2.681	3.055	3.428	3.930	4.318
13	0.694	0.870	1.079	1.350	1.771	2.160	2.282	2.650	3.012	3.372	3.852	4.221
14	0.692	0.868	1.076	1.345	1.761	2.145	2.264	2.624	2.977	3.326	3.787	4.140
15	0.691	0.866	1.074	1.341	1.753	2.131	2.249	2.602	2.947	3.286	3.733	4.073
16	0.690	0.865	1.071	1.337	1.746	2.120	2.235	2.583	2.921	3.252	3.686	4.015
17	0.689	0.863	1.069	1.333	1.740	2.110	2.224	2.567	2.898	3.222	3.646	3.965
18	0.688	0.862	1.067	1.330	1.734	2.101	2.214	2.552	2.878	3.197	3.611	3.922
19	0.688	0.861	1.066	1.328	1.729	2.093	2.205	2.539	2.861	3.174	3.579	3.883
20	0.687	0.860	1.064	1.325	1.725	2.086	2.197	2.528	2.845	3.153	3.552	3.850
21	0.686	0.859	1.063	1.323	1.721	2.080	2.189	2.518	2.831	3.135	3.527	3.819
22	0.686	0.858	1.061	1.321	1.717	2.074	2.183	2.508	2.819	3.119	3.505	3.792
23	0.685	0.858	1.060	1.319	1.714	2.069	2.177	2.500	2.807	3.104	3.485	3.768
24	0.685	0.857	1.059	1.318	1.711	2.064	2.172	2.492	2.797	3.091	3.467	3.745
25	0.684	0.856	1.058	1.316	1.708	2.060	2.167	2.485	2.787	3.078	3.450	3.725
26	0.684	0.856	1.058	1.315	1.706	2.056	2.162	2.479	2.779	3.067	3.435	3.707
27	0.684	0.855	1.057	1.314	1.703	2.052	2.158	2.473	2.771	3.057	3.421	3.690
28	0.683	0.855	1.056	1.313	1.701	2.048	2.154	2.467	2.763	3.047	3.408	3.674
29	0.683	0.854	1.055	1.311	1.699	2.045	2.150	2.462	2.756	3.038	3.396	3.659
30	0.683	0.854	1.055	1.310	1.697	2.042	2.147	2.457	2.750	3.030	3.385	3.646
40	0.681	0.851	1.050	1.303	1.684	2.021	2.123	2.423	2.704	2.971	3.307	3.551
50	0.679	0.849	1.047	1.299	1.676	2.009	2.109	2.403	2.678	2.937	3.261	3.496
60	0.679	0.848	1.045	1.296	1.671	2.000	2.099	2.390	2.660	2.915	3.232	3.460
80	0.678	0.846	1.043	1.292	1.664	1.990	2.088	2.374	2.639	2.887	3.195	3.416
100	0.677	0.845	1.042	1.290	1.660	1.984	2.081	2.364	2.626	2.871	3.174	3.390
1000	0.675	0.842	1.037	1.282	1.646	1.962	2.056	2.330	2.581	2.813	3.098	3.300
z^*	0.674	0.841	1.036	1.282	1.645	1.960	2.054	2.326	2.576	2.807	3.091	3.291
	50%	60%	70%	80%	90%	95%	96%	98%	99%	99.5%	99.8%	99.9%
	Confidence level C											

Table entry for p is the critical value F^* with probability p lying to its right.

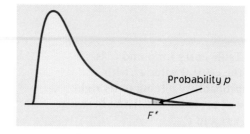

Probability p

F^*

TABLE D F distribution critical values

| | | \multicolumn{8}{c}{Degrees of freedom in the numerator} |
	p	1	2	3	4	5	6	7	8
1	0.100	39.86	49.50	53.59	55.83	57.24	58.20	58.91	59.44
	0.050	161.45	199.50	215.71	224.58	230.16	233.99	236.77	238.88
	0.025	647.79	799.50	864.16	899.58	921.85	937.11	948.22	956.66
	0.010	4052.2	4999.5	5403.4	5624.6	5763.6	5859	5928.4	5981.1
	0.001	405284	500000	540379	562500	576405	585937	592873	598144
2	0.100	8.53	9.00	9.16	9.24	9.29	9.33	9.35	9.37
	0.050	18.51	19.00	19.16	19.25	19.30	19.33	19.35	19.37
	0.025	38.51	39.00	39.17	39.25	39.30	39.33	39.36	39.37
	0.010	98.50	99.00	99.17	99.25	99.30	99.33	99.36	99.37
	0.001	998.50	999.00	999.17	999.25	999.30	999.33	999.36	999.37
3	0.100	5.54	5.46	5.39	5.34	5.31	5.28	5.27	5.25
	0.050	10.13	9.55	9.28	9.12	9.01	8.94	8.89	8.85
	0.025	17.44	16.04	15.44	15.10	14.88	14.73	14.62	14.54
	0.010	34.12	30.82	29.46	28.71	28.24	27.91	27.67	27.49
	0.001	167.03	148.50	141.11	137.10	134.58	132.85	131.58	130.62
4	0.100	4.54	4.32	4.19	4.11	4.05	4.01	3.98	3.95
	0.050	7.71	6.94	6.59	6.39	6.26	6.16	6.09	6.04
	0.025	12.22	10.65	9.98	9.60	9.36	9.20	9.07	8.98
	0.010	21.20	18.00	16.69	15.98	15.52	15.21	14.98	14.80
	0.001	74.14	61.25	56.18	53.44	51.71	50.53	49.66	49.00
5	0.100	4.06	3.78	3.62	3.52	3.45	3.40	3.37	3.34
	0.050	6.61	5.79	5.41	5.19	5.05	4.95	4.88	4.82
	0.025	10.01	8.43	7.76	7.39	7.15	6.98	6.85	6.76
	0.010	16.26	13.27	12.06	11.39	10.97	10.67	10.46	10.29
	0.001	47.18	37.12	33.20	31.09	29.75	28.83	28.16	27.65
6	0.100	3.78	3.46	3.29	3.18	3.11	3.05	3.01	2.98
	0.050	5.99	5.14	4.76	4.53	4.39	4.28	4.21	4.15
	0.025	8.81	7.26	6.60	6.23	5.99	5.82	5.70	5.60
	0.010	13.75	10.92	9.78	9.15	8.75	8.47	8.26	8.10
	0.001	35.51	27.00	23.70	21.92	20.80	20.03	19.46	19.03
7	0.100	3.59	3.26	3.07	2.96	2.88	2.83	2.78	2.75
	0.050	5.59	4.74	4.35	4.12	3.97	3.87	3.79	3.73
	0.025	8.07	6.54	5.89	5.52	5.29	5.12	4.99	4.90
	0.010	12.25	9.55	8.45	7.85	7.46	7.19	6.99	6.84
	0.001	29.25	21.69	18.77	17.20	16.21	15.52	15.02	14.63
8	0.100	3.46	3.11	2.92	2.81	2.73	2.67	2.62	2.59
	0.050	5.32	4.46	4.07	3.84	3.69	3.58	3.50	3.44
	0.025	7.57	6.06	5.42	5.05	4.82	4.65	4.53	4.43
	0.010	11.26	8.65	7.59	7.01	6.63	6.37	6.18	6.03
	0.001	25.41	18.49	15.83	14.39	13.48	12.86	12.40	12.05

Degrees of freedom in the denominator

Table entry for p is the critical value F^* with probability p lying to its right.

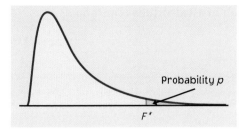

Probability p

F^*

TABLE D *F* distribution critical values (*continued*)

| | p | \multicolumn{8}{c|}{Degrees of freedom in the numerator} |
		9	10	15	20	30	60	120	1000
1	0.100	59.86	60.19	61.22	61.74	62.26	62.79	63.06	63.30
	0.050	240.54	241.88	245.95	248.01	250.10	252.20	253.25	254.19
	0.025	963.28	968.63	984.87	993.10	1001.4	1009.8	1014	1017.7
	0.010	6022.5	6055.8	6157.3	6208.7	6260.6	6313	6339.4	6362.7
	0.001	602284	605621	615764	620908	626099	631337	633972	636301
2	0.100	9.38	9.39	9.42	9.44	9.46	9.47	9.48	9.49
	0.050	19.38	19.40	19.43	19.45	19.46	19.48	19.49	19.49
	0.025	39.39	39.40	39.43	39.45	39.46	39.48	39.49	39.50
	0.010	99.39	99.40	99.43	99.45	99.47	99.48	99.49	99.50
	0.001	999.39	999.40	999.43	999.45	999.47	999.48	999.49	999.50
3	0.100	5.24	5.23	5.20	5.18	5.17	5.15	5.14	5.13
	0.050	8.81	8.79	8.70	8.66	8.62	8.57	8.55	8.53
	0.025	14.47	14.42	14.25	14.17	14.08	13.99	13.95	13.91
	0.010	27.35	27.23	26.87	26.69	26.50	26.32	26.22	26.14
	0.001	129.86	129.25	127.37	126.42	125.45	124.47	123.97	123.53
4	0.100	3.94	3.92	3.87	3.84	3.82	3.79	3.78	3.76
	0.050	6.00	5.96	5.86	5.80	5.75	5.69	5.66	5.63
	0.025	8.90	8.84	8.66	8.56	8.46	8.36	8.31	8.26
	0.010	14.66	14.55	14.20	14.02	13.84	13.65	13.56	13.47
	0.001	48.47	48.05	46.76	46.10	45.43	44.75	44.40	44.09
5	0.100	3.32	3.30	3.24	3.21	3.17	3.14	3.12	3.11
	0.050	4.77	4.74	4.62	4.56	4.50	4.43	4.40	4.37
	0.025	6.68	6.62	6.43	6.33	6.23	6.12	6.07	6.02
	0.010	10.16	10.05	9.72	9.55	9.38	9.20	9.11	9.03
	0.001	27.24	26.92	25.91	25.39	24.87	24.33	24.06	23.82
6	0.100	2.96	2.94	2.87	2.84	2.80	2.76	2.74	2.72
	0.050	4.10	4.06	3.94	3.87	3.81	3.74	3.70	3.67
	0.025	5.52	5.46	5.27	5.17	5.07	4.96	4.90	4.86
	0.010	7.98	7.87	7.56	7.40	7.23	7.06	6.97	6.89
	0.001	18.69	18.41	17.56	17.12	16.67	16.21	15.98	15.77
7	0.100	2.72	2.70	2.63	2.59	2.56	2.51	2.49	2.47
	0.050	3.68	3.64	3.51	3.44	3.38	3.30	3.27	3.23
	0.025	4.82	4.76	4.57	4.47	4.36	4.25	4.20	4.15
	0.010	6.72	6.62	6.31	6.16	5.99	5.82	5.74	5.66
	0.001	14.33	14.08	13.32	12.93	12.53	12.12	11.91	11.72
8	0.100	2.56	2.54	2.46	2.42	2.38	2.34	2.32	2.30
	0.050	3.39	3.35	3.22	3.15	3.08	3.01	2.97	2.93
	0.025	4.36	4.30	4.10	4.00	3.89	3.78	3.73	3.68
	0.010	5.91	5.81	5.52	5.36	5.20	5.03	4.95	4.87
	0.001	11.77	11.54	10.84	10.48	10.11	9.73	9.53	9.36

Degrees of freedom in the denominator

					Degrees of freedom in the numerator					
	p	1	2	3	4	5	6	7	8	9	10
9	0.100	3.36	3.01	2.81	2.69	2.61	2.55	2.51	2.47	2.44	2.42
	0.050	5.12	4.26	3.86	3.63	3.48	3.37	3.29	3.23	3.18	3.14
	0.025	7.21	5.71	5.08	4.72	4.48	4.32	4.20	4.10	4.03	3.96
	0.010	10.56	8.02	6.99	6.42	6.06	5.80	5.61	5.47	5.35	5.26
	0.001	22.86	16.39	13.90	12.56	11.71	11.13	10.70	10.37	10.11	9.89
10	0.100	3.29	2.92	2.73	2.61	2.52	2.46	2.41	2.38	2.35	2.32
	0.050	4.96	4.10	3.71	3.48	3.33	3.22	3.14	3.07	3.02	2.98
	0.025	6.94	5.46	4.83	4.47	4.24	4.07	3.95	3.85	3.78	3.72
	0.010	10.04	7.56	6.55	5.99	5.64	5.39	5.20	5.06	4.94	4.85
	0.001	21.04	14.91	12.55	11.28	10.48	9.93	9.52	9.20	8.96	8.75
12	0.100	3.18	2.81	2.61	2.48	2.39	2.33	2.28	2.24	2.21	2.19
	0.050	4.75	3.89	3.49	3.26	3.11	3.00	2.91	2.85	2.80	2.75
	0.025	6.55	5.10	4.47	4.12	3.89	3.73	3.61	3.51	3.44	3.37
	0.010	9.33	6.93	5.95	5.41	5.06	4.82	4.64	4.50	4.39	4.30
	0.001	18.64	12.97	10.80	9.63	8.89	8.38	8.00	7.71	7.48	7.29
15	0.100	3.07	2.70	2.49	2.36	2.27	2.21	2.16	2.12	2.09	2.06
	0.050	4.54	3.68	3.29	3.06	2.90	2.79	2.71	2.64	2.59	2.54
	0.025	6.20	4.77	4.15	3.80	3.58	3.41	3.29	3.20	3.12	3.06
	0.010	8.68	6.36	5.42	4.89	4.56	4.32	4.14	4.00	3.89	3.80
	0.001	16.59	11.34	9.34	8.25	7.57	7.09	6.74	6.47	6.26	6.08
20	0.100	2.97	2.59	2.38	2.25	2.16	2.09	2.04	2.00	1.96	1.94
	0.050	4.35	3.49	3.10	2.87	2.71	2.60	2.51	2.45	2.39	2.35
	0.025	5.87	4.46	3.86	3.51	3.29	3.13	3.01	2.91	2.84	2.77
	0.010	8.10	5.85	4.94	4.43	4.10	3.87	3.70	3.56	3.46	3.37
	0.001	14.82	9.95	8.10	7.10	6.46	6.02	5.69	5.44	5.24	5.08
25	0.100	2.92	2.53	2.32	2.18	2.09	2.02	1.97	1.93	1.89	1.87
	0.050	4.24	3.39	2.99	2.76	2.60	2.49	2.40	2.34	2.28	2.24
	0.025	5.69	4.29	3.69	3.35	3.13	2.97	2.85	2.75	2.68	2.61
	0.010	7.77	5.57	4.68	4.18	3.85	3.63	3.46	3.32	3.22	3.13
	0.001	13.88	9.22	7.45	6.49	5.89	5.46	5.15	4.91	4.71	4.56
50	0.100	2.81	2.41	2.20	2.06	1.97	1.90	1.84	1.80	1.76	1.73
	0.050	4.03	3.18	2.79	2.56	2.40	2.29	2.20	2.13	2.07	2.03
	0.025	5.34	3.97	3.39	3.05	2.83	2.67	2.55	2.46	2.38	2.32
	0.010	7.17	5.06	4.20	3.72	3.41	3.19	3.02	2.89	2.78	2.70
	0.001	12.22	7.96	6.34	5.46	4.90	4.51	4.22	4.00	3.82	3.67
100	0.100	2.76	2.36	2.14	2.00	1.91	1.83	1.78	1.73	1.69	1.66
	0.050	3.94	3.09	2.70	2.46	2.31	2.19	2.10	2.03	1.97	1.93
	0.025	5.18	3.83	3.25	2.92	2.70	2.54	2.42	2.32	2.24	2.18
	0.010	6.90	4.82	3.98	3.51	3.21	2.99	2.82	2.69	2.59	2.50
	0.001	11.50	7.41	5.86	5.02	4.48	4.11	3.83	3.61	3.44	3.30
200	0.100	2.73	2.33	2.11	1.97	1.88	1.80	1.75	1.70	1.66	1.63
	0.050	3.89	3.04	2.65	2.42	2.26	2.14	2.06	1.98	1.93	1.88
	0.025	5.10	3.76	3.18	2.85	2.63	2.47	2.35	2.26	2.18	2.11
	0.010	6.76	4.71	3.88	3.41	3.11	2.89	2.73	2.60	2.50	2.41
	0.001	11.15	7.15	5.63	4.81	4.29	3.92	3.65	3.43	3.26	3.12
1000	0.100	2.71	2.31	2.09	1.95	1.85	1.78	1.72	1.68	1.64	1.61
	0.050	3.85	3.00	2.61	2.38	2.22	2.11	2.02	1.95	1.89	1.84
	0.025	5.04	3.70	3.13	2.80	2.58	2.42	2.30	2.20	2.13	2.06
	0.010	6.66	4.63	3.80	3.34	3.04	2.82	2.66	2.53	2.43	2.34
	0.001	10.89	6.96	5.46	4.65	4.14	3.78	3.51	3.30	3.13	2.99

Degrees of freedom in the denominator

		Degrees of freedom in the numerator									
	p	12	15	20	25	30	40	50	60	120	1000
9	0.100	2.38	2.34	2.30	2.27	2.25	2.23	2.22	2.21	2.18	2.16
	0.050	3.07	3.01	2.94	2.89	2.86	2.83	2.80	2.79	2.75	2.71
	0.025	3.87	3.77	3.67	3.60	3.56	3.51	3.47	3.45	3.39	3.34
	0.010	5.11	4.96	4.81	4.71	4.65	4.57	4.52	4.48	4.40	4.32
	0.001	9.57	9.24	8.90	8.69	8.55	8.37	8.26	8.19	8.00	7.84
10	0.100	2.28	2.24	2.20	2.17	2.16	2.13	2.12	2.11	2.08	2.06
	0.050	2.91	2.85	2.77	2.73	2.70	2.66	2.64	2.62	2.58	2.54
	0.025	3.62	3.52	3.42	3.35	3.31	3.26	3.22	3.20	3.14	3.09
	0.010	4.71	4.56	4.41	4.31	4.25	4.17	4.12	4.08	4.00	3.92
	0.001	8.45	8.13	7.80	7.60	7.47	7.30	7.19	7.12	6.94	6.78
12	0.100	2.15	2.10	2.06	2.03	2.01	1.99	1.97	1.96	1.93	1.91
	0.050	2.69	2.62	2.54	2.50	2.47	2.43	2.40	2.38	2.34	2.30
	0.025	3.28	3.18	3.07	3.01	2.96	2.91	2.87	2.85	2.79	2.73
	0.010	4.16	4.01	3.86	3.76	3.70	3.62	3.57	3.54	3.45	3.37
	0.001	7.00	6.71	6.40	6.22	6.09	5.93	5.83	5.76	5.59	5.44
15	0.100	2.02	1.97	1.92	1.89	1.87	1.85	1.83	1.82	1.79	1.76
	0.050	2.48	2.40	2.33	2.28	2.25	2.20	2.18	2.16	2.11	2.07
	0.025	2.96	2.86	2.76	2.69	2.64	2.59	2.55	2.52	2.46	2.40
	0.010	3.67	3.52	3.37	3.28	3.21	3.13	3.08	3.05	2.96	2.88
	0.001	5.81	5.54	5.25	5.07	4.95	4.80	4.70	4.64	4.47	4.33
20	0.100	1.89	1.84	1.79	1.76	1.74	1.71	1.69	1.68	1.64	1.61
	0.050	2.28	2.20	2.12	2.07	2.04	1.99	1.97	1.95	1.90	1.85
	0.025	2.68	2.57	2.46	2.40	2.35	2.29	2.25	2.22	2.16	2.09
	0.010	3.23	3.09	2.94	2.84	2.78	2.69	2.64	2.61	2.52	2.43
	0.001	4.82	4.56	4.29	4.12	4.00	3.86	3.77	3.70	3.54	3.40
25	0.100	1.82	1.77	1.72	1.68	1.66	1.63	1.61	1.59	1.56	1.52
	0.050	2.16	2.09	2.01	1.96	1.92	1.87	1.84	1.82	1.77	1.72
	0.025	2.51	2.41	2.30	2.23	2.18	2.12	2.08	2.05	1.98	1.91
	0.010	2.99	2.85	2.70	2.60	2.54	2.45	2.40	2.36	2.27	2.18
	0.001	4.31	4.06	3.79	3.63	3.52	3.37	3.28	3.22	3.06	2.91
50	0.100	1.68	1.63	1.57	1.53	1.50	1.46	1.44	1.42	1.38	1.33
	0.050	1.95	1.87	1.78	1.73	1.69	1.63	1.60	1.58	1.51	1.45
	0.025	2.22	2.11	1.99	1.92	1.87	1.80	1.75	1.72	1.64	1.56
	0.010	2.56	2.42	2.27	2.17	2.10	2.01	1.95	1.91	1.80	1.70
	0.001	3.44	3.20	2.95	2.79	2.68	2.53	2.44	2.38	2.21	2.05
100	0.100	1.61	1.56	1.49	1.45	1.42	1.38	1.35	1.34	1.28	1.22
	0.050	1.85	1.77	1.68	1.62	1.57	1.52	1.48	1.45	1.38	1.30
	0.025	2.08	1.97	1.85	1.77	1.71	1.64	1.59	1.56	1.46	1.36
	0.010	2.37	2.22	2.07	1.97	1.89	1.80	1.74	1.69	1.57	1.45
	0.001	3.07	2.84	2.59	2.43	2.32	2.17	2.08	2.01	1.83	1.64
200	0.100	1.58	1.52	1.46	1.41	1.38	1.34	1.31	1.29	1.23	1.16
	0.050	1.80	1.72	1.62	1.56	1.52	1.46	1.41	1.39	1.30	1.21
	0.025	2.01	1.90	1.78	1.70	1.64	1.56	1.51	1.47	1.37	1.25
	0.010	2.27	2.13	1.97	1.87	1.79	1.69	1.63	1.58	1.45	1.30
	0.001	2.90	2.67	2.42	2.26	2.15	2.00	1.90	1.83	1.64	1.43
1000	0.100	1.55	1.49	1.43	1.38	1.35	1.30	1.27	1.25	1.18	1.08
	0.050	1.76	1.68	1.58	1.52	1.47	1.41	1.36	1.33	1.24	1.11
	0.025	1.96	1.85	1.72	1.64	1.58	1.50	1.45	1.41	1.29	1.13
	0.010	2.20	2.06	1.90	1.79	1.72	1.61	1.54	1.50	1.35	1.16
	0.001	2.77	2.54	2.30	2.14	2.02	1.87	1.77	1.69	1.49	1.22

Degrees of freedom in the denominator

Table entry for *p* is the critical value x^* with probability *p* lying to its right.

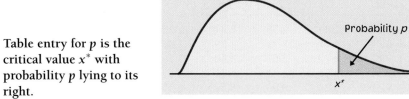

Probability *p*

x^*

TABLE E Chi-square distribution critical values

df	.25	.20	.15	.10	.05	.025	.02	.01	.005	.0025	.001	.0005
											p	
1	1.32	1.64	2.07	2.71	3.84	5.02	5.41	6.63	7.88	9.14	10.83	12.12
2	2.77	3.22	3.79	4.61	5.99	7.38	7.82	9.21	10.60	11.98	13.82	15.20
3	4.11	4.64	5.32	6.25	7.81	9.35	9.84	11.34	12.84	14.32	16.27	17.73
4	5.39	5.99	6.74	7.78	9.49	11.14	11.67	13.28	14.86	16.42	18.47	20.00
5	6.63	7.29	8.12	9.24	11.07	12.83	13.39	15.09	16.75	18.39	20.51	22.11
6	7.84	8.56	9.45	10.64	12.59	14.45	15.03	16.81	18.55	20.25	22.46	24.10
7	9.04	9.80	10.75	12.02	14.07	16.01	16.62	18.48	20.28	22.04	24.32	26.02
8	10.22	11.03	12.03	13.36	15.51	17.53	18.17	20.09	21.95	23.77	26.12	27.87
9	11.39	12.24	13.29	14.68	16.92	19.02	19.68	21.67	23.59	25.46	27.88	29.67
10	12.55	13.44	14.53	15.99	18.31	20.48	21.16	23.21	25.19	27.11	29.59	31.42
11	13.70	14.63	15.77	17.28	19.68	21.92	22.62	24.72	26.76	28.73	31.26	33.14
12	14.85	15.81	16.99	18.55	21.03	23.34	24.05	26.22	28.30	30.32	32.91	34.82
13	15.98	16.98	18.20	19.81	22.36	24.74	25.47	27.69	29.82	31.88	34.53	36.48
14	17.12	18.15	19.41	21.06	23.68	26.12	26.87	29.14	31.32	33.43	36.12	38.11
15	18.25	19.31	20.60	22.31	25.00	27.49	28.26	30.58	32.80	34.95	37.70	39.72
16	19.37	20.47	21.79	23.54	26.30	28.85	29.63	32.00	34.27	36.46	39.25	41.31
17	20.49	21.61	22.98	24.77	27.59	30.19	31.00	33.41	35.72	37.95	40.79	42.88
18	21.60	22.76	24.16	25.99	28.87	31.53	32.35	34.81	37.16	39.42	42.31	44.43
19	22.72	23.90	25.33	27.20	30.14	32.85	33.69	36.19	38.58	40.88	43.82	45.97
20	23.83	25.04	26.50	28.41	31.41	34.17	35.02	37.57	40.00	42.34	45.31	47.50
21	24.93	26.17	27.66	29.62	32.67	35.48	36.34	38.93	41.40	43.78	46.80	49.01
22	26.04	27.30	28.82	30.81	33.92	36.78	37.66	40.29	42.80	45.20	48.27	50.51
23	27.14	28.43	29.98	32.01	35.17	38.08	38.97	41.64	44.18	46.62	49.73	52.00
24	28.24	29.55	31.13	33.20	36.42	39.36	40.27	42.98	45.56	48.03	51.18	53.48
25	29.34	30.68	32.28	34.38	37.65	40.65	41.57	44.31	46.93	49.44	52.62	54.95
26	30.43	31.79	33.43	35.56	38.89	41.92	42.86	45.64	48.29	50.83	54.05	56.41
27	31.53	32.91	34.57	36.74	40.11	43.19	44.14	46.96	49.64	52.22	55.48	57.86
28	32.62	34.03	35.71	37.92	41.34	44.46	45.42	48.28	50.99	53.59	56.89	59.30
29	33.71	35.14	36.85	39.09	42.56	45.72	46.69	49.59	52.34	54.97	58.30	60.73
30	34.80	36.25	37.99	40.26	43.77	46.98	47.96	50.89	53.67	56.33	59.70	62.16
40	45.62	47.27	49.24	51.81	55.76	59.34	60.44	63.69	66.77	69.70	73.40	76.09
50	56.33	58.16	60.35	63.17	67.50	71.42	72.61	76.15	79.49	82.66	86.66	89.56
60	66.98	68.97	71.34	74.40	79.08	83.30	84.58	88.38	91.95	95.34	99.61	102.7
80	88.13	90.41	93.11	96.58	101.9	106.6	108.1	112.3	116.3	120.1	124.8	128.3
100	109.1	111.7	114.7	118.5	124.3	129.6	131.1	135.8	140.2	144.3	149.4	153.2

SOLUTIONS TO SELECTED EXERCISES

CHAPTER 1

1.1 **(a)** Vehicles. **(b)** Vehicle type, transmission type (both categorical); number of cylinders, city MPG, highway MPG (all quantitative).

1.3 **(b)** No: The entries in the table do not represent parts of a single whole.

1.5 Follow the steps in Example 1.2. The data range from 16 to 33 mpg.

1.7 Lightning: centered at or about noon, spread from 7 to 17 (6:30am to 5:30pm). Shakespeare: centered at 4 words, and spread from 1 to 12 words.

1.9 Outlier: 200. Center: between 137 and 140. Spread: 101 to 178.

1.11 **(a)** Baseball players. **(b)** Four other variables; team and position (categorical), age and salary (quantitative). **(c)** Age in years, salary in thousands of dollars per year.

1.13 **(a)** Strongly right-skewed, centered around 4 letters, spread from 1 to 15 letters. **(b)** Shakespeare uses more short words and fewer very long words than does *Popular Science*.

1.15 **(a)** Roughly symmetric (though with two apparent peaks). **(b)** Center: between .265 and .275. Spread: .185 to .355.

1.17 **(a)** For example, $29.3/151.1 = 19.4\%$, $34.9/310.6 = 11.2\%$, etc. **(b)** Children (under 10) are the largest group; percentages trail off gradually after that. **(c)** A much greater proportion in the higher age brackets; now proportions rise gradually up to ages 40–49, then decline.

1.19 The two shortened seasons were 1993 and 1994. Otherwise, McGwire has been a fairly productive home-run hitter; his 1998 number is a possible outlier. Ruth was pretty consistent, and his best year was not an outlier.

1.21 **(b)** Women's times decreased quite rapidly from 1972 until the mid-1980s; since that time, they have been fairly consistent.

1.25 The distribution has two peaks. There are two groups of states: those where less than one-third of all students take the SAT, and those where at least 45% take the SAT. The center of this distribution is 30%, but this could not be considered typical.

1.27 **(a)** $\bar{x} \doteq 141.06$. **(b)** $\bar{x}^* \doteq 137.588$. The outlier pulls the mean up.

1.29 The skewness makes the mean ($675,000) higher than the median ($330,000).

1.31 (a) Male: 20, 27, 34, 50, and 86 C-sections. Female: 5, 10, 18.5, 29, and 33 C-sections. (b) Female doctors generally performed fewer C-sections.

1.33 Both boxplots have similar shapes, but SATM scores have a slightly greater range. Average SATV scores are generally higher than average SATM scores.

1.35 (a) $\bar{x} \doteq 26.1$ and $s \doteq 15.61$ home runs. (b) $\bar{x}^* \doteq 22.2$ and $s^* \doteq 10.24$ home runs. The outlier made both larger.

1.37 $\bar{x} = \$60,000$; seven of the eight employees earned less than the mean. $M = \$22,000$.

1.39 Beef and meat hot dogs are similar, but poultry hot dogs are generally lower in calories than the other two.

1.41 The distribution is fairly symmetrical, with a low outlier of 4.88; $\bar{x} \doteq 5.448$ and $s \doteq 0.22095$ should be reasonable. Our best estimate is $\bar{x}$.

1.43 Because of the outliers, the five-number summary is better: 5.2%, 11.4%, 12.7%, 13.8%, and 18.5%.

1.45 Either a stemplot or a histogram is an acceptable graph. The distribution is clearly skewed to the right, with several clusters. Because of the skewness, the five-number summary should be used: 170, 800, 1663, 3600, and 6495 (in thousands of dollars).

1.47 This is the mean, since half the players should make more than the median.

1.49 (a) Any set of four identical numbers works. (b) 0, 0, 10, 10 is the only possible answer.

1.51 (a) A 1-by-1 square has area 1. (b) 20%. (c) 60%. (d) 50%. (e) 0.5.

1.53 Place 69 at the center of the curve, then mark 66.5 and 71.5 at the change-of-curvature points.

1.55 (a) 50%. (b) 2.5%. (c) 60 to 160 points.

1.57 (a) 0.9978. (b) 0.0022. (c) 0.9515. (d) 0.9493.

1.59 (a) $z \doteq -0.67$. (b) $z \doteq 0.25$.

1.61 Tall one: $\sigma \doteq 0.2$; short one: $\sigma \doteq 0.5$.

1.63 (a) 234 to 298 days. (b) Less than 234 days.

1.65 (a) 0.0122. (b) 0.9878. (c) 0.0384. (d) 0.9494.

1.67 (a) -21% to 45%. (b) About 0.23. (c) About 0.215.

1.69 (a) About 2.5%. (b) About 25%.

1.71 (a) About ± 1.28. (b) 61.3 and 67.7 inches.

1.73 8.8% of all murders were by "other methods." Either a bar chart or a pie chart could be used.

1.75 (a) The distribution is symmetric with no clear outliers. (b) The outliers stand out more clearly in the time plot. (c) $\bar{x} \doteq 8.3628$ and $s \doteq 0.4645$ minutes. (d) $\bar{x} \pm 1s$: 25 (64.1%). $\bar{x} \pm 2s$: 37 (94.9%). $\bar{x} \pm 3s$: 39 (100%).

1.77 (a) Normal: 272, 337, 358, 400.5 and 462 g. New: 318, 383.5, 406.5, 428.5 and 477 g. New corn seems to increase weight gain. (b) Normal: $\bar{x} = 366.3$ and $s \doteq 50.8$ g. New: $\bar{x} = 402.95$ and $s \doteq 42.73$ g. New corn chicks gained an additional 36.65 g.

1.79 (a) Use either a stemplot or histogram. Aside from two high outliers, the distribution is fairly symmetrical; the five-number summaries should be used. Midsize: 16, 25, 26, 29 and 33 mpg. SUV: 16, 19, 19, 20 and 26 mpg. (b) SUVs generally have lower highway gas mileage.

1.81 (a) $-34.04\%, -2.95\%, 3.47\%, 8.45\%$ and 58.68%. (b) Fairly symmetric, with several high and low outliers, but no particular skewness. (c) $1586.78; $659.57.

1.83 (a) After the first two years, the median return is usually above zero, but there is no particular evidence of a trend. (b) The spread of the boxplots is considerably smaller in recent years (with the exception of 1987). (c) High outliers: 58.7% in 1973, 57.9% in 1975, 32% in 1979, and either 42.1% or 41.8% in 1974. The lowest outlier appears in 1973; we see either -26.6% or -27.1% in 1987 (the most striking outlier). Most of the outliers occurred in the early years; later variability is lower.

1.85 The cost per megabyte fell drastically over this period, rapidly at first, then leveling off by the end.

1.89 Those outside the range 22.8 ± 1.81 inches—approximately, less than 21 inches or greater than 24.6 inches.

CHAPTER 2

2.1 (a) Explanatory: time spent studying; response: grade. (b) Explore the relationship. (c) Explanatory: rainfall; response: crop yield. (d) Explore the relationship. (e) Explanatory: father's class; response: son's class.

2.3 Explanatory: treatment (categorical); response: survival time (quantitative).

2.5 (a) Positive association. (b) Linear. (c) A fairly strong relationship, allowing for reasonably accurate prediction. With 716,000 boat registrations, expect about 50 manatee deaths/year.

2.7 (a) Body mass is explanatory variable. (b) A moderately strong, linear, positive association. (c) The men's relationship is basically the same, but with greater scatter (the strength is less). Males typically have larger values for both variables.

2.9 (a) Lowest: about 107 calories (145 mg of sodium); highest: about 195 calories (510 mg of sodium). (b) Positive association: High-calorie hot dogs tend to be high in salt, and low-calorie hot dogs tend to have low sodium. (c) The lower left point in the scatterplot is an outlier. The remaining points show a moderately strong linear relationship.

2.11 (a) Alcohol should be on the x axis. (b) A fairly strong, linear relationship. (c) The association is negative: High wine consumption goes with fewer heart disease deaths, while low wine consumption goes with more deaths. This does not prove causation.

2.13 (a) Planting rate is explanatory. (c) The pattern is curved. The association is not linear, and is neither positive nor negative. (d) 131.025, 143.15, 146.225, 143.07, and 134.75 bushels/acre. At or around 20,000 plants/acre is best.

2.15 (a) A strong, positive association, but slightly curved (not linear). (b) A strong, positive, linear association.

2.17 (a) There appears to be only one species represented. (b) $\bar{x} =$ 58.2 cm and $s_x \doteq 13.20$ cm (femur); $\bar{y} = 66$ cm and $s_y \doteq 15.89$ cm (humerus); $r = 3.97659/4 = 0.994$.

2.19 $r = 1$

2.21 (a) $r \doteq -0.746$, consistent with a moderate negative association. (b) r does not change.

2.23 (a) "Correct calories" is explanatory. (b) $r \doteq 0.8245$, consistent with a positive association. (c) This has no effect on r. Adding 100 to every guess would not change r. (d) $r^* \doteq 0.9837$; without the outliers, the relationship is much stronger.

2.25 (c) Both correlations equal 0.253; scales do not affect r.

2.27 (a) Small-cap stocks. (b) A negative correlation.

2.29 (a) Gender has a nominal scale. (b) $r = 1.09$ is impossible. (c) Correlation has no units.

2.31 (a) Negative association; pH decreases over time. (b) Initial pH: 5.4247; final pH: 4.6350. (c) -0.0053; pH decreased by 0.0053 units per week (on the average).

2.33 (b) We predict $y \doteq 147.4$ bpm, about 4.6 bpm lower than the actual value. (c) With $y =$ time and $x =$ pulse, $y = 43.10 - 0.0574x$, so the predicted time is 34.38 minutes—only 0.08 minutes

(4.8 seconds) too high. **(d)** The results depend on which variable is viewed as explanatory.

2.35 **(a)** Stumps is explanatory; the plot shows a positive linear association. **(b)** The regression line is $\hat{y} = -1.286 + 11.89x$. **(c)** $r^2 \doteq 83.9\%$.

2.37 **(b)** All 10 points: $\hat{y} = 58.588 + 1.3036x$. Without outliers: $\hat{y} = 43.881 + 1.1472x$. **(c)** The two outliers could be considered influential because together they move the line up.

2.39 **(b)** When $x = 20$, $y = \$2500$. **(c)** $y = 500 + 200x$.

2.41 **(a)** $\hat{y} = -3.557 + 0.1010x$. **(b)** $r^2 \doteq 40.16\%$. **(c)** $\hat{y} \doteq 6.85$; the residual is -6.32.

2.43 **(a)** $r \doteq 0.9999$, so recalibration is not necessary. **(b)** $\hat{y} = 1.6571 + 0.1133x$; when $x = 500$ mg/liter, $\hat{y} \doteq 58.31$. The relationship is strong, so the prediction should be very accurate.

2.45 **(b)** $r = 0.463$ and $r^2 = 21.4\%$. A positive association, but not very strong. **(c)** $\hat{y} = 5.683 + 0.6181x$. **(d)** $\hat{y} \doteq 26.3\%$. Since r is not too large, the predictions will not be very reliable. **(e)** The largest residual comes from 1986. There are no points that look influential.

2.47 **(a)** $b = 0.16$, $a = 30.2$. **(b)** $\hat{y} = 78.2$. **(c)** Only $r^2 = 36\%$ of the variability in y is accounted for by the regression.

2.49 **(a)** U.S.: -26.4%, 5.1%, 18.2%, 30.5%, and 37.6%. Overseas: -23.4%, 2.1%, 11.2%, 29.6%, and 69.4%. **(b)** The three middle numbers of the U.S. five-number summary are higher, but the minimum and maximum overseas returns are higher. **(c)** Overseas stocks are more volatile: The boxplot is more widely spread, and the low U.S. return appears to be an outlier, but the low overseas return is not.

2.51 About 4.1 points above the final exam mean.

2.53 **(a)** Scatterplot shows a strong, negative, straight-line association. $\hat{y} = 1166.93 - 0.58679x$. **(b)** About 0.587 million (587,000) per year. About 97.7%. **(c)** $\hat{y} \doteq -0.781$. A population must be greater than or equal to 0; the rate of decrease dropped in the 1980s.

2.55 Less time studying; less parental supervision.

2.57 Both variables increase with the seriousness of the fire.

2.59 Patients with more serious illnesses are more likely to go to larger hospitals, and to require more time to recuperate afterwards.

2.61 Age is the lurking variable.

2.63 **(a)** The straight-line relationship explains about $r^2 \doteq 94.1\%$ of the variation in either variable.

(b) With individual data, there is more variation, so r would be much smaller.

2.65 Explanatory: Whether or not a student has studied a foreign language; response: Score on the test. Lurking variable: English skills before taking (or not taking) the foreign language.

2.67 E.g.: Family history for those who develop leukemia, proximity to power lines, length of residence in this location.

2.69 25 to 34 years old: 24.9%; 35 to 54 years old: 43.9%; 55 or older: 31.3%.

2.71 25 to 34 years old: 12.9%; 35 to 54 years old: 12.5%; 55 or older: 30.8%. The two under-55 age groups are very similar, but a much higher percentage of the older group has not completed high school.

2.73 25 to 34 years old: 27.1%; 35 to 54 years old: 52.0%; 55 or older: 20.9%.

2.75 Start by setting a equal to any number from 10 to 50.

2.77 **(a)** White defendant: 19 yes, 141 no. Black defendant: 17 yes, 149 no. **(b)** Overall death penalty: 11.9% of white defendants, 10.2% of black defendants. For white victims, 12.6% and 17.5%; for black victims, 0% and 5.8%. **(c)** The death penalty is more likely when the victim was white (14%) rather than black (5.4%). Because most convicted killers are of the same race as their victims, whites are more often sentenced to death.

2.79 **(a)** 41.1%. **(b)** 32.5%.

2.81 Older students: 7.1%, 43.2%, 13.5%, and 36.2%. For all students: 13.9%, 24.6%, 43.3%, and 18.1%.

2.83 89.3% of homicides were committed with handguns, 8.2% with long guns, and 2.5% unknown. Among suicides, 70.9% used handguns, 26.3% long guns, and 2.9% unknown. Long guns are more often used in suicides.

2.85 **(a)** No-relapse percentages are 58.3% (desipramine), 25.0% (lithium), and 16.7% (placebo). **(b)** Because random assignment was used, causation is indicated.

2.87 **(a)** Males: 490 admitted, 210 not admitted; females: 280 admitted, 220 not admitted. **(b)** Males: 70% admitted; females: 56% admitted. **(c)** Business school admission rates: 80% of males, 90% of females; law school: 10% of males, 33.3% of females. **(d)** Most male applicants apply to the business school, where admission is easier. A majority of women apply to the law school, which is more selective.

2.89 **(a)** r is negative because the association is negative. The straight-line relationship explains about $r^2 \doteq 71.1\%$ of the variation in death rates. **(b)** $\hat{y} \doteq 168.8$ deaths per 100,000 people. **(c)** $b = rs_y/s_x$, and s_y and s_x are both positive.

2.91 **(b)** There is no clear relationship. **(c)** Generally, those with short incubation periods are more likely to die. **(d)** Person 6— the youngest in the group, and a survivor in spite of a short incubation—merits extra attention. All survivors other than subjects 6 and 17 had incubation periods of 43 hours or more.

2.93 **(a)** $\hat{y} = 0.3531 + 1.1694x$; 27.6%. **(b)** When the S&P monthly return rises (falls) 1 percentage point, Philip Morris stock returns rise (fall) about 1.17 percentage points. **(c)** In a rising market, stocks with beta > 1 rise faster than the overall market. When the market falls, stocks with beta < 1 drop more slowly than the market.

2.95 If women are concentrated in lower-paying disciplines, their overall median salary will be lower than that of men even if salaries within each department are identical.

2.97 The slope is 0.54; $\hat{y} = 33.67 + 0.54x$. We predict the husband's height will be about 69.85 inches.

2.99 **(a)** $\hat{y} = -3.2796 + 1.001226x$; when $x = 455$, we predict $\hat{y} = 452.3$. **(b)** The outlier state is Hawaii, with median SATV score 485 and median SATM score 510—much higher than the predicted median SATM score of 482.2.

2.101 Compute column percents for comparison. Both genders use firearms more than any other method, but they are considerably more common with men (64.5% versus 42.0%). Women are more likely to use poison (34.6% versus 14.0%).

CHAPTER 3

3.1 An observational study: information is gathered without imposing any treatment. Explanatory variable: gender, response variable: political party.

3.3 We can never know how much of the change in attitudes was due to the explanatory variable (reading propaganda) and how much to the historical events of that time.

3.5 **(a)** Adult U.S. residents. **(b)** U.S. households. **(c)** All regulators from the supplier.

3.7 Label down the columns, starting with 01. 04–Bowman, 10–Fleming, 17–Liao, 19–Naber, 12–Goel, 13–Gomez.

3.9 Label from 001 to 440; select 400, 077, 172, 417, 350, 131, 211, 273, 208, and 074.

3.11 Label midsize accounts from 001 to 500, and small accounts from 0001 to 4400. Midsize: 417, 494, 322, 247, and 097. Small: 3698, 1452, 2605, 2480, and 3716.

3.13 (a) Households without telephones or with unlisted numbers. Such households would likely be made up of poor individuals, those who choose not to have phones, and those who do not wish to have their phone number published. (b) Those with unlisted numbers.

3.15 Form A would draw the higher response favoring the ban: It uses loaded language, and only presents one side of the issue.

3.17 An observational study; no treatment is imposed. Explanatory variable: living in public housing or not. Response variable: (some measure of) family stability.

3.19 (a) Individual: a small business; population: "eating and drinking establishments" in the large city. (b) Individual: an adult; desired population: the Congressman's constituents. (c) Individual: auto insurance claim; population: all claims filed in a given month.

3.21 Voluntary response, and the cost of the call.

3.23 Population: Black residents of Miami, probably over 18 years of age. Sample: One adult from each (responsive) black household among the 300 selected addresses. Bias: The sample will underestimate black dissatisfaction because of reluctance to make negative comments about the police to an officer.

3.25 Select three-digit numbers and ignore those that do not appear on the map. This gives 214, 313, 409, 306, and 511.

3.27 (a) Select 35, 75, 115, 155, 195. (Only the first number is from Table B; the others are 40, 60, 120, and 160 places down the list.) (b) Each of the first 40 addresses has a 1/40 chance of being selected; selections from the other four groups of 40 addresses are tied to the first choice. However, the only possible samples have exactly one address from the first 40, one address from the second 40, and so on; while an SRS could contain any five of the 200 addresses in the population.

3.29 Label the women 001 to 500 and the men 0001 to 2000; select 138, 159, 052, 087, 359, then 1369, 0815, 0727, 1025, and 1868.

3.31 Larger sample sizes lead to smaller margins of error (with the same confidence level).

3.33 (a) Pairs of pieces of package liner. (b) Factor: jaw temperature, with four levels: 250°F, 275°F, 300°F, 325°F. (c) Response variable: peeling force.

3.35 (a) Condition of the patient: doctors may avoid surgery on the weakest patients. (b) Make a diagram similar to Figure 3.4.

3.37 **(a)** Make a diagram similar to Figure 3.4. The response variable is the company chosen. **(b)** Numbering from 01 to 40, the first group is 05–Cansico, 32–Roberts, 19–Hwang, 04–Brown, 25–Lippman, 29–Ng, 20–Iselin, 16–Gupta, 37–Turing, 39–Williams, 31–Rivera, 18–Howard, 07–Cortez, 13–Garcia, 33–Rosen, 02–Adamson, 36–Travers, 23–Kim, 27–McNeill, and 35–Thompson.

3.39 In the first design (an observational study), the men who exercise (and those who choose not to) may have other characteristics (lurking variables) which might affect their risk of having a heart attack. Since treatments are assigned to the subjects in the second design, the randomization should "wash out" these factors.

3.41 The experimenter expects meditation to lower anxiety, and probably hopes to show that it does. This may influence his assessment of anxiety.

3.43 For each person, flip a coin to decide which hand they should use first. Record the difference in hand strength for each person.

3.45 **(a)** Ordered by increasing weight, the five blocks are Williams, Deng, Hernandez, Moses; Santiago, Kendall, Mann, Smith; Brunk, Obrach, Rodriguez, Loren; Jackson, Stall, Brown, Cruz; Birnbaum, Tran, Nevesky, Wilansky. **(b)** The simplest method is to number from 1 to 4 within each block, then assign the members of block 1 to a weight-loss treatment, then assign block 2, etc.

3.47 **(a)** The 210 children. **(b)** Factor: the "choice set"; with three levels (2 milk/2 fruit drink, 4 milk/2 fruit drink, and 2 milk/4 fruit drink). Response variable: the choice made by each child. **(d)** Label from 001 to 210; select 119, 033, 199, 192, 148.

3.49 In a controlled scientific study, the effects of factors other than the treatment can be eliminated or accounted for, meaning that the differences in improvement observed between the subjects can be attributed to the differences in treatments.

3.51 E.g.: (1) Imprisonment or fine. (2) Waive punishment if the offender attends AA meetings regularly for 6 weeks. (3) Waive punishment if the offender completes alcohol abuse treatment at a local hospital. Use a completely randomized design with three groups; randomly assign the next 300 (e.g.) convicted drunk drivers to one of the three treatments. Possible response variables: Time to next drunk-driving arrest, number of such arrests in a stated period.

3.53 **(a)** Measure the blood pressure for all subjects, then randomly select half to get a calcium supplement, with the other half getting a placebo. Observe the change in blood pressure. **(b)** Assign labels 01 to 40; give calcium to 18–Howard, 20–Imrani, 26–Maldonado, 35–Tompkins, 39–Willis, 16–Guillen, 04–Bikalis, 21–James, 19–Hruska,

37–Tullock, 29–O'Brian, 07–Cranston, 34–Solomon, 22–Kaplan, 10–Durr, 25–Liang, 13–Fratianna, 38–Underwood, 15–Green, and 05–Chen.

3.55 Placebos do work with real pain, so the placebo response tells nothing about physical basis of the pain.

3.57 (a) Make a diagram similar to Figure 3.4. (b) Have each subject do the task twice, once under each temperature condition, randomly choosing which temperature comes first. Compute the difference in each subject's performances at the two temperatures.

3.59 (a) Use a block design, similar to Figure 3.6. (b) When more subjects are involved, the random differences between individuals have less influence, and we can expect the average of our sample to be a better representation of the whole population.

3.61 It is an observational study—no treatment was imposed.

3.63 (a) Population: Ontario residents; sample: the 61,239 people interviewed. (b) The sample size is very large, so if there were large numbers of both sexes in the sample—a safe assumption, since we are told this is a "random sample"—these samples should accurately represent the whole population.

3.65 (a) Label the students from 0001 to 3478. (b) 2940, 0769, 1481, 2975, and 1315.

3.67 (a) E.g.: All full-time undergraduate students in the fall term on a list provided by the Registrar. (b) E.g.: A stratified sample with 125 students from each year. (c) Mailed questionnaires might have high nonresponse rates. Telephone interviews exclude those without phones, and may mean repeated calling for those that are not home. Face-to-face interviews might be more costly than your funding will allow. Some students might be sensitive about responding to questions about sexual harassment.

3.69 (a) Experimental units: chicks; response variable: weight gain. (b) Two factors (corn type and % protein); nine treatments. The diagram should be a 3×3 table. 90 chicks are needed. (c) Note: This diagram is quite large.

3.71 E.g.: All letters have typed addresses and standard envelope size and are all mailed at 10 a.m. on (say) Tuesday at the same post office to eliminate some variation. Days in transit is the response variable. Mail some letters with ZIP codes and some without. Possible lurking variables: destination; day of the week the letter is mailed. And so on.

3.75 Subjects were randomly chosen to receive treatments (aspirin and/or beta carotene), so that the groups should be substantially identical. Neither the subjects and those who work with them know who is getting what treatment; this prevents a subject's or researcher's expectations from affecting the outcome. Some of the subjects were given placebos. Even though these possess no medical properties, some subjects may show improvement or benefits just as a result of participating in the experiment; the placebos allow researchers to account for this.

CHAPTER 4

4.1   2.5003 cm is a parameter; 2.5009 cm is a statistic.

4.3   Both 335 g and 289 g are statistics.

4.7   (a) We expect probability 1/2. (b) The theoretical probability is 2/3.

4.9   In the long run, of a large number of hands of five cards, about 2% (one out of 50) will contain a three of a kind.

4.15   (a) {all numbers between 0 and 24}. (b) {0, 1, 2, . . . , 11000}. (c) {0, 1, 2, . . . , 12}. (d) {all numbers greater than or equal to 0}, or {0, 0.01, 0.02, 0.03, . . .}. (e) {all positive and negative numbers}.

4.17   0.67; 0.33.

4.19   Models 1, 3, and 4 have probability sums not equal to 1; Model 4 also has individual probabilities greater than 1.

4.21   (a) 0.04, so the sum equals 1. (b) 0.69.

4.23   (a) The density curve is a triangle. (b) 0.5. (c) 0.125.

4.25   The possible values are 2, 3, 4, . . . , 12; with probabilities 1/36, 2/36, 3/36, 4/36, 5/36, 6/36, 5/36, 4/36, 3/36, 2/36, and 1/36.

4.27   {0, 1, 2, . . .}.

4.29   (a) 0.1. (b) 0.3. (c) 0.5; 0.4.

4.31   (a) Just using initials, {(A,D), (A,S), (A,T), (A,R), (D,S), (D,T), (D,R), (S,T), (S,R), (T,R)}. (b) 1/10. (c) 4/10. (d) 3/10.

4.33   (a) 1/38 (all are equally likely). (b) 18/38. (c) 12/38.

4.35   (a) 1%. (b) All probabilities are between 0 and 1, and they add to 1. (c) 0.94. (d) 0.86. (e) Either $X \geq 4$ or $X > 3$; 0.06.

4.37   (a) height $= 1/2$, so the area will be 1. (b) 0.5. (c) 0.4. (d) 0.6.

4.39   On the average, Joe loses 40 cents each time he plays.

4.41   (a) 5.7735 mg. (b) $n = 4$. The average of several measurements is more likely than a single measurement to be close to the mean.

4.43   (a) 0.3409. (b) Mean 18.6, standard deviation 0.8344. (c) 0.0020.

4.45   0.0749; 0.1685.

4.47   (b) 141.847 days. (c) Means will vary. (d) It would be unlikely (though not impossible) for all five $\bar{x}$ values to fall on the same side of μ. (e) The mean of the sampling distribution should be μ.

4.49   (a) Approximately 0. (b) 0.0150. (c) 0.9700.

4.51   (a) 0.1587. (b) 0.0071.

4.53   (a) Approximately $N(2.2, 0.1941)$. (b) 0.1515. (c) 0.0764.

4.55 $L \doteq 12.513$.

4.57   Both 386 and 416 are statistics.

4.59   0.55.

4.61   (a) 75.2%. (b) All probabilities are between 0 and 1, and they add to 1. (c) 0.983. (d) 0.976. (e) Either $X \geq 9$ or $X > 8$; 0.931.

4.63   (a) 0.3707. (b) Mean 100, standard deviation 1.93649. (c) 0.0049. (d) The answer to (a) could be different; (b) would be the same; (c) would be fairly reliable because of the central limit theorem.

4.65   (a) No: A count assumes only whole-number values. (b) Approximately $N(1.5, 0.02835)$. (c) 0.1038.

CHAPTER 5

5.1   1/4; 1/16; 9/16.

5.3   No: Independence is not a reasonable assumption.

5.5   0.02.

5.7   (a) 0.31. (b) 0.16.

5.9   Fight one large battle.

5.11   0.5404.

5.13   (b) 0.3. (c) 0.3.

5.15   (a) 0.57. (b) 0.0481. (c) 0.0962. (d) 0.6864.

5.17   (a) 1/18; 1/5832. (b) The odds against 11 are 17 to 1 (the writer was correct); the odds against three 11s in a row are 5831 to 1 (higher than the writer claimed.)

5.19   No: The number of observations is not fixed.

5.21   (a) 0, 1, . . . , 5. (b) 0.2373, 0.3955, 0.2637, 0.0879, 0.0146, 0.0010.

5.23 0.0074.

5.25 (a) $\mu = 4.5$ Hispanics. (b) $\sigma \doteq 1.7748$ Hispanics. (c) With $p = 0.1$, $\sigma \doteq 1.1619$ Hispanics; with $p = 0.01$, $\sigma \doteq 0.3854$ Hispanics. As p gets close to 0, σ decreases.

5.27 (a) There are 200 independent responses, each with probability $p = 0.4$ of seeking nutritious food. (b) $\mu = 80$ and $\sigma \doteq 6.9282$ households. (c) Depending on the interpretation of the word "between," either 0.5284 or 0.4380.

5.29 (a) $\mu = 180$ and $\sigma \doteq 12.5857$ black adults. (b) 0.2148; $np = 180$ and $n(1 - p) = 1320$, so the normal approximation is safe.

5.31 (a) Yes: It is reasonable that each student's results are independent, and each has the same chance of passing. (b) No: Her probability of success is likely to increase. (c) No: Temperature may affect the outcome of the test.

5.33 (a) $n = 10$, $p = 0.25$. (b) 0.2816. (c) 0.5256. (d) $\mu = 2.5$ and $\sigma \doteq 1.3693$ women.

5.35 (a) $n = 5$, $p = 0.65$. (b) $0, 1, \ldots, 5$. (c) 0.00525, 0.04877, 0.18115, 0.33642, 0.31239, 0.11603. (d) $\mu = 3.25$ and $\sigma \doteq 1.0665$ years.

5.37 (a) 0.1251. (b) 0.0336.

5.39 (a) There are 150 independent observations, each with response probability $p = 0.5$. (b) $\mu = 75$ responses. (c) 0.2061. (d) $n = 200$.

5.41 0.32.

5.43 (a) 0.5264. (b) 0.4054. (c) No: If they were, the answers to (a) and (b) would be the same (or at least close).

5.45 (a) 0.1; 0.9. (b) 9999 switches remain; 999 are bad; $999/9999 \doteq 0.09991$. (c) 9999 switches remain; 1000 are bad; $1000/9999 \doteq 0.10001$.

5.47 (a) 0.4736. (b) 0.6870. (c) 0.3253.

5.49 1/4.

5.51 (a) 0.25. (b) 0.2.

5.53 (a) 5/36. (b) 25/216. (c) $(5/6)^3(1/6)$; $(5/6)^4(1/6)$; $(5/6)^k(1/6)$.

5.55 0.0096.

5.57 0.0901; this is some evidence against the coin being fair, but it is not overwhelming.

5.59 (a) To find $P(A \text{ or } C)$, we would need to know $P(A \text{ and } C)$. (b) To find $P(A \text{ and } C)$, we would need to know $P(A \text{ or } C)$.

5.61 (a) 0.8. (b) 0.5.

5.63 (a) $P(A) = 0.846$, $P(B \mid A) = 0.951$, and $P(B \mid \text{not } A) = 0.919$. (b) No: If A and B were independent, then $P(B \mid A)$ would equal $P(B \mid \text{not } A)$. (c) 0.8045. (d) 0.1415. (e) 0.9461.

CHAPTER 6

6.1 (a) 44% to 50%. (b) The results for our sample will almost certainly not be exactly the same as the population proportion. (c) The method used gives correct results 95% of the time.

6.3 (a) 1.8974. (c) $m \doteq 3.8$. (d) Both intervals should be 7.6 units wide. (e) 95%.

6.5 (a) Aside from two low outliers, the distribution is close to normal. (b) 98.90 to 112.78 IQ points. (c) All the girls at one school are not an SRS.

6.7 (a) 0.8354 to 0.8454 g/liter. (b) 0.8275 to 0.8533 g/liter. (c) Increasing confidence makes the interval longer.

6.9 (a) 271.4 to 278.6. (b) 267.6 to 282.4. (c) 273.1 to 276.9. (d) 7.4, 3.6, and 1.9, respectively. Margin of error decreases with larger samples.

6.11 $n = 68$.

6.13 (a) The computations are correct. (b) No: The numbers are based on a voluntary response, rather than an SRS.

6.15 (a) We can be 95% confident that between 63% and 69% of all adults favor such an amendment. (b) People without telephones (many of whom would be poor), and those living in Alaska and Hawaii, are not included in the sample.

6.17 (a) Intended population: hotel managers. This sample came entirely from Chicago and Detroit, and had a fairly low return rate. (b) 5.101 to 5.691. (c) 4.010 to 4.786. (d) n is large enough that the central limit theorem applies (if we accept the sample as an SRS).

6.19 (a) No marked deviations from normality. (b) 22.57 to 28.77 micrometers/hr. (c) Wider: Higher confidence requires a larger margin of error.

6.21 $n = 174$.

6.23 (a) The method used gives correct results 95% of the time. (b) The confidence interval contains some values of p which give the election to Ford, and the true estimation error could be even larger. (c) Either a majority favors Carter, or they don't. We do not view the parameter as random.

6.25 (a) Approximately $N(115, 6)$. (b) 118.6 is fairly close to the middle of the curve, and would not be too surprising if H_0 were true. Meanwhile, 125.7 lies out toward the high tail of the curve, and would rarely occur when $\mu = 115$.

6.27 $H_0: \mu = 5$ mm; $H_a: \mu \neq 5$ mm.

6.29 $H_0: \mu = 50$; $H_a: \mu < 50$.

6.31 (a) 118.6: $P = 0.2743$. 125.7: $P = 0.0375$. (b) 125.7 is significant at 0.05 but not at 0.01.

6.33 A difference in earnings as large as we observed in our sample would rarely occur if there was no difference in the average earnings of men and women. Meanwhile, the average earnings of blacks and whites in our sample were so close together that such results could easily arise when mean black and white incomes were equal.

6.35 (a) $H_0: \mu = 224$ mm; $H_a: \mu \neq 224$ mm. (b) $z \doteq 0.13$. (c) $P = 0.8966$; no reason to doubt $\mu = 224$ mm.

6.37 (a) $z \doteq -2.20$. (b) Yes. (c) No. (d) $2.054 < |z| < 2.326$; $0.02 < P < 0.04$.

6.39 (a) 100.56 to 111.12 IQ points. (b) $H_0: \mu = 100$; $H_a: \mu \neq 100$. 100 is not in the 95% confidence interval, we reject H_0 at the 5% level (for a two-sided alternative).

6.41 $H_0: \mu = 1250$ ft^2; $H_a: \mu < 1250$ ft^2.

6.43 $H_0: \mu = -0.545°$ C; $H_a: \mu > -0.545°$ C; $z \doteq 1.96$; $P \doteq 0.0250$. The supplier is apparently adding water.

6.45 If church attenders were no more ethnocentric than nonattenders, then sample results like the one observed would rarely occur.

6.47 $\bar{x} = 0.09$: $0.40 < P < 0.50$. $\bar{x} = 0.27$: $0.01 < P < 0.02$.

6.49 Table C gives $0.10 < P < 0.20$; Table A gives $P = 0.1706$.

6.51 If (population) perceived age was the same for all brands, then what was observed in this sample would rarely occur (less than once in 1000 samples). If it was a randomly chosen sample, it should be a fair representation of all ads.

6.53 No; this statement means that if H_0 is true, we have observed outcomes that occur less than 5% of the time.

6.55 (a) $P = 0.3821$. (b) $P = 0.1711$. (c) $P = 0.0013$.

6.57 No: Confidence interval methods do not apply to voluntary response.

6.59 Question (b).

6.61 Because of the large sample size, this significant difference may not indicate a strong preference.

6.63 (a) H_0: Patient is ill; H_a: Patient is healthy. Type I error: clearing a patient who is ill. Type II error: sending a healthy patient to the doctor.

6.65 (a) 0.50. (b) 0.1841. (c) 0.0013.

6.67 (a) 0.2033. (b) 0.9927. (c) Higher, since 290 is further from 300.

6.69 (a) Reject H_0 if $\bar{x} \leq 0.84989$ or $\bar{x} \geq 0.87011$. (b) 0.8944. (c) 0.1056.

6.71 $P(\text{Type I error}) = 0.05$; $P(\text{Type II error}) = 0.0073$.

6.73 A low-power test will often accept H_0 when it is false, simply because it is difficult to distinguish between H_0 and nearby alternatives.

6.75 (a) 141.6 to 148.4 mg/g. (b) H_0: $\mu = 140$ mg/g; H_a: $\mu > 140$ mg/g; $z \doteq 2.42$, $P = 0.0078$. Mean cellulose content is higher than 145 mg/g. (c) We need an SRS from a (nearly) normal population.

6.77 (a) The distribution is right-skewed. (b) 6.20% to 7.95%. (c) H_0: $\mu = 5.5\%$; H_a: $\mu > 5.5\%$; $z \doteq 2.98$; $P = 0.0014$. The mean return on Treasury bills is higher than 5.5%.

6.79 (a) Margin of error decreases. (b) P-value decreases. (c) Power increases.

6.81 No; this statement means that *if H_0 is true,* we have observed outcomes that occur about 3% of the time.

6.83 (a) The observed difference would occur in less than 1% of all samples if the two populations actually have the same proportion remaining on welfare. (b) The method used is correct 95% of the time. (c) No: Treatments were not randomly assigned.

CHAPTER 7

7.1 (a) 1.7898. (b) 0.0173.

7.3 (a) 2.145. (b) 0.688.

7.5 (a) 14. (b) 1.761 ($p = 0.05$) and 2.145 ($p = 0.025$). (c) 0.025 and 0.05. (d) Significant at 5% but not at 1%.

7.7 (a) 1.75 ms; 0.06455 ms. (b) 1.598 to 1.902 ms. (c) $H_0: \mu = 1.3$ ms; $H_a: \mu > 1.3$ ms; $t \doteq 6.97$; $0.0025 < P < 0.005$. DDT slows nerve recovery.

7.9 (a) For each subject, randomly select which knob should be used first. (b) μ is the mean of (right-hand-thread time minus left-hand-thread time); $H_0: \mu = 0$ sec; $H_a: \mu < 0$ sec. (c) $t = -2.90$; $0.0025 < P < 0.005$. Right-hand-threaded times are less.

7.11 (a) Both the subjects and those who work with them do not know who is getting what treatment. (b) Depression scores are reasonably normal. $H_0: \mu = 0$; $H_a: \mu < 0$ (μ is the placebo minus caffeine mean). $t \doteq -3.53$; $0.005 < P < 0.0025$. Caffeine deprivation raises depression scores. (c) The Beats distribution has outliers and skewness; the sample is small.

7.13 (a) The distribution is slightly skewed, but has no apparent outliers. t procedures are acceptable. (b) $H_0: \mu = 224$ mm; $H_a: \mu \neq 224$ mm; $t \doteq 0.13$; $P > 0.50$. There is no reason to doubt that $\mu = 224$ mm.

7.15 (a) 21.46 to 26.54 mixers. (b) A large sample size will overcome the skewness.

7.17 (a) 111.2 to 118.6 mm Hg. (b) The important assumption is that this is an SRS from some population. We also assume a normal distribution, but this is not crucial provided there are no outliers and little skewness.

7.19 (a) Standard error of the mean. (b) About 0.81 to 0.87.

7.21 (a) Work from the set of differences for the given data: $\bar{x} \doteq -5.71$ μm/hr, $s \doteq 10.56$ μm/hr, and $s/\sqrt{14} \doteq 2.82$ μm/hr. (b) $H_0: \mu = 0$; $H_a: \mu < 0$; $t \doteq -2.02$; $0.025 < P < 0.05$; significant at 5% but not at 1%. Altering the electric field appears to reduce the healing rate. (c) -10.70 to -0.72 μm/hr. The method used gives correct results 90% of the time.

7.23 (a) The large sample size will overcome the skewness. (b) df $= 103$. (c) 2.42% to 11.38%. The data must come from an SRS of the population of all corporate CEOs.

7.25 We know about the whole population, not just a sample.

7.27 (a) $t^* = 2.080$. (b) Reject H_0 if $|t| \geq 2.080$; that is, $|\bar{x}| \geq 0.133$. (c) 0.8531.

7.29 (a) Single sample. (b) Two samples.

7.31 (a) With df reasonably large, the t distribution is similar to $N(0, 1)$ distribution. Since 7.36 would almost never come from a $N(0, 1)$ distribution, P is small and the result is significant. (b) df $= 32$.

7.33 (a) $H_0: \mu_1 = \mu_2$; $H_a: \mu_1 > \mu_2$; $t \doteq 1.91$; $0.025 < P < 0.05$; this is significant at 5% and 10%. (b) 0.03 to 6.27.

7.35 (a) We find no particular skewness or outliers. (b) $H_0: \mu_1 = \mu_2$; $H_a: \mu_1 < \mu_2$; $t \doteq -2.47$; $0.01 < P < 0.02$. The high-lysine diet leads to increased weight gain. (c) 5.59 to 67.71 g.

7.37 (a) $H_0: \mu_1 = \mu_2$; $H_a: \mu_1 \neq \mu_2$. (b) $t = 2.9912$; $P = 0.0247$. The means are different. (c) 2.6% to 13.6%.

7.39 E.g.: The difference between average female (55.5) and male (57.9) self concept scores was so small that it can be attributed to chance variation in the samples ($t = -0.83$, df = 62.8, $P = 0.4110$). In other words, based on this sample, we have no evidence that mean self concept scores differ by gender.

7.41 27.91 to 32.09. Normality of SAT scores is not needed since the sample sizes are so large.

7.43 (a) $H_0: \mu_A = \mu_P$; $H_a: \mu_A > \mu_P$; $t \doteq 4.28$; $P < 0.0005$. Active learning results in more correct identifications. (b) 22.2 to 26.6 Blissymbols. (c) We assume (but cannot check) that we have an SRS from the population of learning-impaired children. We also assume (near) normality, while the active score show a high outlier and some skewness. Reasonably large (and equal) sample sizes make t procedures fairly reliable anyway.

7.45 (a) $H_0: \mu_1 = \mu_2$; $H_a: \mu_1 \neq \mu_2$; $t \doteq -8.24$; $P < 0.001$. This is significant at 5% and 1%. (b) A volunteer group of middle-aged college professors is hardly an SRS of all middle-aged men.

7.47 (a) $H_0: \mu_A = \mu_B$; $H_a: \mu_A \neq \mu_B$; $t \doteq -1.48$; $0.10 < P < 0.20$. So the difference is not significant. The bank should choose whichever option costs them less. (b) Large, equal sample sizes will overcome the skewness. (c) It is an experiment, since treatments are imposed, so the conclusions should be reliable.

7.49 (a) $H_0: \mu_1 = \mu_2$; $H_a: \mu_1 \neq \mu_2$; $t \doteq 1.25$; $0.20 < P < 0.30$. The difference is not significant. (b) -15.8 to 54.8 msec (df = 9) or -12.6 to 51.6 msec (df = 25.4). Since $P > 0.20$, 0 would appear in any interval with a confidence level greater than about 80%.

7.51 (a) Either 1.660 (df = 100), or 1.649 (df = 349). The following uses 1.660. (b) $c = 50.19$. (c) Approximately 0.05.

7.53 (a) Significant at 5% but not at 1%. (b) $0.02 < P < 0.05$.

7.55 $H_0: \sigma_1 = \sigma_2$; $H_a: \sigma_1 \neq \sigma_2$; $F \doteq 8.68$; $0.002 < P < 0.02$. This is significant evidence that $\sigma_1 \neq \sigma_2$.

7.57 $H_0: \sigma_1 = \sigma_2$; $H_a: \sigma_1 \neq \sigma_2$; $F \doteq 2.24$; $P > 0.20$. This gives us no reason to believe that $\sigma_1 \neq \sigma_2$.

7.59 (a) $H_0: \mu = 0; H_a: \mu > 0; t \doteq 3.20; 0.01 < P < 0.02$. This is strong evidence that low-rate exercise raises heart rate; the confidence interval for the increase is 2.60 to 13.00 bpm. (b) With the same hypotheses, $t \doteq 10.63$ and $P < 0.0005$. This is stronger evidence that medium-rate exercise raises heart rate; the confidence interval is 14.87 to 22.33 bpm. (c) $H_0: \mu_1 = \mu_2; H_a: \mu_1 < \mu_2; t = -3.60; 0.01 < P < 0.02$. Medium-rate exercise has a greater effect than low-rate exercise.

7.61 (a) A two-sample t test; the two groups are (presumably) independent. (b) df $= 44$. (c) Nonnormality is not a problem because of the large, equal sample sizes.

7.63 (a) $H_0: \mu = 0; H_a: \mu > 0; t \doteq 2.59; 0.025 < P < 0.05$. Average treatment-2 yields are higher. (b) -70.3 to 384.3 lbs/acre.

7.65 (a) The distribution is slightly, but not unreasonably, left-skewed, with no outliers. (b) 54.78% to 64.40%.

7.67 (a) Standard error (of the mean). Drivers: $\bar{x}_1 = 2821$ and $s_1 = 435.58$ cal; $\bar{x}_2 = 0.24$ and $s_2 = 0.59397$ g. Conductors: $\bar{x}_3 = 2844$ and $s_3 = 437.30$ cal; $\bar{x}_4 = 0.39$ and $s_4 = 1.00215$ g. (b) $t \doteq -0.35; P > 0.50$; no significant difference in calorie consumption. (c) $t \doteq -1.20; P > 0.20$; no significant difference in alcohol consumption. (d) 0.207 to 0.573 g. (e) -0.31 to 0.01 g/day.

7.69 (a) $H_0: \mu = 86$ ppm; $H_a: \mu < 86$ ppm; $t \doteq -1.897; 0.025 < P < 0.05$; not significant at 1%. (b) Use a matched-pairs design, with each soil specimen split in half and measured by each method. Test $H_0: \mu_1 = \mu_2; H_a: \mu_1 > \mu_2$.

7.71 The distribution is fairly symmetric, with a slightly low outlier of 4.88. $\bar{x} \doteq 5.4479$ is our best estimate of the earth's density, with margin of error $t^*s/\sqrt{29}$ (which depends on the desired confidence level).

7.73 (a) $H_0: \sigma_1 = \sigma_2; H_a: \sigma_1 \neq \sigma_2; F \doteq 1.16; P > 0.20$. We have no reason to believe that $\sigma_1 \neq \sigma_2$. (b) No: The F test is not robust.

7.75 The distribution of differences (city minus rural) has two high outliers. With the outliers, $t \doteq 2.38$ and $0.01 < P < 0.02$; without them, $t \doteq 2.33$ and $0.01 < P < 0.02$. To estimate the difference, use a confidence interval for the mean ($\bar{x}$ equals either 2.19 or 1.00 g).

CHAPTER 8

8.1 (a) Population: the 175 residents of Tonya's dorm; p is the proportion who like the food. (b) $\hat{p} = 0.28$.

8.3 (a) Population: the 15,000 alumni; p is the proportion who support the president's decision. (b) $\hat{p} = 0.38$.

8.5 (a) $N(0.14, 0.0155)$. (b) $P(\hat{p} > 0.20) < 0.0002$. $P(\hat{p} > 0.15) \doteq 0.2611$.

8.7 (a) No: The population-to-sample ratio is too small. (b) Yes: We have an SRS, the population is 48 times as large as the sample, and the success and failure counts are both greater than 10. (c) No: There were only 5 or 6 successes in the sample.

8.9 (a) 0.51 to 0.57; the margin of error is about 3%. (b) We weren't given sample sizes for each gender. (c) Greater than 0.03, since the sample size is smaller.

8.11 (a) 0.594 to 0.726. (b) $H_0: p = 0.73$; $H_a: p \neq 0.73$; $z \doteq -2.23$; $P = 0.0258$—fairly strong evidence of a difference. (c) We have an SRS; the population-to-sample ratio is large enough (assuming that there are at least 2000 new students); we had 132 successes and 68 failures.

8.13 $n = 451$.

8.15 (a) Approximately $N(0.14, 0.003613)$. (b) $H_0: p = 0.14$; $H_a: p > 0.14$; $z \doteq 36$, so $P \doteq 0$; overwhelming evidence that Harleys are more likely to be stolen.

8.17 $H_0: p = 1/3$; $H_a: p > 1/3$; $z \doteq 2.72$; $P = 0.0033$. This is strong evidence that more than one-third of this group never use condoms.

8.19 (a) 0.3901 to 0.4503. (b) $H_0: p = 0.5$; $H_a: p < 0.5$; $z \doteq -6.75$; $P < 0.0002$. This is very strong evidence that less than half the population attended church or synagogue in the preceding week. (c) $n = 16{,}590$. Our confidence interval shows that p is in the range 0.3 to 0.7, so that this conservative approach will not greatly inflate the sample size.

8.21 (a) $H_0: p = 0.5$; $H_a: p > 0.5$; $z \doteq 1.70$; $P = 0.0446$. This is strong enough evidence (at the 5% level) to conclude that a majority prefer fresh-brewed coffee. (b) 0.5071 to 0.7329. (c) The coffee should be presented in random order.

8.23 (a) 0.0588, 0.0784, 0.0898, 0.0960, 0.0980, 0.0960, 0.0898, 0.0784, 0.0588. (b) 0.0263, 0.0351, 0.0402, 0.0429, 0.0438, 0.0429, 0.0402, 0.0351, 0.0263. The new margins of error are less than half their former size.

8.25 -0.0518 to 0.3753. The populations of mice are certainly more than 10 times as large as the samples, and the counts of successes and failures are more than 5 in both samples.

8.27 $H_0: p_1 = p_2$; $H_a: p_1 > p_2$; $z \doteq 0.98$; $P = 0.1635$; we cannot conclude that death rates are different.

8.29 Home computer: H_0: $p_1 = p_2$; H_a: $p_1 \neq p_2$; $z \doteq 1.01$; $P = 0.3124$. Access at work: Same hypotheses; $z \doteq 3.90$; $P < 0.0004$. There is no evidence of a difference in home computers, but have very strong evidence of a difference at work.

8.31 (a) 0.0341 to 0.0757. (b) All counts are larger than 5; the populations are much larger than the samples (assuming that the population of [potential] complainers is fairly large).

8.33 (a) H_0: $p_1 = p_2$; H_a: $p_1 \neq p_2$; $z \doteq 2.99$; $P = 0.0028$—strong evidence of a difference. (b) 0.1172 to 0.3919.

8.35 H_0: $p_1 = p_2$; H_a: $p_1 \neq p_2$; $z \doteq 0.02$; $P = 0.9840$—no evidence of a difference.

8.37 (a) 0.1626 to 0.2398. (b) The confidence interval does not even come close to 0, so the P-value against the two-sided alternative will be (much) smaller than 0.01. (c) The counts are 158 (January to April) and 313 (July and August). H_0: $p_1 = p_2$; H_a: $p_1 \neq p_2$; $z \doteq -4.39$; $P < 0.0004$; very strong evidence that other nonresponse rates also differ between the seasons.

8.39 (a) 0.2465 to 0.3359; since 0 is not in this interval, the difference is significant at the 1% level. (b) No: $t \doteq -0.8658$; P is close to 0.4.

8.41 (a) $\hat{p}_1 \doteq 0.1415$, $\hat{p}_2 \doteq 0.1667$; $z \doteq -0.39$; $P = 0.6966$. (b) $z \doteq 2.12$ and $P = 0.0340$. (c) For (a): -0.1559 to 0.1056. For (b): -0.04904 to -0.001278. Larger samples make the margin of error smaller.

8.43 (a) The samples should be randomly chosen from a variety of schools. (b) 0.0135 to 0.0270. (c) $z \doteq 0.54$ and $P = 0.5892$—no evidence of a difference.

8.45 No: The numbers are based on a voluntary response sample.

8.47 H_0: $p = 0.488$; H_a: $p > 0.488$; $z \doteq 0.18$; $P = 0.4286$; no evidence that the chemists are more likely to have girls.

8.49 (a) H_0: $p_1 = p_2$; H_a: $p_1 \neq p_2$. We have random samples from large populations, and the smallest count is 44. (b) $z \doteq -3.80$, $P < 0.0004$. This is very strong evidence of a difference in the proportion experiencing adverse symptoms.

CHAPTER 9

9.1 (a) $r = 2$, $c = 3$. (b) 0.55, 0.747, and 0.375. Some (but not too much) time spent in extracurricular activities seems to be beneficial. (d) 13.78, 62.71, 5.51; 6.22, 28.29, 2.49. (e) The first and last columns have lower numbers than we expect in the passing row (and higher numbers in the failing row), while the middle column has this reversed.

9.3 (a) $0.5614 + 0.4470 + 1.1452 + 1.2442 + 0.9906 + 2.5381 = 6.926$. (b) $P = 0.031$. We conclude that there *is* a relationship between hours spent in extracurricular activities and performance in the course. (c) Column 3, row 2. Too much time spent on these activities seems to hurt academic performance. (d) No: This was a study, not an experiment.

9.5 (b) 5.99 and 7.38; $0.025 < P < 0.05$. (c) The mean would be 2; this value is larger.

9.7 (a) Cardiac event: 3, 7, 12; No cardiac event: 30, 27, 28. (b) 0.9091, 0.7941, 0.7000. (c) 6.79, 6.99, 8.22; 26.21, 27.01, 31.78. All are greater than 5. (d) $P = 0.089$; we don't have significant evidence of a difference in success rates.

9.9 (a) $H_0: p_1 = p_2$; $H_a: p_1 \neq p_2$; $z \doteq -0.57$; $P = 0.5686$. (b) Improved: 28, 30; no improvement: 54, 48. $X^2 = 0.322 \doteq z^2$. Table E gives $P > 0.25$. (c) Gastric freezing is not significantly more (or less) effective than a placebo treatment.

9.11 (a) No: Treatments were not imposed on the subjects. (b) It doesn't seem to be; roughly one-third of each group of patients fell into each olive oil consumption category. (c) $X^2 = 1.552$. The mean is df $= 4$, so this gives no evidence of a relationship between olive oil consumption and cancer. $P = 0.817$. (d) A high participation rate means that nonresponse had little effect on the results.

9.13 (a) $X^2 = 10.827$; $P = 0.013$. We have fairly strong evidence that gender and choice of major are related. (b) The biggest difference is that a higher percentage of women chose Administration, while greater proportions of men chose other fields, especially Finance. (c) The two from the Administration row: Many more women (and fewer men) than expected chose this major. (d) Yes: Only one (12.5%) count is less than 5. (e) 46.5%.

9.15 (a) $H_0: p_1 = p_2$; $\hat{p}_1 \doteq 0.8423$, $\hat{p}_2 \doteq 0.6881$; $z \doteq 3.92$; $P < 0.0004$. (b) $X^2 = 15.334 = z^2$. Table E gives $P < 0.0005$. (c) 0.0774 to 0.2311.

9.17 (a) 7.01%, 14.02%, 13.05%, respectively. (b) The table entries are 172, 2283; 167, 1024; 86, 573. (c) Expected counts are all much bigger than 5, so the chi-square test is safe. H_0: there is no relationship between worker class and race; H_a: there is some relationship. (d) df $= 2$; $P < 0.0005$. (e) Black female child-care workers are more likely to work in non-household or preschool positions.

9.19 (a) $X^2 = 14.863$; $P < 0.001$. The differences are significant. (b) The data should come from independent SRSs of the (unregulated) child-care providers in each city.

9.21 (a) df $= 9$. (b) The entries in the two-way table are 1067, 491; 833, 756; 901, 1174; 1583, 1055. $X^2 = 256.8$, and P is very small. The highest response rates occur from September to mid-April; the lowest occur in the summer months, when more people are likely to be on vacation.

9.23 (a) Use a 2×2 table formed by the four entries on the "Total" line. $X^2 = 332.205$; P is tiny, so the evidence is very strong. Possibly many sacrificed themselves out of a sense of chivalry. (b) $X^2 = 103.767$—a very significant result. The probability of dying decreased with increasing social status. (c) $X^2 = 34.621$—another very significant result; again, the probability of dying decreased with increasing social status.

9.25 (a) $X^2 = 11.141$; $0.0025 < P < 0.005$. The two cells in the second row contribute 8.507 to the value of X^2. (b) The table entries are 488, 327; 102, 113. $X^2 = 10.751$; $0.0005 < P < 0.001$—a significant result. A comparison of actual and expected counts agrees with our expectations.

CHAPTER 10

10.1 (a) The stemplots show no extreme outliers or skewness. (b) The means suggest that a dog reduces heart rate, but being with a friend appears to raise it. (c) $F = 14.08$; $P < 0.0005$. H_0: $\mu_P = \mu_F = \mu_C$; H_a: at least one mean is different. It appears that the mean heart rate is lowest with a pet is present, and highest when a friend is present.

10.3 (a) $I = 3$ (number of populations); $n_1 = n_2 = n_3 = 15$ (sample sizes from each population); $N = 45$ (total sample size). (b) $I - 1 = 2$ and $N - I = 42$. (c) Since $F > 9.22$, $P < 0.001$.

10.5 (a) Populations: tomato varieties; response: yield. $I = 4$; $n_1 = \cdots = n_4 = 10$; $N = 40$; 3 and 36 degrees of freedom. (b) Populations: consumers (responding to different package designs); response: attractiveness rating. $I = 6$; $n_1 = \cdots = n_6 = 120$; $N = 720$; 5 and 714 degrees of freedom. (c) Populations: dieters (under different diet programs); response: weight change after six months. $I = 3$; $n_1 = n_2 = 10$, $n_3 = 12$; $N = 32$; 2 and 29 degrees of freedom.

10.7 (a) Single men earn considerably less than the other groups; widowed and married men earn the most. (b) Yes: $8119/5731 \doteq 1.42$. (c) 3 and 8231. (d) With large sample sizes, even *small* differences would be significant. (e) No: Age is the lurking variable.

10.9 (a) 32.32, 34.58, 35.51; 35.51/32.32 $\doteq$ 1.10. (b) MSG $\doteq$ 96.41, MSE $\doteq$ 1216, $F \doteq$ 0.08; df 2 and 126; $P >$ 0.100. This is not enough evidence that nest temperature affects mean weight.

10.11 (a) $\bar{x} \doteq$ 140.02. MSE and MSG agree with computer output (except for rounding). (b) 130.80 to 161.64 bushels/acre.

10.13 (a) 123.8, 123.6, 132.6, and 116.4 lb. The means vary over time, but do not consistently decrease. (b) 4.6043, 6.5422, 9.0443, 16.0873 lb. 16.0873/4.6043 $\doteq$ 3.49; ANOVA should not be used. (c) The two-sample t test does not require that the standard deviations be equal, but the ANOVA assumes that they are, and is not reliable if there is evidence that they are different.

10.15 (a) The stemplots show no extreme outliers or strong skewness (given small sample sizes). (b) 5.761/4.981 $\doteq$ 1.16 is acceptable. The means suggest that logging reduces the number of trees per plot, and that recovery is slow. $F = $ 11.43; df 2 and 30; $P <$ 0.001; these differences are significant.

10.17 (a) Means: 10.65, 10.425, 5.60, 5.45 cm; standard deviations: 2.053, 1.486, 1.244, 1.771 cm. The means and the stemplots suggest that the presence of too many nematodes reduces growth. (b) $H_0: \mu_1 = \cdots = \mu_4$; H_a: not all means are the same. We test whether nematodes affect mean plant growth. (c) $F = $ 12.08, df 3 and 12, $P = $ 0.001. The first two levels are similar, as are the last two. Somewhere between 1000 and 5000 nematodes, the tomato plants are hurt by the worms.

10.19 MSG $\doteq$ 102.2, MSE $\doteq$ 16.98, $F \doteq$ 6.02; df 2 and 30; 0.001 $< P <$ 0.010.

CHAPTER 11

11.1 (a) $r = $ 0.994, $\hat{y} = -3.660 + 1.1969x$. (b) β (estimated by 1.1969) represents how much we can expect the humerus length to increase with a 1-cm increase in femur length. The estimate of α is -3.660. (c) The residuals are -0.8226, -0.3668, 3.0425, -0.9420, and -0.9110. $s \doteq$ 1.982.

11.3 $\hat{y} = $ 560.65 $-$ 3.0771x; generally, the longer a child remains at the table, the fewer calories he or she will consume. -4.8625 to -1.2917 calories per minute.

11.5 -12.9454 to -6.4444 bpm per minute. For each one-minute increase in swimming time, pulse rate drops by 6 to 13 bpm.

11.7 $b = -22.969$, $t = -6.46$, $P <$ 0.0005; we have strong evidence that $\beta < 0$, and hence that the correlation is negative.

11.9   (a) Use the prediction interval: 135.79 to 159.01 bpm. (b) Use df $= 21$ and $t^* = 1.721$.

11.11   (a) One residual may be a high outlier, but the stemplot does not show any other deviations from normality. (b) The scatterplot shows no striking features (other than the outlier).

11.13   (a) The relationship is reasonably linear, with no outliers or influential points. (b) 88.6% of the variation in manatee deaths is explained by the linear relationship with powerboat registrations. (c) β is the number of additional manatee deaths we expect for each additional thousand powerboat registrations. The interval is 0.1019 to 0.1478 deaths per thousand registrations.

11.15   44.00 to 51.94 manatee deaths.

11.17   (a) $\bar{x} \doteq 0.00174$, $s \doteq 1.0137$. Standardized data should have $\bar{x} \doteq 0$ and $s \doteq 1$. (b) A stemplot is not strikingly non-normal (for such a small sample). About 95% of the observations should be between -2 and 2, so -1.99 is quite reasonable. (c) The scatterplot gives no cause for concern.

11.19   (a) Variability about the line may not be constant for all x; it seems to increase from left to right. (b) The stemplot suggests that the distribution is right-skewed. The outlier is from 1986.

11.21   A scatterplot shows a positive association; $\hat{y} = 113.2 + 26.88x$. The confidence interval is 20.29 to 33.47 cal/kg; for each additional kilogram of mass, metabolic rate increases by about 20 to 33 calories. A stemplot of the residuals suggests that the distribution is right-skewed, and the largest residual may be an outlier.

11.23   (a) The association is roughly linear and positive; SUVs generally have lower mileage than midsize cars. (b) With the Passat, $r = 0.914$; without it, $r^* = 0.882$. Without it, the relative scatter about the regression line is greater, reducing the strength of the relationship. (c) $\hat{y} = 0.465 + 1.3026x$. (d) The largest residuals correspond to the two cars with the most extreme (smallest and largest) differences between city and highway mileage. (e) β is the average increase in highway mileage for each 1-mpg increase in city mileage. The confidence interval is 1.109 to 1.497 highway mpg per city mpg.

11.25   (a) The plot shows a fairly strong curved pattern (weight increases with length). Two fish stray from the curve, but would not particularly be considered outliers. (b) Weight should be roughly proportional to volume; when all dimensions change by a factor of x, the volume increases by a factor of x^3. (c) This plot shows a strong, positive, linear association, with no particular outliers. (d) The correlations reflect the increased linearity of the second

plot: With weight, $r^2 = 0.9207$; with weight$^{1/3}$, $r^2 = 0.9851$.
(e) $\hat{y} = -0.3283 + 0.2330x$; $\hat{y} = 5.9623$ when $x = 27$ cm; 5.886 to 6.039 g$^{1/3}$. **(f)** The stemplot shows no gross violations of the assumptions, except for the high outlier for fish #143. The scatterplot suggests that variability in weight may be greater for larger lengths. Dropping fish #143 changes the regression line only slightly, and seems to alleviate both these problems (to some degree, at least).

INDEX

TABLE B Random Digits

Line								
101	19223	95034	05756	28713	96409	12531	42544	82853
102	73676	47150	99400	01927	27754	42648	82425	36290
103	45467	71709	77558	00095	32863	29485	82226	90056
104	52711	38889	93074	60227	40011	85848	48767	52573
105	95592	94007	69971	91481	60779	53791	17297	59335
106	68417	35013	15529	72765	85089	57067	50211	47487
107	82739	57890	20807	47511	81676	55300	94383	14893
108	60940	72024	17868	24943	61790	90656	87964	18883
109	36009	19365	15412	39638	85453	46816	83485	41979
110	38448	48789	18338	24697	39364	42006	76688	08708
111	81486	69487	60513	09297	00412	71238	27649	39950
112	59636	88804	04634	71197	19352	73089	84898	45785
113	62568	70206	40325	03699	71080	22553	11486	11776
114	45149	32992	75730	66280	03819	56202	02938	70915
115	61041	77684	94322	24709	73698	14526	31893	32592
116	14459	26056	31424	80371	65103	62253	50490	61181
117	38167	98532	62183	70632	23417	26185	41448	75532
118	73190	32533	04470	29669	84407	90785	65956	86382
119	95857	07118	87664	92099	58806	66979	98624	84826
120	35476	55972	39421	65850	04266	35435	43742	11937
121	71487	09984	29077	14863	61683	47052	62224	51025
122	13873	81598	95052	90908	73592	75186	87136	95761
123	54580	81507	27102	56027	55892	33063	41842	81868
124	71035	09001	43367	49497	72719	96758	27611	91596
125	96746	12149	37823	71868	18442	35119	62103	39244
126	96927	19931	36809	74192	77567	88741	48409	41903
127	43909	99477	25330	64359	40085	16925	85117	36071
128	15689	14227	06565	14374	13352	49367	81982	87209
129	36759	58984	68288	22913	18638	54303	00795	08727
130	69051	64817	87174	09517	84534	06489	87201	97245
131	05007	16632	81194	14873	04197	85576	45195	96565
132	68732	55259	84292	08796	43165	93739	31685	97150
133	45740	41807	65561	33302	07051	93623	18132	09547
134	27816	78416	18329	21337	35213	37741	04312	68508
135	66925	55658	39100	78458	11206	19876	87151	31260
136	08421	44753	77377	28744	75592	08563	79140	92454
137	53645	66812	61421	47836	12609	15373	98481	14592
138	66831	68908	40772	21558	47781	33586	79177	06928
139	55588	99404	70708	41098	43563	56934	48394	51719
140	12975	13258	13048	45144	72321	81940	00360	02428
141	96767	35964	23822	96012	94591	65194	50842	53372
142	72829	50232	97892	63408	77919	44575	24870	04178
143	88565	42628	17797	49376	61762	16953	88604	12724
144	62964	88145	83083	69453	46109	59505	69680	00900
145	19687	12633	57857	95806	09931	02150	43163	58636
146	37609	59057	66967	83401	60705	02384	90597	93600
147	54973	86278	88737	74351	47500	84552	19909	67181
148	00694	05977	19664	65441	20903	62371	22725	53340
149	71546	05233	53946	68743	72460	27601	45403	88692
150	07511	88915	41267	16853	84569	79367	32337	03316

Table entry for p and C is the critical value t^* with probability p lying to its right and probability C lying between $-t^*$ and t^*.

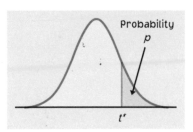

WITHDRAWN

TABLE C t distribution critical values

| df | \multicolumn{12}{c}{Upper tail probability p} |
	.25	.20	.15	.10	.05	.025	.02	.01	.005	.0025	.001	.0005
1	1.000	1.376	1.963	3.078	6.314	12.71	15.89	31.82	63.66	127.3	318.3	636.6
2	0.816	1.061	1.386	1.886	2.920	4.303	4.849	6.965	9.925	14.09	22.33	31.60
3	0.765	0.978	1.250	1.638	2.353	3.182	3.482	4.541	5.841	7.453	10.21	12.92
4	0.741	0.941	1.190	1.533	2.132	2.776	2.999	3.747	4.604	5.598	7.173	8.610
5	0.727	0.920	1.156	1.476	2.015	2.571	2.757	3.365	4.032	4.773	5.893	6.869
6	0.718	0.906	1.134	1.440	1.943	2.447	2.612	3.143	3.707	4.317	5.208	5.959
7	0.711	0.896	1.119	1.415	1.895	2.365	2.517	2.998	3.499	4.029	4.785	5.408
8	0.706	0.889	1.108	1.397	1.860	2.306	2.449	2.896	3.355	3.833	4.501	5.041
9	0.703	0.883	1.100	1.383	1.833	2.262	2.398	2.821	3.250	3.690	4.297	4.781
10	0.700	0.879	1.093	1.372	1.812	2.228	2.359	2.764	3.169	3.581	4.144	4.587
11	0.697	0.876	1.088	1.363	1.796	2.201	2.328	2.718	3.106	3.497	4.025	4.437
12	0.695	0.873	1.083	1.356	1.782	2.179	2.303	2.681	3.055	3.428	3.930	4.318
13	0.694	0.870	1.079	1.350	1.771	2.160	2.282	2.650	3.012	3.372	3.852	4.221
14	0.692	0.868	1.076	1.345	1.761	2.145	2.264	2.624	2.977	3.326	3.787	4.140
15	0.691	0.866	1.074	1.341	1.753	2.131	2.249	2.602	2.947	3.286	3.733	4.073
16	0.690	0.865	1.071	1.337	1.746	2.120	2.235	2.583	2.921	3.252	3.686	4.015
17	0.689	0.863	1.069	1.333	1.740	2.110	2.224	2.567	2.898	3.222	3.646	3.965
18	0.688	0.862	1.067	1.330	1.734	2.101	2.214	2.552	2.878	3.197	3.611	3.922
19	0.688	0.861	1.066	1.328	1.729	2.093	2.205	2.539	2.861	3.174	3.579	3.883
20	0.687	0.860	1.064	1.325	1.725	2.086	2.197	2.528	2.845	3.153	3.552	3.850
21	0.686	0.859	1.063	1.323	1.721	2.080	2.189	2.518	2.831	3.135	3.527	3.819
22	0.686	0.858	1.061	1.321	1.717	2.074	2.183	2.508	2.819	3.119	3.505	3.792
23	0.685	0.858	1.060	1.319	1.714	2.069	2.177	2.500	2.807	3.104	3.485	3.768
24	0.685	0.857	1.059	1.318	1.711	2.064	2.172	2.492	2.797	3.091	3.467	3.745
25	0.684	0.856	1.058	1.316	1.708	2.060	2.167	2.485	2.787	3.078	3.450	3.725
26	0.684	0.856	1.058	1.315	1.706	2.056	2.162	2.479	2.779	3.067	3.435	3.707
27	0.684	0.855	1.057	1.314	1.703	2.052	2.158	2.473	2.771	3.057	3.421	3.690
28	0.683	0.855	1.056	1.313	1.701	2.048	2.154	2.467	2.763	3.047	3.408	3.674
29	0.683	0.854	1.055	1.311	1.699	2.045	2.150	2.462	2.756	3.038	3.396	3.659
30	0.683	0.854	1.055	1.310	1.697	2.042	2.147	2.457	2.750	3.030	3.385	3.646
40	0.681	0.851	1.050	1.303	1.684	2.021	2.123	2.423	2.704	2.971	3.307	3.551
50	0.679	0.849	1.047	1.299	1.676	2.009	2.109	2.403	2.678	2.937	3.261	3.496
60	0.679	0.848	1.045	1.296	1.671	2.000	2.099	2.390	2.660	2.915	3.232	3.460
80	0.678	0.846	1.043	1.292	1.664	1.990	2.088	2.374	2.639	2.887	3.195	3.416
100	0.677	0.845	1.042	1.290	1.660	1.984	2.081	2.364	2.626	2.871	3.174	3.390
1000	0.675	0.842	1.037	1.282	1.646	1.962	2.056	2.330	2.581	2.813	3.098	3.300
z^*	0.674	0.841	1.036	1.282	1.645	1.960	2.054	2.326	2.576	2.807	3.091	3.291
	50%	60%	70%	80%	90%	95%	96%	98%	99%	99.5%	99.8%	99.9%
	\multicolumn{12}{c}{Confidence level C}